D0944939

# PROBABILITY, STATISTICS, AND STOCHASTIC PROCESSES

# PROBABILITY, STATISTICS, AND STOCHASTIC PROCESSES

## Second Edition

PETER OLOFSSON
MIKAEL ANDERSSON

A JOHN WILEY & SONS, INC., PUBLICATION

Published by John Wiley & Sons, Inc., Hoboken, New Jersey
Published simultaneously in Canada



***Library of Congress Cataloging-in-Publication Data:***

Olofsson, Peter, 1963–
Probability, statistics, and stochastic processes / Peter Olofsson, Mikael Andersson. – 2nd ed.
p. cm.
ISBN 978-0-470-88974-9 (hardback)
1. Stochastic processes–Textbooks. 2. Probabilities–Textbooks. 3. Mathematical statistics–Textbooks. I. Andersson, Mikael. II. Title.
QA274.O46 2012
519.2′3–dc23

2011040205

Printed in the United States of America

ISBN: 9780470889749

10 9 8 7 6 5 4 3 2 1

# CONTENTS

# PREFACE

The second edition was motivated by comments from several users and readers that the chapters on statistical inference and stochastic processes would benefit from substantial extensions. To accomplish such extensions, I decided to bring in Mikael Andersson, an old friend and colleague from graduate school. Being five days my junior, he brought a vigorous and youthful perspective to the task and I am very pleased with the outcome. Below, Mikael will outline the major changes and additions introduced in the second edition.

PETER OLOFSSON

*San Antonio, Texas, 2011*

The chapter on statistical inference has been extended, reorganized, and split into two new chapters. Chapter 6 introduces the principles and concepts behind standard methods of statistical inference in general, while the important case of normally distributed samples is treated separately in Chapter 7. This is a somewhat different structure compared to most other textbooks in statistics since common methods such as $t$ tests and linear regression come rather late in the text. According to my experience, if methods based on normal samples are presented too early in a course, they tend to overshadow other approaches such as nonparametric and Bayesian methods and students become less aware that these alternatives exist.

New additions in Chapter 6 include consistency of point estimators, large sample theory, bootstrap simulation, multiple hypothesis testing, Fisher's exact test, Kolmogorov–Smirnov test and nonparametric confidence intervals, as well as a discussion of informative versus noninformative priors and credibility intervals in Section 6.8.

Chapter 7 starts with a detailed treatment of sampling distributions, such as the $t$, chi-square, and $F$ distributions, derived from the normal distribution. There are also new sections introducing one-way analysis of variance and the general linear model.

Chapter 8 has been expanded to include three new sections on martingales, renewal processes, and Brownian motion. These areas are of great importance in probability theory and statistics, but since they are based on quite extensive and advanced mathematical theory, we offer only a brief introduction here.

It has been a great privilege, responsibility, and pleasure to have had the opportunity to work with such an esteemed colleague and good friend. Finally, the joint project that we dreamed about during graduate school has come to fruition!

I also have a victim of preoccupation and absentmindedness, my beloved Eva whom I want to thank for her support and all the love and friendship we have shared and will continue to share for many days to come.

MIKAEL ANDERSSON

*Stockholm, Sweden, 2011*

# PREFACE TO THE FIRST EDITION

## THE BOOK

In November 2003, I was completing a review of an undergraduate textbook in probability and statistics. In the enclosed evaluation sheet was the question "Have you ever considered writing a textbook?" and I suddenly realized that the answer was "Yes," and had been for quite some time. For several years I had been teaching a course on calculus-based probability and statistics mainly for mathematics, science, and engineering students. Other than the basic probability theory, my goal was to include topics from two areas: statistical inference and stochastic processes. For many students this was the only probability/statistics course they would ever take, and I found it desirable that they were familiar with confidence intervals and the maximum likelihood method, as well as Markov chains and queueing theory. While there were plenty of books covering one area or the other, it was surprisingly difficult to find one that covered both in a satisfying way and on the appropriate level of difficulty. My solution was to choose one textbook and supplement it with lecture notes in the area that was missing. As I changed texts often, plenty of lecture notes accumulated and it seemed like a good idea to organize them into a textbook. I was pleased to learn that the good people at Wiley agreed.

It is now more than a year later, and the book has been written. The first three chapters develop probability theory and introduce the axioms of probability, random variables, and joint distributions. The following two chapters are shorter and of an "introduction to" nature: Chapter 4 on limit theorems and Chapter 5 on simulation. Statistical inference is treated in Chapter 6, which includes a section on Bayesian statistics, too often a neglected topic in undergraduate texts. Finally, in Chapter 7, Markov chains in discrete and continuous time are introduced. The reference list at the end of the book is by no means intended to be comprehensive; rather, it is a subjective selection of the useful and the entertaining.

Throughout the text I have tried to convey an intuitive understanding of concepts and results, which is why a definition or a proposition is often preceded by a short discussion or a motivating example. I have also attempted to make the exposition entertaining by choosing examples from the rich source of fun and thought-provoking probability problems. The data sets used in the statistics chapter are of three different kinds: real, fake but realistic, and unrealistic but illustrative.

## THE PEOPLE

Most textbook authors start by thanking their spouses. I know now that this is far more than a formality, and I would like to thank $A\lambda\kappa\mu\acute{\eta}\nu\eta$ not only for patiently putting up with irregular work hours and an absentmindedness greater than usual but also for valuable comments on the aesthetics of the manuscript.

A number of people have commented on various parts and aspects of the book. First, I would like to thank Olle Häggström at Chalmers University of Technology, Göteborg, Sweden for valuable comments on all chapters. His remarks are always accurate and insightful, and never obscured by unnecessary politeness. Second, I would like to thank Kjell Doksum at the University of Wisconsin for a very helpful review of the statistics chapter. I have also enjoyed the Bayesian enthusiasm of Peter Müller at the University of Texas MD Anderson Cancer Center.

Other people who have commented on parts of the book or been otherwise helpful are my colleagues Dennis Cox, Kathy Ensor, Rudy Guerra, Marek Kimmel, Rolf Riedi, Javier Rojo, David W. Scott, and Jim Thompson at Rice University; Prof. Dr. R.W.J. Meester at Vrije Universiteit, Amsterdam, The Netherlands; Timo Seppäläinen at the University of Wisconsin; Tom English at Behrend College; Robert Lund at Clemson University; and Jared Martin at Shell Exploration and Production. For help with solutions to problems, I am grateful to several bright Rice graduate students: Blair Christian, Julie Cong, Talithia Daniel, Ginger Davis, Li Deng, Gretchen Fix, Hector Flores, Garrett Fox, Darrin Gershman, Jason Gershman, Shu Han, Shannon Neeley, Rick Ott, Galen Papkov, Bo Peng, Zhaoxia Yu, and Jenny Zhang. Thanks to Mikael Andersson at Stockholm University, Sweden for contributions to the problem sections, and to Patrick King at ODS–Petrodata, Inc. for providing data with a distinct Texas flavor: oil rig charter rates. At Wiley, I would like to thank Steve Quigley, Susanne Steitz, and Kellsee Chu for always promptly answering my questions. Finally, thanks to John Haigh, John Allen Paulos, Jeffrey E. Steif, and an anonymous Dutchman for agreeing to appear and be mildly mocked in footnotes.

PETER OLOFSSON

*Houston, Texas, 2005*

# 1

# BASIC PROBABILITY THEORY

## 1.1 INTRODUCTION

Probability theory is the mathematics of randomness. This statement immediately invites the question "What is randomness?" This is a deep question that we cannot attempt to answer without invoking the disciplines of philosophy, psychology, mathematical complexity theory, and quantum physics, and still there would most likely be no completely satisfactory answer. For our purposes, an informal definition of randomness as "what happens in a situation where we cannot predict the outcome with certainty" is sufficient. In many cases, this might simply mean lack of information. For example, if we flip a coin, we might think of the outcome as random. It will be either heads or tails, but we cannot say which, and if the coin is fair, we believe that both outcomes are equally likely. However, if we knew the force from the fingers at the flip, weight and shape of the coin, material and shape of the table surface, and several other parameters, we would be able to predict the outcome with certainty, according to the laws of physics. In this case we use randomness as a way to describe uncertainty due to lack of information.[1]

Next question: "What is probability?" There are two main interpretations of probability, one that could be termed "objective" and the other "subjective." The first is

[1]To quote the French mathematician Pierre-Simon Laplace, one of the first to develop a mathematical theory of probability: "Probability is composed partly of our ignorance, partly of our knowledge."

*Probability, Statistics, and Stochastic Processes*, Second Edition. Peter Olofsson and Mikael Andersson.

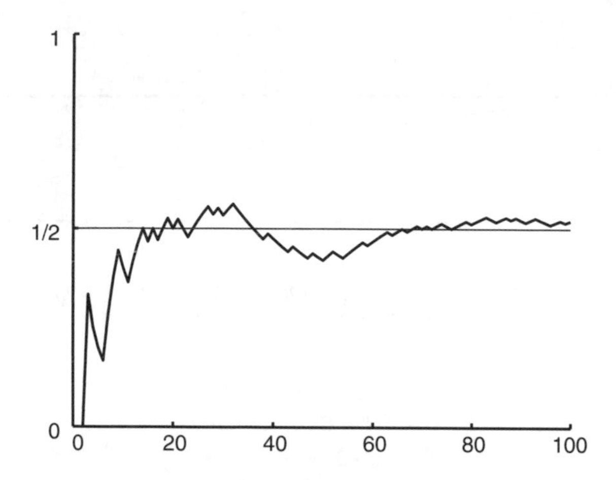


**FIGURE 1.1** Consecutive relative frequencies of heads in 100 coin flips.

the interpretation of a probability as a *limit of relative frequencies*; the second, as a *degree of belief*. Let us briefly describe each of these.

For the first interpretation, suppose that we have an experiment where we are interested in a particular outcome. We can repeat the experiment over and over and each time record whether we got the outcome of interest. As we proceed, we count the number of times that we got our outcome and divide this number by the number of times that we performed the experiment. The resulting ratio is the *relative frequency* of our outcome. As it can be observed empirically that such relative frequencies tend to stabilize as the number of repetitions of the experiment grows, we might think of the limit of the relative frequencies as the probability of the outcome. In mathematical notation, if we consider $n$ repetitions of the experiment and if $S_n$ of these gave our outcome, then the relative frequency would be $f_n = S_n/n$, and we might say that the probability equals $\lim_{n\to\infty} f_n$. Figure 1.1 shows a plot of the relative frequency of heads in a computer simulation of 100 hundred coin flips. Notice how there is significant variation in the beginning but how the relative frequency settles in toward $\frac{1}{2}$ quickly.

The second interpretation, probability as a degree of belief, is not as easily quantified but has obvious intuitive appeal. In many cases, it overlaps with the previous interpretation, for example, the coin flip. If we are asked to quantify our degree of belief that a coin flip gives heads, where 0 means "impossible" and 1 means "with certainty," we would probably settle for $\frac{1}{2}$ unless we have some specific reason to believe that the coin is not fair. In some cases it is not possible to repeat the experiment in practice, but we can still imagine a sequence of repetitions. For example, in a weather forecast you will often hear statements like "there is a 30% chance of rain tomorrow." Of course, we cannot repeat the experiment; either it rains tomorrow or it does not. The 30% is the meteorologist's measure of the chance of rain. There is still a connection to the relative frequency approach; we can imagine a sequence of days with similar weather conditions, same time of year, and so on, and that in roughly 30% of the cases, it rains the following day.

The "degree of belief" approach becomes less clear for statements such as "the Riemann hypothesis is true" or "there is life on other planets." Obviously, these are statements that are either true or false, but we do not know which, and it is not unreasonable to use probabilities to express how strongly we believe in their truth. It is also obvious that different individuals may assign completely different probabilities.

How, then, do we actually *define* a probability? Instead of trying to use any of these interpretations, we will state a strict mathematical definition of probability. The interpretations are still valid to develop intuition for the situation at hand, but instead of, for example, *assuming* that relative frequencies stabilize, we will be able to *prove* that they do, within our theory.

## 1.2 SAMPLE SPACES AND EVENTS

As mentioned in the introduction, probability theory is a mathematical theory to describe and analyze situations where randomness or uncertainty are present. Any specific such situation will be referred to as a *random experiment*. We use the term "experiment" in a wide sense here; it could mean an actual physical experiment such as flipping a coin or rolling a die, but it could also be a situation where we simply observe something, such as the price of a stock at a given time, the amount of rain in Houston in September, or the number of spam emails we receive in a day. After the experiment is over, we call the result an *outcome*. For any given experiment, there is a set of possible outcomes, and we state the following definition.

***Definition 1.1.*** The set of all possible outcomes in a random experiment is called the *sample space*, denoted $S$.

Here are some examples of random experiments and their associated sample spaces.

***Example 1.1.*** Roll a die and observe the number.

Here we can get the numbers 1 through 6, and hence the sample space is

$$S = \{1, 2, 3, 4, 5, 6\}$$

□

***Example 1.2.*** Roll a die repeatedly and count the number of rolls it takes until the first 6 appears.

Since the first 6 may come in the first roll, 1 is a possible outcome. Also, we may fail to get 6 in the first roll and then get 6 in the second, so 2 is also a possible outcome. If we continue this argument we realize that any positive integer is a possible outcome and the sample space is

$$S = \{1, 2, \ldots\}$$

the set of positive integers. □

***Example 1.3.*** Turn on a lightbulb and measure its lifetime, that is, the time until it fails.

Here it is not immediately clear what the sample space should be since it depends on how accurately we can measure time. The most convenient approach is to note that the lifetime, at least in theory, can assume any nonnegative real number and choose as the sample space

$$S = [0, \infty)$$

where the outcome 0 means that the lightbulb is broken to start with. □

In these three examples, we have sample spaces of three different kinds. The first is *finite*, meaning that it has a finite number of outcomes, whereas the second and third are infinite. Although they are both infinite, they are different in the sense that one has its points separated, $\{1, 2, \ldots\}$ and the other is an entire continuum of points. We call the first type *countable infinity* and the second *uncountable infinity*. We will return to these concepts later as they turn out to form an important distinction.

In the examples above, the outcomes are always numbers and hence the sample spaces are subsets of the real line. Here are some examples of other types of sample spaces.

***Example 1.4.*** Flip a coin twice and observe the sequence of heads and tails.

With $H$ denoting heads and $T$ denoting tails, one possible outcome is $HT$, which means that we get heads in the first flip and tails in the second. Arguing like this, there are four possible outcomes and the sample space is

$$S = \{HH, HT, TH, TT\}$$

□

***Example 1.5.*** Throw a dart at random on a dartboard of radius $r$.

If we think of the board as a disk in the plane with center at the origin, an outcome is an ordered pair of real numbers $(x, y)$, and we can describe the sample space as

$$S = \{(x, y) : x^2 + y^2 \leq r^2\}$$

□

Once we have described an experiment and its sample space, we want to be able to compute probabilities of the various things that may happen. What is the probability that we get 6 when we roll a die? That the first 6 does not come before the fifth roll? That the lightbulb works for at least 1500 h? That our dart hits the bull's eye? Certainly, we need to make further assumptions to be able to answer these questions, but before that, we realize that all these questions have something in common. They all ask for probabilities of either single outcomes or groups of outcomes. Mathematically, we can describe these as subsets of the sample space.

***Definition 1.2.*** A subset of $S$, $A \subseteq S$, is called an *event*.

Note the choice of words here. The terms "outcome" and "event" reflect the fact that we are describing things that may happen in real life. Mathematically, these are described as elements and subsets of the sample space. This duality is typical for probability theory; there is a verbal description and a mathematical description of the same situation. The verbal description is natural when real-world phenomena are described and the mathematical formulation is necessary to develop a consistent theory. See Table 1.1 for a list of set operations and their verbal description.

***Example 1.6.*** If we roll a die and observe the number, two possible events are that we get an odd outcome and that we get at least 4. If we view these as subsets of the sample space, we get

$$A = \{1, 3, 5\} \quad \text{and} \quad B = \{4, 5, 6\}$$

If we want to use the verbal description, we might write this as

$$A = \{\text{odd outcome}\} \quad \text{and} \quad B = \{\text{at least } 4\}$$

□

We always use "or" in its nonexclusive meaning; thus, "$A$ or $B$ occurs" includes the possibility that both occur. Note that there are different ways to express combinations of events; for example, $A \setminus B = A \cap B^c$ and $(A \cup B)^c = A^c \cap B^c$. The latter is known as one of *De Morgan's laws*, and we state these without proof together with some other basic set theoretic rules.

**TABLE 1.1 Basic Set Operations and Their Verbal Description**

| Notation | Mathematical Description | Verbal Description |
|---|---|---|
| $A \cup B$ | The union of $A$ and $B$ | $A$ or $B$ (or both) occurs |
| $A \cap B$ | The intersection of $A$ and $B$ | Both $A$ and $B$ occur |
| $A^c$ | The complement of $A$ | $A$ does not occur |
| $A \setminus B$ | The difference between $A$ and $B$ | $A$ occurs but not $B$ |
| $\emptyset$ | The empty set | Impossible event |

**Proposition 1.1.** Let $A$, $B$, and $C$ be events. Then

**(a)** ***(Distributive Laws)*** 
$$(A \cap B) \cup C = (A \cup C) \cap (B \cup C)$$
$$(A \cup B) \cap C = (A \cap C) \cup (B \cap C)$$

**(b)** ***(De Morgan's Laws)*** 
$$(A \cup B)^c = A^c \cap B^c$$
$$(A \cap B)^c = A^c \cup B^c$$

As usual when dealing with set theory, *Venn diagrams* are useful. See Figure 1.2 for an illustration of some of the set operations introduced above. We will later return to how Venn diagrams can be used to calculate probabilities. If $A$ and $B$ are such that $A \cap B = \emptyset$, they are said to be *disjoint* or *mutually exclusive*. In words, this means that they cannot both occur simultaneously in the experiment.

As we will often deal with unions of more than two or three events, we need more general versions of the results given above. Let us first introduce some notation. If $A_1, A_2, \ldots, A_n$ is a sequence of events, we denote

$$\bigcup_{k=1}^{n} A_k = A_1 \cup A_2 \cup \cdots \cup A_n$$

the union of all the $A_k$ and

$$\bigcap_{k=1}^{n} A_k = A_1 \cap A_2 \cap \cdots \cap A_n$$

the intersection of all the $A_k$. In words, these are the events that *at least one* of the $A_k$ occurs and that *all* the $A_k$ occur, respectively. The distributive and De Morgan's laws extend in the obvious way, for example

$$\left(\bigcup_{k=1}^{n} A_k\right)^c = \bigcap_{k=1}^{n} A_k^c$$

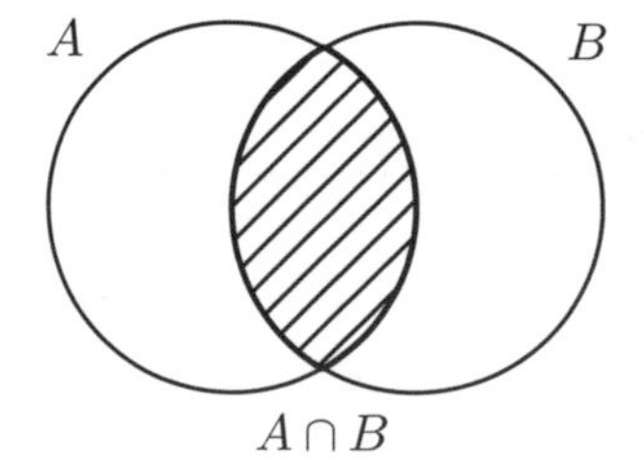


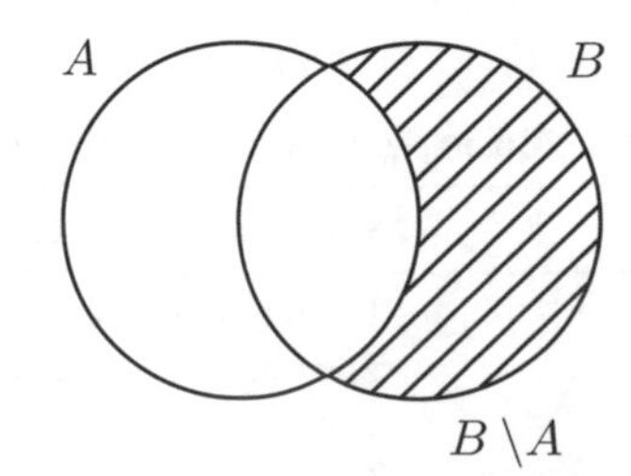


**FIGURE 1.2** Venn diagrams of the intersection and the difference between events.

It is also natural to consider infinite unions and intersections. For example, in Example 1.2, the event that the first 6 comes in an odd roll is the infinite union $\{1\} \cup \{3\} \cup \{5\} \cup \cdots$ and we can use the same type of notation as for finite unions and write

$$\{\text{first 6 in odd roll}\} = \bigcup_{k=1}^{\infty} \{2k-1\}$$

For infinite unions and intersections, distributive and De Morgan's laws still extend in the obvious way.

## 1.3 THE AXIOMS OF PROBABILITY

In the previous section, we laid the basis for a theory of probability by describing random experiments in terms of the sample space, outcomes, and events. As mentioned, we want to be able to compute probabilities of events. In the introduction, we mentioned two different interpretations of probability: as a limit of relative frequencies and as a degree of belief. Since our aim is to build a consistent mathematical theory, as widely applicable as possible, our definition of probability should not depend on any particular interpretation. For example, it makes intuitive sense to require a probability to always be less than or equal to one (or equivalently, less than or equal to 100%). You cannot flip a coin 10 times and get 12 heads. Also, a statement such as "I am 150% sure that it will rain tomorrow" may be used to express extreme pessimism regarding an upcoming picnic but is certainly not sensible from a logical point of view. Also, a probability should be equal to one (or 100%), when there is absolute certainty, regardless of any particular interpretation.

Other properties must hold as well. For example, if you think there is a 20% chance that Bob is in his house, a 30% chance that he is in his backyard, and a 50% chance that he is at work, then the chance that he is at home is 50%, the sum of 20% and 30%. Relative frequencies are also *additive* in this sense, and it is natural to demand that the same rule apply for probabilities.

We now give a mathematical definition of probability, where it is defined as a real-valued function of the events, satisfying three properties, which we refer to as the *axioms of probability*. In the light of the discussion above, they should be intuitively reasonable.

***Definition 1.3* (*Axioms of Probability*).** A *probability measure* is a function $P$, which assigns to each event $A$ a number $P(A)$ satisfying

**(a)** $0 \leq P(A) \leq 1$

**(b)** $P(S) = 1$

**(c)** If $A_1, A_2, \ldots$ is a sequence of *pairwise disjoint* events, that is, if $i \neq j$, then $A_i \cap A_j = \emptyset$, then

$$P\left(\bigcup_{k=1}^{\infty} A_k\right) = \sum_{k=1}^{\infty} P(A_k)$$

We read $P(A)$ as "the probability of $A$." Note that a probability in this sense is a real number between 0 and 1 but we will occasionally also use percentages so that, for example, the phrases "The probability is 0.2" and "There is a 20% chance" mean the same thing.[2]

The third axiom is the most powerful assumption when it comes to deducing properties and further results. Some texts prefer to state the third axiom for finite unions only, but since infinite unions naturally arise even in simple examples, we choose this more general version of the axioms. As it turns out, the finite case follows as a consequence of the infinite. We next state this in a proposition and also that the empty set has probability zero. Although intuitively obvious, we must prove that it follows from the axioms. We leave this as an exercise.

**Proposition 1.2.** Let $P$ be a probability measure. Then

**(a)** $P(\emptyset) = 0$

**(b)** If $A_1, \ldots, A_n$ are pairwise disjoint events, then

$$P(\bigcup_{k=1}^{n} A_k) = \sum_{k=1}^{n} P(A_k)$$

In particular, if $A$ and $B$ are disjoint, then $P(A \cup B) = P(A) + P(B)$. In general, unions need not be disjoint and we next show how to compute the probability of a union in general, as well as prove some other basic properties of the probability measure.

**Proposition 1.3.** Let $P$ be a probability measure on some sample space $S$ and let $A$ and $B$ be events. Then

**(a)** $P(A^c) = 1 - P(A)$

**(b)** $P(A \setminus B) = P(A) - P(A \cap B)$

**(c)** $P(A \cup B) = P(A) + P(B) - P(A \cap B)$

**(d)** If $A \subseteq B$, then $P(A) \leq P(B)$

[2]If the sample space is very large, it may be impossible to assign probabilities to *all* events. The class of events then needs to be restricted to what is called a $\sigma$*-field*. For a more advanced treatment of probability theory, this is a necessary restriction, but we can safely disregard this problem.

*Proof.* We prove (b) and (c), and leave (a) and (d) as exercises. For (b), note that $A = (A \cap B) \cup (A \setminus B)$, which is a disjoint union, and Proposition 1.2 gives

$$P(A) = P(A \cap B) + P(A \setminus B)$$

which proves the assertion. For (c), we write $A \cup B = A \cup (B \setminus A)$, which is a disjoint union, and we get

$$P(A \cup B) = P(A) + P(B \setminus A) = P(A) + P(B) - P(A \cap B)$$

by part (b). ■

Note how we repeatedly used Proposition 1.2(b), the finite version of the third axiom. In Proposition 1.3(c), for example, the events $A$ and $B$ are not necessarily disjoint but we can represent their union as a union of other events that are disjoint, thus allowing us to apply the third axiom.

***Example 1.7.*** Mrs Boudreaux and Mrs Thibodeaux are chatting over their fence when the new neighbor walks by. He is a man in his sixties with shabby clothes and a distinct smell of cheap whiskey. Mrs B, who has seen him before, tells Mrs T that he is a former Louisiana state senator. Mrs T finds this very hard to believe. "Yes," says Mrs B, "he is a former state senator who got into a scandal long ago, had to resign and started drinking." "Oh," says Mrs T, "that sounds more probable." "No," says Mrs B, "I think you mean less probable."

Actually, Mrs B is right. Consider the following two statements about the shabby man: "He is a former state senator" and "He is a former state senator who got into a scandal long ago, had to resign, and started drinking." It is tempting to think that the second is more probable because it gives a more exhaustive explanation of the situation at hand. However, this is precisely why it is a *less* probable statement. To explain this with probabilities, consider the experiment of observing a person and the two events

$$A = \{\text{he is a former state senator}\}$$
$$B = \{\text{he got into a scandal long ago, had to resign, and started drinking}\}$$

The first statement then corresponds to the event $A$ and the second to the event $A \cap B$, and since $A \cap B \subseteq A$, we get $P(A \cap B) \leq P(A)$. Of course, what Mrs T meant was that it was easier to believe that the man was a former state senator once she knew more about his background.

In their book *Judgment under Uncertainty*, Kahneman et al. [5], show empirically how people often make similar mistakes when asked to choose the most probable among a set of statements. With a strict application of the rules of probability, we get it right. □

***Example 1.8.*** Consider the following statement: "I heard on the news that there is a 50% chance of rain on Saturday and a 50% chance of rain on Sunday. Then there must be a 100% chance of rain during the weekend."

This is, of course, not true. However, it may be harder to point out precisely where the error lies, but we can address it with probability theory. The events of interest are

$$A = \{\text{rain on Saturday}\} \quad \text{and} \quad B = \{\text{rain on Sunday}\}$$

and the event of rain during the weekend is then $A \cup B$. The percentages are reformulated as probabilities so that $P(A) = P(B) = 0.5$ and we get

$$\begin{aligned} P(\text{rain during the weekend}) &= P(A \cup B) \\ &= P(A) + P(B) - P(A \cap B) \\ &= 1 - P(A \cap B) \end{aligned}$$

which is less than 1, that is, the chance of rain during the weekend is less than 100%. The error in the statement lies in that we can add probabilities only when the events are disjoint. In general, we need to subtract the probability of the intersection, which in this case is the probability that it rains both Saturday and Sunday. □

***Example 1.9.*** A dartboard has an area of 143 in.$^2$ (square inches). In the center of the board, there is the "bulls eye," which is a disk of area 1 in.$^2$. The rest of the board is divided into 20 sectors numbered $1, 2, \ldots, 20$. There is also a triple ring that has an area of 10 in.$^2$ and a double ring of area 15 in.$^2$ (everything rounded to nearest integers). Suppose that you throw a dart at random on the board. What is the probability that you get **(a)** double 14, **(b)** 14 but not double, **(c)** triple or the bull's eye, and **(d)** an even number or a double?

Introduce the events $F = \{14\}$, $D = \{\text{double}\}$, $T = \{\text{triple}\}$, $B = \{\text{bull's eye}\}$, and $E = \{\text{even}\}$. We interpret "throw a dart at random" to mean that any region is hit with a probability that equals the fraction of the total area of the board that region occupies. For example, each number has area $(143 - 1)/20 = 7.1$ in.$^2$ so the corresponding probability is $7.1/143$. We get

$$P(\text{double 14}) = P(D \cap F) = \frac{0.75}{143} \approx 0.005$$

$$\begin{aligned} P(\text{14 but not double}) &= P(F \setminus D) = P(F) - P(F \cap D) \\ &= \frac{7.1}{143} - \frac{0.75}{143} \approx 0.044 \end{aligned}$$

$$P(\text{triple or bulls eye}) = P(T \cup B) = P(T) + P(B) = \frac{10}{143} + \frac{1}{143} \approx 0.077$$

$$P(\text{even or double}) = P(E \cup D) = P(E) + P(D) - P(E \cap D) = \frac{71}{143} + \frac{15}{143} - \frac{7.5}{143} \approx 0.55$$

□

Let us say a word here about the interplay between logical statements and events. In the previous example, consider the events $E = \{\text{even}\}$ and $F = \{14\}$. Clearly, if we get 14, we also get an even number. As a logical relation between statements, we would express this as

$$\text{the number is 14} \Rightarrow \text{the number is even}$$

and in terms of events, we would say "If $F$ occurs, then $E$ must also occur." But this means that $F \subseteq E$ and hence

$$\{\text{the number is 14}\} \subseteq \{\text{the number is even}\}$$

and thus the set-theoretic analog of "$\Rightarrow$" is "$\subseteq$" that is useful to keep in mind.

Venn diagrams turn out to provide a nice and useful interpretation of probabilities. If we imagine the sample space $S$ to be a rectangle of area 1, we can interpret the probability of an event $A$ as the area of $A$ (see Figure 1.3). For example, Proposition 1.3(c) says that $P(A \cup B) = P(A) + P(B) - P(A \cap B)$. With the interpretation of probabilities as areas, we thus have

$$\begin{aligned} P(A \cup B) &= \text{area of } A \cup B \\ &= \text{area of } A + \text{area of } B - \text{area of } A \cap B \\ &= P(A) + P(B) - P(A \cap B) \end{aligned}$$

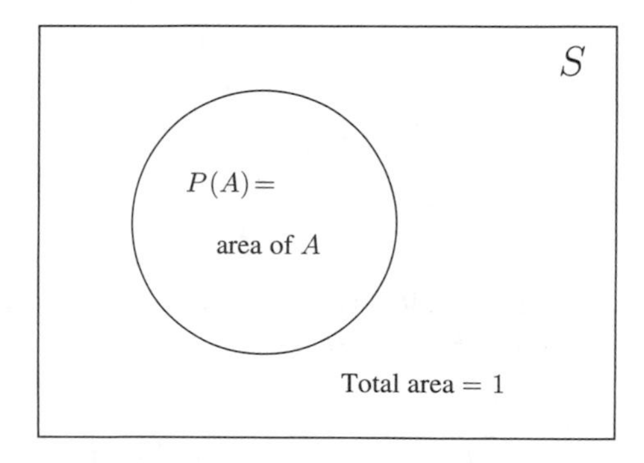


**FIGURE 1.3** Probabilities with Venn diagrams.

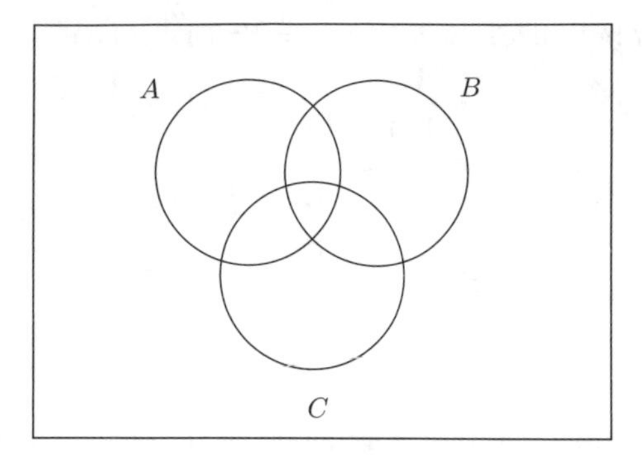


**FIGURE 1.4** Venn diagram of three events.

since when we add the areas of $A$ and $B$, we count the area of $A \cap B$ twice and must subtract it (think of $A$ and $B$ as overlapping pancakes where we are interested only in how much area they cover). Strictly speaking, this is not a proof but the method can be helpful to find formulas that can then be proved formally. In the case of three events, consider Figure 1.4 to argue that

$$\begin{aligned}\text{Area of } A \cup B \cup C = \; & \text{area of } A + \text{ area of } B + \text{ area of } C \\ & - \text{ area of } A \cap B - \text{ area of } A \cap C - \text{ area of } B \cap C \\ & + \text{ area of } A \cap B \cap C\end{aligned}$$

since the piece in the middle was first added three times and then removed three times, so in the end we have to add it again. Note that we must draw the diagram so that we get all possible combinations of intersections between the events. We have argued for the following proposition, which we state and prove formally.

**Proposition 1.4.** Let $A$, $B$, and $C$ be three events. Then

$$\begin{aligned}P(A \cup B \cup C) = \; & P(A) + P(B) + P(C) \\ & - P(A \cap B) - P(A \cap C) - P(B \cap C) \\ & + P(A \cap B \cap C)\end{aligned}$$

*Proof.* By applying Proposition 1.3(c) twice—first to the two events $A \cup B$ and $C$ and second to the events $A$ and $B$—we obtain

$$\begin{aligned}P(A \cup B \cup C) &= P(A \cup B) + P(C) - P((A \cup B) \cap C) \\ &= P(A) + P(B) - P(A \cap B) + P(C) - P((A \cup B) \cap C)\end{aligned}$$

The first four terms are what they should be. To deal with the last term, note that by the distributive laws for set operations, we obtain

$$(A \cup B) \cap C = (A \cap C) \cup (B \cap C)$$

and yet another application of Proposition 1.3(c) gives

$$\begin{aligned} P((A \cup B) \cap C) &= P((A \cap C) \cup (B \cap C)) \\ &= P(A \cap C) + P(B \cap C) - P(A \cap B \cap C) \end{aligned}$$

which gives the desired result. ■

***Example 1.10.*** Choose a number at random from the numbers 1, ..., 100. What is the probability that the chosen number is divisible by either 2, 3, or 5?

Introduce the events

$$A_k = \{\text{divisible by } k\} \text{ for } k = 1, 2, \ldots$$

We interpret "at random" to mean that any set of numbers has a probability that is equal to its relative size, that is, the number of elements divided by 100. We then get

$$P(A_2) = 0.5, \;\; P(A_3) = 0.33, \;\text{ and } P(A_5) = 0.2$$

For the intersection, first note that, for example, $A_2 \cap A_3$ is the event that the number is divisible by both 2 and 3, which is the same as saying it is divisible by 6. Hence $A_2 \cap A_3 = A_6$ and

$$P(A_2 \cap A_3) = P(A_6) = 0.16$$

Similarly, we get

$$P(A_2 \cap A_5) = P(A_{10}) = 0.1, \quad P(A_3 \cap A_5) = P(A_{15}) = 0.06$$

and

$$P(A_2 \cap A_3 \cap A_5) = P(A_{30}) = 0.03$$

The event of interest is $A_2 \cup A_3 \cup A_5$, and Proposition 1.4 yields

$$P(A_2 \cup A_3 \cup A_5) = 0.5 + 0.33 + 0.2 - (0.16 + 0.1 + 0.06) + 0.03 = 0.74$$

□

It is now easy to believe that the general formula for a union of $n$ events starts by adding the probabilities of the events, then subtracting the probabilities of the pairwise intersections, adding the probabilities of intersections of triples, and so on, finishing with either adding or subtracting the intersection of all the $n$ events, depending on whether $n$ is odd or even. We state this in a proposition that is sometimes referred to as the *inclusion–exclusion formula*. It can, for example, be proved by induction, but we leave the proof as an exercise.

**Proposition 1.5.** Let $A_1, A_2, \ldots, A_n$ be a sequence of $n$ events. Then

$$\begin{aligned}P\left(\bigcup_{k=1}^{n} A_k\right) &= \sum_{k=1}^{n} P(A_k) \\ &\quad - \sum_{i<j} P(A_i \cap A_j) \\ &\quad + \sum_{i<j<k} P(A_i \cap A_j \cap A_k) \\ &\quad \vdots \\ &\quad + (-1)^{n+1} P(A_1 \cap A_2 \cap \cdots \cap A_n)\end{aligned}$$

We finish this section with a theoretical result that will be useful from time to time. A sequence of events is said to be *increasing* if

$$A_1 \subseteq A_2 \subseteq \cdots$$

and *decreasing* if

$$A_1 \supseteq A_2 \supseteq \cdots$$

In each case we can define the *limit* of the sequence. If the sequence is increasing, we define

$$\lim_{n\to\infty} A_n = \bigcup_{k=1}^{\infty} A_k$$

and if the sequence is decreasing

$$\lim_{n\to\infty} A_n = \bigcap_{k=1}^{\infty} A_k$$

Note how this is similar to limits of sequences of numbers, with $\subseteq$ and $\supseteq$ corresponding to $\leq$ and $\geq$, respectively, and union and intersection corresponding to supremum and infimum. The following proposition states that the probability measure is a *continuous set function*. The proof is outlined in Problem 18.

**Proposition 1.6.** If $A_1, A_2, \ldots$ is either increasing or decreasing, then

$$P(\lim_{n\to\infty} A_n) = \lim_{n\to\infty} P(A_n)$$

## 1.4 FINITE SAMPLE SPACES AND COMBINATORICS

The results in the previous section hold for an arbitrary sample space $S$. In this section, we will assume that $S$ is finite, $S = \{s_1, \ldots, s_n\}$, say. In this case, we can always define the probability measure by assigning probabilities to the individual outcomes.

**Proposition 1.7.** Suppose that $p_1, \ldots, p_n$ are numbers such that

**(a)** $p_k \geq 0, \quad k = 1, \ldots, n$

**(b)** $\sum_{k=1}^{n} p_k = 1$

and for any event $A \subseteq S$, define

$$P(A) = \sum_{k:s_k \in A} p_k$$

Then $P$ is a probability measure.

*Proof.* Clearly, the first two axioms of probability are satisfied. For the third, note that in a finite sample space, we cannot have infinitely many disjoint events, so we only have to check this for a disjoint union of two events $A$ and $B$. We get

$$P(A \cup B) = \sum_{k:s_k \in A \cup B} p_k = \sum_{k:s_k \in A} p_k + \sum_{k:s_k \in B} p_k = P(A) + P(B)$$

and we are done. (Why are two events enough?) ∎

Hence, when dealing with finite sample spaces, we do not need to explicitly give the probability of every event, only for each outcome. We refer to the numbers $p_1, \ldots, p_n$ as a *probability distribution* on $S$.

***Example 1.11.*** Consider the experiment of flipping a fair coin twice and counting the number of heads. We can take the sample space

$$S = \{HH, HT, TH, TT\}$$

and let $p_1 = \cdots = p_4 = \frac{1}{4}$. Alternatively, since all we are interested in is the number of heads and this can be 0, 1, or 2, we can use the sample space

$$S = \{0, 1, 2\}$$

and let $p_0 = \frac{1}{4}$, $p_1 = \frac{1}{2}$, $p_2 = \frac{1}{4}$. □

Of particular interest is the case when all outcomes are equally likely. If $S$ has $n$ equally likely outcomes, then $p_1 = p_2 = \cdots = p_n = \frac{1}{n}$, which is called a *uniform distribution* on $S$. The formula for the probability of an event $A$ now simplifies to

$$P(A) = \sum_{k:s_k \in A} \frac{1}{n} = \frac{\#A}{n}$$

where $\#A$ denotes the number of elements in $A$. This formula is often referred to as the *classical definition of probability* since historically this was the first context in which probabilities were studied. The outcomes in the event $A$ can be described as *favorable* to $A$ and we get the following formulation.

**Corollary 1.1.** *In a finite sample space with uniform probability distribution*

$$P(A) = \frac{\#\textit{ favorable outcomes}}{\#\textit{ possible outcomes}}$$

In daily language, the term "at random" is often used for something that has a uniform distribution. Although our concept of randomness is more general, this colloquial notion is so common that we will also use it (and already have). Thus, if we say "pick a number at random from 1, . . . , 10," we mean "pick a number according to a uniform probability distribution on the sample space $\{1, 2, \ldots, 10\}$."

***Example 1.12.*** Roll a fair die three times. What is the probability that all numbers are the same?

The sample space is the set of the 216 ordered triples $(i, j, k)$, and since the die is fair, these are all equally probable and we have a uniform probability distribution. The event of interest is

$$A = \{(1, 1, 1), (2, 2, 2), \ldots, (6, 6, 6)\}$$

which has six outcomes and probability

$$P(A) = \frac{\#\text{ favorable outcomes}}{\#\text{ possible outcomes}} = \frac{6}{216} = \frac{1}{36}$$

□

***Example 1.13.*** Consider a randomly chosen family with three children. What is the probability that they have exactly one daughter?

There are eight possible sequences of boys and girls (in order of birth), and we get the sample space

$$S = \{bbb, bbg, bgb, bgg, gbb, gbg, ggb, ggg\}$$

where, for example, *bbg* means that the oldest child is a boy, the middle child a boy, and the youngest child a girl. If we assume that all outcomes are equally likely, we get a uniform probability distribution on $S$, and since there are three outcomes with one girl, we get

$$P(\text{one daughter}) = \frac{3}{8}$$

□

***Example 1.14.*** Consider a randomly chosen girl who has two siblings. What is the probability that she has no sisters?

Although this seems like the same problem as in the previous example, it is not. If, for example, the family has three girls, the chosen girl can be any of these three, so there are three different outcomes and the sample space needs to take this into account. Let $g^*$ denote the chosen girl to get the sample space

$$S = \{g^*gg, gg^*g, ggg^*, g^*gb, gg^*b, g^*bg, gbg^*, bg^*g, bgg^*, g^*bb, bg^*b, bbg^*\}$$

and since 3 out of 12 equally likely outcomes have no sisters we get

$$P(\text{no sisters}) = \frac{1}{4}$$

which is smaller than the $\frac{3}{8}$ we got above. On average, 37.5% of families with three children have a single daughter and 25% of girls in three-children families are single daughters. □

### 1.4.1 Combinatorics

Combinatorics, "the mathematics of counting," gives rise to a wealth of probability problems. The typical situation is that we have a set of objects from which we draw repeatedly in such a way that all objects are equally likely to be drawn. It is often tedious to list the sample space explicitly, but by counting combinations we can find the total number of cases and the number of favorable cases and apply the methods from the previous section.

The first problem is to find general expressions for the total number of combinations when we draw $k$ times from a set of $n$ distinguishable objects. There are different ways to interpret this. For example, we can draw *with* or *without replacement*, depending on whether the same object can be drawn more than once. We can also draw *with* or *without regard to order*, depending on whether it matters in which order the objects are drawn. With these distinctions, there are four different cases, illustrated in the following simple example.

***Example 1.15.*** Choose two numbers from the set $\{1, 2, 3\}$ and list the possible outcomes.

Let us first choose with regard to order. If we choose with replacement, the possible outcomes are

$$(1, 1), (1, 2), (1, 3), (2, 1), (2, 2), (2, 3), (3, 1), (3, 2), (3, 3)$$

and if we choose without replacement

$$(1, 2), (1, 3), (2, 1), (2, 3), (3, 1), (3, 2)$$

Next, let us choose without regard to order. This means that, for example, the outcomes $(1, 2)$ and $(2, 1)$ are regarded as the same and we denote it by $\{1, 2\}$ to stress that this is the *set* of 1 and 2, *not* the ordered pair. If we choose with replacement, the possible cases are

$$\{1, 1\}, \{1, 2\}, \{1, 3\}, \{2, 2\}, \{2, 3\}, \{3, 3\}$$

and if we choose without replacement

$$\{1, 2\}, \{1, 3\}, \{2, 3\}$$

□

To find expressions in the four cases for arbitrary values of $n$ and $k$, we first need the following result. It is intuitively quite clear, and we state it without proof.

**Proposition 1.8.** If we are to perform $r$ experiments in order, such that there are $n_1$ possible outcomes of the first experiment, $n_2$ possible outcomes of the second experiment, ..., $n_r$ possible outcomes of the $r$th experiment, then there is a total of $n_1 n_2 \cdots n_r$ outcomes of the sequence of the $r$ experiments.

This is called the *fundamental principle of counting* or the *multiplication principle*. Let us illustrate it by a simple example.

***Example 1.16.*** A Swedish license plate consists of three letters followed by three digits. How many possible license plates are there?

Although there are 28 letters in the Swedish alphabet, only 23 are used for license plates. Hence we have $r = 6, n_1 = n_2 = n_3 = 23$, and $n_4 = n_5 = n_6 = 10$. This gives a total of $23^3 \times 10^3 \approx 12.2$ million different license plates. □

We can now address the problem of drawing $k$ times from a set of $n$ objects. It turns out that choosing with regard to order is the simplest, so let us start with this and first consider the case of choosing with replacement. The first object can be chosen in $n$ ways, and for each such choice, we have $n$ ways to choose also the second object, $n$ ways to choose the third, and so on. The fundamental principle of counting gives

$$n \times n \times \cdots \times n = n^k$$

ways to choose with replacement and with regard to order.

If we instead choose without replacement, the first object can be chosen in $n$ ways, the second in $n-1$ ways, since the first object has been removed, the third in $n-2$ ways, and so on. The fundamental principle of counting gives

$$n(n-1)\cdots(n-k+1)$$

ways to choose without replacement and with regard to order. Sometimes, the notation

$$(n)_k = n(n-1)\cdots(n-k+1)$$

will be used for convenience, but this is not standard.

***Example 1.17.*** From a group of 20 students, half of whom are female, a student council president and vice president are chosen at random. What is the probability of getting a female president and a male vice president?

The set of objects is the 20 students. Assuming that the president is drawn first, we need to take order into account since, for example, (Brenda, Bruce) is a favorable outcome but (Bruce, Brenda) is not. Also, drawing is done without replacement. Thus, we have $k = 2$ and $n = 20$ and there are $20 \times 19 = 380$ equally likely different ways to choose a president and a vice president. The sample space is the set of these 380 combinations and to find the probability, we need the number of favorable cases. By the fundamental principle of counting, this is $10 \times 10 = 100$. The probability of getting a female president and male vice president is $\frac{100}{380} \approx 0.26$. □

***Example 1.18.*** A human gene consists of nucleotide base pairs of four different kinds, $A$, $C$, $G$, and $T$. If a particular region of interest of a gene has 20 base pairs, what is the probability that a randomly chosen individual has no base pairs in common with a particular reference sequence in a database?

The set of objects is $\{A, C, G, T\}$, and we draw 20 times with replacement and with regard to order. Thus $k = 20$ and $n = 4$, so there are $4^{20}$ possible outcomes, and let us, for the sake of this example, assume that they are equally likely (which would not be true in reality). For the number of favorable outcomes, $n = 3$ instead of 4 since we need to avoid one particular letter in each choice. Hence, the probability is $3^{20}/4^{20} \approx 0.003$. □

***Example 1.19* (*The Birthday Problem*).** This problem is a favorite in the probability literature. In a group of 100 people, what is the probability that at least two have the same birthday?

To simplify the solution, we disregard leap years and assume a uniform distribution of birthdays over the 365 days of the year. To assign birthdays to 100 people, we choose 100 out of 365 with replacement and get $365^{100}$ different combinations. The sample space is the set of those combinations, and the event of interest is

$$A = \{\text{at least two birthdays are equal}\}$$

and as it turns out, it is easier to deal with its complement

$$A^c = \{\text{all 100 birthdays are different}\}$$

To find the probability of $A^c$, note that the number of cases favorable to $A^c$ is obtained by choosing 100 days out of 365 *without* replacement and hence

$$P(A) = 1 - P(A^c) = 1 - \frac{365 \times 364 \times \cdots \times 266}{365^{100}} \approx 0.9999997$$

Yes, that is a sequence of six 9s followed by a 7! Hence, we can be almost certain that any group of 100 people has at least two people sharing birthdays. A similar calculation reveals the probability of a shared birthday already exceeds $\frac{1}{2}$ at 23 people, a quite surprising result. About 50% of school classes thus ought to have kids who share birthdays, something that those with idle time on their hands can check empirically. □

A check of real-life birthday distributions will reveal that the assumption of birthdays being uniformly distributed over the year is not true. However, the already high probability of shared birthdays only gets higher with a nonuniform distribution. Intuitively, this is because the less uniform the distribution, the more difficult it becomes to avoid birthdays already taken. For an extreme example, suppose that everybody was born in January, in which case there would be only 31 days to choose from instead of 365. Thus, in a group of 100 people, there would be absolute certainty of shared birthdays. Generally, it can be shown that the uniform distribution minimizes the probability of shared birthdays (we return to this in Problems 55 and 56).

***Example 1.20* (*The Birthday Problem Continued*).** A while ago I was in a group of exactly 100 people and asked for their birthdays. It turned out that nobody had the same birthday as I do. In the light of the previous problem, would this not be a very unlikely coincidence?

No, because here we are only considering the case of avoiding one particular birthday. Hence, with

$$B = \{\text{at least 1 out of 99 birthdays is the same as mine}\}$$

we get

$$B^c = \{99 \text{ birthdays are different from mine}\}$$

and the number of cases favorable to $B^c$ is obtained by choosing with replacement from the 364 days that do not match my birthday. We get

$$P(B) = 1 - P(B^c) = 1 - \frac{364^{99}}{365^{99}} \approx 0.24$$

Thus, it is actually quite likely that nobody shares my birthday, and it is at the same time almost certain that at least somebody shares somebody else's birthday. □

Next, we turn to the case of choosing without regard to order. First, suppose that we choose without replacement and let $x$ be the number of possible ways, in which this can be done. Now, there are $n(n-1)\cdots(n-k+1)$ ways to choose with regard to order and each such ordered set can be obtained by first choosing the objects and then order them. Since there are $x$ ways to choose the unordered objects and $k!$ ways to order them, we get the relation

$$n(n-1)\cdots(n-k+1) = x \times k!$$

and hence there are

$$x = \frac{n(n-1)\cdots(n-k+1)}{k!} \tag{1.1}$$

ways to choose without replacement, without regard to order. In other words, this is the number of subsets of size $k$ of a set of size $n$, called the *binomial coefficient*, read "$n$ choose $k$" and usually denoted and defined as

$$\binom{n}{k} = \frac{n!}{(n-k)!k!}$$

but we use the expression in Equation (1.1) for computations. By convention,

$$\binom{n}{0} = 1$$

and from the definition it follows immediately that

$$\binom{n}{k} = \binom{n}{n-k}$$

which is useful for computations. For some further properties, see Problem 24.

***Example 1.21.*** In Texas Lotto, you choose five of the numbers $1, \ldots, 44$ and one bonus ball number, also from $1, \ldots, 44$. Winning numbers are chosen randomly. Which is more likely: that you match the first five numbers but not the bonus ball or that you match four of the first five numbers and the bonus ball?

Since we have to match five of our six numbers in each case, are the two not equally likely? Let us compute the probabilities and see. The set of objects is $\{1, 2, \ldots, 44\}$ and the first five numbers are drawn without replacement and without regard to order. Hence, there are $\binom{44}{5}$ combinations and for each of these there are then 44 possible choices of the bonus ball. Thus, there is a total of $\binom{44}{5} \times 44 = 47,784,352$ different combinations. Introduce the events

$$A = \{\text{match the first five numbers but not the bonus ball}\}$$
$$B = \{\text{match four of the first five numbers and the bonus ball}\}$$

For $A$, the number of favorable cases is $1 \times 43$ (only one way to match the first five numbers, 43 ways to avoid the winning bonus ball). Hence

$$P(A) = \frac{1 \times 43}{\binom{44}{5} \times 44} \approx 9 \times 10^{-7}$$

To find the number of cases favorable to $B$, note that there are $\binom{5}{4} = 5$ ways to match four out of five winning numbers and then $\binom{39}{1} = 39$ ways to avoid the fifth winning number. There is only one choice for the bonus ball and we get

$$P(B) = \frac{5 \times 39 \times 1}{\binom{44}{5} \times 44} \approx 4 \times 10^{-6}$$

so $B$ is more than four times as likely as $A$. □

***Example 1.22.*** You are dealt a poker hand (5 cards out of 52 without replacement). **(a)** What is the probability that you get no hearts? **(b)** What is the probability that you get exactly $k$ hearts? **(c)** What is the most likely number of hearts?

We will solve this by disregarding order. The number of possible cases is the number of ways in which we can choose 5 out of 52 cards, which equals $\binom{52}{5}$. In (a), to get a favorable case, we need to choose all 5 cards from the 39 that are not hearts. Since this can be done in $\binom{39}{5}$ ways, we get

$$P(\text{no hearts}) = \frac{\binom{39}{5}}{\binom{52}{5}} \approx 0.22$$

In (b), we need to choose $k$ cards among the 13 hearts, and for each such choice, the remaining $5-k$ cards are chosen among the remaining 39 that are not hearts. This gives

$$P(k \text{ hearts}) = \frac{\binom{13}{k}\binom{39}{5-k}}{\binom{52}{5}}, \quad k = 0, 1, \ldots, 5$$

and for (c), direct computation gives the most likely number as 1, which has probability 0.41. □

The problem in the previous example can also be solved by taking order into account. Hence, we imagine that we get the cards one by one and list them in order and note that there are $(52)_5$ different cases. There are $(13)_k(39)_{5-k}$ ways to choose so that we get $k$ hearts and $5-k$ nonhearts in a particular order. Since there are $\binom{5}{k}$ ways to choose position for the $k$ hearts, we get

$$P(k \text{ hearts}) = \frac{\binom{5}{k}(13)_k(39)_{5-k}}{(52)_5}$$

which is the same as we got when we disregarded order above. It does not matter to the solution of the problem whether we take order into account, but we must be consistent and count the same way for the total and the favorable number of cases. In this particular example, it is probably easier to disregard order.

***Example 1.23.*** An urn contains 10 white balls, 10 red balls, and 10 black balls. You draw five balls at random without replacement. What is the probability that you do not get all colors?

Introduce the events

$$R = \{\text{no red balls}\}, \quad W = \{\text{no white balls}\}, \quad B = \{\text{no black balls}\}$$

The event of interest is then $R \cup W \cup B$, and we will apply Proposition 1.4. First note that by symmetry, $P(R) = P(W) = P(B)$. Also, each intersection of any two events has the same probability and finally $R \cap W \cap B = \emptyset$. We get

$$P(\text{not all colors}) = 3P(R) - 3P(R \cap W)$$

In order to get no red balls, the 5 balls must be chosen among the 20 balls that are not red and hence

$$P(R) = \binom{20}{5} \Big/ \binom{30}{5}$$

Similarly, to get neither red nor white balls, the five balls must be chosen among the black balls and

$$P(R \cap W) = \binom{10}{5} \Big/ \binom{30}{5}$$

We get

$$P(\text{not all colors}) = 3\left(\binom{20}{5} - \binom{10}{5}\right) \Big/ \binom{30}{5} \approx 0.32$$

□

***Example 1.24.*** The final case, choosing with replacement and without regard to order, turns out to be the trickiest. As we noted above, when we choose without replacement, each unordered set of $k$ objects corresponds to exactly $k!$ ordered sets. The relation is not so simple when we choose with replacement. For example, the unordered set $\{1, 1\}$ corresponds to one ordered set $(1, 1)$, whereas the unordered set $\{1, 2\}$ corresponds to two ordered sets $(1, 2)$ and $(2, 1)$. To find the general expression, we need to take a less direct route.

Imagine a row of $n$ slots, numbered from 1 to $n$ and separated by single walls where slot number $j$ represents the $j$th object. Whenever object $j$ is drawn, a ball is put in slot number $j$. After $k$ draws, we will thus have $k$ balls distributed over the $n$ slots (and slots corresponding to objects never drawn are empty). The question now reduces to how many ways there are to distribute $k$ balls over $n$ slots. This is equivalent to rearranging the $n - 1$ inner walls and the $k$ balls, which in turn is equivalent to choosing positions for the $k$ balls from a total of $n - 1 + k$ positions. But this can be done in $\binom{n-1+k}{k}$ ways, and hence this is the number of ways to choose with replacement and without regard to order. □

***Example 1.25.*** The Texas Lottery game "Pick 3" is played by picking three numbers with replacement from the numbers $0, 1, \ldots, 9$. You can play "exact order" or "any order." With the "exact order" option, you win when your numbers match the winning numbers in the exact order they are drawn. With the "any order" option, you win whenever your numbers match the winning numbers in any order. How many possible winning combinations are there with the "any order" option?

We have $n = 10$, $k = 3$, and the winning numbers are chosen with replacement and without regard to order and hence there are

$$\binom{10 - 1 + 3}{3} = \binom{12}{3} = 220$$

possible winning combinations. □

***Example 1.26.*** Draw twice from the set $\{1, \ldots, 9\}$ at random with replacement. What is the probability that the two drawn numbers are equal?

We have $n = 9$ and $k = 2$. Taking order into account, there are $9 \times 9 = 81$ possible cases, 9 of which are favorable. Hence the probability is $\frac{9}{81} = \frac{1}{9}$. If we disregard order, we have $\binom{9-1+2}{2} = 45$ possible cases and still 9 favorable and the probability is $\frac{9}{45} = \frac{1}{5}$. Since whether we draw with or without regard to order does not seem to matter to the question, why do we get different results?

The problem is that in the second case, when we draw without regard to order, the *distribution is not uniform*. For example, the outcome $\{1, 2\}$ corresponds to the two equally likely ordered outcomes $(1, 2)$ and $(2, 1)$ and is thus twice as likely as the outcome $\{1, 1\}$, which corresponds to only one ordered outcome $(1, 1)$. Thus, the first solution $\frac{1}{9}$ is correct. □

Thus, when we draw with replacement but without regard to order, we must be careful when we compute probabilities, since the distribution is not uniform, as it is in the other three cases. Luckily, this case is far more uncommon in applications than are the other three cases. There is one interesting application, though, that has to do with the number of integer solutions to a certain type of equation. If we look again at the way in which we arrived at the formula and let $x_j$ denote the number of balls in slot $j$, we realize that we must have $x_1 + \cdots + x_n = k$ and get the following observation.

**Corollary 1.2.** *There are* $\binom{n-1+k}{k}$ *nonnegative integer solutions* $(x_1, \ldots, x_n)$ *to the equation* $x_1 + \cdots + x_n = k$.

The four different ways of choosing $k$ out of $n$ objects are summarized in Table 1.2. Note that when we choose without replacement, $k$ must be less than or equal to $n$, but when we choose with replacement, there is no such restriction.

We finish with another favorite problem from the probability literature. It combines combinatorics with previous results concerning the probability of a union.

***Example 1.27*** **(*The Matching Problem*).** The numbers $1, 2, \ldots, n$ are listed in random order. Whenever a number remains in its original position in the permutation, we call this a "match." For example, if $n = 5$, then there are two matches in the

**TABLE 1.2 Choosing *k* Out of *n* Objects**

| | With Replacement | Without Replacement |
|---|---|---|
| With regard to order | $n^k$ | $n(n-1)\cdots(n-k+1)$ |
| Without regard to order | $\binom{n-1+k}{k}$ | $\binom{n}{k}$ |

permutation 32541 and none in 23451. **(a)** What is the probability that there are no matches? **(b)** What happens to the probability in (a) as $n \to \infty$?

Before we solve this, let us try to think about part (b). Does it get easier or harder to avoid matches when $n$ is large? It seems possible to argue for both. With so many choices, it is easy to avoid a match in each particular position. On the other hand, there are many positions to try, so it should not be too hard to get at least one match. It is not easy to have good intuition for what happens here.

To solve the problem, we first consider the complement of no matches and introduce the events

$$\begin{aligned} A &= \{\text{at least one match}\} \\ A_k &= \{\text{match in the } k\text{th draw}\}, \quad k = 1, 2, \ldots, n \end{aligned}$$

so that

$$A = \bigcup_{k=1}^{n} A_k$$

We will apply Proposition 1.5, so we need to figure out the probabilities of the events $A_k$ as well as all intersections of two events, three events, and so on.

First, note that there are $n!$ different permutations of the numbers $1, 2, \ldots, n$. To get a match in position $k$, there is only one choice for that number and the rest can be ordered in $(n-1)!$ different ways. We get the probability

$$P(A_k) = \frac{\#\text{ favorable outcomes}}{\#\text{ possible outcomes}} = \frac{(n-1)!}{n!} = \frac{1}{n}$$

which means that the first sum in Proposition 1.5 equals 1. To get a match in both the $i$th and the $j$th positions, we have only one choice for each of these two positions and the remaining $n-2$ numbers can be ordered in $(n-2)!$ ways and

$$P(A_i \cap A_j) = \frac{(n-2)!}{n!} = \frac{1}{n(n-1)}$$

Since there are $\binom{n}{2}$ ways to select two events $A_i$ and $A_j$, we get the following equation for the second sum in Proposition 1.5:

$$\begin{aligned} \sum_{i<j} P(A_i \cap A_j) &= \binom{n}{2} \frac{1}{n(n-1)} \\ &= \frac{n(n-1)}{2!} \times \frac{1}{n(n-1)} = \frac{1}{2!} \end{aligned}$$

Proceeding to the third sum, a similar argument gives that, for fixed $i < j < k$,

$$\sum_{i<j<k} P(A_i \cap A_j \cap A_k) = \binom{n}{3} \times \frac{1}{n(n-1)(n-2)} = \frac{1}{3!}$$

and the pattern emerges. The $j$th sum in Proposition 1.5 equals $1/j!$, and with the alternating signs we get

$$P(\text{at least one match}) = 1 - \sum_{j=2}^{n} \frac{(-1)^j}{j!} = 1 - \sum_{j=0}^{n} \frac{(-1)^j}{j!}$$

which finally gives

$$P(\text{no matches}) = \sum_{j=0}^{n} \frac{(-1)^j}{j!}$$

This is interesting. First, the probability is not monotone in $n$, so we cannot say that it gets easier or harder to avoid matches as $n$ increases. Second, as $n \to \infty$, we recognize the limit as the Taylor expansion of $e^{-1}$ and hence the probability of no matches converges to $e^{-1} \approx 0.37$ as $n \to \infty$. We can also note how rapid the convergence is; already for $n = 4$, the probability is 0.375. Thus, for all practical purposes, the probability to get no matches is 0.37 regardless of $n$. In Problem 36, you are asked to find the probability of exactly $j$ matches. □

## 1.5 CONDITIONAL PROBABILITY AND INDEPENDENCE

In this section, we introduce the important notion of *conditional probability*. The idea behind this concept is that the value of a probability can change if we get additional information. For example, the probability of contracting lung cancer is higher among smokers than nonsmokers and the probability of voting Republican is higher in Texas than in Massachusetts.

To arrive at a formal definition of conditional probabilities, we consider the example with the dartboard from Example 1.9. Suppose you throw darts repeatedly at random on a dartboard and consider only those darts that hit the number 14. In the long run, what proportion of those will also be doubles? Since the area of 14 is $142/20 = 7.1$ in.$^2$ and the area of the double ring inside 14 is $15/20 = 0.75$ in.$^2$, in the long run we expect the proportion $0.75/7.1 \approx 0.11$ of hits of 14 to also be doubles. To express this as a statement about probabilities, we can say that if we know that a dart hits 14, the probability that it is also a double is 0.11. Since the probability of 14 is $P(F) = 7.1/143$ and of both double and 14 is $P(F \cap D) = 0.75/143$, we see that the probability that a dart hits a double if we know that it hits 14 is the ratio $P(F \cap D)/P(F)$.

Now, consider a sample space in general and let $A$ and $B$ be two events. If we know that $B$ occurred in an experiment, what is the probability that $A$ also occurred? We can draw a Venn diagram and apply the same reasoning as above. Since the fraction of area of $A$ inside $B$ is $P(A \cap B)/P(B)$, it seems reasonable that this is the probability we seek. This is the intuition behind the following definition.

***Definition 1.4.*** Let $B$ be an event such that $P(B) > 0$. For any event $A$, denote and define the *conditional probability of A given B* as

$$P(A|B) = \frac{P(A \cap B)}{P(B)}$$

We think of this as the probability of $A$ if we know that $B$ has occurred. Hence, to compute a conditional probability means to compute a probability given additional information.

***Example 1.28.*** Let us revisit Mrs B and Mrs T from Example 1.7. If we introduce a third event

$$C = \{\text{he is shabby-looking}\}$$

then one way to interpret Mrs T's comment "that sounds more probable" is that

$$P(A|B \cap C) > P(A|C)$$

that is, given that more of the background is known, it seems more likely that the person is who Mrs B says he is. □

***Example 1.29.*** Roll a die and observe the number. Let

$$A = \{\text{odd outcome}\} \text{ and } B = \{\text{at least } 4\}$$

What is $P(A|B)$?

We solve this in two different ways: (1) by using the definition and (2) by intuitive reasoning. Since $P(A \cap B) = P(\{5\}) = \frac{1}{6}$ and $P(B) = \frac{1}{2}$, the definition gives

$$P(A|B) = \frac{P(A \cap B)}{P(B)} = \frac{1/6}{1/2} = \frac{1}{3}$$

If we think about this intuitively, to condition on the event $B$ means that we get the additional information that the outcome is at least 4. Since one of these three outcomes is also odd and outcomes are equally likely, the conditional probability of odd is $\frac{1}{3}$. □

There is no general rule for whether it is easier to use the definition or intuitive reasoning. In the previous example, the "one out of three" approach works since outcomes are equally likely, but this is not always the case.

Conditional probabilities can make it easier to compute probabilities of intersections. Say that we want to compute $P(A \cap B)$ but that it is tricky to do so directly. However, if we can find $P(B)$ and $P(A|B)$, then the definition tells us that $P(A \cap B) = P(A|B)P(B)$ and we are done. Let us look at some examples of this.

***Example 1.30.*** In Example 1.8, we had the events $A = \{\text{rain on Saturday}\}$ and $B = \{\text{rain on Sunday}\}$, where $P(A) = P(B) = 0.5$. Now, suppose that a rainy day is followed by another rainy day with probability 0.7. What is the probability of rain during the weekend?

We already know that the probability of a rainy weekend is

$$P(A \cup B) = 1 - P(A \cap B)$$

where we can now compute $P(A \cap B)$ as

$$P(A \cap B) = P(B|A)P(A) = 0.7 \times 0.5 = 0.35$$

and we get

$$P(A \cup B) = 0.65$$

as the probability of rain during the weekend. □

***Example 1.31.*** From a deck of cards, draw four cards at random, without replacement. If you get $j$ aces, draw $j$ cards from another deck. What is the probability of getting exactly two aces from each deck?

With

$$A = \{\text{two aces from the first deck}\}$$
$$B = \{\text{two aces from the second deck}\}$$

the event of interest is $A \cap B$, and it is not that easy to figure out its probability directly. However, if we use conditional probabilities, it is simple. We get

$$P(A) = \frac{\binom{4}{2}\binom{48}{2}}{\binom{52}{4}} \quad \text{and} \quad P(B|A) = \frac{\binom{4}{2}}{\binom{52}{2}}$$

and hence

$$P(A \cap B) = P(B|A)P(A) = \frac{\binom{4}{2}\binom{48}{2}}{\binom{52}{4}} \times \frac{\binom{4}{2}}{\binom{52}{2}} \approx 0.0001$$

□

***Example 1.32.*** The online bookseller amazon.com has a feature called the "Gold Box." When you enter this, you are presented with 10 special offers to buy various merchandise, anything from books and DVDs, to kitchenware and the "Panasonic ER411NC nose and ear hair groomer." The offers are presented one at a time and each time you have to decide whether to take it or to pass. If you take it, you are done and will not get to see the rest of the offers. If you pass, that offer is gone and cannot be retrieved. Suggest a strategy that gives you at least 25% chance to win the best offer.

Let us assume that the offers are presented in random order. If your strategy is to always take the first offer or if you choose at random, your chance to win is 10%. How can this be improved?

A better strategy is to let five offers pass, remember the best thus far, and take the next offer that is better. If this never happens, you are forced to take the last offer. One case in which you will certainly win is if the second best offer is among the first five and the best is among the remaining five. Thus, let

$$A = \{\text{second best offer is among the first five}\}$$
$$B = \{\text{best offer is among the last five}\}$$

so that the event of interest is $A \cap B$, which has probability

$$P(A \cap B) = P(A|B)P(B)$$

Since the offers are randomly ordered, the best offer is equally likely to be in any position and hence $P(B) = \frac{5}{10}$. Given that the best is among the last five, the second best is equally likely to be any of the remaining nine, so the probability that it is among the first five is $P(A|B) = \frac{5}{9}$ and we get

$$P(\text{get the best offer}) = \frac{5}{9} \times \frac{5}{10} \approx 0.28$$

which is larger than 0.25. Note that $A \cap B$ is not the only way in which you can get the best offer, so the true probability is in fact higher than 0.28.

Generally, if there are $n$ offers, the same strategy gives a probability to get the best offer that is at least

$$P(A \cap B) = P(A|B)P(B) = \frac{n/2}{n-1} \times \frac{n/2}{n} = \frac{n}{4(n-1)}$$

which is greater than $\frac{1}{4}$ regardless of $n$ [if $n$ is odd, we can replace $n/2$ by $(n+1)/2$]. It is quite surprising that we can do so well and for example have at least 25% chance to find the best of 10 million offers. It can be shown that an even better strategy is to first discard roughly $ne^{-1}$ offers and then take the next that is better. The probability to win is then approximately $e^{-1} \approx 0.37$ (a number that also showed up in Example 1.27). □

The way in which we have defined conditional probability makes good intuitive sense. However, remember that a probability is defined as something that satisfies the three axioms in Definition 1.3. We must therefore show that whenever we condition on an event $B$, the definition of conditional probability does not violate any of the axioms. We state this in a proposition.

**Proposition 1.9.** For fixed $B$, $P(A|B)$ satisfies the probability axioms:

**(a)** $0 \leq P(A|B) \leq 1$

**(b)** $P(S|B) = 1$

**(c)** If $A_1, A_2, \ldots$ is a sequence of pairwise disjoint events, then

$$P\left(\bigcup_{k=1}^{\infty} A_k \,\middle|\, B\right) = \sum_{k=1}^{\infty} P(A_k|B)$$

*Proof.* Since $A \cap B \subseteq B$, we get $0 \leq P(A \cap B) \leq P(B)$ and part (a) follows. For (b), note that $B \subseteq S$ so that $P(S \cap B) = P(B)$ and hence $P(S|B) = 1$. Finally, for (c) first note that

$$\left(\bigcup_{k=1}^{\infty} A_k\right) \cap B = \bigcup_{k=1}^{\infty} (A_k \cap B)$$

and since $A_1, A_2, \ldots$ are pairwise disjoint, so are the events $A_1 \cap B, A_2 \cap B, \ldots$, and we get

$$P\left(\left(\bigcup_{k=1}^{\infty} A_k\right) \cap B\right) = \sum_{k=1}^{\infty} P(A_k \cap B)$$

Divide both sides with $P(B)$ to conclude the proof. ■

It is easily realized that $P(B|B) = 1$, and with this in mind, we can think of conditioning on $B$ as viewing $B$ as the new sample space. The nice thing about the proposition is that we now know that conditional probabilities have all the properties of probabilities that we stated in Proposition 1.3. We restate these properties for conditional probabilities in a corollary.

**Corollary 1.3.** *Provided that the conditional probabilities are defined, the following properties hold:*

**(a)** $P(A^c|B) = 1 - P(A|B)$

**(b)** $P(B \setminus A|C) = P(B|C) - P(A \cap B|C)$

**(c)** $P(A \cup B|C) = P(A|C) + P(B|C) - P(A \cap B|C)$
**(d)** *If* $A \subseteq B$, *then* $P(A|C) \leq P(B|C)$

It is important to keep in mind that properties of probabilities hold for events to the left of the conditioning bar and that the event to the right is fixed (see Problem 37).

If we think of probability as a measure of degree of belief, we can think of conditional probability as an update of that degree, in the light of new information. Here is an example of a logical oddity that philosophers of science love to toss around to confuse the rest of us.

***Example 1.33.*** Consider the hypothesis "all swans are white." We can say that each observation of a white swan strengthens our belief in, or *corroborates*, the hypothesis. Also, since the two statements "all swans are white" and "all nonwhite objects are nonswans" are logically equivalent, the hypothesis is also corroborated by the observation of something that is neither white nor a swan. Thus, every sighting of a yellow dog corroborates the hypothesis that all swans are white.[3]

Weird, isn't it? A zoologist trying to prove the hypothesis would certainly decide to examine swans for whiteness, rather than checking various red, green, and blue objects to make sure that they are not swans. Still, there is certainly nothing wrong with the logic, so how can the paradox be resolved? Let us try a probabilistic approach.

Suppose that we have all examinable objects in a big urn. Suppose that there are $n$ such objects, $k$ of which are white, and that the other $n - k$ are black (representing "nonwhite"). Suppose further that $j$ of the objects are swans, and call the remaining objects "ravens," another favorite bird among philosophers of science. If we do not know anything about the whiteness of swans, we may assume that the $j$ swans are randomly spread among the $n$ objects. Thus, when we choose a swan, the probability that it is white is $\frac{k}{n}$ (if we have very strong belief in the hypothesis to begin with, we can just introduce a lot of white "dummy objects" to make this probability anything we want). The probability that the hypothesis is true can now be thought of as the probability to get only white objects when we draw without replacement $j$ times (assign the "swan property" to $j$ objects). Our hypothesis is then the event

$$H = \{\text{all swans are white}\} = \{\text{get } j \text{ white objects}\}$$

Let us choose with regard to order (which does not matter to the problem, but expressions get less messy). Thus, the probability that all swans are white is

$$P(H) = \frac{k(k-1)\cdots(k-j+1)}{n(n-1)\cdots(n-j+1)}$$

[3]For ornithologists: This has nothing to do with *Cygnus atratus*.

We now follow two different strategies: (a) to examine swans and (b) to examine black objects. Suppose that we get a corroborating observation. How does this affect the probability of $H$, now pertaining to the remaining $n-1$ objects? Let $C_a$ and $C_b$ be the events to get corroborating observations with the two strategies, respectively. With strategy (a), a corroborating observation means that one white swan has been removed, and the conditional probability of $H$ becomes

$$P(H|C_a) = \frac{(k-1)(k-2)\cdots(k-j+1)}{(n-1)(n-2)\cdots(n-j+1)}$$

With strategy (b), one black raven has been removed, and we get

$$P(H|C_b) = \frac{k(k-1)\cdots(k-j+1)}{(n-1)(n-2)\cdots(n-j)}$$

Both these are larger than the original $P(H)$, so each corroborating observation indeed strengthens belief in the hypothesis. But do they do so to equal extents? Let us compare the two conditional probabilities. We get

$$\frac{P(H|C_a)}{P(H|C_b)} = \frac{n-j}{k}$$

If we now assume that the number of swans is less than the number of black objects, certainly a reasonable assumption, we have that $j < n-k$, which gives $k < n-j$, and hence

$$\frac{P(H|C_a)}{P(H|C_b)} > 1$$

so that the observation of a black raven does corroborate the hypothesis but *not as much* as the sighting of a white swan. The intuition is simple: since there are fewer swans than black objects, it is easier to check the swans. If instead $j > n-k$, strategy (b) would be preferable. If we, for example, were to corroborate the hypothesis "All Volvo drivers live outside the Vatican," it would be better to ask a thousand Vaticanos what they drive, than to track down Volvo drivers in London and Paris to check if they happen to be vacationing Swiss Guardsmen. □

### 1.5.1 Independent Events

In the previous section, we dealt with conditional probabilities and learned to interpret them as probabilities that are computed given additional information. It is easy to think of cases when such additional information is irrelevant and does not change the probability. For example, if we are about to flip a fair coin, the probability to get heads is $\frac{1}{2}$. Now suppose that we get the additional information that the coin was flipped once yesterday and showed heads. Since our upcoming coin flip is not affected by what happened yesterday and we know that the coin is fair, the conditional probability given this information is still $\frac{1}{2}$. With $A = \{\text{heads in next flip}\}$ and $B = \{\text{heads yesterday}\}$ we thus have $P(A) = P(A|B)$; the unconditional and conditional probabilities are the same. Since $P(A|B) = P(A \cap B)/P(B)$, this means that $P(A \cap B) = P(A)P(B)$, and we call two events with this property *independent*.

***Definition 1.5.*** If $A$ and $B$ are two events such that

$$P(A \cap B) = P(A)P(B)$$

then they are said to be *independent*.

Not surprisingly, events that are not independent are called *dependent*. In the introductory motivation for the definition, we talked about conditional and unconditional probabilities being equal. We could take this as the definition of independence, but since conditional probabilities are not always defined, we use the definition of independence above and get the following consequence.

**Corollary 1.4.** *If $P(A|B)$ is defined, then the events $A$ and $B$ are independent if and only if $P(A) = P(A|B)$.*

When checking for independence, it might sometimes be easier to condition on the event $B^c$ instead of $B$, that is, by supposing that $B$ did not occur. Intuitively, information regarding $B$ and information on $B^c$ are equivalent since saying that one occurred is the same as saying that the other one did not occur. This is stated formally as follows.

**Proposition 1.10.** If $A$ and $B$ are independent, then $A$ and $B^c$ are also independent.

*Proof.* By Proposition 1.3(b), we get

$$P(A \cap B^c) = P(A \setminus B) = P(A) - P(A \cap B)$$

and if $A$ and $B$ are independent, this equals

$$P(A) - P(A)P(B) = P(A)(1 - P(B)) = P(A)P(B^c)$$

and $A$ and $B^c$ are independent. ∎

***Example 1.34.*** In Example 1.30, suppose that a rainy Saturday and a rainy Sunday are independent events. What is the probability of rain during the weekend?

In this case

$$P(A \cap B) = P(A)P(B) = 0.25$$

and hence

$$P(A \cup B) = 0.75$$

which we note is higher than the 0.65 we obtained if rainy Saturdays are more likely to be followed by rainy Sundays. The reason is that under this assumption, rainy Saturdays and Sundays tend to come together more often than under the independence assumption. □

***Example 1.35.*** A card is chosen at random from a deck of cards. Consider the events

$$A = \{\text{the card is an ace}\} \quad \text{and} \quad H = \{\text{the card is a heart}\}$$

Are $A$ and $H$ independent?

Let us first solve this by using the definition. We have $P(A) = \frac{4}{52}$, $P(H) = \frac{1}{4}$, and $P(A \cap H) = P(\text{ace of hearts}) = \frac{1}{52}$ and hence

$$P(A \cap H) = P(A)P(H)$$

so that $A$ and $H$ are independent. Intuitively, the events give no information about each other. The probability of drawing an ace is $\frac{4}{52} = \frac{1}{13}$ and if we are given the information that the chosen card is a heart, the probability of an ace is still $\frac{1}{13}$. The proportion of aces is the same in the deck as within the suit of hearts. □

***Example 1.36.*** Consider the previous example but suppose that we have removed the 2 of spades from the deck. Are the events $A$ and $H$ still independent?

At first glance, we might think that the answer is "Yes" since the 2 of spades has nothing to do with either hearts or aces. However, the probabilities are now $P(A) = \frac{4}{51}$, $P(H) = \frac{13}{51}$, and $P(A \cap H) = P(\text{ace of hearts}) = \frac{1}{51}$ and hence

$$P(A \cap H) \neq P(A)P(H)$$

and $A$ and $H$ are no longer independent. Intuitively, although the 2 of spades has nothing to do with hearts or aces, its removal changes the proportion of aces in the deck from $\frac{4}{52}$ to $\frac{4}{51}$ but does not change the proportion within the suit of hearts, where it remains at $\frac{1}{13}$. Formulated as a statement about conditional probabilities, we have that

$$P(A) = \frac{4}{51} \quad \text{and} \quad P(A|H) = \frac{1}{13}$$

which are not equal. □

***Example 1.37.*** Are disjoint events independent?

It seems that disjoint events have nothing to do with each other and should thus be independent. However, this reasoning is faulty. The correct reasoning is that if we condition on one event having occurred, then the other cannot have occurred, and hence its conditional probability drops to 0. We can also see this from the definition of independence since if $A$ and $B$ are disjoint, then $A \cap B = \emptyset$ and hence $P(A \cap B) = P(\emptyset) = 0$, which does not equal the product $P(A)P(B)$ (assuming that neither of these probabilities equal 0). Hence, the answer in general is "absolutely not." □

In Example 1.36, the events {ace} and {hearts} are dependent. Computation yields that $P(A) = 0.078$ and $P(A|H) = 0.077$, so the difference is negligible from a practical point of view. We could say that although the events are dependent, the dependence is not strong. Compare this with the case of disjoint events where the conditional probability drops down to 0, which indicates a much stronger dependence. Dependence could also go in different directions; $P(A|B)$ could be either larger or smaller than $P(A)$. We will later return to the problem of measuring the degree of dependence in a more general context (see also Problem 42).

The following two examples illustrate how it is not always obvious which event to condition on and how it is important to find the correct such event.

***Example 1.38.*** You know that your new neighbors have two children. Given that they have at least one daughter, what is the conditional probability that they have two daughters?

The sample space is

$$S = \{bb, bg, gb, gg\}$$

where $b$ represents boy, $g$ represents girl, and the order is birth order. If we assume that genders are equally likely and that genders of different children are independent, each outcome has probability $\frac{1}{4}$. Since the outcome $bb$ is out of the question and one out of the other three outcomes has two girls, the conditional probability is $\frac{1}{3}$. Formally

$$P(gg|bg, gb, gg) = \frac{P(gg)}{P(bg, gb, gg)} = \frac{1/4}{3/4} = \frac{1}{3}$$

□

***Example 1.39.*** You know that your new neighbors have two children. One day you see the mother taking a walk with a girl. What is the probability that the other child is also a girl?

This looks like the same problem. On the basis of your observation, you rule out the outcome $bb$ and the conditional probability of another girl is $\frac{1}{3}$. On the other hand, since we assume that genders of different children are independent, the probability ought to be $\frac{1}{2}$.

Confusing? Let us clear it up. The first solution is incorrect, but why? While it is true that the probability of two girls, given at least one girl, is $\frac{1}{3}$, this is *not the correct event* on which to condition in this case. We are not just observing "at least one girl;" we are observing the mother walking with a *particular girl*. This distinction is important but quite subtle, and requires that we extend the sample space to be able to also describe how the mother chooses which child to walk with.[4] Thus, we split each outcome into two, and if we denote the child that goes for the walk by an asterisk, the new sample space is

$$S = \{b^*b, bb^*, b^*g, bg^*, g^*b, gb^*, g^*g, gg^*\}$$

where, for example, $b^*g$ means that the older child is a boy, and the younger, a girl, and that the mother takes a walk with the boy. If the mother chooses child at random, each outcome has probability $\frac{1}{8}$. It is now easy to see that four outcomes have the mother walking with a girl and that two of these have another girl, and we arrive at the solution $\frac{1}{2}$ once more (see also Problem 80). □

We also want to define independence of more than two events. To arrive at a reasonable definition, let us first examine an example that highlights one of the problems that must be addressed.

***Example 1.40.*** Flip two fair coins and consider the events

$$A = \{\text{heads in first flip}\} = \{HH, HT\}$$
$$B = \{\text{heads in second flip}\} = \{HH, TH\}$$
$$C = \{\text{different in first and second flip}\} = \{HT, TH\}.$$

Then, for example, $P(A \cap B) = P(HH) = \frac{1}{4} = P(A)P(B)$, so $A$ and $B$ are independent. Similarly, it is easy to show that any two of the events are independent. Hence, these events are *pairwise independent*. However, it does not seem quite right to say that the three events $A$, $B$, and $C$ are independent since, for example, $C$ is not independent of the event $A \cap B$. Indeed, $P(C) = \frac{1}{2}$ but $P(C|A \cap B) = 0$, since if $A \cap B$ has occurred, both flips showed heads and $C$ is impossible. □

This example indicates that in order to call three events independent, we want each event to be independent of any combination of the other two. It turns out that the following definition guarantees this (see Problem 53).

[4]Ironically, in the first edition of his excellent book *Innumeracy: Mathematical Illiteracy and Its Consequences*, John Allen Paulos described this problem a bit obscurely [4]. His terse formulation was "Consider now some randomly selected family of four. Given that Myrtle has a sibling, what is the conditional probability that her sibling is a brother?" and he went on to claim that the probability is $\frac{2}{3}$. This ambiguity was clarified in the 2001 edition.

*Definition 1.6.* Three events $A$, $B$, and $C$ are called independent if the following two conditions hold:

**(a)** They are pairwise independent

**(b)** $P(A \cap B \cap C) = P(A)P(B)P(C)$

For more than three events, the definition is analogous and can also be extended to infinitely many events.

*Definition 1.7.* The events $A_1, A_2, \ldots$ are called independent if

$$P(A_{i_1} \cap A_{i_2} \cap \cdots \cap A_{i_k}) = P(A_{i_1})P(A_{i_2}) \cdots P(A_{i_k})$$

for all sequences of integers $i_1 < i_2 < \cdots < i_k$, $k = 2, 3, \ldots$

Sometimes events satisfying this definition are called *mutually independent*, to distinguish from *pairwise independent*, which, as we have seen, is a weaker property.

***Example 1.41.*** Recall the experiment of rolling a die repeatedly until the first 6 appears. What is the probability that this occurs in the $n$th roll for $n = 1, 2, \ldots$?

The event of interest is

$$B_n = \{\text{first 6 in } n\text{th roll}\}, \quad n = 1, 2, \ldots$$

and let us also introduce the events

$$A_k = \{6 \text{ in } k\text{th roll}\}, \quad k = 1, 2, \ldots$$

Note the difference: $B_n$ is the event that the *first* 6 comes in the $n$th roll; $A_k$, the event that we get 6 in the $k$th roll but not necessarily for the first time. How do the events relate to each other? Obviously, $B_1 = A_1$. For $n = 2$, note that $B_2$ is the event that we do not get 6 in the first roll and that we do get 6 in the second roll. In terms of the $A_k$, this is $A_1^c \cap A_2$. In general

$$B_n = A_1^c \cap A_2^c \cap \cdots \cap A_{n-1}^c \cap A_n$$

To compute the probability of $B_n$, we make two reasonable assumptions: that the die is fair and that rolls are independent. The first assumption means that $P(A_k) = \frac{1}{6}$ for all $k$ and the second that probabilities of intersections equal products of probabilities. Since independence carries over to complements, we get

$$\begin{aligned} P(B_n) &= P(A_1^c \cap A_2^c \cap \cdots \cap A_{n-1}^c \cap A_n) \\ &= P(A_1^c)P(A_2^c) \cdots P(A_{n-1}^c)P(A_n) \\ &= \frac{5}{6} \times \frac{5}{6} \times \cdots \times \frac{5}{6} \times \frac{1}{6} \end{aligned}$$

and we conclude that

$$P(B_n) = \frac{1}{6}\left(\frac{5}{6}\right)^{n-1}, \quad n = 1, 2, \ldots$$

□

More generally, consider independent repetitions of a trial where the event $A$ occurs with probability $p > 0$ and let $E$ be the event that we *never* get $A$. With

$$B_n = \{\text{first occurrence of} A \text{ comes after the } n\text{th trial}\}$$

we have

$$E = \bigcap_{n=1}^{\infty} B_n$$

where

$$P(B_n) = P(\text{the first } n \text{ trials give } A^c) = (1 - p)^n$$

by independence. The $B_n$ are clearly decreasing (why?), so by Proposition 1.6 we get

$$P(E) = \lim_{n \to \infty} P(B_n) = 0$$

and we summarize in the following corollary.

**Corollary 1.5.** *In independent repetitions of a trial, any event with positive probability occurs sooner or later.*

From Example 1.21, we can compute the probability to win the Texas Lotto jackpot (match all numbers including bonus ball) as $1/47,784,352 = 2.1 \times 10^{-8}$. This is very small, but if you keep playing, the last result tells you that you will win eventually.[5] It may take some time, though; there are two draws a week and if you play every time for 50 years, the probability that you never win is

$$(1 - 2.1 \times 10^{-8})^{5200} \approx 0.9999$$

[5]The subtle difference between *certain* occurrence and occurrence *with probability one* is important in a more advanced study of probability theory but not for us at this point.

The probability that you win in a drawing is very low, but since there are millions of players in each draw, the probability that *somebody* wins is much higher. Suppose that 5 million number combinations are played independently and at random for a drawing. The probability that somebody wins is

$$1 - (1 - 2.1 \times 10^{-8})^{5{,}000{,}000} \approx 0.10$$

which is not that low, and it could be you.

***Example 1.42*** **(*Reliability Theory*).** Consider a system of two electronic components connected in series. Each component functions with probability $p$ and the components function independent of each other. What is the probability that the system functions?

If we interpret "functions" as the natural "lets current through," then the system functions if and only if both components function. Hence, with the events

$$\begin{aligned} A &= \{\text{system functions}\} \\ A_1 &= \{\text{first component functions}\} \\ A_2 &= \{\text{second component functions}\} \end{aligned}$$

we get $A = A_1 \cap A_2$ and by independence

$$P(A) = P(A_1)P(A_2) = p^2$$

If the components are instead connected in parallel, the system functions as long as at least one of the components function, and we have

$$A = A_1 \cup A_2$$

which gives

$$\begin{aligned} P(A) = P(A_1 \cup A_2) \; &= \; 1 - P(A_1^c \cap A_2^c) \\ = 1 - (1 - P(A_1))(1 - P(A_2)) \; &= \; 1 - (1 - p)^2 \end{aligned}$$

These are simple examples from the discipline of *reliability theory* where the probability of functioning is referred to as the *reliability* of a system. Hence, we have seen that the reliability of a series system is $p^2$ and that of a parallel system is $1 - (1 - p)^2$. An obvious generalization is to $n$ components, where the reliability of a series system is $p^n$ and that of a parallel system is $1 - (1 - p)^n$. This does not have to be about electronic components but applies to any situation where a complex system is dependent on its individual parts to function. The series system is sometimes referred to as a *weakest-link model*. □

## 1.6 THE LAW OF TOTAL PROBABILITY AND BAYES' FORMULA

In this section, we will address one of the most important uses of conditional probabilities. The basic idea is that if a probability is hard to compute directly, it might help to break the problem up in special cases, where in each special case the conditional probability is easier to compute. For example, suppose that you buy a used car in a city where street flooding due to heavy rainfall is a common problem. You know that roughly 5% of all used cars have previously been flood-damaged and estimate that 80% of such cars will later develop serious engine problems, whereas only 10% of used cars that are not flood-damaged develop the same problems. What is the probability that your car will later run into this kind of trouble?

Here is a situation where you can compute the probability in each of two different cases, flood-damaged or not flood-damaged (and no used-car dealer worth his salt would ever let you know which).

Let us first think about this in terms of proportions. Out of every 1000 cars sold, 50 are previously flood-damaged and of those, 80%, or 40 cars, will develop serious engine problems. Among the 950 that are not flood-damaged, we expect 10%, or 95 cars, to develop the same problems. Hence, we get a total of $40 + 95 = 135$ cars out of a 1000, and the probability of future problems is 0.135.

If we introduce the events $F = \{\text{flood-damaged}\}$ and $T = \{\text{trouble}\}$, we have argued that $P(T) = 0.135$. We also know that $P(F) = 0.05$, $P(F^c) = 0.95$, $P(T|F) = 0.80$, and $P(T|F^c) = 0.10$ and the probability we computed is in fact $0.80 \times 0.05 + 0.10 \times 0.95 = 0.135$. Our probability is a weighted average of the probability in the two different cases, flood-damaged or not, and the weights are the corresponding probabilities of the cases. The example illustrates the idea behind the following important result.

***Theorem 1.1 (Law of Total Probability).*** *Let $B_1, B_2, \ldots$ be a sequence of events such that*

**(a)** $P(B_k) > 0$ *for* $k = 1, 2, \ldots$

**(b)** $B_i$ *and* $B_j$ *are disjoint whenever* $i \neq j$

**(c)** $S = \bigcup_{k=1}^{\infty} B_k$

*Then, for any event A, we have*

$$P(A) = \sum_{k=1}^{\infty} P(A|B_k)P(B_k)$$

Condition (a) is a technical requirement to make sure that the conditional probabilities are defined, and you may recall that a collection of sets satisfying (b) and (c) is called a *partition* of $S$.

*Proof.* First note that

$$A = A \cap S = \bigcup_{k=1}^{\infty} (A \cap B_k)$$

by the distributive law for infinite unions. Since $A \cap B_1, A \cap B_2, \ldots$ are pairwise disjoint, we get

$$P(A) = \sum_{k=1}^{\infty} P(A \cap B_k) = \sum_{k=1}^{\infty} P(A|B_k)P(B_k)$$

which proves the theorem. ■

By virtue of Proposition 1.2, we realize that the law of total probability is also true for a finite union of events, $B_1, \ldots, B_n$. In particular, if we choose $n = 2$ and $B_1$ equal to some event $B$, then $B_2$ must equal $B^c$, and we get the following corollary.

**Corollary 1.6.** *If* $0 < P(B) < 1$, *then*

$$P(A) = P(A|B)P(B) + P(A|B^c)P(B^c)$$

The verbal description of conditions (b) and (c) in Theorem 1.1 is that we are able to find different cases that *exclude each other* and *cover all possibilities*. This way of thinking about it is often sufficient to solve problems and saves us the effort to explicitly find the sample space and the partitioning events.

***Example 1.43.*** A sign reads HOUSTON. Two letters are removed at random and then put back together again at random in the empty spaces. What is the probability that the sign still reads HOUSTON?

There are two different cases to consider: the case where two Os are chosen, in which case the text will always be correct and the case when two different letters are chosen, in which case the text will be correct when they are put back in their original order. Clearly, these two cases exclude each other and cover all possibilities and the assumptions in the law of total probability are satisfied. Hence, without spelling out exactly what the sample space is, we can define the events

$$A = \{\text{the sign still reads HOUSTON}\}$$
$$B = \{\text{two Os are chosen}\}$$

and obtain

$$P(A) = P(A|B)P(B) + P(A|B^c)P(B^c)$$

If the two letters are different, they are put back in their original order with probability $\frac{1}{2}$. Hence, the conditional probabilities are

$$P(A|B) = 1 \quad \text{and} \quad P(A|B^c) = \frac{1}{2}$$

and $P(B)$ is obtained by noting that we are choosing two letters out of seven without replacement and without regard to order. The total number of ways to do this is $\binom{7}{2} = 21$, and since there is only one way to choose the two Os, we get $P(B) = \frac{1}{21}$. This gives $P(B^c) = \frac{20}{21}$ and we get

$$P(A) = 1 \times \frac{1}{21} + \frac{1}{2} \times \frac{20}{21} = \frac{11}{21}$$

which is slightly larger than $\frac{1}{2}$, as was to be expected. □

***Example 1.44.*** In the United States, the overall risk of developing lung cancer is about 0.1%. Among the 20% of the population who are smokers, the risk is about 0.4%. What is the risk that a nonsmoker will develop lung cancer?

Introduce the events $C = \{\text{cancer}\}$ and $S = \{\text{smoker}\}$. The percentages above give $P(C) = 0.001$, $P(S) = 0.20$, and $P(C|S) = 0.004$, and we wish to compute $P(C|S^c)$. The law of total probability gives

$$P(C) = P(C|S)P(S) + P(C|S^c)P(S^c)$$

which with our numbers becomes

$$0.001 = 0.004 \times 0.20 + P(C|S^c) \times 0.80$$

which we solve for $P(C|S^c)$ to get

$$P(C|S^c) = 0.00025$$

in other words, a 250 in a million risk. □

***Example 1.45.*** Here is an example of a simple game of dice that does not seem to be to your advantage but turns out to be so.

Consider three dice, $A$, $B$, and $C$, numbered on their six sides as follows:

Die $A$ 1, 1, 5, 5, 5, 5
Die $B$ 3, 3, 3, 4, 4, 4
Die $C$ 2, 2, 2, 2, 6, 6

The game now goes as follows. You and your opponent bet a dollar each, and you offer your opponent to choose any die and roll it. Next, you choose one of the remaining dice and roll it, and whoever gets the higher number wins the money. It seems that

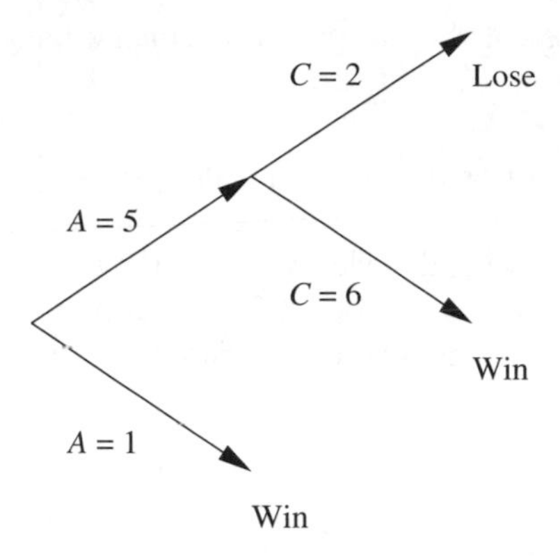


**FIGURE 1.5** Tree diagram when you roll die C against die A in Example 1.45.

your opponent will have an edge since he gets to choose first. However, it turns out that once you know his choice, you can always choose so that your probability to win is more than one half! The reason for this is that, when rolled two by two against each other, these dice are such that on average $A$ beats $B$, $B$ beats $C$, and $C$ beats $A$. The probabilities are (using $A$ and $C$ also to denote the numbers on dice $A$ and $C$)

$$P(A \text{ beats } B) = P(A = 5) = \frac{2}{3}$$
$$P(B \text{ beats } C) = P(C = 2) = \frac{2}{3}$$

For the third case, we need to use the law of total probability and get

$$\begin{aligned} P(C \text{ beats } A) &= P(C \text{ beats } A | A = 1) \times \frac{1}{3} + P(C \text{ beats } A | A = 5) \times \frac{2}{3} \\ &= 1 \times \frac{1}{3} + P(C = 6) \times \frac{2}{3} = \frac{1}{3} + \frac{1}{3} \times \frac{2}{3} = \frac{5}{9} \end{aligned}$$

which is also greater than $\frac{1}{2}$. Although you appear generous to let your opponent choose first, this is precisely what gives you the advantage.[6] □

Tree diagrams provide a nice way to illustrate the law of total probability. We represent each different case with a branch and look at the leaves to see which cases are of interest. We then compute the probability by first multiplying along each branch, then adding across the branches. See Figure 1.5 for an illustration of the situation in Example 1.45, where you roll die $C$ against die $A$.

Sometimes, we need to condition repeatedly. For example, to compute $P(A | B)$, it may be necessary to condition further on some event $C$. Since a conditional probability is

[6] A completely deterministic version of this is the game "rock, paper, scissors," in which you would always win if your opponent were to choose first. Games like these are called *nontransitive*.

a probability, this is nothing new, but the formula looks more complicated. We get

$$P(A|B) = P(A|B \cap C)P(C|B) + P(A|B \cap C^c)P(C^c|B) \tag{1.2}$$

where we note that every probability has the event $B$ to the right of the conditioning bar. In Problem 81 you are asked to prove this.

***Example 1.46* (*Simpson's Paradox*).** In a by now famous study of gender bias at the University of California, Berkeley, it was noted that men were more likely than women to be admitted to graduate school. In 1 year, in the six largest majors, 45% of male applicants but only 30% of the female ones were admitted. To further study the bias, we divide the majors into two groups "difficult" and "easy," referring to whether it is relatively difficult or easy to be admitted, not to the subjects themselves. It then turns out that in the "difficult" category, 26% of both men and women were admitted (actually even slightly above 26% for women and slightly below for men), so the bias obviously has to be in the other category. However, in the "easy" category, 80% of women but only 62% of men were admitted. Thus, there was no bias for difficult majors, a bias against men in easy majors, and an overall bias against women! Clearly there must be an error somewhere?

Consider a randomly chosen applicant. Let $A$ be the event that the applicant is admitted, and let $M$ and $W$ be the events that the applicant is a man and a woman, respectively. We then have $P(A|M) = 0.45$ and $P(A|W) = 0.30$. Now also introduce the events $D$ and $E$, for "difficult" and "easy" majors. By Table 1.3 we have, for men

$$P(A|M \cap D) = \frac{334}{1306} \approx 0.26 \quad \text{and} \quad P(A|M \cap E) = \frac{864}{1385} \approx 0.62$$

and for women

$$P(A|W \cap D) \approx 0.26 \quad \text{and} \quad P(A|W \cap E) \approx 0.80$$

and hence

$$P(A|M \cap D) = P(A|W \cap D) \quad \text{and} \quad P(A|M \cap E) < P(A|W \cap E)$$

but

$$P(A|M) > P(A|W)$$

**TABLE 1.3 Numbers of Admitted, and Total Numbers (in Parentheses) of Male and Female Applicants in the Two Categories "Easy" and "Difficult" at UC Berkeley**

| | Male | Female |
|---|---|---|
| Easy major | 864 (1385) | 106 (133) |
| Difficult major | 334 (1306) | 451 (1702) |

Thus, the conditional probabilities of being admitted are equal or higher for women in both categories but the overall probability for a woman to be admitted is lower than that of a man. Apparently, there was no error, but it still seems paradoxical. To resolve this, recall Equation 1.43, by which

$$P(A|W) = P(A|W \cap D)P(D|W) + P(A|W \cap E)P(E|W)$$

and

$$P(A|M) = P(A|M \cap D)P(D|M) + P(A|M \cap E)P(E|M)$$

and we realize that the explanation lies in the conditional probabilities $P(D|W)$, $P(E|W)$, $P(D|M)$, and $P(E|M)$, which reflect how men and women choose their majors. The probabilities that a man chooses a difficult major and an easy major, respectively, are

$$P(D|M) = \frac{1306}{2691} \approx 0.49 \quad \text{and} \quad P(E|M) \approx 0.51$$

and the corresponding probabilities for women are

$$P(D|W) = \frac{1702}{1835} \approx 0.93 \quad \text{and} \quad P(E|W) \approx 0.07$$

Thus, women almost exclusively applied for difficult majors, whereas men applied equally for difficult and easy majors, and this is the resolution of the paradox. Was it harder for women to be admitted? Yes. Was this due to gender discrimination? No. The effect on admission rates that was initially attributed to gender bias was really due to choice of major, an example of what statisticians call *confounding of factors*. The effect of gender on choice of major is a completely different issue. □

The last example is a version of what is known as *Simpson's paradox*. If we formulate it as a mathematical problem, it completely loses its charm. The question then becomes if it is possible to find numbers $A$, $a$, $B$, $b$, $p$, and $q$, all between 0 and 1, such that

$$A > a \quad \text{and} \quad B > b$$

and

$$pA + (1 - p)B < qa + (1 - q)b$$

No problems here. Let $A > a > B > b$, and choose $p$ sufficiently close to 0 and $q$ sufficiently close to 1. Ask your mathematician friends this question, and also if there is something strange about the Berkeley admissions data, and don't be surprised if you get the answer "Yes" to both questions!

### 1.6.1 Bayes' Formula

We next turn to the situation when we know conditional probabilities in one direction but want to compute conditional probabilities "backward." The following result is helpful.

**Proposition 1.11 (Bayes' Formula).** Under the same assumptions as in the law of total probability and if $P(A) > 0$, then for any event $B_j$, we have

$$P(B_j|A) = \frac{P(A|B_j)P(B_j)}{\sum_{k=1}^{\infty} P(A|B_k)P(B_k)}$$

*Proof.* Note that, by the law of total probability, the denominator is nothing but $P(A)$, and hence we must show that

$$P(B_j|A) = \frac{P(A|B_j)P(B_j)}{P(A)}$$

which is to say that

$$P(B_j|A)P(A) = P(A|B_j)P(B_j)$$

which is true since both sides equal $P(A \cap B_j)$, by the definition of conditional probability. ■

Again, the obvious analog for finitely many conditioning events holds, and in particular we state the case of two such events, $B$ and $B^c$, as a corollary.

**Corollary 1.7.** *If* $0 < P(B) < 1$ *and* $P(A) > 0$, *then*

$$P(B|A) = \frac{P(A|B)P(B)}{P(A|B)P(B) + P(A|B^c)P(B^c)}$$

***Example 1.47.*** The polygraph is an instrument used to detect physiological signs of deceptive behavior. Although it is often pointed out that the polygraph is not a lie detector, this is probably the way most of us think of it. For the purpose of this example, let us retain this notion. It is debated how accurate a polygraph test is, but there are several reports of accuracies above 95% (and as a counterweight, a Web site that gladly claims "Don't worry, the polygraph can be beaten rather easily!"). Let us assume that the polygraph test is indeed very accurate and that it decides "lie" or "truth" correctly with probability 0.95. Now consider a randomly chosen individual who takes the test and is determined to be lying. What is the probability that this person did indeed lie?

First, the probability is *not* 0.95. Introduce the events

$$L = \{\text{the person tells a lie}\}$$
$$L_P = \{\text{the polygraph reading says the person is lying}\}$$

and let $T = L^c$ and $T_P = L_P^c$. We are given the conditional probabilities $P(L_P|L) = P(T_P|T) = 0.95$, but what we want is $P(L|L_P)$. By Bayes' formula

$$P(L|L_P) = \frac{P(L_P|L)P(L)}{P(L_P|L)P(L) + P(L_P|T)P(T)} = \frac{0.95P(L)}{0.95P(L) + 0.05(1 - P(L))}$$

and to be able to finish the computation we need to know the probability that a randomly selected person would lie on the test. Suppose that we are dealing with a largely honest population; let us say that one out of a thousand would tell a lie in the given situation. Then $P(L) = 0.001$, and we get

$$P(L|L_P) = \frac{0.95 \times 0.001}{0.95 \times 0.001 + 0.05 \times 0.999} \approx 0.02$$

and the probability that the person actually lied is only 0.02. Since lying is so rare, most detected lies actually stem from errors, not actual lies. One way to understand this is to imagine that a large number of, say, 100,000, people are tested. We then expect 100 liars and of those, 95 will be discovered. Among the remaining 99,900 truthtellers, we expect 5%, or 4995 individuals to be misclassified as liars. Hence, out of a total of $95 + 4995 = 5090$ individuals who are classified as liars, only 95, or 2% actually are liars. A truth misclassified as a lie is called a "false-positive" and in this case, we say that the *false-positive rate* is 98%. □

In the last example, there are two types of errors we can make: classifying a lie as truth, and vice versa. The probability $P(L_P|L)$ to correctly classify a lie as a lie is called the *sensitivity* of the procedure. Obviously, we want the sensitivity to be high but with increased sensitivity we may risk to misclassify more truths as lies as well. Another probability of interest is therefore the *specificity*, namely, the probability $P(T_P|T)$ that a truth is correctly classified as truth. For an extreme but illustrative example, we can achieve maximum sensitivity by classifying all statements as lies; however, the specificity is then 0. Likewise, we can achieve maximum specificity by classifying all statements as truths but then instead getting sensitivity 0. The terms are borrowed from the field of medical testing for illnesses where good procedures should be both sensitive to detect an illness but also specific for that illness. For example, using high fever to diagnose measles would have high sensitivity (not many cases of measles will go undetected) but low specificity (many other diseases cause high fever and will be misclassified as measles).

Another probability of interest in any kind of testing situation is the false-positive rate, mentioned above. In the lie-detector example, it is $P(T|L_P)$, the probability that a detected lie is actually a truth. Also, the *false-negative rate* is $P(L|T_P)$, the probability that a detected truth is actually a lie. The sensitivity, specificity, false-positive rate, and false-negative rate are related via Bayes' formula where we also need to know the *base rate*, namely, the unconditional probability $P(T)$ of telling a lie (or having a disease). For typical examples from medical testing, see Problem 92 and subsequent problems.

***Example 1.48* (*The Monty Hall Problem*).** This problem has become a modern classic and was hotly discussed after it first appeared in the column "Ask Marilyn" in *Parade Magazine* in 1991. The problem was inspired by the game show "Let's Make a Deal" with host Monty Hall, and it goes like this. You are given the choice of three doors. Behind one door is a car; behind the others are goats. You pick a door without opening it, and the host opens another door that reveals a goat. He then gives you the choice to either open your door and keep what is behind it, or switch to the remaining door and take what is there. Is it to your advantage to switch?

At first glance, it would not seem to make a difference whether you stay or switch since the car is either behind your door or behind the remaining door. However, this is incorrect, at least if we make some reasonable assumptions. To solve the problem, we assume that the car and goats are placed at random behind the doors and that the host *always opens a door and shows a goat*. Let us further assume that in the case where you have chosen the car, he chooses which door to open at random. Now introduce the two events

$$C = \{\text{you chose the car}\}$$
$$G = \{\text{he shows a goat}\}$$

so that the probability to win after switching is $1 - P(C|G)$. But

$$P(C|G) = \frac{P(G|C)P(C)}{P(G|C)P(C) + P(G|C^c)P(C^c)} = P(C) = \frac{1}{3}$$

since $P(G|C) = P(G|C^c) = 1$. Thus, if you switch, you win the car with probability $\frac{2}{3}$, so switching is to your advantage. Note that the events $G$ and $C$ are in fact independent.

Intuitively, since you know that the host will always show you a goat, there is no additional information when he does. Since there are two goats and the host will always show one of them, to choose a door and then switch is equivalent to choosing the *two other doors* and telling the host to open one of them and show you a goat. Your chance of winning the car is then $\frac{2}{3}$.

One variant of the problem that has been suggested to make it easier to understand is to assume that there are not 3 but 1000 doors. One has a car, and 999 have goats.

Once you have chosen, the host opens 998 doors and shows you 998 goats. Given how unlikely it is that you found the car in the first pick, is it not obvious that you should now switch to the remaining door? You could also use one of the several computer simulations that are available online, or write your own. Still not convinced? Ask Marilyn. □

***Example 1.49*** **(*The Monty Hall Problem Continued*).** Suppose that you are playing "Let's Make a Deal" and have made your choice when the host suddenly realizes that he has forgotten where the car is. Since the show must go on, he keeps a straight face, takes a chance, and opens a door that reveals a goat. Is it to your advantage to switch?

Although the situation looks the same from your perspective, it is actually different since it could have happened that the host revealed the car. With $C$ and $G$ as above, Bayes' formula now gives

$$\begin{aligned} P(C|G) &= \frac{P(G|C)P(C)}{P(G|C)P(C) + P(G|C^c)P(C^c)} \\ &= \frac{1 \times (1/3)}{1 \times (1/3) + (1/2) \times (2/3)} = \frac{1}{2} \end{aligned}$$

so it makes no difference whether you stay or switch. In this case, the showing of a goat behind the open door actually does give some additional information, and $G$ and $C$ are no longer independent. □

***Example 1.50*** **(*The Island Problem*).** Probability theory is frequently used in courts of law, especially when DNA evidence is considered. As an example, consider the following situation. A person is murdered on an island, and the murderer must be one of the $n$ remaining islanders. DNA evidence on the scene reveals that the murderer has a particular genotype that is known to exist in a proportion $p$ in the general population, and we assume that the islanders' genotypes are independent. Crime investigators start screening all islanders for their genotypes. The first one who is tested is Mr Joe Bloggs, who turns out to have the murderer's genotype. What is the probability that he is guilty?

To solve this, we introduce the events

$$\begin{aligned} G &= \{\text{Mr Bloggs is guilty}\} \\ B &= \{\text{Mr Bloggs' genotype is found at the scene of the crime}\} \end{aligned}$$

so that we are asking for the probability $P(G|B)$. By Bayes' formula

$$P(G|B) = \frac{P(B|G)P(G)}{P(B|G)P(G) + P(B|G^c)P(G^c)}$$

Here, $P(G)$ is the probability that Mr Bloggs is guilty before any genotyping has been done, and if we assume that there is no reason to suspect any particular person more than anyone else, it is reasonable to let $P(G) = \frac{1}{n}$. If Mr Bloggs is guilty, then his genotype is certain to show up at the scene of the crime, and we have $P(B|G) = 1$. If Mr Bloggs is innocent, his genotype can still show up by chance, which gives $P(B|G^c) = p$, the proportion of his genotype in the population. All put together, we get

$$P(G|B) = \frac{1 \times 1/n}{1 \times 1/n + p \times (n-1)/n} = \frac{1}{1 + (n-1)p}$$

as the probability that Mr Bloggs is guilty. □

The last problem is a simple example of the general problem of how to quantify the weight of evidence in forensic identification. This "island problem" has been analyzed and discussed by lawyers and probabilists and different approaches have shown to give different results (not all correct).[7] We will return to this in more detail in Section 2.5. For now, let us present a simple example that demonstrates how calculations can go agley.

***Example 1.51.*** You know that your new neighbors have two children. One night you hear a stone thrown at your window and you see a child running from your yard into the neighbor's house. It is dark, and the only thing you can see for certain is that the child is a boy. The next day you walk over to the neighbor's house and ring the doorbell. A boy opens the door. What is the probability that he is guilty?

We will do this in two different ways. First approach: If the other child is a girl, you know that the boy is guilty and if the other child is a boy, the boy who opened the door is equally likely to be guilty or not guilty. Thus, with

$$G = \{\text{child who opened the door is guilty}\}$$

we condition on the gender of the other child and recall Example 1.39 to obtain

$$\begin{aligned} P(G) &= P(G|\text{boy})P(\text{boy}) + P(G|\text{girl})P(\text{girl}) \\ &= \frac{1}{2} \times \frac{1}{2} + 1 \times \frac{1}{2} = \frac{3}{4}. \end{aligned}$$

Second approach: Note how the situation is similar to that in the previous example, with genotype replaced by gender and Mr Bloggs replaced by the child who opened

[7]The island problem is made up (yes, really!), but there is a famous real case, *People* versus *Collins*, in which a couple in Los Angeles was first convicted of a crime, on the basis of circumstantial evidence, and later acquitted by the California Supreme Court. Both the initial verdict and the appeal were based on (questionable) probability arguments.

the door. In that formulation, we have $n = 2$ and $p = \frac{1}{2}$ and we get

$$P(\text{child who opened the door is guilty}) = \frac{1}{1 + 1 \times (1/2)} = \frac{2}{3}$$

There we go again; different methods give different results! As usual, we need to be very careful with which events we condition on. Let us assume that each child is equally likely to decide to go out and throw a stone at your window and that each child is equally likely to open the door. For each gender combination of two children, we thus choose at random who is guilty and who opens the door, so that each gender combination is split up into four equally likely cases. Let us use the subscript $d$ for the child who opened the door and the superscript $g$ for the child who is guilty. The sample space consists of the 16 equally likely outcomes

$$\begin{aligned} S = \{&b_d^g b, b_d b^g, b^g b_d, b b_d^g, b_d^g g, b_d g^g, b^g g_d, b g_d^g, \\ &g_d^g b, g_d b^g, g^g b_d, g b_d^g, g_d^g g, g_d g^g, g^g g_d, g g_d^g\} \end{aligned}$$

and the event that the child who opened the door is guilty is

$$G = \{b_d^g b, b b_d^g, b_d^g g, g b_d^g, b g_d^g, g_d^g b, g_d^g g, g g_d^g\}$$

What event do we condition on? We know two things: that the guilty child is a boy and that a boy opened the door. These events are

$$\begin{aligned} A &= \{b_d^g b, b_d b^g, b^g b_d, b b_d^g, b_d^g g, b^g g_d, g_d b^g, g b_d^g\} \\ B &= \{b_d^g b, b_d b^g, b^g b_d, b b_d^g, b_d^g g, b_d g^g, g^g b_d, g b_d^g\} \end{aligned}$$

and we condition on their intersection

$$A \cap B = \{b_d^g b, b_d b^g, b^g b_d, b b_d^g, b_d^g g, g b_d^g\}$$

Since four of these six outcomes are in $G$ and the distribution on $S$ is uniform, we get

$$P(\text{child who opened the door is guilty}) = P(G | A \cap B) = \frac{2}{3}$$

in accordance with the previous example.

The first approach gives the wrong solution but why? When we computed the probabilities $P(\text{boy})$ and $P(\text{girl})$, we implicitly conditioned on event $B$ above but forgot to also condition on $A$. What we need to do is to compute $P(\text{boy})$ as

$$P(\text{other child is a boy} \,| A \cap B) = \frac{2}{3}$$

and not $\frac{1}{2}$. Note how the conditional probability that the other child is a boy is higher now that we also know that the guilty child is a boy. This is quite subtle and resembles the situation in Example 1.39, in the sense that we need to be careful to condition on precisely the information we have, no more and no less. We can now state the correct version of the first solution. Everything must be computed conditioned on the event $A \cap B$, but for ease of notation let us not write this conditioning out explicitly. We get

$$\begin{aligned} P(G) &= P(G|\text{boy})P(\text{boy}) + P(G|\text{girl})P(\text{girl}) \\ &= \frac{1}{2} \times \frac{2}{3} + 1 \times \frac{1}{3} = \frac{2}{3} \end{aligned}$$

just as we should. For a variant, see Problem 99. □

***Example 1.52.*** Consider the previous example and also assume that on your way over to the neighbor's, you meet another neighbor who tells you that she saw the mother of the family take a walk with a boy a few days ago. If a boy opens the door, what is the probability that he is guilty?

By now you know how to solve this. In the previous sample space, split each outcome further into two, marking who the mother took a walk with, and proceed. The sample space now has 32 outcomes, and we will suggest a more convenient approach. We can view the various sightings of a boy as repeated sampling with replacement from a randomly chosen family. Let us convert this into a problem about black and white balls in urns.

Consider three urns, containing two balls each, such that the $k$th urn contains $k$ black balls, $k = 0, 1, 2$. We first choose an urn according to the probabilities $\frac{1}{4}$, $\frac{1}{2}$, and $\frac{1}{4}$ (think of the gender combinations above) and then pick balls with replacement and note their colors. If we do this $j$ times and get only black balls, what is the probability that we have chosen the urn with only black balls? Let

$$\begin{aligned} B &= \{\text{get only black balls}\} \\ U_k &= \{\text{the } k\text{th urn chosen}\}, \quad k = 0, 1, 2 \end{aligned}$$

and compute $P(U_2|B)$. The reversed probabilities are

$$P(B|U_0) = 0, \quad P(B|U_1) = \frac{1}{2^j}, \quad P(B|U_2) = 1$$

and Bayes' formula gives

$$\begin{aligned} P(U_2|B) &= \frac{P(B|U_2)P(U_2)}{P(B|U_1)P(U_1) + P(B|U_2)P(U_2)} \\ &= \frac{1 \times (1/4)}{(1/2^j) \times (1/2) + 1 \times (1/4)} = \frac{2^{j-1}}{2^{j-1} + 1} \end{aligned}$$

In our examples with families and their children, we let urns represent families and black and white balls represent genders. Consider the probability that the other child has the same gender as the observed child. In Example 1.39, we have $j = 1$, which gives probability $\frac{1}{2}$ and in Example 1.51 we have $j = 2$, which gives probability $\frac{2}{3}$. Finally, in this example we have $j = 3$ and probability $\frac{4}{5}$. The more observations we have on boys, the stronger our belief that both children are boys. □

From mother:

| | $A$ | $a$ |
|---|---|---|
| **From father:** $A$ | $AA$ | $Aa$ |
| $a$ | $aA$ | $aa$ |

**FIGURE 1.6** A Punnett square illustrating possible genotypes.

### 1.6.2 Genetics and Probability

Genetics is a science where probability theory is extremely useful. Recall that genes occur in pairs where one copy is inherited from the mother and one from the father. Suppose that a particular gene has two different *alleles* (variants) called $A$ and $a$. An individual can then have either of the three *genotypes* $AA$, $Aa$, and $aa$. If the parents both have genotype $Aa$, what is the probability that their child gets the same genotype?

We assume that each of the two gene copies from each parent is equally likely to be passed on to the child and that genes from the father and the mother are inherited independently. There are then the four equally likely outcomes illustrated in Figure 1.6, and the probability that the child also has genotype $Aa$ is $\frac{1}{2}$ (order has no meaning here, so $Aa$ and $aA$ are the same). Each of the genotypes $AA$ and $aa$ has probability $\frac{1}{4}$. The square in the figure is an example of a *Punnett square*.

***Example 1.53.*** An allele is said to be *recessive* if it is required to exist in two copies to be expressed and *dominant* if one copy is enough. For example, the hereditary disease *cystic fibrosis* (CF) is caused by a recessive allele of a particular gene. Let us denote this allele $C$ and the healthy allele $H$ so that only individuals with genotype $CC$ get the disease. Individuals with genotype $CH$ are *carriers*, that is, they have the disease-causing allele but are healthy. It is estimated that approximately 1 in 25 individuals are carriers (among people of central and northern European descent; it is much less common in other ethnic groups). Given this information, what is the probability that a newborn of healthy parents has CF?

Introduce the events

$$D = \{\text{newborn has CF}\}$$
$$B = \{\text{both parents are carriers}\}$$

so that

$$P(D) = P(D|B)P(B)$$

since $B^c$ is the event that at least one parent has genotype $HH$, in which case the baby will also be healthy. Assuming that the mother's and father's genotypes are independent, we get

$$P(B) = \frac{1}{25} \times \frac{1}{25} = \frac{1}{625}$$

and since the child will get the disease only if it inherits the $C$ allele from each parent, we get $P(D|B) = \frac{1}{4}$, which gives

$$P(D) = \frac{1}{625} \times \frac{1}{4} = \frac{1}{2500}$$

In other words, the *incidence* of CF among newborns is 1 in 2500, or 0.04%.

Now consider a family with one child where we know that both parents are healthy, that the mother is a carrier of the disease allele and nothing is known about the father's genotype. What is the probability that the child neither is a carrier nor has the disease?

Let $E$ be the event we are interested in. The mother's genotype is $CH$, and we condition on the father's genotype to obtain

$$\begin{aligned} P(E) &= P(E|CH)P(CH) + P(E|HH)P(HH) \\ &= \frac{1}{4} \times \frac{1}{25} + \frac{1}{2} \times \frac{24}{25} = 0.49 \end{aligned}$$

where we figure out the conditional probabilities with Punnett squares. See the problem section at the end of the chapter for more on genetics. □

### 1.6.3 Recursive Methods

Certain probability problems can be solved elegantly with recursive methods, involving the law of total probability. The general idea is to condition on a number of cases that can either be solved explicitly or lead back to the original problem. We will illustrate this in a number of examples.

***Example 1.54.*** In the final scene of the classic 1966 Sergio Leone movie *The Good, the Bad, and the Ugly*, the three title characters, also known as "Blondie," "Angel Eyes," and "Tuco," stand in a cemetery, guns in holsters, ready to draw. Let us interfere slightly with the script and assume that Blondie always hits his target, Angel Eyes hits with probability 0.9, and Tuco with probability 0.5. Let us also suppose that they take turns in shooting, that whomever is shot at shoots next (unless he is hit), and that Tuco starts. What strategy maximizes his probability of survival?

Introduce the events

$$\begin{aligned} S &= \{\text{Tuco survives}\} \\ H &= \{\text{Tuco hits his target}\} \end{aligned}$$

Let us first suppose that Tuco tries to kill Blondie. If he fails, Blondie kills Angel Eyes, and Tuco gets one shot at Blondie. We thus have

$$P(S) = P(S|H)P(H) + P(S|H^c)P(H^c) = P(S|H)\frac{1}{2} + \frac{1}{4}$$

where we need to find $P(S|H)$, the probability that Tuco survives a shootout with Angel Eyes, who gets the first shot. If we assume an infinite supply of bullets (hey, it's a Clint Eastwood movie!), we can solve this recursively. Note how this is repeated conditioning, as in Equation (1.43), but let us ease the notation and rename the event that Tuco survives the shootout $T$. Now let $p = P(T)$ and condition on the three events

$$A = \{\text{Angel Eyes hits}\}$$
$$B = \{\text{Angel Eyes misses, Tuco hits}\}$$
$$C = \{\text{Angel Eyes misses, Tuco misses}\}$$

to obtain

$$p = P(T|A)P(A) + P(T|B)P(B) + P(T|C)P(C)$$

where $P(A) = 0.9$, $P(B) = 0.1 \times 0.5 = 0.05$, $P(C) = 0.1 \times 0.5 = 0.05$, $P(T|A) = 0$, and $P(T|B) = 1$. To find $P(T|C)$, note that if both Angel Eyes and Tuco have missed their shots, they start over from the beginning and hence $P(T|C) = p$. This gives

$$p = 0.05 + 0.05p$$

which gives $p = 0.05/0.95$, and with this strategy, Tuco has survival probability

$$P(S) = \frac{0.05}{0.95} \times 0.5 + 0.25 \approx 0.28$$

Next, suppose that Tuco tries to kill Angel Eyes. If he succeeds, he faces certain death as Blondie shoots him. If he fails, Angel Eyes will try to kill Blondie to maximize his own probability of survival. If Angel Eyes fails, Blondie kills him for the same reason and Tuco again gets one last shot at Blondie. Tuco surviving this scenario has probability $0.5 \times 0.1 \times 0.5 = 0.025$. If Angel Eyes succeeds and kills Blondie, Tuco must again survive a shootout with Angel Eyes but this time, Tuco gets to start. By an argument similar to that stated above, his probability to survive the shootout is $p = 0.5 + 0.05p$ that gives $p = 0.5/0.95$ and Tuco's survival probability is

$$P(S) = 0.025 + 0.5 \times 0.9 \times \frac{0.5}{0.95} \approx 0.26$$

not quite as good as with the first strategy.

Notice, however, that Tuco really gains from missing his shot, letting the two better shots fight it out first. The smartest thing he can do is to miss on purpose! If he aims at Blondie and misses, Blondie kills Angel Eyes and Tuco gets one last shot at Blondie.

His survival probability is 0.5. An even better strategy is to aim at Angel Eyes, miss on purpose, and give Angel Eyes a chance to kill Blondie. If Angel Eyes fails, he is a dead man and Tuco gets one last shot at Blondie. If Angel Eyes succeeds, Tuco again needs to survive the shootout which, as we just saw, has probability $p = 0.5/0.95$ and his overall survival probability is

$$P(S) = 0.1 \times 0.5 + 0.9 \times \frac{0.5}{0.95} \approx 0.52$$

When Fredric Mosteller presents a similar problem in his 1965 book *Fifty Challenging Problems in Probability* [1], he expresses some worry over the possibly unethical dueling conduct to miss on purpose. In the case of Tuco, we can safely disregard any such ethical considerations. □

***Example 1.55.*** The shootout between Tuco and Angel Eyes in the previous example is a special case of the following situation: Consider an experiment where the events $A$ and $B$ are disjoint and repeat the experiment until either $A$ or $B$ occurs. What is the probability that $A$ occurs before $B$?

First, by Corollary 1.5, we will sooner or later get either $A$ or $B$. Let $C$ be the event that $A$ occurs before $B$, let $p = P(C)$, and condition on the first trial. If we get $A$, we have $A$ before $B$ for certain and if we get $B$, we do not. If we get neither, that is, get $(A \cup B)^c$, we start over. The law of total probability now gives

$$\begin{aligned} p &= P(C|A)P(A) + P(C|B)P(B) + P(C|(A \cup B)^c)P(A \cup B)^c) \\ &= P(A) + p(1 - P(A \cup B)) \\ &= P(A) + p(1 - (P(A) + P(B))) \end{aligned}$$

and we have established an equation for $p$. Solving it gives

$$p = \frac{P(A)}{P(A) + P(B)}$$

□

***Example 1.56.*** Recall that a single *game* in tennis is won by the first player to win four points but that it must also be won by a margin of at least two points. If no player has won after six played points, they are at *deuce* and the first to get two points ahead wins the game. Suppose that Ann is the server and has probability $p$ of winning a single point against Bob, and suppose that points are won independent of each other. If the players are at deuce, what is the probability that Ann wins the game?

We are waiting for the first player to win two consecutive points from deuce, so let us introduce the events

$$\begin{aligned} A &= \{\text{Ann wins two consecutive points from deuce}\} \\ B &= \{\text{Bob wins two consecutive points from deuce}\} \end{aligned}$$

with the remaining possibility that they win a point each, in which case they are back at deuce. By independence of consecutive points, $P(A) = p^2$ and $P(B) = (1 - p)^2$, and by Example 1.55 we get

$$P(\text{Ann wins}) = \frac{p^2}{p^2 + (1 - p)^2}$$

□

***Example 1.57.*** The next sports application is to the game of badminton. The scoring system is such that you can score a point only when you win a rally as the server. If you win a rally as the receiver, the score is unchanged, but you get to serve and thus the opportunity to score. Suppose that Ann wins a rally against Bob with probability $p$, regardless of who serves (a reasonable assumption in badminton but would, of course, not be so in tennis, where the server has a big advantage). What is the probability that Ann scores the next point if she is the server?

If Ann is the server, the next point is scored either when she wins a rally as server or loses two consecutive rallies starting from being server. In the remaining case, the players will start over from Ann being the server with no points scored yet. Hence, we can apply the formula from Example 1.55 to the events

$$A = \{\text{Ann wins a rally as server}\}$$
$$B = \{\text{Ann loses two consecutive rallies as server}\}$$

to obtain

$$P(\text{Ann scores next point}) = \frac{P(A)}{P(A) + P(B)} = \frac{p}{p + (1 - p)^2}$$

If the players are equally good, so that $p = \frac{1}{2}$, the server thus has a $\frac{2}{3}$ probability to score the next point. □

***Example 1.58 (Gambler's Ruin).*** Next, Ann and Bob play a game where a fair coin is flipped repeatedly. If it shows heads, Ann pays Bob one dollar, otherwise Bob pays Ann one dollar. If Ann starts with $a$ dollars and Bob with $b$ dollars, what is the probability that Ann ends up winning all the money and Bob is ruined?

Introduce the event

$$A = \{\text{Ann wins all the money}\}$$

and let $p_a$ be the probability of $A$ if Ann's initial fortune is $a$. Thinking a few minutes makes us realize that it is quite complicated to compute $p_a$ directly. Instead, let us condition on the first flip and note that if it is heads, the game starts over with the new initial fortunes $a - 1$ and $b + 1$, and if it is tails, the new fortunes are $a + 1$ and $b - 1$. Introduce the events

$$H = \{\text{heads in first flip}\} \quad \text{and} \quad T = \{\text{tails in first flip}\}$$

and apply the law of total probability to get

$$p_a = P(A|H)\frac{1}{2} + P(A|T)\frac{1}{2} = \frac{1}{2}(p_{a-1} + p_{a+1})$$

or equivalently

$$p_{a+1} = 2p_a - p_{a-1}$$

First note that $p_0 = 0$ and let $a = 1$ to obtain

$$p_2 = 2p_1$$

With $a = 2$ we get

$$p_3 = 2p_2 - p_1 = 3p_1$$

and we find the general relation

$$p_a = ap_1$$

Now, $p_{a+b} = 1$, and hence

$$p_1 = \frac{1}{a+b}$$

which finally gives the solution

$$P(\text{Ann wins all the money}) = \frac{a}{a+b}$$

By symmetry, the same argument applies to give

$$P(\text{Bob wins all the money}) = \frac{b}{a+b}$$

Note that this means that the probability that *somebody* wins is 1, which excludes the possibility that the game goes on forever, something we cannot immediately rule out.

This *gambler's ruin* problem is an example of a *random walk*. We may think of a particle that in each step decides to go up or down (or, if you prefer, left/right), and does so independent of its previous path.[8] We can view the position after $n$ steps as Ann's total gain, so if the walk starts in 0, Ann has won the game when it hits $b$, and she has lost when it hits $-a$. We refer to $-a$ and $b$ as *absorbing barriers* (see Figure 1.7). □

[8]A more romantic allegory is that of a drunken Dutchman who staggers back and forth until he either is back in his favorite *bruine cafe* or falls into the canal.

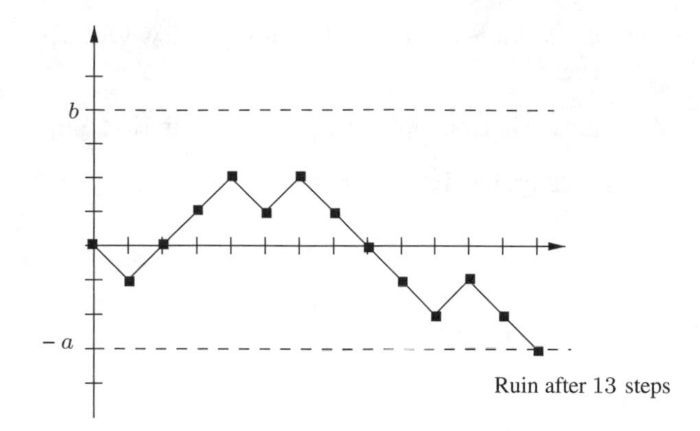


**FIGURE 1.7** Gambler's ruin as a random walk.

***Example 1.59.*** Consider the gambler's ruin problem from the previous example, but suppose that Ann has only one dollar and Bob is infinitely wealthy. What is the probability that Ann eventually goes broke?

Since the range is infinite, we cannot use the technique from above, but let us still condition on the first coin flip. If it shows heads, Ann's fortune drops to zero and she is ruined. If it shows tails, Ann's fortune goes up to \$2, and the game continues. If Ann is to go broke, her fortune must eventually hit 0, and before it does so, it must first hit 1. Now, the probability to eventually hit 1 starting from 2 is the same as the probability to eventually hit 0 starting from 1, and once her fortune is back at 1, the game starts over from the beginning. If we let $B = \{\text{Ann goes broke eventually}\}$ and condition on the first flip being heads or tails, we thus get

$$P(B) = P(B|H)P(H) + P(B|T)P(T) = \frac{1}{2} + P(B|T)\frac{1}{2}$$

Now let $q = P(B)$. By the argument above, $P(B|T) = q^2$, and we get the equation

$$q = \frac{1}{2} + \frac{q^2}{2}$$

which we solve for $q$ to get $q = 1$, so Ann will eventually go broke. Since the game is fair, there is no trend that drags her fortune down toward ruin, only inevitable bad luck. □

***Example 1.60.*** Consider the gambler's ruin problem but suppose that the game is unfair, so that Ann wins with probability $p \neq \frac{1}{2}$ in each round. If her initial fortune is $a$ and Bob's initial fortune is $b$, what is the probability that she wins?

The solution method is the same as in the original gambler's ruin: to condition on the first flip and apply the law of total probability. Again, let $A$ be the event that Ann wins and $p_a$ the probability of $A$ if she starts with $a$ dollars. For ease of notation, let $q = 1 - p$, the probability that Ann loses a round. We get

$$p_a = P(A|H)P(H) + P(A|T)P(T) = p_{a-1}q + p_{a+1}p$$

which gives

$$p_{a+1} = \frac{1}{p}(p_a - qp_{a-1})$$

First let $a = 1$. Since $p_0 = 0$, we get

$$p_2 = \frac{1}{p}p_1$$

which we rewrite as

$$p_2 = \left(1 + \frac{q}{p}\right)p_1$$

For $a = 2$, we get

$$\begin{aligned} p_3 &= \frac{1}{p}(p_2 - qp_1) \\ &= \frac{1}{p}\left(1 + \frac{q}{p} - q\right)p_1 = \left(1 + \frac{q}{p} + \left(\frac{q}{p}\right)^2\right)p_1 \end{aligned}$$

and the general formula emerges as

$$p_a = \left(1 + \frac{q}{p} + \left(\frac{q}{p}\right)^2 + \cdots + \left(\frac{q}{p}\right)^{a-1}\right)p_1 = \frac{1-(q/p)^a}{1-(q/p)}p_1$$

Finally, we use $p_{a+b} = 1$ to obtain

$$\frac{1-(q/p)^{a+b}}{1-(q/p)}p_1 = 1$$

which gives

$$p_1 = \frac{1-(q/p)}{1-(q/p)^{a+b}}$$

and the probability that Ann wins, starting from a fortune of $a$ dollars, is thus

$$p_a = \frac{1-(q/p)^a}{1-(q/p)^{a+b}}$$

if $p \neq \frac{1}{2}$. The game is unfair to Ann if $p < \frac{1}{2}$ and to Bob if $p > \frac{1}{2}$. It is interesting to note that the change in winning probabilities can be dramatic for small changes of $p$.

For example, if the players start with 20 dollars each and $p = \frac{1}{2}$, they are equally likely to win in the end. Now change $p$ to 0.55, so that Ann has a slight edge. Then

$$p_{20} = \frac{1 - (0.45/0.55)^{20}}{1 - (0.45/0.55)^{40}} \approx 0.98$$

so Ann is almost certain to win. See Problem 106 for an interesting application to roulette. □

***Example 1.61* (*Penney-ante*).** We finish the section with a game named for Walter Penney, who in 1969 described it in an article in *Journal of Recreational Mathematics*. If a fair coin is flipped three times, there are eight outcomes

$$HHH, HHT, HTH, THH, HTT, THT, TTH, TTT$$

each of which has probability $\frac{1}{8}$. You now suggest the following game to your friend John.[9] You bet \$1 each, he gets to choose one of the eight patterns, and you choose another. A coin is flipped repeatedly, and the sequence of heads and tails is recorded. Whoever first sees his sequence come up wins. Since all patterns are equally likely to come up in a sequence of three flips, this game seems fair. However, it turns out that after John has chosen his pattern, you can always choose so that your chance of winning is at least $\frac{2}{3}$!

The idea is to always let your sequence end with the two symbols that his begins with. Intuitively, this means that whenever his pattern is about to come up, there is a good chance that yours has come up already. For example, if he chooses $HHH$, you choose $THH$, and the only way in which he can win is if the first three flips give heads. Otherwise, the sequence $HHH$ cannot appear without having a $T$ before it, and thus your pattern $THH$ has appeared. With these choices, your probability to win is $\frac{7}{8}$.

The general strategy is to let his first two be your last two, and never choose a palindrome. Suppose that John chooses $HTH$ so that according to the strategy, you choose $HHT$. Let us calculate your probability of winning.

Let $A$ be the event that you win, and let $p$ be the probability of $A$. To find $p$, we condition on the first flip. If this is $T$, the game starts over, and hence $P(A|T) = p$. If it is $H$, we condition further on the second flip. If this is $H$, you win (if we start with $HH$, then $HHT$ must come before $HTH$), and if it is $T$, we condition further on the third flip. If this is $H$, the full sequence is $HTH$, and you have lost. If it is $T$, the full sequence is $HTT$ and the game starts over. See the tree diagram in Figure 1.8 for

[9]Named after John Haigh, who in his splendid book *Taking Chances: Winning with Probability* [3] describes this game and names the loser Doyle after Doyle Lonnegan, victim in the 1973 movie *The Sting*. I feel that Doyle has now lost enough and in this way let him get a small revenge. Hopefully, John's book has sold so well that he is able to take the loss.

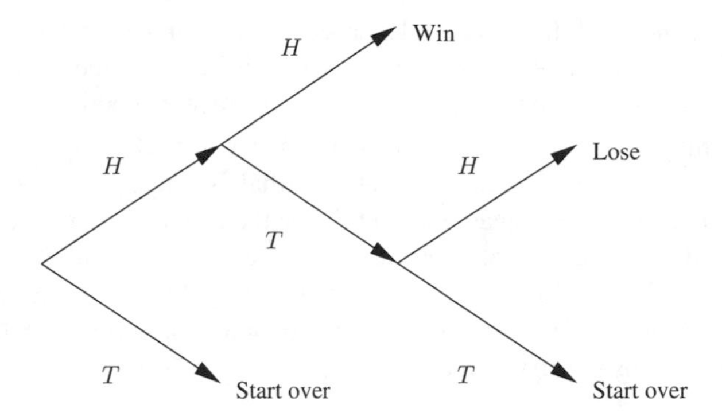


**FIGURE 1.8** The four possible cases in Penney-ante when *HHT* competes with *HTH*.

an illustration of the possible cases. The law of total probability gives (ignoring the case in which you lose)

$$\begin{aligned} p &= P(A|T)P(T) + P(A|HH)P(HH) + P(A|HTT)P(HTT) \\ &= p \times \frac{1}{2} + 1 \times \frac{1}{4} + p \times \frac{1}{8} = \frac{2+5p}{8} \end{aligned}$$

which we solve for $p$ to get $p = \frac{2}{3}$. Just as in the dice game in Example 1.45, your apparent generosity to let your opponent choose first is precisely what gives you the advantage. See also Problem 108. □

## PROBLEMS

### Section 1.2. Sample Spaces and Events

1. Suggest sample spaces for the following experiments: **(a)** Three dice are rolled and their sum computed. **(b)** Two real numbers between 0 and 1 are chosen. **(c)** An American is chosen at random and is classified according to gender and age. **(d)** Two different integers are chosen between 1 and 10 and are listed in increasing order. **(e)** Two points are chosen at random on a yardstick and the distance between them is measured.
2. Suggest a sample space for Example 1.8.
3. Consider the experiment to toss a coin three times and count the number of heads. Which of the following sample spaces can be used to describe this experiment?
   **(a)** $S = \{H, T\}$
   **(b)** $S = \{HHH, TTT\}$
   **(c)** $S = \{0, 1, 2, 3\}$
   **(d)** $S = \{1, 2, 3\}$
   **(e)** $S = \{HHH, HHT, HTH, THH, HTT, THT, TTH, TTT\}$

4. Let $A$, $B$, and $C$ be three events. Express the following events in terms of $A$, $B$, and $C$: **(a)** exactly one of the events occurs. **(b)** None of the events occurs. **(c)** At least one of the events occurs. **(d)** All of the events occur.
5. The Stanley Cup final is played in best of seven games. Suppose that the good old days are brought back and that the final is played between the Boston Bruins and Montreal Canadiens. Let $B_k$ be the event that Boston wins the $k$th game and describe the following events in terms of the $B_k$: **(a)** Boston wins game 1, **(b)** Boston loses game 1 and wins games 2 and 3, **(c)** Boston wins the series without losing any games, **(d)** Boston wins the series with one loss, and **(e)** Boston wins the first three games and loses the series.

## Section 1.3. The Axioms of Probability

6. A certain thick and asymmetric coin is tossed and the probability that it lands on the edge is 0.1. If it does not land on the edge, it is twice as likely to show heads as tails. What is the probability that it shows heads?
7. Let $A$ and $B$ be two events such that $P(A) = 0.3$, $P(A \cup B) = 0.5$, and $P(A \cap B) = 0.2$. Find **(a)** $P(B)$, **(b)** the probability that $A$ but not $B$ occurs, **(c)** $P(A \cap B^c)$, **(d)** $P(A^c)$, **(e)** the probability that $B$ does not occur, and **(f)** the probability that neither $A$ nor $B$ occurs.
8. Let $A$ be the event that it rains on Saturday and $B$ the event that it rains on Sunday. Suppose that $P(A) = P(B) = 0.5$. Furthermore, let $p$ denote the probability that it rains on both days. Express the probabilities of the following events as functions of $p$: **(a)** it rains on Saturday but not Sunday. **(b)** It rains on one day but not the other. **(c)** It does not rain at all during the weekend.
9. The probability in Problem 8(b) is a decreasing function of $p$. Explain this intuitively.
10. People are asked to assign probabilities to the events "rain on Saturday," "rain on Sunday," "rain both days," and "rain on at least one of the days." Which of the following suggestions are consistent with the probability axioms: **(a)** 70%, 60%, 40%, and 80%, **(b)** 70%, 60%, 40%, and 90%, **(c)** 70%, 60%, 80%, and 50%, and **(d)** 70%, 60%, 50%, and 90%?
11. Two fish are caught and weighed. Consider the events $A = \{$the first weighs more than 10 pounds$\}$, $B = \{$the second weighs more than 10 pounds$\}$, and $C = \{$the sum of the weights is more than 20 pounds$\}$. Argue that $C \subseteq A \cup B$.
12. Let $A$, $B$, and $C$ be three events, such that each event has probability $\frac{1}{2}$, each intersection of two has probability $\frac{1}{4}$, and $P(A \cap B \cap C) = \frac{1}{8}$. Find the probability that **(a)** exactly one of the events occurs, **(b)** none of the events occurs, **(c)** at least one of the events occurs, **(d)** all of the events occur, and **(e)** exactly two of the events occur.
13. **(a)** Let $A$ and $B$ be two events. Show that

$$P(A) + P(B) - 1 \leq P(A \cup B) \leq P(A) + P(B)$$

**(b)** Let $A_1, \ldots, A_n$ be a sequence of events. Show that

$$\sum_{k=1}^{n} P(A_k) - (n-1) \leq P\left(\bigcup_{k=1}^{n} A_k\right) \leq \sum_{k=1}^{n} P(A_k)$$

14. A particular species of fish is known to weigh more than 10 pounds with probability 0.01. Suppose that 10 such fish are caught and weighed. Show that the probability that the total weight of the 10 fish is above 100 pounds is at most 0.1.

15. Consider the Venn diagram of four events below. If we use the "area method" to find the probability of $A \cup B \cup C \cup D$, we get

$$\begin{aligned} P(A \cup B \cup C \cup D) = &\ P(A) + P(B) + P(C) + P(D) \\ &- \ P(A \cap B) - P(A \cap C) - P(B \cap D) - P(C \cap D) \\ &+ \ P(A \cap B \cap C \cap D) \end{aligned}$$

However, this does not agree with Proposition 1.5 for $n = 4$. Explain!

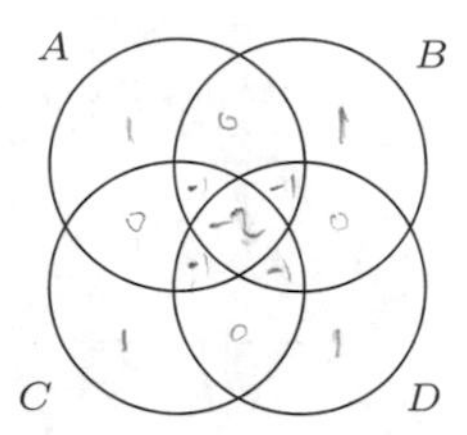


16. Choose a number at random from the integers 1,...,100. What is the probability that it is divisible by **(a)** 2, 3, or 4, **(b)** $i$, $j$, or $k$?

17. Consider Example 1.9 where you throw a dart at random. Find the probability that you get **(a)** 14 or double, **(b)** 14, double, or triple, **(c)** even, double, a number higher than 10, or bull's eye.

18. Prove Proposition 1.6 by considering disjoint events $B_1, B_2, \ldots$ defined by $B_1 = A_1,\ \ B_2 = A_2 \setminus B_1, \ldots, B_k = A_k \setminus B_{k-1}, \ldots$

## Section 1.4. Finite Sample Spaces and Combinatorics

19. You are asked to select a password for a Web site. It must consist of five lowercase letters and two digits in any order. How many possible such passwords are there if **(a)** repetitions are allowed, and **(b)** repetitions are not allowed?

20. Consider the Swedish license plate from Example 1.16. Find the probability that a randomly selected plate has **(a)** no duplicate letters, **(b)** no duplicate digits, **(c)** all letters the same, **(d)** only odd digits, and **(e)** no duplicate letters and all digits equal.

21. "A thousand monkeys, typing on a thousand typewriters will eventually type the entire works of William Shakespeare" is a statement often heard in one

form or another. Suppose that one monkey presses 10 keys at random. What is the probability that he types the word HAMLET if he is **(a)** allowed to repeat letters, and **(b)** not allowed to repeat letters?

22. Four envelopes contain four different amounts of money. You are allowed to open them one by one, each time deciding whether to keep the amount or discard it and open another envelope. Once an amount is discarded, you are not allowed to go back and get it later. Compute the probability that you get the largest amount under the following different strategies: **(a)** You take the first envelope. **(b)** You open the first envelope, note that it contains the amount $x$, discard it and take the next amount which is larger than $x$ (if no such amount shows up, you must take the last envelope). **(c)** You open the first two envelopes, call the amounts $x$ and $y$, and discard both and take the next amount that is larger than both $x$ and $y$.
23. In the early 1970s, four talented Scandinavians named Agneta, Annifrid, Benny, and Björn put a band together and decided to name it using their first name initials. **(a)** How many possible band names were there? What if a reunion is planned and the reclusive Agneta is replaced by some guy named Robert? **(b)** A generalization of (a): You are given $n$ uppercase letters such that the numbers of $A, B, \ldots, Z$ are $n_A, n_B, \ldots, n_Z$, respectively (these numbers may be 0). Show that you can create

$$\frac{n!}{n_A!n_B!\cdots n_Z!}$$

different possible words. Compare with your answers in part (a).
24. Prove the following identities (rather than using the definition, try to give combinatorial arguments):

**(a)** $\binom{n+1}{k+1} = \binom{n}{k+1} + \binom{n}{k}$ **(b)** $k\binom{n}{k} = n\binom{n-1}{k-1}$

**(c)** $\binom{2n}{n} = \sum_{k=1}^{n} \binom{n}{k}^2$ **(d)** $\sum_{k=0}^{n} \binom{n}{k} = 2^n$

25. On a chessboard ($8 \times 8$ squares, alternating black and white), you place three chess pieces at random. What is the probability that they are all **(a)** in the first row, **(b)** on black squares, **(c)** in the same row, and **(d)** in the same row and on the same color?
26. In a regular coordinate system, you start at $(0, 0)$ and flip a fair coin to decide whether to go sideways to $(1, 0)$ or up to $(0, 1)$. You continue in this way, and after $n$ flips you have reached the point $(j, k)$, where $j + k = n$. What is the probability that **(a)** all the $j$ steps sideways came before the $k$ steps up, **(b)** all the $j$ steps sideways came either before or after the $k$ steps up, and **(c)** all the $j$ steps sideways came in a row?

27. An urn contains $n$ red balls, $n$ white balls, and $n$ black balls. You draw $k$ balls at random without replacement (where $k \leq n$). Find an expression for the probability that you do not get all colors.
28. You are dealt a bridge hand (13 cards). What is the probability that you do not get cards in all suits?
29. Recall Texas Lotto from Example 1.21, where five numbers are chosen among 1,...,44 and one bonus ball number from the same set. Find the probability that you match **(a)** four of the first five numbers but not the bonus ball, **(b)** three of the first five numbers and the bonus ball.
30. You are dealt a poker hand. What is the probability of getting **(a)** royal flush, **(b)** straight flush, **(c)** four of a kind, **(d)** full house, **(e)** flush, **(f)** straight, **(g)** three of a kind, **(h)** two pairs, and **(i)** one pair? (These are listed in order of descending value in poker, not in order of difficulty!)
31. From the integers $1, \ldots, 10$, three numbers are chosen at random without replacement. **(a)** What is the probability that the smallest number is 4? **(b)** What is the probability that the smallest number is 4 and the largest is 8? **(c)** If you choose three numbers from $1, \ldots, n$, what is the probability that the smallest number is $j$ and the largest is $k$ for possible values of $j$ and $k$?
32. An urn contains $n$ white and $m$ black balls. You draw repeatedly at random and without replacement. What is the probability that the first black ball comes in the $k$th draw, $k = 1, 2, \ldots, n+1$?
33. In the "Pick 3" game described in Example 1.25, suppose that you choose the "any order" option and play the numbers 111. Since there are a total of 220 cases and 1 favorable case, you think that your chance of winning is $1/220$. However, when playing this repeatedly, you notice that you win far less often than once every 220 times. Explain!
34. How many strictly positive, integer-valued solutions $(x_1, \ldots, x_n)$ are there to the equation $x_1 + \cdots + x_n = k$?
35. Ann and Bob shuffle a deck of cards each. Ann wins if she can find a card that has the same position in her deck as in Bob's. What is the (approximate) probability that Ann wins?
36. Consider the matching problem in Example 1.4.17 and let $n_j$ be the number of permutations with exactly $j$ matches for $j = 0, 1, \ldots, n$. **(a)** Find an expression for $n_0$. *Hint:* How does $n_0/n!$ relate to the probability computed in the example? **(b)** Find the probability of exactly $j$ matches, for $j = 0, 1, \ldots, n$ and its limit as $n \to \infty$. *Hint:* You need to find $n_j$. First fix a particular set of $j$ numbers, for example, $\{1, 2, \ldots, j\}$ and note that the number of ways to match exactly those equals the number of ways to have no matches among the remaining $n - j$ numbers, which you can obtain from part (a).

### Section 1.5. Conditional Probability and Independence

37. Let $A$ and $B$ be two events. Is it then true that $P(A|B) + P(A|B^c) = 1$? Give proof or counterexample.

38. Let $A$ and $B$ be disjoint events. Show that

$$P(A|A \cup B) = \frac{P(A)}{P(A) + P(B)}$$

39. Let $A$, $B$, and $C$ be three events such that $P(B \cap C) > 0$. Show that

$$P(A \cap B \cap C) = P(A|B \cap C)P(B|C)P(C)$$

and that

$$P(A|B \cap C) = \frac{P(A \cap B|C)}{P(B|C)}$$

40. Let $A$ and $B$ be events, with $P(A) = \frac{1}{2}$ and $P(B) = \frac{1}{3}$. Compute both $P(A \cup B)$ and $P(A \cap B)$ if **(a)** $A$ and $B$ are independent, **(b)** $A$ and $B$ are disjoint, **(c)** $A^c$ and $B$ are independent, and **(d)** $A^c$ and $B$ are disjoint.
41. A politician considers running for election and has decided to give it two tries. He figures that the current conditions are favorable and that he has about a 60% chance of winning this election as opposed to a 50–50 chance in the next election. However, if he does win this election, he estimates that there ought to be a 75% chance of being reelected. **(a)** Find the probability that he wins both elections. **(b)** Find the probability that he wins the first election and loses the second. **(c)** If you learn that he won the second election, what is the probability that he won the first election? **(d)** If he loses the first election, what is the probability that he wins the second?
42. Consider two events $A$ and $B$. We say that $B$ gives *positive information* about $A$, denoted $B \nearrow A$, if $P(A|B) > P(A)$, that is if knowing $B$ increases the probability of $A$. Similarly, we say that $B$ gives negative information about $A$, denoted $B \searrow A$, if $P(A|B) < P(A)$. Are the following statements true or false? **(a)** If $B \nearrow A$, then $A \nearrow B$. **(b)** If $A \nearrow B$ and $B \nearrow C$, then $A \nearrow C$. **(c)** If $B \nearrow A$, then $B \searrow A^c$. **(d)** $A \searrow A^c$.
43. Show that both $\emptyset$ and the sample space $S$ are independent of any event. Explain intuitively.
44. Let $S$ be a sample space with $n$ equally likely outcomes where $n$ is a prime number. Show that there are no independent events (unless one of them is $S$ or $\emptyset$).
45. A coin has probability $p$ of showing heads. Flip it three times and consider the events $A = \{\text{at most one tails}\}$ and $B = \{\text{all flips are the same}\}$. For which values of $p$ are $A$ and $B$ independent?
46. A fair coin is flipped twice. Explain the difference between the following: **(a)** the probability that both flips give heads, and **(b)** the conditional probability that the second flip gives heads given that the first flip gave heads.
47. In December 1992, a small airplane crashed in a residential area near Bromma Airport outside Stockholm, Sweden. In an attempt to calm the residents, the airport manager claimed that they should now feel safer than before since the

probability of two crashes is much smaller than the probability of one crash and hence it has now become less likely that another crash will occur in the future.[10] What do you think of his argument?

48. Bob and Joe are working on a project. They each have to finish their individual tasks to complete the project and work independent of each other. When Bob is asked about the chances of him getting his part done, Joe getting his part done, and then both getting the entire project done, he estimates these to be 99%, 90%, and 95%, respectively. Is this reasonable?
49. You roll a die and consider the events $A$: get an even outcome, and $B$: get at least 2. Find $P(B|A)$ and $P(A|B)$.
50. You roll a die twice and record the largest number (if the two rolls give the same outcome, this is the largest number). **(a)** Given that the first roll gives 1, what is the conditional probability that the largest number is 3? **(b)** Given that the first roll gives 3, what is the conditional probability that the largest number is 3?
51. Roll two fair dice. Let $A_k$ be the event that the first die gives $k$, and let $B_n$ be the event that the sum is $n$. For which values of $n$ and $k$ are $A_k$ and $B_n$ independent?
52. The distribution of blood types in the United States according to the "ABO classification" is O:45%, A:40%, B:11%, and AB:4%. Blood is also classified according to Rh type, which can be negative or positive and is independent of the ABO type (the corresponding genes are located on different chromosomes). In the U.S. population, about 84% are Rh positive. Sample two individuals at random and find the probability that **(a)** both are A negative, **(b)** one of them is O and Rh positive, while the other is not, **(c)** at least one of them is O positive, **(d)** one is Rh positive and the other is not AB, **(e)** they have the same ABO type, and **(f)** they have the same ABO type and different Rh types.
53. Let $A$, $B$, and $C$ be independent events. Show that $A$ is independent of both $B \cap C$ and $B \cup C$.
54. You are offered to play the following game: A roulette wheel is spun eight times. If any of the 38 numbers (0,00,1–36) is repeated, you lose \$10, otherwise you win \$10. Should you accept to play this game? Argue by computing the relevant probability.
55. Consider the following simplified version of the birthday problem in Example 1.19. Divide the year into "winter half" and "summer half." Suppose that the probability is $p$ that an individual is born in the winter half. What is the probability that two people are born in the same half of the year? For which value of $p$ is this minimized?
56. Consider the birthday problem with two people and suppose that the probability distribution of birthdays is $p_1, \ldots, p_{365}$. **(a)** Express the probability that they have the same birthday as a function of the $p_k$. **(b)** Show that the probability in (a) is minimized for $p_k = \frac{1}{365}, k = 1, 2, \ldots, 365$. If you are familiar with Lagrange

[10]True story!

multipliers, you can use these. Alternatively, first show that $\sum_{k=1}^{365}(p_k - \frac{1}{365} + \frac{1}{365})^2 \geq \frac{1}{365}$.

57. A certain text has one-third vowels and two-thirds consonants. Five letters are chosen at random and you are asked to guess the sequence. Find the probability that all guesses are correct if for each letter you **(a)** guess vowel or consonant with equal probabilities, **(b)** guess vowel with probability $\frac{1}{3}$ and consonant with probability $\frac{2}{3}$, and **(c)** always guess consonant.
58. Two events $A$ and $B$ are said to be *conditionally independent* given the event $C$ if

$$P(A \cap B|C) = P(A|C)P(B|C)$$

**(a)** Give an example of events $A$, $B$, and $C$ such that $A$ and $B$ are independent but not conditionally independent given $C$. **(b)** Give an example of events $A$, $B$, and $C$ such that $A$ and $B$ are not independent but conditionally independent given $C$. **(c)** Suppose that $A$ and $B$ are independent events. When are they conditionally independent given their union $A \cup B$? **(d)** Since the information in $C$ and $C^c$ is equivalent (remember Proposition 1.10 and the preceding discussion), we might suspect that if $A$ and $B$ are independent given $C$, they are also independent given $C^c$. However, this is not true in general. Give an example of three events $A$, $B$, and $C$ such that $A$ and $B$ are independent given $C$ but not given $C^c$.

59. Roll a die twice and consider the events $A = \{\text{first roll gives at least 4 }\}$, $B = \{\text{second roll gives at most 4 }\}$, and $C = \{\text{the sum of the rolls is 10 }\}$. **(a)** Find $P(A)$, $P(B)$, $P(C)$, and $P(A \cap B \cap C)$. **(b)** Are $A$, $B$, and $C$ independent?
60. Roll a die $n$ times and let $A_{ij}$ be the event that the $i$th and $j$th rolls give the same number, where $1 \leq i < j \leq n$. Show that the events $A_{ij}$ are pairwise independent but not independent.
61. You throw three darts independently and at random at a dartboard. Find the probability that you get **(a)** no bull's eye, **(b)** at least one bull's eye, **(c)** only even numbers, and **(d)** exactly one triple and at most one double.
62. Three fair dice are rolled. Given that there are no 6s, what is the probability that there are no 5s?
63. You have three pieces of string and tie together the ends two by two at random. **(a)** What is the probability that you get one big loop? **(b)** Generalize to $n$ pieces.
64. Choose $n$ points independently at random on the perimeter of a circle. What is the probability that all points are on the same half-circle?
65. Do Example 1.23 assuming that balls are instead drawn with replacement.
66. A fair coin is flipped $n$ times. Let $A_k = \{\text{heads in } k\text{th flip}\}$, $k = 1, 2, \ldots, n$, and $B = \{\text{the total number of heads is even}\}$. Show that $A_1, \ldots, A_n, B$ are not independent but that if any one of them is removed, the remaining $n$ events are independent (from Stoyanov, *Counterexamples in Probability* [9]).
67. Compute the reliability of the two systems below given each component functioning independently with probability $p$.

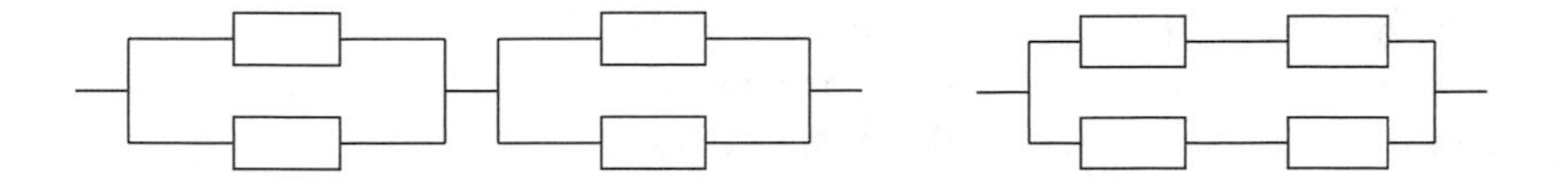

68. A system is called a "$k$-out-of-$n$ system" if it functions whenever at least $k$ of the $n$ components function. Suppose that components function independent of each other with probability $p$ and find an expression for the reliability of the system.
69. You play the following game: You bet \$1, a fair die is rolled and if it shows 6 you win \$4, otherwise you lose your dollar. If you must choose the number of rounds in advance, how should you choose it to maximize your chance of being ahead (having won more than you have lost) when you quit, and what is the probability of this?
70. Suppose that there is a one-in-a-million chance that a person is struck by lightning and that there are $n$ people in a city during a thunderstorm. **(a)** If $n$ is 2 million, what is the probability that somebody is struck? **(b)** How large must $n$ be for the probability that somebody is struck to be at least $\frac{1}{2}$?
71. A fair die is rolled $n$ times. Once a number has come up, it is called *occupied* (e.g., if $n = 5$ and we get 2, 6, 5, 6, 2, the numbers 2, 5, and 6 are occupied). Let $A_k$ be the event that $k$ numbers are occupied. Find the probability of $A_1$ (easy) and $A_2$ (trickier).

### Section 1.6. The Law of Total Probability and Bayes' Formula

72. In the United States, the overall chance that a baby survives delivery is 99.3%. For the 15% that are delivered by cesarean section, the chance of survival is 98.7%. If a baby is not delivered by cesarean section, what is its survival probability?
73. You roll a die and flip a fair coin a number of times determined by the number on the die. What is the probability that you get no heads?
74. In a blood transfusion, you can always give blood to somebody of your own ABO type (see Problem 52). Also, type O can be given to anybody and those with type AB can receive from anybody (people with these types are called *universal donors* and *universal recipients*, respectively). Suppose that two individuals are chosen at random. Find the probability that **(a)** neither can give blood to the other, **(b)** one can give to the other but not vice versa, **(c)** at least one can give to the other, and **(d)** both can give to each other.
75. You have two urns, 10 white balls, and 10 black balls. You are asked to distribute the balls in the urns, choose an urn at random, and then draw a ball at random from the chosen urn. How should you distribute the balls in order to maximize the probability to get a black ball?
76. A sign reads ARKANSAS. Three letters are removed and then put back into the three empty spaces again, at random. What is the probability that the sign still reads ARKANSAS?

77. A sign reads IDAHO. Two letters are removed and put back at random, each equally likely to be put upside down as in the correct orientation. What is the probability that the sign still reads IDAHO?
78. In the "Pick 3" game from Example 1.25, play the "any order" options and choose your three numbers at random. What is the probability that you win?
79. From a deck of cards, draw four cards at random without replacement. If you get $k$ aces, draw $k$ cards from another deck. What is the probability to get exactly $k$ aces from the first deck and exactly $n$ aces from the second deck?
80. Recall Example 1.39, where you observe a mother walking with a girl. Find the conditional probability that the other child is also a girl in the following cases: **(a)** The mother chooses the older child with probability $p$. **(b)** If the children are of different genders, the mother chooses the girl with probability $p$. **(c)** When do you get the second solution in the example that the probability equals $\frac{1}{2}$?
81. Let $A$, $B$, and $C$ be three events. Assuming that all conditional probabilities are defined, show that

$$P(A|B) = P(A|B \cap C)P(C|B) + P(A|B \cap C^c)P(C^c|B).$$

82. Graduating students from a particular high school are classified as "weak" or "strong." Among those who apply to college, it turns out that 56% of the weak students but only 39% of the strong students are accepted at their first choice. Does this indicate a bias against strong students?
83. In Example 1.45, if all three dice are rolled at once, which is the most likely to win?
84. Consider the introduction to Section 1.6. If your car develops engine problems, how likely is it that the dealer sold you a flood-damaged car?
85. Consider the Monty Hall problem in Example 1.48. **(a)** What is the relevance of the assumption that Monty opens a door at random in the case where you chose the car? **(b)** Suppose that there are $n$ doors and $k$ cars, everything else being the same. What is your probability of winning a car with the switching strategy?
86. The three prisoners Shadrach, Mesach, and Abednego learn that two of them will be set free but not who. Later, Mesach finds out that he is one of the two, and, excited, he runs to Shadrach to share his good news. When Shadrach finds out, he gets upset and complains "Why did you tell me? Now that there are only me and Abednego left, my chance to be set free is only $\frac{1}{2}$, but before it was $\frac{2}{3}$." What do you think of his argument? What assumptions do you make?
87. A box contains two regular quarters and one fake two-headed quarter. **(a)** You pick a coin at random. What is the probability that it is the two-headed quarter? **(b)** You pick a coin at random, flip it, and get heads. What is the probability that it is the two-headed quarter?
88. Two cards are chosen at random without replacement from a deck and inserted into another deck. This deck is shuffled, and one card is drawn. If this card is an ace, what is the probability that no ace was moved from the first deck?

89. A transmitter sends 0s and 1s to a receiver. Each digit is received correctly (0 as 0, 1 as 1) with probability 0.9. Digits are received correctly independent of each other and on the average twice as many 0s as 1s are being sent. **(a)** If the sequence 10 is sent, what is the probability that 10 is received? **(b)** If the sequence 10 is received, what is the probability that 10 was sent?
90. Consider two urns, one with 10 balls numbered 1 through 10 and one with 100 balls numbered 1 through 100. You first pick an urn at random, then pick a ball at random, which has number 5. **(a)** What is the probability that it came from the first urn? **(b)** What is the probability in (a) if the ball was instead chosen randomly from all the 110 balls?
91. Smoking is reported to be responsible for about 90% of all lung cancer. Now consider the risk that a smoker develops lung cancer. Argue why this is not 90%. In order to compute the risk, what more information is needed?
92. The serious disease $D$ occurs with a frequency of 0.1% in a certain population. The disease is diagnosed by a method that gives the correct result (i.e., positive result for those with the disease and negative for those without it) with probability 0.99. Mr Smith goes to test for the disease and the result turns out to be positive. Since the method seems very reliable, Mr Smith starts to worry, being "99% sure of actually having the disease." Show that this is not the relevant probability and that Mr Smith may actually be quite optimistic.
93. You test for a disease that about 1 in 500 people have. If you have the disease, the test is always positive. If you do not have the disease, the test is 95% accurate. If you test positive, what is the probability that you have the disease?
94. **(a)** Ann and Bob each tells the truth with probability 1/3 and lies otherwise, independent of each other. If Bob tells you something and Ann tells you Bob told the truth, what is the probability Bob told you the truth? **(b)** Add a third person, Carol, who is as prone to lying as Ann and Bob one. If Ann says that Bob claims that Carol told the truth, what is the probability Carol told the truth?
95. A woman witnesses a hit-and-run accident one night and reports it to the police that the escaping car was black. Since it was dark, the police test her ability to distinguish black from dark blue (other colors are ruled out) under similar circumstances and she is found to be able to pick the correct color about 90% of the time. One police officer claims that they can now be 90% certain that the escaping car was black, but his more experienced colleague says that they need more information. In order to determine the probability that the car was indeed black, what additional information is needed, and how is the probability computed?
96. Joe and Bob are about to drive home from a bar. Since Joe is sober and Bob is not, Joe takes the wheel. Bob has recently read in the paper that drunk drivers are responsible for 25% of car accidents, that about 95% of drivers are sober, and that the overall risk of having an accident is 10%. "You sober people cause 75% of the accidents," slurs Bob, "and there are so many of you too! You should let me drive!" Joe who knows his probability theory has his answer ready. How does he respond?

97. Consider Example 1.50, where the murderer must be one of $n$ individuals. Suppose that Joe Bloggs is initially considered the main suspect and that the detectives judge that there is a 50–50 chance that he is guilty. If his DNA matches the DNA found at the scene of the crime, what is then the probability that he is guilty?
98. Consider a parallel system of two components. The first component functions with probability $p$ and if it functions, the second also functions with probability $p$. If the first has failed, the second functions with probability $r < p$, due to heavier load on the single component. **(a)** What is the probability that the second component functions? **(b)** What is the reliability of the system? **(c)** If the second component does not function, what is the probability that the first does?
99. Recall Example 1.51, where you know that the guilty child is a boy, a boy opens the door, and he has one sibling. Compute the probability that the child who opened the door is guilty if the guilty child opens the door with probability $p$.
100. Your new neighbors have three children. **(a)** If you are told about three independent observations of a boy, what is the probability that they have three boys? **(b)** If you get two confirmations of an observed boy and one of an observed girl, what is the probability that they have two boys and a girl? **(c)** If you get $j \geq 1$ confirmations of an observed boy and $n - j \geq 1$ of an observed girl, what is the probability that they have two boys and a girl?
101. Consider Example 1.53 about cystic fibrosis. **(a)** What is the probability that two healthy parents have a child who neither is a carrier nor has the disease? **(b)** Given that a child is healthy, what is the probability that both parents are carriers (you may disregard parents with the disease)?
102. A genetic disease or condition is said to be *sex-linked* if the responsible gene is located on either of the sex chromosomes, $X$ and $Y$ (recall that women have two $X$ chromosomes and men have one each of $X$ and $Y$). One example is red-green color-blindness for which the responsible gene is located on the $X$ chromosome. The allele for color-blindness is recessive, so that one normal copy of the gene is sufficient for normal vision. **(a)** Consider a couple where the woman is color-blind and the man has normal vision. If they have a daughter, what is the probability that she is color-blind? If they have a son? **(b)** Compute the probabilities in (a) under the assumption that both parents have normal vision and the woman's father was color-blind. **(c)** It is estimated that about 7% of men are color-blind but only about 0.5% of women. Explain!
103. *Tay–Sachs Disease* is a serious genetic disease that usually leads to death in early childhood. The allele for the disease is recessive and *autosomal* (not located on any of the sex chromosomes). **(a)** In the general population, about 1 in 250 is a carrier of the disease. What incidence among newborns does this give? **(b)** Certain subpopulations are at greater risk for the disease. For example, the incidence among newborns in the Cajun population of Louisiana is 1 in 3600. What proportion of carriers does this give? **(c)** Generally, if a serious recessive disease has a carrier frequency of one in $n$ and an incidence among newborns

of one in $N$, what is the relation between $n$ and $N$? (Why is it relevant that the disease is "serious?")

104. Consider the game of badminton in Example 1.57. **(a)** Find the probability that Ann scores the next point if she is the receiver. **(b)** Now suppose that Ann wins a rally as server with probability $p_A$ and let the corresponding probability for Bob be $p_B$. If Ann serves, what is the probability that she is the next player to score?

105. In table tennis, a set is won by the first player to reach 11 points, unless the score is 10–10, in which case serves are alternated and the player who first gets ahead by two points wins. Suppose that Ann wins a point as server with probability $p_A$ and Bob wins a point as server with probability $p_B$. If the score is 10–10 and Ann serves, what is the probability that she wins the set?

106. You are playing roulette, each time betting on "odd," which occurs with probability $\frac{18}{38}$ and gives you even money back. You start with \$10 and decide to play until you have either doubled your fortune or gone broke. Compute the probability that you manage to double your fortune if in each round you bet \$10, \$5, \$2, and \$1 dollar, respectively. After you have found the best strategy, give an intuitive explanation of why it is the best and why it is called "bold play."

107. In Example 1.59, suppose that Ann wins each round with probability $p > \frac{1}{2}$. What is the probability that she eventually goes broke?

108. The game of Penney-ante can be played with patterns of any length $n$. In the case $n = 1$, the game is fair (this is trivial); if $n = 2$, it can be fair or to your advantage, depending on the patterns chosen, and if $n \geq 3$, you can always choose a winning strategy. **(a)** Let $n = 2$, so that the possible patterns are *HH*, *HT*, *TH*, and *TT*. Suggest a strategy and compute your winning probability in the different cases. **(b)** Let $n = 4$ and suppose that your opponent chooses *HHHH*. Suggest how you should choose your best pattern and compute the winning probability.

109. In the game of craps, you roll two dice and add the numbers. If you get 7 or 11 (a *natural*) you win, if you roll 2, 3, or 12 (*craps*) you lose. Any other roll establishes your *point*. You then roll the dice repeatedly until you get either 7 or your point. If you get your point first you win, otherwise you lose. Starting a new game of craps, what is the probability that you win?

# 2

# RANDOM VARIABLES

## 2.1 INTRODUCTION

We saw in the previous chapter that many random experiments have numerical outcomes. Even if the outcome itself is not numerical, such as the case is Example 1.4, where a coin is flipped twice, we often consider events that can be described in terms of numbers, for example, {the number of heads equals 2}. It would be convenient to have some mathematical notation to avoid the need to spell out all events in words. For example, instead of writing {the number of heads equals 1} and {the number of heads equals 2}, we could start by denoting the number of heads by $X$ and consider the events $\{X = 1\}$ and $\{X = 2\}$. The quantity $X$ is then something whose value is not known before the experiment but becomes known after.

> ***Definition 2.1.*** A *random variable* is a real-valued variable that gets its value from a random experiment.

There is a more formal definition that defines a random variable as a real-valued function on the sample space. If $X$ denotes the number of heads in two coin flips, we would thus, for example, have $X(HH) = 2$. In a more advanced treatment of probability theory, this formal definition is necessary, but for our purposes, Definition 2.1 is enough.

*Probability, Statistics, and Stochastic Processes*, Second Edition. Peter Olofsson and Mikael Andersson.

A random variable $X$ is thus something that does not have a value until after the experiment. Before the experiment, we can only describe the set of possible values, that is, the *range* of $X$ and the associated probabilities. Let us look at a simple example.

***Example 2.1.*** Flip a coin twice and let $X$ denote the number of heads. Then $X$ has range $\{0, 1, 2\}$ and the associated probabilities are

$$P(X = 0) = \frac{1}{4}, \quad P(X = 1) = \frac{1}{2}, \quad P(X = 2) = \frac{1}{4}$$

and we refer to these probabilities as the *distribution* of $X$. □

In the last example, any three numbers between 0 and 1 that sum to 1 is a possible distribution (recall Section 1.4), and the particular choice in the example indicates that the coin is fair. Let us next restate some of the examples from Section 1.2 in terms of random variables. In each case, we define $X$ and find its range.

***Example 2.2.*** Let $X$ be the number of dots when a die is rolled. The range of $X$ is $\{1, 2, \ldots, 6\}$. □

***Example 2.3.*** Let $X$ be the number of rolls of a die until the first 6 appears. The range of $X$ is $\{1, 2, \ldots\}$. □

***Example 2.4.*** Let $X$ be the lifetime of a lightbulb. The range of $X$ is $[0, \infty)$, the nonnegative real numbers. □

As noted in Section 1.2, the three sets above are different in nature. The first is finite, the second is countably infinite, and the third is uncountable. The formal definition of a countably infinite set is one that can be put in a one-to-one correspondence with the natural numbers. Examples of such sets are the natural numbers themselves $\{1, 2, 3, \ldots\}$, the odd natural numbers $\{1, 3, 5, \ldots\}$, and the integers $\{\ldots, -2, -1, 0, 1, 2, \ldots\}$. Countably infinite sets need not be sets of integers only; for example, the set $\{\frac{1}{n} : n = 1, 2, \ldots\} = \{1, \frac{1}{2}, \frac{1}{3}, \ldots\}$ is countably infinite. Hopefully it is intuitively clear what a countably infinite set is, and we will not discuss the complications and subtleties that arise in the study of cardinality of sets.

We use the term *countable* to refer to a set that is either finite or countably infinite. The reason for this is that in the study of random variables, the important distinction turns out to be not between finite and infinite ranges, but between countable and uncountable ranges.

## 2.2 DISCRETE RANDOM VARIABLES

We distinguish primarily between random variables that have countable range and those that have uncountable range. Let us examine the first case and start with a definition.

***Definition 2.2.*** If the range of $X$ is countable, then $X$ is called a *discrete random variable*.

For a discrete random variable $X$, we are interested in computing probabilities of the type $P(X = x_k)$ for various values of $x_k$ in the range of $X$. As we vary $x_k$, the probability $P(X = x_k)$ changes, so it is natural to view $P(X = x_k)$ as a function of $x_k$. We now formally define and name this function.

***Definition 2.3.*** Let $X$ be a discrete random variable with range $\{x_1, x_2, \ldots\}$ (finite or countably infinite). The function

$$p(x_k) = P(X = x_k), \quad k = 1, 2, \ldots$$

is called the *probability mass function* (pmf) of $X$.

Sometimes, we also use the notation $p_X$ for the pmf, if it is needed to stress which the random variable is. When we represent a pmf as a bar chart, the height of a bar equals the probability of the corresponding value on the $x$ axis. Since $X$ cannot take on values other than $x_1, x_2, \ldots$, we can imagine bars of height 0 at all other values and could thus view the pmf as a function on all of $R$ (the real numbers) if we wish. The numbers $x_1, x_2, \ldots$ do not have to be integers, but they often are.

***Example 2.5.*** Let $X$ be the number of daughters in a family with three children. By Example 1.13, the range of $X$ is $\{0, 1, 2, 3\}$ and the values of the pmf are $p(0) = \frac{1}{8}$, $p(1) = \frac{3}{8}$, $p(2) = \frac{3}{8}$, $p(3) = \frac{1}{8}$. The pmf is illustrated in Figure 2.1. □

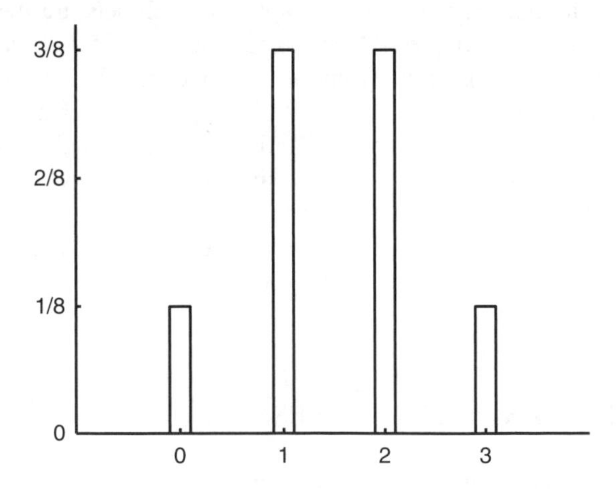


**FIGURE 2.1** Bar chart for the number of daughters in Example 2.5. The probability of 0 daughters is $\frac{1}{8}$ and so on.

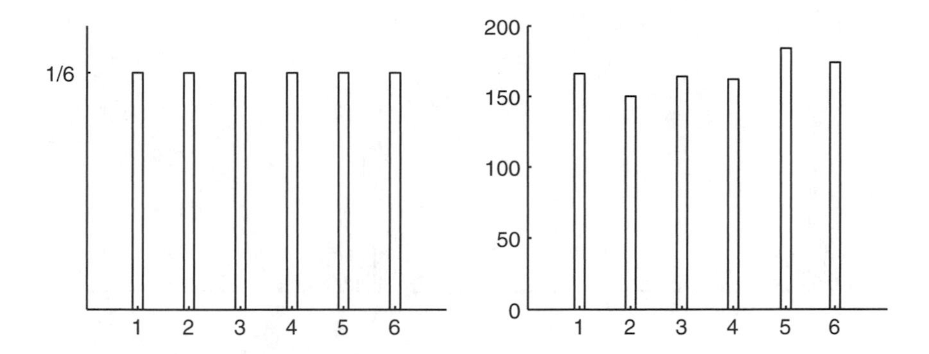


**FIGURE 2.2** Pmf and histogram of the roll of a fair die.

Now suppose that we have a discrete random variable $X$ with pmf $p$ and repeat the experiment over and over to get a number, say, $n$, of outcomes of $X$. We then expect the *absolute frequency*, that is, the number of times that we get the outcome $x_k$, to be approximately equal to $np(x_k)$. We can plot the absolute frequencies in a *histogram*, which would then be expected to look similar in shape to the pmf (but different in scale). In other words, we can think of the pmf as representing the "perfect histogram."

***Example 2.6.*** If we roll a die and let $X$ denote the outcome, then $X$ has pmf

$$p(k) = P(X = k) = \frac{1}{6}, \quad k = 1, 2, \ldots, 6$$

which is plotted in Figure 2.2, together with a histogram of 1000 simulated rolls of the die. This is an example of a *uniform distribution* (compare with Section 1.4), characterized by the pmf having the same value for all the possible outcomes of $X$. Note that it is not possible to have a uniform distribution on an infinite set (why?), so there is no immediately clear interpretation of a statement such as "choose an integer at random." □

***Example 2.7.*** If we roll a die and let $X$ denote the number of rolls until the first 6 appears, then Example 1.41 reveals that the pmf of $X$ is

$$p(k) = P(X = k) = \frac{1}{6}\left(\frac{5}{6}\right)^{k-1}, \quad k = 1, 2, \ldots$$

which is plotted in Figure 2.3 together with a histogram of 1000 simulated observations on $X$. The pmf has a bar at every positive integer, but the histogram must end somewhere. In this case, the largest observed value of $X$ was 44. □

By the properties of probability measures, we have the following proposition.

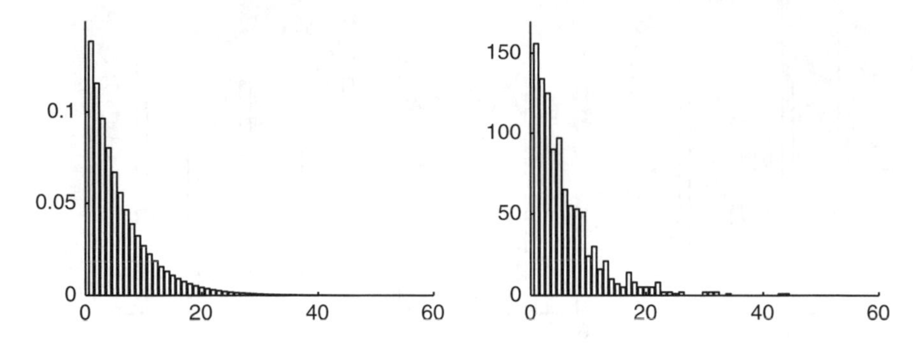


**FIGURE 2.3** Pmf and histogram of the number of rolls until the first 6 appears.

**Proposition 2.1.** A function $p$ is a possible pmf of a discrete random variable on the range $\{x_1, x_2, \ldots\}$ if and only if

**(a)** $p(x_k) \geq 0 \quad$ for $k = 1, 2, \ldots$

**(b)** $\sum_{k=1}^{\infty} p(x_k) = 1$

If the range of $X$ is finite, the sum in (b) is finite. So far we have considered events of the type $\{X = k\}$, the event that $X$ equals a particular value $k$. We could also look at events of the type $\{X \leq k\}$, the event that $X$ is less than or equal to $k$. For example, when we roll a die we might ask for the probability that we get at most 1, at most 2, and so on. This leads to another function to be defined next.

***Definition 2.4.*** Let $X$ be any random variable. The function

$$F(x) = P(X \leq x), \quad x \in R$$

is called the *(cumulative) distribution function* (cdf) of $X$.

The word "cumulative" appears in parentheses because it is rarely used other than for the acronym cdf. Note that the cdf is a function on the entire real line. To get an idea of what it typically looks like, let us return to Example 2.5.

***Example 2.8.*** In a family with three children, let $X$ be the number of daughters. The range of $X$ is $\{0, 1, 2, 3\}$, so let us start by computing $F(k)$ for these values. We get

$$F(0) = P(X \leq 0) = p(0) = \frac{1}{8}$$

since the only way to be less than or equal to 0 is to be equal to 0. For $k = 1$, we first note that being less than or equal to 1 means being equal to 0 or 1. In terms of events, we have

$$\{X \leq 1\} = \{X = 0\} \cup \{X = 1\}$$

and since the events are disjoint, we get

$$\begin{aligned} F(1) = P(X \leq 1) &= P(X = 0) + P(X = 1) \\ = p(0) + p(1) &= \frac{1}{8} + \frac{3}{8} = \frac{1}{2} \end{aligned}$$

Continuing like this, we also get $F(2) = \frac{7}{8}$ and $F(3) = 1$. Now we have the values of $F$ at the points in the range of $X$. What about other values? Let us, for example, consider the point 0.5. Noting that $X \leq 0.5$ means that $X \leq 0$, we get

$$F(0.5) = F(0) = \frac{1}{8}$$

and we realize that $F(x) = F(0)$ for all points $x \in [0, 1)$. Similarly, all points $x \in [1, 2)$ have $F(x) = F(1)$ and all points $x \in [2, 3)$ have $F(x) = F(2)$. Finally, since $X$ cannot take on any negative values, we have $F(x) = 0$ for $x < 0$, and since $X$ is always at most 3, we have $F(x) = 1$ for $x \geq 3$. This gives the final form of $F$, which is illustrated in Figure 2.4. □

The graph in Figure 2.4 is typical for the cdf of a discrete random variable. It jumps at each value that $X$ can assume, and the size of the jump is the probability of that value. Between points in the range of $X$, the cdf is constant. Note how it is always nondecreasing and how it ranges from 0 to 1. Also note in which way the points are filled where $F$ jumps, which can be expressed by saying that $F$ is *right-continuous*.

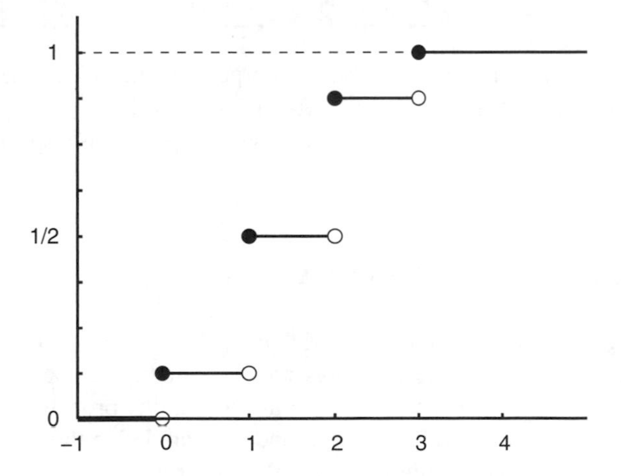


**FIGURE 2.4** Graph of the cdf of the number of daughters in Example 2.8.

Some of these observed properties turn out to be true for any cdf, not only for that of a discrete random variable. In the next section, we shall return to this.

In Example 2.8, $F$ assumes the values 0 and 1. If the range of $X$ is finite, this is always the case but not necessarily if the range is countably infinite, as the following example shows.

***Example 2.9.*** When $X$ is the number of rolls until the first 6 appears, the range is $\{1, 2, \ldots\}$ and the pmf is given in Example 1.41. The value of the cdf at an integer $n$ is

$$\begin{aligned} F(n) = \sum_{k=1}^{n} p(k) &= \frac{1}{6} \sum_{k=1}^{n} \left(\frac{5}{6}\right)^{k-1} \\ &= \frac{1}{6} \times \frac{1-(5/6)^n}{1-5/6} = 1 - \left(\frac{5}{6}\right)^n, \quad n = 1, 2, \ldots \end{aligned}$$

and the cdf approaches but never reaches 1. □

The cdf and the pmf are related according to the following proposition, which we state without proof.

**Proposition 2.2.** Let $X$ be a discrete random variable with range $\{x_1, x_2, \ldots\}$, pmf $p$, and cdf $F$. Then

**(a)** $F(x) = \sum_{k:x_k \leq x} p(x_k), \quad x \in R$

**(b)** $p(x_k) = F(x_k) - \lim_{y \uparrow x_k} F(y), \quad k = 1, 2, \ldots$

**(c)** For $B \subseteq R, \quad P(X \in B) = \sum_{k:x_k \in B} p(x_k)$

Note how part (b) says that the probability of a point is the size of the jump of $F$ in that point. If $F$ does not jump at $x$, the left-hand side limit $\lim_{y \uparrow x} F(y)$ equals $F(x)$ and the probability that $X$ equals $x$ is 0. Also note that part (a) is a special case of part (c) with the particular choice $B = (-\infty, x]$.

## 2.3 CONTINUOUS RANDOM VARIABLES

To introduce another type of random variable, let $X$ be the lifetime of a lightbulb so that it has range $[0, \infty)$. Since this is an uncountable set, $X$ is not a discrete random variable. However, we can still imagine computing probabilities of events of the type $\{X \leq x\}$ for different values of $x$, and we can define the cdf as previously, $F(x) = P(X \leq x)$. Indeed, Definition 2.4 is stated for an arbitrary random variable, not necessarily discrete. What does $F$ look like in this case?

Recall the cdf for a discrete random variable. This is a nondecreasing function, ranging from 0 to 1, which jumps precisely at the points that $X$ can assume. For our lightbulb, the cdf should be nondecreasing (why?), range from 0 to 1, and have $F(x) = 0$ for $x < 0$. Finally, since $X$ can assume *every* positive value, there must be an "infinitely small jump at every point," and the only way for this to be possible is if $F$ is a continuous function. With the definition of cdf remaining the same as in the previous section, we define the following.

***Definition 2.5.*** If the cdf $F$ is a continuous function, then $X$ is said to be a *continuous random variable*.

The definitions of discrete and continuous random variables are qualitatively different. A discrete random variable is defined through its range; a continuous random variable, through its cdf. Although a continuous random variable must have an uncountable range, this is not the definition and as we will see later, a random variable can have an uncountable range and still have discontinuities in its cdf.

In Example 2.8 and Figure 2.4, we observed some properties of the cdf. These turn out to hold in general, and we state the following proposition.

**Proposition 2.3.** If $F$ is the cdf of any random variable, $F$ has the following properties:

**(a)** It is nondecreasing.

**(b)** It is right-continuous.

**(c)** It has the limits $F(-\infty) = 0$ and $F(\infty) = 1$ (where the limits may or may not be attained at finite $x$).

Since these properties are intuitively reasonable, we omit the technical proof. It should be noted that $F(\infty)$ is defined as $\lim_{x\to\infty} F(x)$ and is *not* the same as $P(X \leq \infty)$ but rather $P(X < \infty)$. This is so because, for any increasing sequence $\{x_n\}$ of real numbers that is such that $x_n \to \infty$, we have

$$\{X < \infty\} = \bigcup_{n=1}^{\infty} \{X \leq x_n\}$$

since if $X$ is finite, it must be less than *some* $x_n$. By continuity of probabilities (Proposition 1.6), we get

$$P(X < \infty) = \lim_{n\to\infty} P(X \leq x_n) = \lim_{n\to\infty} F(x_n) = F(\infty)$$

by definition (and similarly for $-\infty$, using instead decreasing sequences). Right now, this is no big deal since we do not allow random variables to be infinite and hence both $P(X < \infty)$ and $P(X \leq \infty)$ equal 1. However, there are situations where it is natural to let random variables also take on the value $\infty$. Consider, for example, the gambler's ruin problem from Example 1.58, and let the random variable $T$ be the time until Ann is ruined. It is then natural to let $T = \infty$ mean that Bob was ruined first, that is, Ann's ruin never occurred. We will see more examples of infinite-valued random variables in Chapter 8.

The properties listed in Proposition 2.3 are useful for a quick "sanity check" as to whether a function is a cdf. If you have attempted to find a cdf and your candidate violates any of (a) through (c) above, you have made an error somewhere. For computations, the cdf can be used as follows.

**Proposition 2.4.** Let $X$ be any random variable with cdf $F$. Then

**(a)** $P(a < X \leq b) = F(b) - F(a), \quad a \leq b$

**(b)** $P(X > x) = 1 - F(x), \quad x \in R$

*Proof.* To prove (a), take $a \leq b$ and first note that

$$\{X \leq b\} = \{X \leq a\} \cup \{a < X \leq b\}$$

which is a disjoint union. Hence

$$P(X \leq b) = P(a < X \leq b) + P(X \leq a)$$

which is to say that

$$P(a < X \leq b) = P(X \leq b) - P(X \leq a) = F(b) - F(a)$$

by the definition of cdf. For part (b), note that the event $\{X > x\}$ is the complement of the event $\{X \leq x\}$, so this follows from Proposition 1.3 together with the definition of cdf. Alternatively, we can view this as a special case of (a) with $b = \infty$. ■

The cdf is thus defined in the same way for discrete and continuous random variables. For discrete random variables, we also defined the pmf, $p(x_k) = P(X = x_k)$, for values $x_k$ in the range of $X$ where $p(x_k)$ measures the size of the jump of the cdf at $x_k$. In the continuous case, there are no such jumps, so there is no exact analogy. However, instead of considering the jump size, let us consider the *slope* of the cdf. A large jump size in the discrete case then corresponds to a steep slope in the continuous case. Since the size of the slope of a function is measured by its derivative, we present the following definition.

***Definition 2.6.*** The function $f(x) = F'(x)$ is called the *probability density function* (pdf) of $X$.

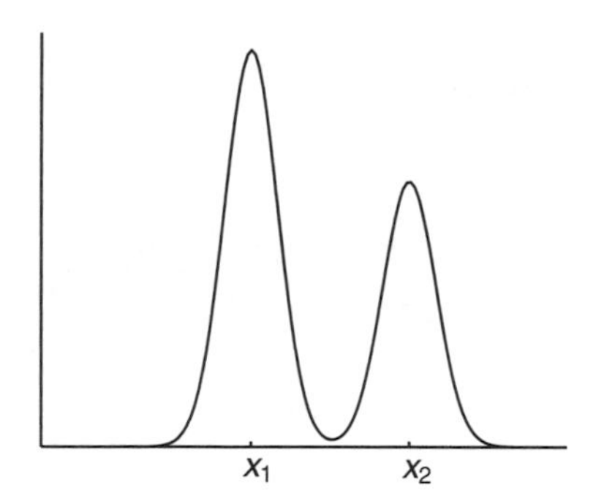


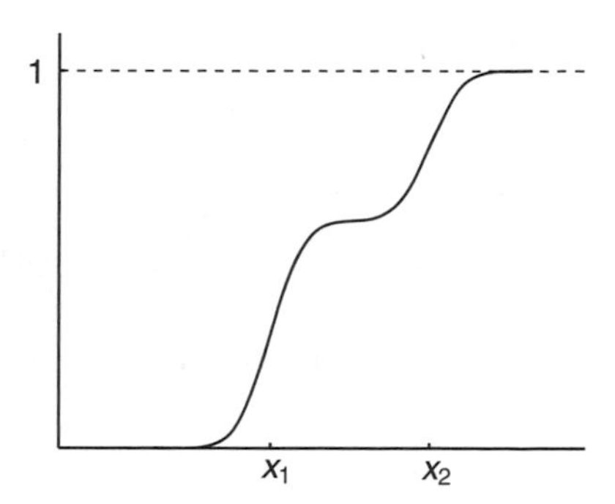


**FIGURE 2.5** A pdf and corresponding cdf of a continuous random variable. Note how high values of the pdf correspond to sharp increases in the cdf.

For a continuous random variable, the pdf plays the role of a discrete random variable's pmf. Since the pdf is the derivative of the cdf, the fundamental theorem of calculus gives that we can obtain the cdf from the pdf as

$$F(x) = \int_{-\infty}^{x} f(t)dt$$

(see Figure 2.5). Compare with the discrete case where the relation between cdf and pmf is

$$F(x) = \sum_{k:x_k \leq x} p(x_k)$$

We get the following analog of Proposition 2.2, stated without proof.

**Proposition 2.5.** Let $X$ be a continuous random variable with pdf $f$ and cdf $F$. Then

**(a)** $F(x) = \int_{-\infty}^{x} f(t)dt, \quad x \in R$

**(b)** $f(x) = F'(x), \quad x \in R$

**(c)** For $B \subseteq R, \ P(X \in B) = \int_B f(x)dx$

The notation $\int_B$ means that we compute the integral over the set $B$. Often $B$ is simply an interval but it could also be a union of intervals or possibly some more complicated set.[1] The special case of an interval, $B = (a, b]$, gives, for a

[1]It turns out that $B$ cannot be any set but must be chosen among the so-called "Borel sets." This is a necessary restriction in a more advanced treatment of probability theory but is of no concern to us. The sets that are not Borel sets are very bizarre and never arise in practice. It should also be pointed out that the existence of a function $f$ satisfying part (c) of Proposition 2.5 is often taken as the *definition* of a continuous

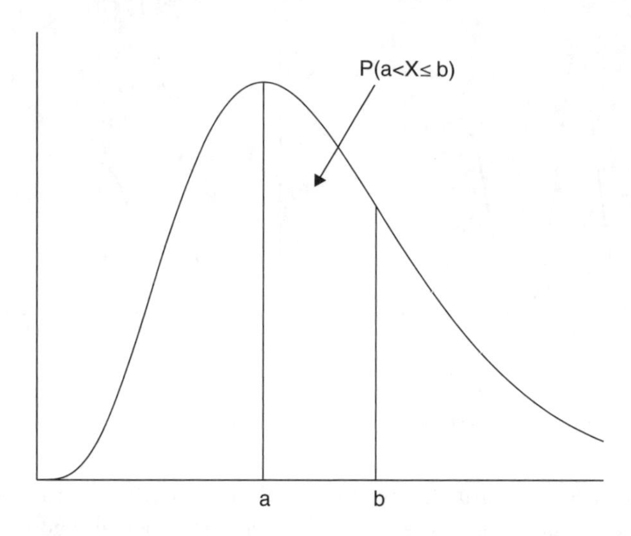


**FIGURE 2.6** Illustration of Proposition 2.5(c); the probability that $X$ is between $a$ and $b$ is the area under the pdf.

continuous random variable

$$P(a < X \leq b) = \int_a^b f(x)dx$$

and combining this with Proposition 2.4(a) gives

$$F(b) - F(a) = \int_a^b f(x)dx$$

The probability that $X$ falls in the interval $(a, b]$ is thus the area under the pdf between $a$ and $b$ (see Figure 2.6).

Since $P(X \in R) = 1$, we also get the following analog of Proposition 2.1.

**Proposition 2.6.** A function $f$ is a possible pdf of some continuous random variable if and only if

**(a)** $f(x) \geq 0, \quad x \in R$

**(b)** $\int_{-\infty}^{\infty} f(x)dx = 1$

random variable. In fact, our definition in terms of the continuity of the cdf is slightly more restrictive since there exist continuous functions that are not differentiable anywhere. However, such functions are of no practical use, so we stick to our definition.

Since the value of an integral does not change by the inclusion or exclusion of a single point, we realize that we can alter the inequalities between strict and nonstrict, and, for instance, we get

$$P(a \le X \le b) = \int_a^b f(x)dx = F(b) - F(a) = P(a < X \le b)$$

and so on. Let us state this important fact as a corollary.

**Corollary 2.1.** *For a continuous random variable X, probabilities do not change if strict and nonstrict inequalities are interchanged in events.*

This is not true for discrete random variables. For example, if we roll a die there is a difference between $P(X \le 6)$ and $P(X < 6)$. The intuition behind Corollary 2.1 is that the probabilities are so "smeared out" over the range of $X$ that it does not matter if we add or remove single points.

***Example 2.10.*** Let $X$ be the lifetime (in hours) of a lightbulb and suppose that $X$ has pdf

$$f(x) = \begin{cases} 0 & \text{if } x < 0 \\ 0.001e^{-0.001x} & \text{if } x \ge 0 \end{cases}$$

Verify that this is a possible pdf and find **(a)** $P(X < 1000)$, **(b)** $P(100 \le X \le 1000)$, **(c)** a number $x$ such that a lightbulb survives the age $x$ with probability 0.5.

First note that the fact that $X$ can assume only nonnegative values is reflected in the pdf being 0 for negative $x$. Clearly $f$ is nonnegative and also

$$\begin{aligned}\int_{-\infty}^{\infty} f(x)dx &= \int_0^{\infty} 0.001e^{-0.001x}dx \\ &= \left[-e^{-0.001x}\right]_0^{\infty} = 1\end{aligned}$$

and $f$ is a possible pdf. To compute the probabilities, let us first find the cdf:

$$\begin{aligned}F(x) &= \int_{-\infty}^{x} f(t)dt = \int_0^{x} 0.001\, e^{-0.001t}dt \\ &= 1 - e^{-0.001x}, \quad x \ge 0\end{aligned}$$

For (a) we get

$$\begin{aligned}P(X < 1000) &= P(X \le 1000) = F(1000) \\ &= 1 - e^{-0.001 \times 1000} \approx 0.63\end{aligned}$$

and for (b)

$$P(100 \leq X \leq 1000) = F(1000) - F(100)$$
$$= 1 - e^{-1} - (1 - e^{-0.1}) \approx 0.54$$

and finally for (c), note that "surviving age $x$" is the event $\{X > x\}$, and we get

$$0.5 = P(X > x) = 1 - F(x) = e^{-0.001x}$$

which gives

$$x = \frac{\log 0.5}{-0.001} \approx 693$$

This number is called the *median* of $X$. If we have a large number of lightbulbs, about half of them function longer than 693 h. □

In the example, we formally integrate the pdf from $-\infty$ to $\infty$, but the negative part vanishes since $f$ equals 0 there. From now on, we will give the functional expression for a pdf only where it is positive and understand that it is 0 everywhere else.

Now let $X$ be a continuous random variable and consider a point $x$ in the range of $X$. If $X$ is less than or equal to $x$, then it is either strictly less than $x$ or exactly equal to $x$, or in terms of events

$$\{X \leq x\} = \{X < x\} \cup \{X = x\}$$

This union is clearly disjoint and hence

$$P(X \leq x) = P(X < x) + P(X = x)$$

which gives

$$P(X = x) = P(X \leq x) - P(X < x)$$

and by Corollary 2.1 this equals 0! The probability that a continuous random variable equals any particular value is thus 0, which at first sight may seem surprising since we must get *some* value of $X$ in the experiment. If we think about the underlying mathematics, this becomes less mysterious since we know that the integral over an interval can be positive even though the interval over each single *point* is 0. We will not delve any deeper into this; just remember that the range of a continuous random variable is so large that we cannot assign positive probabilities to individual points, only to intervals.

One consequence of the discussion above is that the interpretation of the pdf is different from its discrete counterpart, the pmf. Recall that if $X$ is discrete and has pmf $p$, then $p(x) = P(X = x)$. If $X$ is continuous, however, $P(X = x) = 0$ for all $x$ and is thus not equal to the pdf $f(x)$. How, then, should we interpret $f(x)$?

To get some feeling for what a pdf is, consider a continuous random variable $X$ with pdf $f$ and take two points, $x_1$ and $x_2$, in the range of $X$ such that $f(x_1) > f(x_2)$.

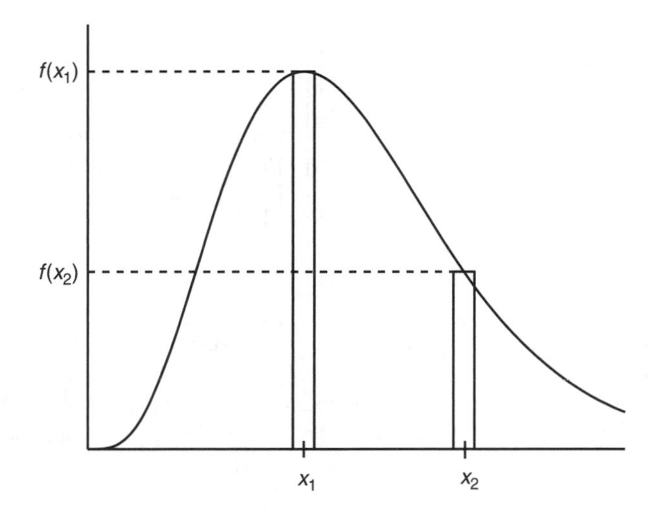


**FIGURE 2.7** Interpretation of the pdf. Since $f(x_1) > f(x_2)$, it is more likely that $X$ is close to $x_1$ than to $x_2$.

Now take a small $\epsilon > 0$ and place symmetric intervals of lengths $\epsilon$ around $x_1$ and $x_2$ (see Figure 2.7). Let us call the intervals $I_1$ and $I_2$, respectively. We get

$$P(X \in I_1) = \int_{x_1-\epsilon/2}^{x_1+\epsilon/2} f(t)dt \approx \epsilon f(x_1)$$

and similarly

$$P(X \in I_2) \approx \epsilon f(x_2)$$

Now, since $f(x_1) > f(x_2)$, the probability that $X$ belongs to $I_1$ is the larger of the two. This suggests an interpretation of the pdf in a point $x$; it measures how likely it is that $X$ is *in the neighborhood of* $x$, not exactly equal to $x$. Although not exactly the same as the pmf, the shape of the pdf still gives information on where $X$ is most likely to assume its values.

Consider again the lightbulb with the pdf given in Example 2.10 above. In Figure 2.8, the pdf and a histogram of 1000 simulated values of $X$ are plotted. To construct a histogram for a continuous distribution, we divide the $x$-axis into *bins* and count the number of observations in each bin. The number of bins depends on how many observations we have and has to be decided in each case. (Think about what the histogram would look like if we had too many or too few bins.) Note how the shape of the histogram mirrors the shape of the pdf.

Whenever the pdf of $X$ is given, we say that it defines the *distribution* of $X$. By virtue of Proposition 2.5, we realize that the pdf is uniquely determined by the cdf and vice versa, so either of the two specifies the distribution of $X$. The advantage

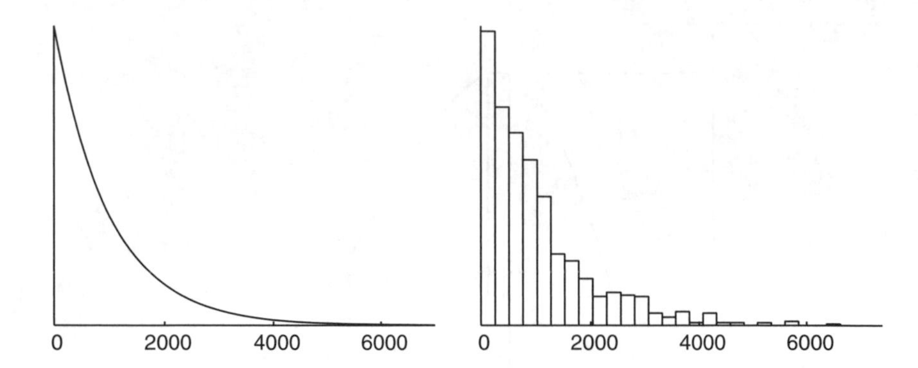


**FIGURE 2.8** Pdf of the lifetime of the lightbulb in Example 2.10 and histogram of 1000 simulated such lifetimes. Since we are interested only in comparing the shape, the scale on the $y$-axis has been omitted.

of stating the pdf is that it has the graphical interpretation mentioned above, and the advantage of the cdf is that it is directly interpretable as a probability.

### 2.3.1 The Uniform Distribution

Consider the experiment to choose a real number at random between 0 and 1. As usual, the interpretation of "at random" is that no number is more likely than any other to be chosen. Since the range is an uncountable set, we now need to specify what this means in terms of probabilities of intervals rather than single points. Call the number $X$, and consider an interval $I_h \subseteq [0, 1]$ of length $h$. The interpretation of "at random" is then that

$$P(X \in I_h) = h$$

regardless of the value of $h$ and where the interval $I_h$ is located. This means that the pdf must be *constant* between 0 and 1 (and 0 otherwise), and since it must also integrate to one, we realize that the pdf of $X$ is

$$f(x) = 1, \quad 0 \leq x \leq 1$$

and we say that $X$ has a *uniform distribution* on [0, 1]. To find the cdf of $X$, take $x \in [0, 1]$ to obtain

$$F(x) = \int_{-\infty}^{x} f(t)dt = x, \quad 0 \leq x \leq 1$$

and $F(x) = 0$ for $x < 0$, and $F(x) = 1$ for $x > 1$. Note that the cdf is a continuous function, so $X$ is by definition a continuous random variable. With a similar argument, we can define the uniform distribution on any interval $[a, b]$ as follows.

***Definition 2.7.*** If the pdf of $X$ is

$$f(x) = \frac{1}{b-a}, \quad a \leq x \leq b$$

then $X$ is said to have a *uniform distribution* on $[a, b]$, written $X \sim \text{unif}[a, b]$.

The corresponding cdf is

$$F(x) = \int_a^x f(t)dt = \frac{x-a}{b-a}, \quad a \leq x \leq b$$

Since this is a continuous distribution, it does not matter if we include or exclude the endpoints of the interval, since these have probability 0. Thus, a uniform distribution on the open interval $(a, b)$ is, from this perspective, the same as a uniform distribution on the closed interval $[a, b]$. Also note how the cdf $F$ is not differentiable at the points $a$ and $b$, which for the same reason does not matter.

Recall that we previously talked about the uniform distribution on a finite set (Section 1.4 and Example 2.6). This was an example of the *discrete uniform distribution* characterized by a constant pmf on a finite set. What we have defined here is the continuous analog, characterized by a constant pdf on a finite interval. It is impossible to have a uniform distribution on an infinite interval (why?). Hence, a statement such as "choose a real number at random" does not have a clear interpretation.

The uniform distribution on $[0, 1]$ is often referred to as the *standard* uniform distribution. There is a simple relation between the general and the standard uniform distribution, which we state next.

**Proposition 2.7.** Let $X \sim \text{unif}[0, 1]$ and define the random variable $Y$ by $Y = a + (b-a)X$. Then $Y \sim \text{unif}[a, b]$.

*Proof.* Note that $Y$ has range $[a, b]$. We need to show that the pdf of $Y$ is of the form in Definition 2.7. To do this, let us first consider the cdf of $Y$, $F_Y$. Take $x \in [a, b]$ to obtain

$$\begin{aligned} F_Y(x) &= P(a + (b-a)X \leq x) \\ &= P\left(X \leq \frac{x-a}{b-a}\right) = F_X\left(\frac{x-a}{b-a}\right) \end{aligned}$$

the cdf of the random variable $X$ evaluated at the point $(x-a)/(b-a)$. Now we can use the fact that $X \sim \text{unif}[0, 1]$ and thus $F_X(t) = t$, to obtain

$$F_Y(x) = \frac{x-a}{b-a}, \quad a \leq x \leq b$$

from which we get the pdf

$$f_Y(x) = F'_Y(x) = \frac{1}{b-a}, \quad a \le x \le b$$

which we recognize as the pdf of a uniform distribution on $[a, b]$. ■

Thus we can transform a standard uniform distribution into any other uniform distribution. Conversely, if $X \sim \text{unif}[a, b]$, then the random variable $(X - a)/(b - a)$ is standard uniform. As we will see later, the standard uniform distribution is a fundamental tool for computer simulation of random variables. In that context, observations generated from the standard uniform distribution are called *random numbers*, and we see here that if we are given such random numbers, we can transform them to observations from a uniform distribution on $[a, b]$ by multiplying each with $b - a$ and adding $a$ (scaling and translating).

Finally, a word on terminology. The term used for the distribution may also be applied directly to the random variable. Hence, we mean the same thing by saying "$X$ is a uniform random variable" and "$X$ has a uniform distribution" and the phrase "generate a couple of standard uniforms" is a quicker way to say "generate a couple of random variables that have the uniform distribution on $[0, 1]$."

### 2.3.2 Functions of Random Variables

In Proposition 2.7, we started from a random variable $X$ and defined a new random variable $Y$ through the relation $Y = a + (b - a)X$. We then managed to find the distribution of $Y$ from knowledge of the distribution of $X$. More generally, suppose that we are given a random variable $X$ with known distribution and a function $g$, and define a new random variable $Y = g(X)$. Can we find the distribution of $Y$ on the basis of our knowledge of the distribution of $X$?

We need to distinguish between the discrete and the continuous cases. The discrete case is easy and can be treated with methods that we already know. Suppose that $X$ is discrete with range $\{x_1, x_2, \ldots\}$ and pmf $p_X$ and let $Y = g(X)$. The range of $Y$ is then $\{y_1, y_2, \ldots\}$, the possible function values. Note that this range may be strictly smaller than the range of $X$, in the case that $g$ maps several $x$ values to the same $y$ value. In the extreme case that $g$ is constant, $Y$ can only assume one value, regardless of what $X$ is. Now consider the value of the pmf of $Y$ in a point $y_k$, $p_Y(y_k) = P(Y = y_k)$. We can express the event $\{Y = y_k\}$ as a subset of the range of $X$. Thus, let

$$B_k = \{x_j : g(x_j) = y_k\}$$

the set of values that are mapped to $y_k$ and apply Proposition 2.2(c) to obtain

$$p_Y(y_k) = P(Y = y_k) = P(X \in B_k) = \sum_{j:x_j \in B_k} p_X(x_j)$$

The continuous case is more complicated since a distribution is defined through its pdf. The trick here is to start with the cdf, and we illustrate this in a few examples.

***Example 2.11.*** Suppose that $X \sim \text{unif}(-\pi/2, \pi/2)$ and let $Y = \sin(X)$. Find the pdf of $Y$.

First note that the range of $Y$ is $(-1, 1)$. Now take a $y$ in this range and note that $\{Y \leq y\} = \{X \leq \sin^{-1}(y)\}$ to obtain

$$\begin{aligned} F_Y(y) = P(Y \leq y) &= P(X \leq \sin^{-1}(y)) \\ = F_X(\sin^{-1}(y)) &= \frac{1}{\pi}\sin^{-1}(y) + \frac{1}{2} \end{aligned}$$

where we have used that the cdf of $X$ is $F_X(x) = (x + \pi/2)/\pi$, $-\pi/2 \leq x \leq \pi/2$. This gives the pdf

$$f_Y(y) = F_Y'(y) = \frac{1}{\pi}\frac{d}{dy}\sin^{-1}(y) = \frac{1}{\pi\sqrt{1-y^2}}, \quad -1 < y < 1$$

We chose to define the uniform distribution on the open interval rather than the closed interval to avoid infinite values of the pdf, but as noted previously, this is merely a cosmetic issue. □

***Example 2.12.*** Road salt is spread on a road to melt snow and ice. Suppose that the resulting water has a temperature of $X$°C, which has pdf

$$f_X(x) = \frac{1}{25}(10 - 2x), \quad 0 \leq x \leq 5$$

What is the pdf of the temperature if we convert it to degrees Fahrenheit?

The relation between Fahrenheit $Y$ and Celsius $X$ is

$$Y = 1.8X + 32$$

so the range of $Y$ is $[32, 41]$. Start with the cdf to obtain

$$\begin{aligned} F_Y(y) &= P(1.8X + 32 \leq y) \\ &= P\left(X \leq \frac{y-32}{1.8}\right) = F_X\left(\frac{y-32}{1.8}\right) \end{aligned}$$

which we differentiate to get the pdf

$$\begin{aligned} f_Y(y) = F_Y'(y) &= \frac{d}{dy}F_X\left(\frac{y-32}{1.8}\right) \\ &= f_X\left(\frac{y-32}{1.8}\right) \times \frac{1}{1.8} = \frac{1}{45}\left(10 - 2 \times \frac{y-32}{1.8}\right) \\ &= \frac{1}{81}(82 - 2y), \quad 32 \leq y \leq 41. \end{aligned}$$

□

These examples illustrate how to arrive at the desired pdf. Start with the cdf of one random variable, express it in terms of the cdf of the other, and finally differentiate to get the pdf. The last example deals with a linear transformation of the form $Y = aX + b$, where $a > 0$. By going through the same steps as in the example, we realize that the pdf of $Y$ is

$$f_Y(y) = \frac{1}{a} f_X\left(\frac{y-b}{a}\right)$$

for $y$ in the range of $Y$. Generally, if $Y = g(X)$ where $g$ is a strictly increasing and differentiable function, the event $\{g(X) \leq y\}$ is the same as $\{X \leq g^{-1}(y)\}$, and we have the relation

$$F_Y(y) = F_X(g^{-1}(y))$$

which gives the pdf

$$f_Y(y) = f_X(g^{-1}(y)) \frac{d}{dy} g^{-1}(y)$$

If $g$ is instead strictly decreasing and differentiable, the event $\{g(X) \leq y\}$ is the same as $\{X \geq g^{-1}(y)\}$ and we get

$$F_Y(y) = 1 - F_X(g^{-1}(y))$$

and

$$f_Y(y) = f_X(g^{-1}(y)) \left(-\frac{d}{dy} g^{-1}(y)\right)$$

We summarize in the following proposition.

**Proposition 2.8.** Let $X$ be a continuous random variable with pdf $X$, let $g$ be a strictly increasing or strictly decreasing, differentiable function, and let $Y = g(X)$. Then $Y$ has pdf

$$f_Y(y) = \left|\frac{d}{dy} g^{-1}(y)\right| f_X(g^{-1}(y))$$

for $y$ in the range of $Y$.

Functions that are not strictly increasing or decreasing can usually be broken up into pieces that are. In practice, it is often just as easy to start with the cdf of $Y$ and work through the steps in the examples, as it is to apply the proposition.

***Example 2.13.*** Let $X \sim \text{unif}[-1, 1]$, and let $A$ be the area of a square that has one corner at the origin and the next at the point $X$ on the $x$-axis. Find the pdf of $A$.

The range of $A$ is $[0, 1]$ and since $A = X^2$, we get the cdf

$$\begin{aligned} F_A(a) &= P(X^2 \le a) \;=\; P(-\sqrt{a} \le X \le \sqrt{a}) \\ &= F_X(\sqrt{a}) - F_X(-\sqrt{a}) \\ &= \frac{\sqrt{a}+1}{2} - \frac{-\sqrt{a}+1}{2} \;=\; \sqrt{a}, \quad 0 \le a \le 1 \end{aligned}$$

which gives the pdf

$$f_A(a) = F_A'(a) = \frac{1}{2\sqrt{a}}, \quad 0 \le a \le 1$$

It would be more cumbersome to try to apply Proposition 2.8 by considering the cases $X > 0$ and $X \le 0$ separately. □

We finish with an example that shows that applying a function $g$ can turn a continuous random variable into a discrete random variable.

***Example 2.14.*** Let $X \sim \text{unif}[0, 1]$ and let $Y = [6X] + 1$, where $[\,\cdot\,]$ denotes integer part. What is the distribution of $Y$?

First note that the range of $Y$ is $\{1, 2, \ldots, 6\}$ so $Y$ is a discrete random variable. Second, for $k$ in the range of $Y$, we obtain

$$\{Y = k\} = \{k - 1 \le [6X] < k\} = \left\{\frac{k-1}{6} \le X < \frac{k}{6}\right\}$$

which gives

$$P(Y = k) = P\left(\frac{k-1}{6} \le X < \frac{k}{6}\right) = \frac{1}{6}, \quad k = 1, 2, \ldots, 6$$

and $Y$ has the discrete uniform distribution on $\{1, 2, \ldots, 6\}$. Thus, if we generate random numbers and apply the function $g(x) = [6x] + 1$ to each, we can simulate rolls of a fair die. We will return to this in Chapter 5. □

## 2.4 EXPECTED VALUE AND VARIANCE

In both daily language and scientific reporting, quantitative properties are often described in terms of averages. For example, a tourist guide for a travel destination may state average monthly temperatures. The stock market indices are weighted averages of several individual stock prices. We talk about averages when we discuss salaries, home prices, amounts of rainfall, and so on. Common for all these examples is that instead of presenting the entire data set, a single number is given to summarize the data. This is convenient, and if I plan to visit Honolulu in December, the information that the average daily high temperature is 81°F is enough for me to plan my trip.

It probably would not give me any extra useful information to see the entire data set of all the temperature measurements from which this number is computed.

With this in mind, it would be convenient to have a similar way to describe random variables. That is, instead of giving the entire distribution, we could give a single number that summarizes it and gives useful information. How should we choose this number? Let us consider the experiment of rolling a die. If we roll it repeatedly and denote the consecutive outcomes $X_1, X_2, \ldots$, we can consider the average $(X_1 + \cdots + X_n)/n$ of the $n$ first rolls. If we further denote the number of $k$'s by $N_k$ for $k = 1, 2, \ldots, 6$, we get

$$\frac{X_1 + \cdots + X_n}{n} = \sum_{k=1}^{6} k \frac{N_k}{n}$$

and if $n$ is large, we expect the relative frequencies $N_k/n$ to be approximately equal to the corresponding probabilities, which in this case are all $\frac{1}{6}$. But this means that we expect

$$\frac{X_1 + \cdots + X_n}{n} \approx \frac{1}{6}(1 + 2 + \cdots + 6) = 3.5$$

and we expect the approximation to be better the larger the value of $n$. Hence, 3.5 is a long-term average of the rolls of a die, and we can give this number as a summary of the experiment. Inspired by this we state the following definition.

***Definition 2.8.*** Let $X$ be a discrete random variable with range $\{x_1, x_2, \ldots\}$ (finite or countably infinite) and probability mass function $p$. The *expected value* of $X$ is defined as

$$E[X] = \sum_{k=1}^{\infty} x_k p(x_k)$$

The expected value is thus a weighted average of the possible values of $X$, with the corresponding probabilities as weights. It can be thought of as a theoretical spatial average over the range of $X$, whereas we can describe the consecutive averages mentioned above as time averages. Note that the expected value is a number computed from the distribution, whereas time averages are computed from experiments and are thus random. We will return to this discussion and make it strict in Chapter 4.

The terms *expectation* and *mean* are used as synonyms of expected value, and the letter $\mu$ is often used to denote it. Let us look at some examples.

***Example 2.15.*** An American roulette table has the numbers 1–36, plus 0 and 00. Suppose that you bet \$1 on "odd." What is your expected gain?

If the winning number is odd, you win \$1; if it is not, you lose \$1, which we can describe as a negative gain. Hence, if $X$ is your gain, we have

$$X = \begin{cases} 1 & \text{with probability } 18/38 \\ -1 & \text{with probability } 20/38 \end{cases}$$

and get

$$E[X] = 1 \times \frac{18}{38} + (-1) \times \frac{20}{38} = -\frac{1}{19}$$

so on average you lose about 5 cents for each dollar you bet. □

***Example 2.16.*** In the previous example, consider instead a *straight bet*, which means that you bet on a single number. Which payout should the casino choose for you to have the same expected loss as when you bet on odd?

Again denote your gain by $X$. We have

$$X = \begin{cases} a & \text{with probability } 1/38 \\ -1 & \text{with probability } 37/38 \end{cases}$$

which has expected value

$$E[X] = a \times \frac{1}{38} + (-1) \times \frac{37}{38} = \frac{a - 37}{38}$$

which we set equal to $-\frac{1}{19}$ to obtain $a = 35$. □

As you have noticed by now, the expected value is not necessarily a value that $X$ can assume, so in this sense, the term "expected" is slightly misleading.

If the random variable is nonnegative, there is an alternative way to compute the expected value according to the following proposition.

**Proposition 2.9.** Let $X$ be a discrete random variable with range $\{0, 1, \ldots\}$. Then

$$E[X] = \sum_{n=0}^{\infty} P(X > n)$$

*Proof.* Note that $k = \sum_{n=1}^{k} 1$, and use the definition of expected value to obtain

$$\begin{aligned}
E[X] &= \sum_{k=1}^{\infty} kP(X = k) = \sum_{k=1}^{\infty}\sum_{n=1}^{k} P(X = k) \\
&= \sum_{n=1}^{\infty}\sum_{k=n}^{\infty} P(X = k) = \sum_{n=1}^{\infty} P(X \geq n) \\
&= \sum_{n=0}^{\infty} P(X > n)
\end{aligned}$$

■

***Example 2.17.*** What is the expected number of rolls of a fair die until we get the first 6?

If we let the number be $X$, Example 2.9 gives

$$P(X > n) = 1 - F(n) = \left(\frac{5}{6}\right)^n, \quad n = 0, 1, \ldots$$

and the expected value is

$$E[X] = \sum_{n=0}^{\infty} P(X > n) = \sum_{n=0}^{\infty} \left(\frac{5}{6}\right)^n = \frac{1}{1 - 5/6} = 6$$

so on average it takes 6 rolls to get the first 6. We leave it to the reader to ponder whether this is intuitively clear. □

We proceed to the definition of expected value in the continuous case. Remember how sums are replaced by integrals, and the following definition should not come as a surprise.

***Definition 2.9.*** Let $X$ be a continuous random variable with pdf $f$. The *expected value* of $X$ is defined as

$$E[X] = \int_{-\infty}^{\infty} xf(x)dx$$

The formal integral limits are $-\infty$ and $\infty$, but in reality the limits are determined by the range of $X$, which is where $f(x)$ is positive.

***Example 2.18.*** Let $X \sim \text{unif}[a, b]$. Find $E[X]$.

By Definition 2.9,

$$E[X] = \int_{-\infty}^{\infty} xf(x)dx = \frac{1}{b-a}\int_a^b x\,dx = \frac{a+b}{2}$$

which is the midpoint of the interval $[a, b]$. □

For nonnegative continuous random variables, there is an analog of Proposition 2.9. We leave the proof as an exercise.

**Proposition 2.10.** Let $X$ be a continuous random variable with range $[0, \infty)$. Then

$$E[X] = \int_0^{\infty} P(X > x)dx$$

***Example 2.19.*** Find the expected lifetime of the lightbulb in Example 2.10.

The pdf is

$$f(x) = 0.001e^{-0.001x}, \quad x \geq 0$$

so we can find the expected value according to the definition as

$$E[X] = \int_0^{\infty} xf(x)dx$$

Since the range of $X$ is $[0, \infty)$, we can also use Proposition 2.10. We know from the example that

$$P(X > x) = e^{-0.001x}, \quad x \geq 0$$

and get

$$E[X] = \int_0^{\infty} e^{-0.001x}dx = \left[-\frac{e^{-0.001x}}{0.001}\right]_0^{\infty} = 1000$$

which, if done by hand, is a little bit simpler than using the definition directly since it avoids the partial integration. □

One important property of the expected value is that it is linear, in the following sense.

**Proposition 2.11.** Let $X$ be any random variable, and let $a$ and $b$ be real numbers. Then

$$E[aX+b] = aE[X]+b$$

*Proof.* We prove this in the continuous case, for $a > 0$. The discrete is analogous, replacing integrals by sums. Let $Y = aX + b$, and note that $Y$ is a continuous random variable, which by Proposition 2.8 has pdf

$$f_Y(y) = \frac{1}{a} f_X\left(\frac{y-b}{a}\right)$$

and by definition, the expected value of $Y$ is

$$E[Y] = \int_{-\infty}^{\infty} y f_Y(y) dy = \frac{1}{a}\int_{-\infty}^{\infty} y f_X\left(\frac{y-b}{a}\right) dy$$

where the variable substitution $y = ax + b$ gives $dy = a\,dx$ and hence

$$\begin{aligned} E[Y] &= \int_{-\infty}^{\infty} (ax+b) f_X(x) dx \\ &= a\int_{-\infty}^{\infty} x f_X(x) dx + b\int_{-\infty}^{\infty} f_X(x) dx \; = \; aE[X]+b \end{aligned}$$

and we are done. ■

In the proof we discovered the identity

$$E[aX+b] = \int_{-\infty}^{\infty} (ax+b) f_X(x) dx$$

which, as we shall see next, is a special case of a more general result.

### 2.4.1 The Expected Value of a Function of a Random Variable

In the previous section, we considered the expected value of a linear transformation $aX + b$. It is natural to ask what can be said about the expected value of an arbitrary function $g$ applied to $X$, $E[g(X)]$. In Section 2.3.2, we learned how to find the pdf of $g(X)$, at least for strictly monotone functions $g$, and this can be used to find the expected value of $g(X)$ according to the definition, but there is a quicker way. Let us start with an example.

***Example 2.20.*** Let $X \sim \text{unif}[0, 2]$. What is the expected area of a square with side-length $X$?

The random variable of interest is $A = X^2$. To find $E[A]$, we compute it according to the definition of expected value. Let us first find the pdf of $A$, $f_A$. Since we are dealing with a continuous random variable, we need to start with the cdf, $F_A$. Note that the range of $A$ is $[0, 4]$, and take an $a$ in this range to obtain

$$F_A(a) = P(A \leq a) = P(X \leq \sqrt{a}) = \frac{\sqrt{a}}{2}, \quad 0 \leq a \leq 4$$

since $X \sim \text{unif}[0, 2]$. We get the pdf

$$f_A(a) = F'_A(a) = \frac{1}{4\sqrt{a}}$$

and expected value

$$E[A] = \int_0^4 a f_A(a) da = \frac{1}{4}\int_0^4 \sqrt{a}\, da = \frac{1}{4}\left[\frac{2}{3}a\sqrt{a}\right]_0^4 = \frac{4}{3}$$

□

This example is illuminating in several ways. First, note that the expected area is *not* equal to the square of the expected sidelength. At first it might seem paradoxical that the "typical square has side 1 and area $\frac{4}{3}$," but the situation clears up if we think of expected values as averages of large numbers of observations. For the sidelength, small values (near 0) and large values (near 2) average each other out around 1. However, when the values are squared, values near 0 stay near 0 but those near 2 end up closer to 4. Thus, larger values have more impact when squared, and push the expected area up.

Second, note that the distribution of $A$ is not uniform on its range $[0, 4]$. To understand why, consider, for example, the two intervals $[0, 0.25]$ and $[3.75, 4]$, both of length 0.25. The probability that $A$ is in the first is the probability that $X$ is in $[\sqrt{0}, \sqrt{0.25}] = [0, 0.5]$, which is 0.25, and the probability that $A$ is in the second is the probability that $X$ is in $[\sqrt{3.75}, \sqrt{4}] = [1.94, 2]$, which is only 0.03.

Finally, if we make the variable substitution $x = \sqrt{a}$ in the expression for $E[A]$ above, we get $da = 2x\,dx$, the new integral limits $[0, 2]$, and

$$E[A] = \int_0^4 a f_A(a) da = \frac{1}{4}\int_0^4 \sqrt{a}\, da = \frac{1}{2}\int_0^2 x^2 dx$$

Since the pdf of $X$ is $f_X(x) = \frac{1}{2}$, $0 \leq x \leq 2$, we have shown that

$$E[X^2] = \int_0^2 x^2 f_X(x) dx$$

that is, in this case, we could have computed the expected value of $A$ without first finding its pdf. Compare the two expressions

$$E[A] = \int_0^4 a f_A(a) da$$
$$E[X^2] = \int_0^2 x^2 f_X(x) dx$$

where the first is according to the definition, and uses the range and pdf of $A$. In contrast, the second expression uses the range and pdf of the original random variable $X$ and plugs in the function $x^2$ in the integral. It turns out that this is no coincidence, and we have the following result.

**Proposition 2.12.** Let $X$ be a random variable with pmf $p_X$ or pdf $f_X$, and let $g : R \to R$ be any function. Then

$$E[g(X)] = \begin{cases} \displaystyle\sum_{k=1}^{\infty} g(x_k) p_X(x_k) & \text{if } X \text{ is discrete with range } \{x_1, x_2, \ldots\} \\ \displaystyle\int_{-\infty}^{\infty} g(x) f_X(x) dx & \text{if } X \text{ is continuous} \end{cases}$$

*Proof.* The discrete case is straightforward. We will do the proof only for the continuous case in the special case when $g$ is strictly increasing. By Proposition 2.8, the pdf of $Y = g(X)$ is

$$f_Y(y) = \left(\frac{d}{dy} g^{-1}(y)\right) f_X(g^{-1}(y))$$

which gives

$$E[Y] = \int_{-\infty}^{\infty} y f_Y(y) dy = \int_{-\infty}^{\infty} y \left(\frac{d}{dy} g^{-1}(y)\right) f_X(g^{-1}(y)) dy$$

where we make the change of variables $x = g^{-1}(y)$, which gives $y = g(x)$, $dx = \frac{d}{dy} g^{-1}(y) dy$, and

$$E[Y] = \int_{-\infty}^{\infty} g(x) f_X(x) dx$$

as desired. ■

The general case of continuous $X$ is more complicated. For instance, depending on the function $g$, $g(X)$ may be discrete even if $X$ is continuous (recall Example 2.14). Proposition 2.18 is true for any function $g$, but the general proof is more involved and we skip it here. We are fortunate to have this proposition, since it is generally much easier to compute $E[g(X)]$ according to this method than it is to go through the procedure of first finding the distribution of $g(X)$.

In Example 2.20, we saw that the expected area did not equal the square of the expected sidelength. In other words, with $\mu = E[X]$, we had $E[X^2] \neq \mu^2$. Generally, if $g$ is a function and $\mu = E[X]$, it is most often the case that

$$E[g(X)] \neq g(\mu)$$

with one important exception: the linear case $g(x) = ax + b$, stated in Proposition 2.11.

***Example 2.21.*** Recall Example 2.14, where $X \sim \text{unif}[0, 1]$ and $Y = [6X] + 1$. Find $E[Y]$.

The function applied to $X$ is $g(x) = [6x] + 1$ so it has expected value

$$E[g(X)] = \int_0^1 ([6x] + 1) f_X(x) dx = \int_0^1 [6x] dx + 1$$

and since $[6x] = k$ if $\frac{k}{6} \leq x < \frac{k+1}{6}$ for $k = 0, 1, \ldots, 5$, we get

$$\int_0^1 [6x] dx = \frac{0 + 1 + \cdots + 5}{6} = 2.5$$

which gives $E[X] = 3.5$, the mean of a fair die. □

***Example 2.22.*** A chemical reaction in a solution in a test tube produces a certain chemical compound. The amount $X$ in grams thus created has pdf

$$f(x) = 2x, \quad 0 \leq x \leq 1$$

Amounts below 0.5 g are considered too low, and in that case the solution is discarded. After the reaction, what is the expected amount of the compound that is kept?

We are looking for $E[g(X)]$, where $g$ is the function

$$g(x) = \begin{cases} 0 & \text{if } x \leq 0.5 \\ x & \text{if } x > 0.5 \end{cases}$$

and Proposition 2.12 gives

$$E[g(X)] = \int_0^1 g(x) f(x) dx = 2 \int_{0.5}^1 x^2 dx \approx 0.58$$

□

Expected values do not have to be finite, as the following famous example shows.

***Example 2.23* (*St Petersburg Paradox*).** Consider the following roulette strategy. You bet \$1 on odd, and if you win, you quit. If you lose, you bet \$2 on odd, and in each subsequent round either quit if you win or double your bet if you lose. **(a)** What is the expected number of rounds until you win, and what is then your net gain? **(b)** What is your expected loss before the first win?

In each round, you win with probability $\frac{18}{38}$, and in an analogy with Example 2.17, the expected number of rounds until your first win is $\frac{38}{18} \approx 2.1$. If you have had $n-1$ consecutive losses prior to the first win in the $n$th round, you have lost $1 + 2 + \cdots + 2^{n-2} = 2^{n-1} - 1$, and since you bet $2^{n-1}$ in the $n$th round, your net gain is one dollar. Hence, each time you quit, you have gained a dollar, and each dollar takes on average 2.1 rounds to gain. This sounds almost too good to be true, does it not?

The problem with the strategy is revealed by solving (b). Let the time of your first win be $T$. In an analogy with Example 2.7, the pmf of $T$ is

$$P(T = n) = \frac{18}{38}\left(\frac{20}{38}\right)^{n-1}, \quad n = 1, 2, \ldots$$

and your accumulated loss before the first win is the random variable $L = 2^{T-1} - 1$, which has expected value

$$E[L] = \sum_{n=1}^{\infty}(2^{n-1} - 1)P(T = n) = \frac{18}{38}\sum_{n=1}^{\infty}\left(\left(\frac{40}{38}\right)^{n-1} - \left(\frac{20}{38}\right)^{n-1}\right) = \infty$$

so your expected loss before the first win is infinite! In practice, this means that whatever fortune you start with, eventually you will go broke if you play this strategy. It was indeed too good to be true, and Donald Trump is still wealthier than all the world's probabilists.

This example is one version of the *St Petersburg paradox*, which dates back to the eighteenth century, and occurs in many variants in the probability literature. □

There is nothing strange about the random variable $L$ in the last example. Each outcome of $T$ gives an outcome of $L$, and the range of $L$ is the set $\{0, 1, 3, 7, \ldots\}$. Thus, $L$ itself is always finite but its expected value is infinite. If we were to get consecutive observations on $L$, the consecutive averages would tend to grow beyond all bounds.

### 2.4.2 Variance of a Random Variable

We introduced the expected value as a number that summarizes the distribution of a random variable $X$. Since it gives an idea of where $X$ is on the average, the expected value is often referred to as a *location parameter*. The expected value does, however, not give any information of the variability of $X$, as the following example illustrates.

***Example 2.24.*** Suppose that we are about to weigh a piece of metal, and have at our disposal two different scales, one that gives the correct weight with an error of $\pm 0.01$ g, and the second with an error of $\pm 0.1$ g. If the true weight of the piece is $w$ g, we thus assume that the scales give the weights $X \sim \text{unif}[w - 0.01, w + 0.01]$ and $Y \sim \text{unif}[w - 0.1, w + 0.1]$, respectively. The expected values are

$$E[X] = w \quad \text{and} \quad E[Y] = w$$

that is, both scales give the correct weight on average. However, it is clear that the first scale is preferred because of its better precision. This is not reflected in the expected values. □

If we, together with the expected value, also had a measure of variability, the two numbers together would give more information about the distribution than would the expected value alone. Since the expected value $\mu$ measures location, we want to somehow measure the random variable's variability relative to the location, that is, measure the average behavior of $X - \mu$. This is a random quantity, and we would like to describe it by a single number. We could try to take the expected value of $X - \mu$, but this would always give 0 since $E[X - \mu] = E[X] - \mu = 0$. Intuitively, values to the left of $\mu$ cancel values to the right. Another attempt would be to take the expected value of $|X - \mu|$, which would give us a nonnegative number that measures the extent to which $X$ deviates from $\mu$ on average. As it turns out, the following definition gives a more convenient measure than using the absolute value.

***Definition 2.10.*** Let $X$ be a random variable with expected value $\mu$. The *variance* of $X$ is defined as

$$\text{Var}[X] = E\left[(X - \mu)^2\right]$$

The variance is often denoted $\sigma^2$. Notice that $(X - \mu)^2 \geq 0$, so the variance is a nonnegative measure of variability where large values indicate that $X$ tends to fluctuate a lot around $\mu$. Just like the mean, the variance can be infinite. Since we have squared the values, the variance is not on the same scale as $X$ and $\mu$. For example, if $X$ and $\mu$ are weights in grams, the unit of measure of Var$[X]$ is square grams that does not have a clear meaning. For this reason, we often use the following definition.

***Definition 2.11.*** Let $X$ be a random variable with variance $\sigma^2 = \text{Var}[X]$. The *standard deviation* of $X$ is then defined as $\sigma = \sqrt{\text{Var}[X]}$.

To compute the variance, we can use Proposition 2.12 with $g(x) = (x - \mu)^2$, but it is often easier to use the following formula.

**Corollary 2.2.**

$$Var[X] = E[X^2] - (E[X])^2$$

*Proof.* We prove the continuous case. The discrete is the obvious analog. By Proposition 2.12, we have

$$\begin{aligned}
\text{Var}[X] = E[(X-\mu)^2] &= \int_{-\infty}^{\infty} (x-\mu)^2 f(x)dx \\
&= \int_{-\infty}^{\infty} (x^2 - 2x\mu + \mu^2) f(x)dx \\
&= \int_{-\infty}^{\infty} x^2 f(x)dx - 2\mu \int_{-\infty}^{\infty} xf(x)dx + \mu^2 \int_{-\infty}^{\infty} f(x)dx \\
&= E[X^2] - 2\mu E[X] + \mu^2 = E[X^2] - (E[X])^2
\end{aligned}$$

■

***Example 2.25.*** Let $X$ be the number when a die is rolled. Find Var$[X]$.

The pmf is $p(k) = \frac{1}{6}$, $k = 1, 2, \ldots, 6$, which gives expected value

$$E[X] = \sum_{k=1}^{6} kp(k) = \frac{1}{6}\sum_{k=1}^{6} k = \frac{7}{2}$$

and by Proposition 2.12,

$$E[X^2] = \sum_{k=1}^{6} k^2 p(k) = \frac{1}{6}\sum_{k=1}^{6} k^2 = \frac{91}{6}$$

which gives

$$\text{Var}[X] = E[X^2] - (E[X])^2 = \frac{35}{12}$$

□

***Example 2.26.*** In Example 2.15, we calculated the expected gain in roulette. Compute the variance of your gain if you **(a)** bet on odd and **(b)** make straight bets.

For (a), your gain $X$ has distribution

$$X = \begin{cases} 1 & \text{with probability } 18/38 \\ -1 & \text{with probability } 20/38 \end{cases}$$

which has mean $E[X] = -\frac{1}{19}$. For the variance we need $E[X^2]$, which equals 1 since $X^2 \equiv 1$. Hence

$$\text{Var}[X] = E[X^2] - (E[X])^2 = 1 - \left(-\frac{1}{19}\right)^2 = \frac{360}{361} \approx 0.997$$

For (b) denote the gain by $Y$, which has distribution

$$Y = \begin{cases} 35 & \text{with probability } 1/38 \\ -1 & \text{with probability } 37/38 \end{cases}$$

which has mean $E[Y] = -\frac{1}{19}$. Furthermore,

$$E[Y^2] = 35^2 \times \frac{1}{38} + (-1)^2 \times \frac{37}{38} = \frac{631}{19}$$

which gives

$$\text{Var}[Y] = E[Y^2] - (E[Y])^2 = \frac{631}{19} - \left(-\frac{1}{19}\right)^2 = \frac{11988}{361} \approx 33.2$$

Thus, $X$ and $Y$ have the same mean, so on average you lose just as much with either strategy. The variance of $Y$ is much larger than the variance of $X$, which reflects the fact that you are more likely to lose a round with a straight bet, but when you win, you win more. □

***Example 2.27.*** Let $X \sim \text{unif}[a, b]$. Find the variance of $X$.

We already know that $E[X] = (a+b)/2$. For the variance, we need $E[X^2]$ and get

$$E[X^2] = \int_a^b x^2 f(x)dx = \frac{1}{b-a}\left[\frac{x^3}{3}\right]_a^b = \frac{a^2 + ab + b^2}{3}$$

which gives

$$\text{Var}[X] = \frac{a^2 + ab + b^2}{3} - \left(\frac{a+b}{2}\right)^2 = \frac{(b-a)^2}{12}$$

after some elementary algebra. □

Notice how these expressions for mean and variance are similar to those of the discrete uniform distribution in Example 2.25. Let us state them separately.

**Proposition 2.13.** If $X \sim \text{unif}[a, b]$, then

$$E[X] = \frac{a+b}{2} \quad \text{and} \quad \text{Var}[X] = \frac{(b-a)^2}{12}$$

In mechanical engineering, these can be interpreted as the center of gravity and moment of inertia, respectively, of a solid bar with endpoints in $a$ and $b$. The variance formula in Corollary 2.2 can then be interpreted as Steiner's theorem about moment of inertia.

***Example 2.28.*** Recall Example 2.24 about the two scales, where one gives a measurement $X \sim \text{unif}[w - 0.01, w + 0.01]$ and the other, $Y \sim \text{unif}[w - 0.1, w + 0.1]$. They both have expected value $w$, and by the formula given above, the variances are

$$\text{Var}[X] = 3.3 \times 10^{-5} \quad \text{and} \quad \text{Var}[Y] = 3.3 \times 10^{-3}$$

and the first scale's higher precision is reflected in its lower variance. The standard deviations are 0.006 and 0.06, respectively, so we could say that the first scale has 10 times better precision than the second. □

Unlike the expected value, there is no immediate intuition behind the value of the variance. In the previous example, we can use the variances to compare the two scales, but the actual numbers are not immediately interpretable. Even if we calculate the standard deviations, it is not clear what these numbers mean. Some help is given by the following result.

**Proposition 2.14 (Chebyshev's Inequality).** Let $X$ be any random variable with mean $\mu$ and variance $\sigma^2$. For any constant $c > 0$, we have

$$P(|X - \mu| \geq c\,\sigma) \leq \frac{1}{c^2}$$

*Proof.* Let us prove the continuous case. Fix $c$ and let $B$ be the set $\{x \in R : |x - \mu| \geq c\,\sigma\}$. We get

$$\sigma^2 = E[(X - \mu)^2] \;=\; \int_{-\infty}^{\infty} (x - \mu)^2 f(x)dx$$
$$\geq \int_B (x - \mu)^2 f(x)dx \;\geq\; c^2\sigma^2 \int_B f(x)dx = c^2\sigma^2 P(X \in B)$$

which gives the desired inequality. ■

Chebyshev's inequality tells us that the probability is at least $1 - 1/c^2$ that a random variable is within $c$ standard deviations of its mean, regardless of what the distribution of $X$ is. For example, $X$ is within two standard deviations of its mean with probability at least 0.75, and within three standard deviations with probability at least 0.89.

***Example 2.29.*** The IQ of a randomly sampled individual can be viewed as a random variable $X$. It is known that this has mean 100 and standard deviation 15. The highest recorded IQ is 228, belonging to Marilyn vos Savant, mentioned in Example 1.48. What is the probability that a randomly sampled individual has an IQ that is at least as high as Marilyn's?

We apply Chebyshev's inequality with $\mu = 100$, $\sigma = 15$. Since

$$\{|X - \mu| \geq c\sigma\} = \{X \leq \mu - c\sigma\} \cup \{X \geq \mu + c\sigma\}$$

we get

$$P(X \geq \mu + c\sigma) \leq P(|X - \mu| \geq c\sigma) \leq \frac{1}{c^2}$$

Setting $\mu + c\sigma = 228$ gives $c \approx 8.5$, which in turn gives

$$P(X \geq 228) \leq 0.014$$

In words, at most 1.4% of the population has an IQ that is at least as high as Marilyn's. Note that this is only an upper bound, and the true number is likely to be smaller. See also Problem 47. □

Since Chebyshev's inequality holds for all random variables, it is fairly coarse (as you can see by plugging in $c = 1$) and we can get much better bounds if we know the distribution of $X$. This is, for example, the case in the IQ example above, and we will return to it later (with better news for Marilyn). Chebyshev's inequality is of mostly theoretical use.

In the previous section we saw that the expected value is linear. This is not true for the variance, but the following proposition holds.

**Proposition 2.15.** Let $X$ be any random variable, and let $a$ and $b$ be real numbers. Then

$$\text{Var}[aX + b] = a^2\text{Var}[X]$$

*Proof.* By the definition of variance applied to the random variable $aX + b$, we obtain

$$\begin{aligned}\text{Var}[aX + b] &= E\left[(aX + b - E[aX + b])^2\right] \\ &= E\left[(aX + b - aE[X] - b)^2\right] = E\left[a^2(X - E[X])^2\right] \\ &= a^2E\left[(X - E[X])^2\right] = a^2\text{Var}[X]\end{aligned}$$

where we have used Proposition 2.11 repeatedly. ■

***Example 2.30.*** Consider Example 2.12, where the temperature $X$ in degrees Celsius has pdf

$$f_X(x) = \frac{10 - 2x}{25}, \quad 0 \le x \le 5$$

Find the mean and variance of $X$, and the mean and variance of $Y$, which is the temperature in degrees Fahrenheit.

The mean of $X$ is

$$E[X] = \frac{1}{25} \int_0^5 x(10 - 2x)dx = \frac{5}{3} \approx 1.7$$

and since

$$E[X^2] = \frac{1}{25} \int_0^5 x^2(10 - 2x)dx = \frac{25}{6}$$

we get

$$\mathrm{Var}[X] = \frac{25}{6} - \frac{25}{9} = \frac{25}{18} \approx 1.4$$

Since $Y = 1.8X + 32$, we get

$$E[Y] = 1.8E[X] + 32 = 35$$

and

$$\mathrm{Var}[Y] = 1.8^2 \mathrm{Var}[X] = 4.5$$

□

Note how adding the constant $b$ does not affect the variance. This is intuitively clear since adding $b$ simply shifts the entire distribution, including the mean, $b$ units but does not affect how $X$ varies around the mean. The following proposition is intuitively obvious.

**Proposition 2.16.** Let $\mu = E[X]$. Then $\mathrm{Var}[X] = 0$ if and only if $X \equiv \mu$.

In words, if a random variable has variance 0, it must be constant.[2] The "if" part follows immediately from the definition of variance; the "only if" part turns out to be trickier, and we leave it for Problem 48.

[2]Strictly speaking, it must be constant *with probability one*, but this is a subtle distinction that we need not worry about.

Above we learned how to compute $E[g(X)]$ for an arbitrary function $g$. To find the variance of $g(X)$, simply use the variance formula applied to $g(X)$, that is

$$\mathrm{Var}[g(X)] = E[g(X)^2] - (E[g(X)])^2$$

where $E[g(X)]$ and $E[g(X)^2]$ are computed by applying Proposition 2.12 to the functions $g(x)$ and $g(x)^2$, respectively.

***Example 2.31.*** Let $X \sim \mathrm{unif}[0, 2]$, and let $A = X^2$, the area of a square with sidelength $X$. Find $\mathrm{Var}[A]$.

We already know that $E[A] = \frac{4}{3}$ and need to find $E[A^2]$. Proposition 2.12 gives

$$E[A^2] = E[X^4] = \frac{1}{2}\int_0^2 x^4 dx = \frac{16}{5}$$

which gives

$$\mathrm{Var}[A] = E[A^2] - (E[A])^2 = \frac{16}{5} - \frac{16}{9} = \frac{64}{45}$$

□

## 2.5 SPECIAL DISCRETE DISTRIBUTIONS

In this section, we examine some special discrete distributions. First, the term "discrete" may be used to describe the random variable itself or its distribution. The same goes for the name of the distribution. For example, if we roll a die, we may say that "$X$ has a uniform distribution on $1, \ldots, 6$" or "$X$ is uniform on $1, \ldots, 6$."

### 2.5.1 Indicators

This is the simplest type of discrete random variable. Consider an experiment where the event $A$ may occur.

***Definition 2.12.*** Let $A$ be an event. The random variable $I_A$ defined by

$$I_A = \begin{cases} 1 & \text{if } A \text{ occurs} \\ 0 & \text{otherwise} \end{cases}$$

is called the *indicator* of the event $A$.

This type of random variable is also called a *Bernoulli* random variable. If the event $A$ has probability $p$, the pmf of $I_A$ is

$$p(k) = \begin{cases} p & \text{for } k = 1 \\ 1 - p & \text{for } k = 0 \end{cases}$$

and the mean and variance are

$$E[I_A] = 1 \times p + 0 \times (1 - p) = p$$
$$\text{Var}[I_A] = E[I_A^2] - (E[I_A])^2 = p(1 - p)$$

since $I_A^2 = I_A$. Indicators are much more useful than they may seem at first glance. In more advanced treatments of probability theory they function as "building blocks" that are used to define random variables and expected values in greater generality than we have done. We will later see how indicators are also very useful in problem solving.

### 2.5.2 The Binomial Distribution

Consider an experiment where we are interested in some particular event $A$, where the probability of $A$ is $p$. Suppose that we repeat the experiment independently $n$ times and count how many times we get $A$. Denote this number by $X$, which is then a discrete random variable with range $0, 1, \ldots, n$. What is the pmf of $X$?

Let us call an occurrence of $A$ a "success" ($S$) and a nonoccurrence a "failure" ($F$). The event $\{X = k\}$ then means that we have a sequence of $k$ successes and $n - k$ failures, which can, for example, be

$$S\ F\ S\ S\ F\ \cdots\ F\ S\ F$$

By independence, this particular sequence has probability

$$p \times (1 - p) \times p \times p \times (1 - p) \times \cdots \times (1 - p) \times p \times (1 - p) = p^k (1 - p)^{n-k}$$

But this is only one possible configuration of $k$ successes and $n - k$ failures, and since the position for the $k$ $S$s can be chosen in $\binom{n}{k}$ ways, we get

$$P(X = k) = \binom{n}{k} p^k (1 - p)^{n-k}, \quad k = 0, 1, \ldots, n$$

and we give this probability distribution a name.

***Definition 2.13.*** If $X$ has probability mass function

$$p(k) = \binom{n}{k} p^k (1 - p)^{n-k}, \quad k = 0, 1, \ldots, n$$

it is said to have a *binomial distribution* with parameters $n$ and $p$, and we write $X \sim \text{bin}(n, p)$.

The binomial distribution thus describes the experiment to count successes in independent trials and is defined through its pmf. The parameter $p$ is often called the *success*

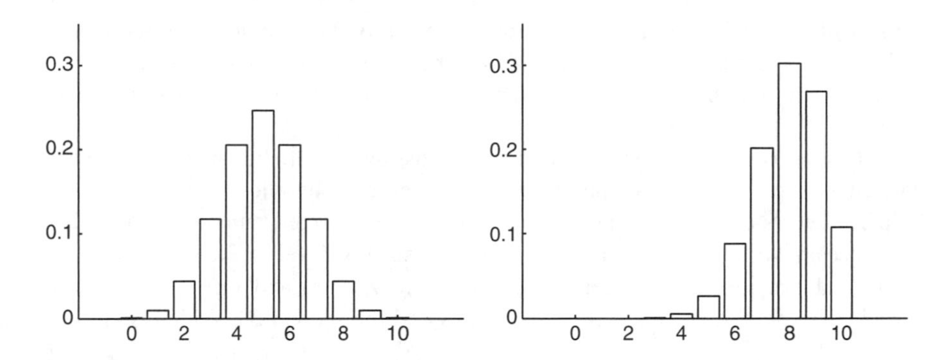

**FIGURE 2.9** The pmf's of two binomial distributions with parameters $n = 10$, $p = 0.5$ and $n = 10$, $p = 0.8$, respectively.

*probability*. In Figure 2.9, a binomial pmf with $n = 10$ and $p = 0.5$ and another with $n = 10$ and $p = 0.8$ are plotted. Note how the pmf with $p = 0.5$ is symmetric, which for this value of $p$ is the case for any value of $n$.

**Proposition 2.17.** If $X \sim \text{bin}(n, p)$, then

$$E[X] = np \quad \text{and} \quad \text{Var}[X] = np(1-p)$$

*Proof.* First note that, by Problem 24(b) in Chapter 1

$$kp(k) = np\binom{n-1}{k-1}p^{k-1}(1-p)^{n-k}$$

and we get

$$E[X] = \sum_{k=0}^{n} kp(k) = np\sum_{k=1}^{n}\binom{n-1}{k-1}p^{k-1}(1-p)^{n-k} = np$$

since the terms in the sum are the probabilities in a binomial distribution with parameters $n - 1$, and $p$ and hence the sum equals 1. The proof of the variance formula is left for Problem 65. ■

***Example 2.32.*** Consider the following two sequences of heads and tails. Only one of them was obtained by repeatedly flipping a fair coin. Let us call this sequence "random." Which one is it? For better readability, we represent heads by X and tails by O:

XOOXXOOXXXOXOOXXOOXXOOOXOOOOXX
XXOXOXOOXOXOXOXXOOXOXOXXOXOXXO

This is, of course, impossible to answer with certainty. The second sequence seems a little better "mixed," whereas the first has some suspicious occurrences of long runs of tails at the end, but we need to come up with some more formal way to make a decision.

Let us try to figure out which sequence is most unlikely to be random. Since each particular sequence of heads and tails has the same probability $(\frac{1}{2})^{30}$, this will not help us. Let us consider the number of heads. The expected number is 15, and the sequences have 14 and 16 heads, respectively, so this does not help us, either. Let us instead consider the number of changes from X to O and vice versa. A count reveals that the first sequence has 14 changes and the second 23. Which is closer to the expected?

Since each change has probability $\frac{1}{2}$, we can denote the number of changes by $X$ and thus have

$$X \sim \text{bin}\left(29, \frac{1}{2}\right)$$

which has $E[X] = 14.5$. Thus, the second sequence has far more changes than what is expected. But how extreme is it? The probability to get a values as large as 23 or larger, is

$$P(X \geq 23) = \sum_{k=23}^{29} \binom{29}{k} \left(\frac{1}{2}\right)^{29} \approx 0.001$$

which is a very small probability. In other words, only once in 1000 experiments of 30 independent coin flips would we get an outcome this extreme, and we decide that this sequence is not random.

So, what is the answer? The first sequence is random, and the second was obtained by letting the probability of a change be 0.7. The number of changes is then bin(29, 0.7), which has expected value $29 \times 0.7 = 20.3$, and our outcome of 23 is not at all extreme.

That the second sequence looks "better mixed" is precisely why we should *not* believe that it is random. Considerations like these are important when testing random number generators, which use deterministic algorithms to produce numbers that seem to be random. It is quite typical that people, when asked to write down a random sequence, are likely to come up with something similar to the second rather than the first sequence. Most human brains are not forgetful enough to be good random number generators. □

***Example 2.33.*** Let us revisit Mr Bloggs from Example 1.50 and give an alternative computation of the probability of his guilt.

Suppose that each person on the island has the genotype with probability $p$, independent of other people. We know that Mr Bloggs has the genotype, so if there are a total of $k$ people with it, he is guilty with probability $\frac{1}{k}$, for $k = 1, 2, \ldots, n$. Let $X$ be the total number of individuals with the genotype. We already know that

Mr Bloggs has the genotype, and among the remaining $n-1$ individuals, the number with the genotype should thus be binomial with parameters $n-1$ and $p$. Let $G$ be the event that Mr Bloggs is guilty and condition on the events $\{X=k\}$, $k=1,2,\ldots,n$ to obtain

$$P(G) = \sum_{k=1}^{n} P(G|X=k)P(X=k)$$
$$= \sum_{k=1}^{n} \frac{1}{k}\binom{n-1}{k-1} p^{k-1}(1-p)^{n-k} = \frac{1-(1-p)^n}{np}$$

after some algebra. In Example 1.50, we found that the probability that he is guilty is $1/(1+(n-1)p)$, which is not the same answer. What is wrong this time? Just as in the case of the stone-throwing boys from Example 1.51, we forgot to condition on all the information we have. We know not only that Mr Bloggs has the genotype but also that the murderer has it and under these two conditions, the proposed binomial distribution is not correct. What, then, is the distribution of $X$?

Initially and unconditionally, the number of people on the island with the genotype is $X \sim \text{bin}(n, p)$. Given the event $A = \{$both the murderer and Mr Bloggs have the genotype$\}$, what is the conditional distribution of $X$? We can think of observations on genotypes as sampling with replacement, just as we did in Example 1.52. Thus, if there are $k$ individuals with the genotype, then the conditional probability of $A$ is $k^2/n^2$, since we have sampled twice and gotten the genotype both times. We can now apply Bayes' formula to obtain

$$P(X=k|A) = \frac{P(A|X=k)P(X=k)}{\sum_{j=0}^{n} P(A|X=j)P(X=j)}$$
$$= \frac{\frac{k^2}{n^2}\binom{n}{k} p^k(1-p)^{n-k}}{\sum_{j=0}^{n}\frac{j^2}{n^2}\binom{n}{j} p^j(1-p)^{n-j}} = \frac{k^2\binom{n}{k} p^k(1-p)^{n-k}}{E[X^2]}$$

and the probability that Mr Bloggs is guilty becomes

$$P(G|A) = \sum_{k=1}^{n} P(G|A\cap\{X=k\})P(X=k|A)$$
$$= \sum_{k=1}^{n} \frac{1}{k} \times \frac{k^2\binom{n}{k} p^k(1-p)^{n-k}}{E[X^2]} = \frac{E[X]}{E[X^2]}$$
$$= \frac{np}{n^2p^2+np(1-p)} = \frac{1}{1+(n-1)p}$$

where we used the variance formula $E[X^2] = (E[X])^2 + \text{Var}[X]$. Thus, we got the same solution as in Example 1.50 once more. The conditional distribution of $X$ we computed is an example of a *size-biased* distribution; for more on this topic, see Problem 32. □

### 2.5.3 The Geometric Distribution

Consider again the situation of successive independent trials where the event $A$ occurs with probability $p$. This time let $X$ be the number of trials until $A$ first occurs. Then $X$ is discrete with range $1, 2, \ldots$, and the event $\{X = k\}$ means that we have the sequence

$$F\ F\ \cdots\ F\ S$$

of $k - 1$ failures and success in the $k$th trial. By independence, this has probability

$$(1 - p)(1 - p)\cdots(1 - p)p = p(1 - p)^{k-1}$$

and we state the following definition.

***Definition 2.14.*** If $X$ has probability mass function

$$p(k) = p(1 - p)^{k-1}, \quad k = 1, 2, \ldots$$

it is said to have a *geometric distribution* with parameter $p$, and we write $X \sim \text{geom}(p)$.

The parameter $p$ is again called the success probability. If we instead count the number of trials *before* the first occurrence of $A$ and denote this by $Y$, then $Y$ has pmf

$$p(k) = p(1 - p)^k, \quad k = 0, 1, \ldots$$

The difference is that the range includes 0. We refer to this as a "geometric distribution including 0" and note that $Y = X - 1$. In Example 2.9, we found the cdf of a geometric distribution with $p = \frac{1}{6}$ by using the definition of cdf. There is, however, a quicker way, if we instead consider the probability that $X$ is strictly greater than $n$. Since this is equivalent to saying that the first $n$ trials resulted in failure, we get

$$P(X > n) = P(n \text{ consecutive failures}) = (1 - p)^n$$

which gives the cdf

$$F(n) = 1 - (1 - p)^n, \quad n = 1, 2, \ldots$$

which is also what we got in the special case in the example.

**Proposition 2.18.** If $X \sim \text{geom}(p)$, then

$$E[X] = \frac{1}{p} \quad \text{and} \quad \text{Var}[X] = \frac{1 - p}{p^2}$$

*Proof.* From Proposition 2.9, we get

$$E[X] = \sum_{n=0}^{\infty} P(X > n) = \sum_{n=0}^{\infty} (1-p)^n = \frac{1}{p}$$

which is certainly quicker than using the definition directly. We will prove the variance formula in Section 3.7.4. ■

The expression for the mean makes good sense. In a large number of trials, the proportion $p$ are successes, and hence they come on average every $(1/p)$th trial. The mean in the geometric distribution including 0 is $1/p - 1 = (1-p)/p$.

***Example 2.34.*** If you play the Texas Lotto twice a week, how long is your expected wait to win the jackpot?

According to Example 1.21, the probability to win is $1/47,784,352$, so your expected wait is $47,784,352$ draws or about 459,000 years. Good luck. □

***Example 2.35.*** The government of a war-torn country decides that the proportion of males in the population needs to be increased. It therefore declares that families are allowed to keep having children as long as the newborn babies are boys, but as soon as they have a daughter, they are not allowed any more children. Will the goal be achieved?

Consider a family that has no children yet, and let $X$ be the number of children they will have. With the rule to stop after the first daughter, $X \sim \text{geom}(\frac{1}{2})$, and hence the expected number of children is 2. Given $Y$ as the number of sons, we have $Y = X - 1$ and hence $E[Y] = E[X] - 1 = 1$, so the expected number of sons is 1, the same as the expected number of daughters. The suggested policy will not change the long-term sex ratio. See also Problem 60. □

### 2.5.4 The Poisson Distribution

The distribution we investigate in this section is different from those discussed previously, in the sense that it does not primarily describe a particular experiment. Rather, it is a distribution that has been observed empirically in many different applications.

***Definition 2.15.*** If $X$ has probability mass function

$$p(k) = e^{-\lambda}\frac{\lambda^k}{k!}, \quad k = 0, 1, \ldots$$

it is said to have a *Poisson distribution* with parameter $\lambda > 0$, and we write $X \sim \text{Poi}(\lambda)$.

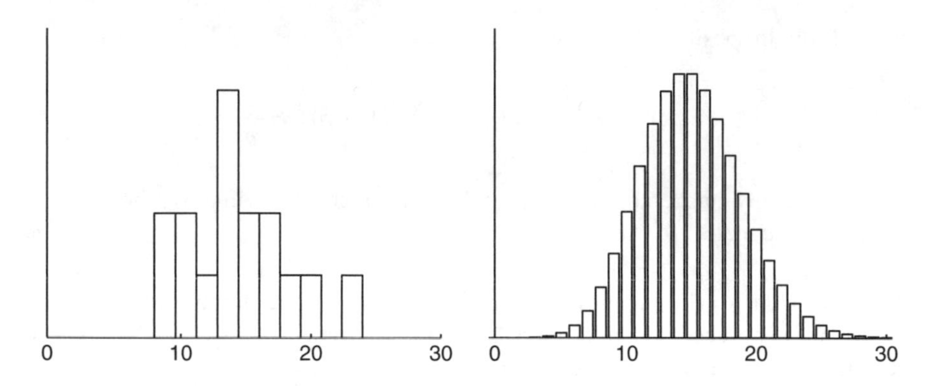


**FIGURE 2.10** Histogram of Pacific tropical cyclones 1988–2003. (*Source*: U.S. National Oceanic and Atmospheric Administration, www.noaa.gov), and pmf of a Poisson distribution with mean 15. Since we are interested only in the shapes, the scales on the $y$-axes have been omitted.

Since we have not derived this expression from any particular experiment, let us first check that it is indeed a pmf according to the two criteria in Proposition 2.1. Clearly it is nonnegative and by Taylor's theorem

$$\sum_{k=0}^{\infty} \frac{\lambda^k}{k!} = e^{\lambda}$$

so the $p(k)$ sum to one. The Poisson distribution tends to arise when we count the number of occurrences of some unpredictable event over a period of time. Typical examples are earthquakes, car accidents, incoming phone calls, misprints in a newspaper, radioactive decay, and hits of a Web site.[3] These all have in common the fact that they are rare on a short timescale but more frequent if we count over a longer period of time. For some real data, consider Figure 2.10, which is a histogram over the annual numbers of tropical cyclones (tropical storms or hurricanes) that were formed off the U.S. Pacific coast during the years 1988–2003. The average number is 15, and we also plot the pmf of a Poisson distribution with $\lambda = 15$ (we will soon see that $\lambda$ is the mean in the distribution). With so few observations, we cannot expect the histogram to look exactly like the pmf, but there is certainly nothing that contradicts the assumption of a Poisson distribution. Thus, both the physical nature of cyclones (unpredictable, rare on a short timescale) and actual observations support the assumption.

***Example 2.36.*** Suppose that the annual number of tropical cyclones that are formed off the U.S. Pacific coast has a Poisson distribution with mean 15. What is the probability that a given year has at most five cyclones?

[3]It would be a shame not to also mention Siméon Poisson's original use of the distribution that would later bear his name, the classical application to the number of Prussian soldiers annually kicked to death by their horses in the nineteenth century. The fit to a Poisson distribution is remarkably good (and the term "success probability" takes on an interesting meaning).

Let $X$ be the number of cyclones, so that $X \sim \text{Poi}(15)$. The probability of at most 5 cyclones is

$$P(X \leq 5) = e^{-15}\left(1 + 15 + \frac{15^2}{2!} + \frac{15^3}{3!} + \frac{15^4}{4!} + \frac{15^5}{5!}\right) \approx 0.003$$

□

It should be stressed that the events must be not only rare but also random. Compare the two statements "large asteroids hit Earth every thousand years" and "a millennium year occurs every thousand years." The first statement is about the average behavior of a random phenomenon, whereas there is no randomness at all in the second. Even with randomness, we must make sure that there is "enough" of it. For example, if buses are supposed to arrive at a bus stop every 10 min, there will be some random variation in their arrival times. However, since the bus drivers attempt to follow a schedule, there is too much regularity to assume a Poisson distribution for the number of arriving buses in a given time interval.

**Proposition 2.19.** If $X \sim \text{Poi}(\lambda)$, then

$$E[X] = \lambda \quad \text{and} \quad \text{Var}[X] = \lambda$$

*Proof.* First note that the pmf satisfies the relation

$$kp(k) = \lambda p(k-1)$$

for $k \geq 1$, which gives

$$E[X] = \sum_{k=0}^{\infty} kp(k) = \sum_{k=1}^{\infty} kp(k) = \lambda \sum_{k=1}^{\infty} p(k-1) = \lambda$$

The proof of the variance expression is left for Problem 65. ■

***Example 2.37.*** A microliter of normal human blood contains on average about 7000 white blood cells and 5 million red blood cells, as well as platelets, plasma, and other matter. In a diluted blood sample, distributed in hundreds of test tubes, it was observed that about 1% of the test tubes contained no white blood cells. What is the mean number of white blood cells per test tube?

Since white blood cells are comparatively rare, we assume a Poisson distribution (and here we are describing events in space rather than time). More specifically, let $X$ denote the number of white blood cells in a test tube and assume that $X \sim \text{Poi}(\lambda)$, and the question is what $\lambda = E[X]$ equals. That 1% of the tubes had no white blood cells at all means that we estimate $p(0) = P(X = 0)$ to be 0.01. Since

$$p(0) = e^{-\lambda}$$

we get

$$\lambda = -\log p(0)$$

which in our case equals $-\log 0.01 \approx 4.6$. Hence, the mean is 4.6 white blood cells per test tube. □

The Poisson distribution can be used as an approximation to the binomial distribution. To illustrate why, consider a newspaper page with $n$ letters and assume that each letter is misprinted with probability $p$, independent of other letters. The number of misprints is then binomial with parameters $n$ and $p$. On the other hand, since $n$ is large and $p$ is small, we are dealing with a rare and unpredictable event, and this fits the situation for a Poisson distribution. Since the mean in the binomial distribution is $np$, we can argue that the number of misprints is approximately Poi($np$). It is also possible to argue directly that

$$\binom{n}{k} p^k (1-p)^{n-k} \approx e^{-np} \frac{(np)^k}{k!}$$

for large $n$ and small $p$ and in Section 4.4, we will state a limit result that warrants the use of the approximation.

How good is the approximation? One obvious difference between the binomial and the Poisson distributions is that the former has finite range $\{0, 1, \ldots, n\}$ and the latter, infinite range $\{0, 1, \ldots\}$. This is also reflected in the variances, which for the binomial distribution is $np(1-p)$ and for the approximating Poisson distribution $np$, slightly larger because of the wider range. There are various rules of thumb for when approximation is allowed, but since this is really a numerical problem, depending on how many correct decimals we are satisfied with, we will not address it here. Let us, however, point out that the approximation is good enough to be of practical use. In Figure 2.11, pmfs are plotted for $n = 10$ and $p = 0.1$.

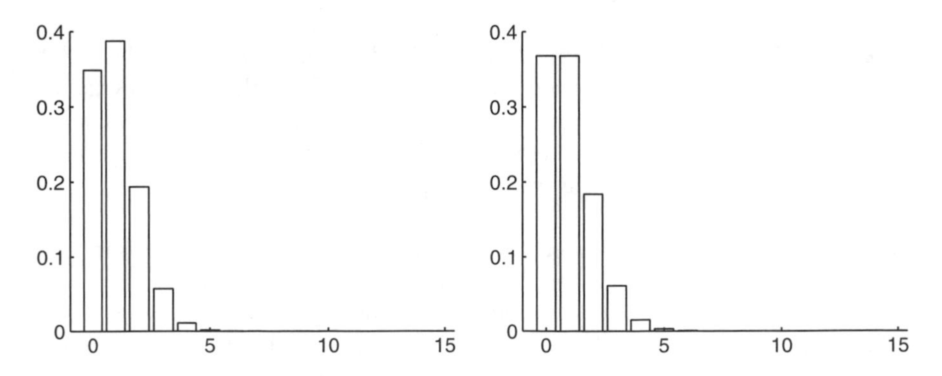


**FIGURE 2.11** Pmfs of a binomial distribution with $n = 10$, $p = 0.1$ (left) and the approximating Poisson distribution with $\lambda = 1$ (right).

### 2.5.5 The Hypergeometric Distribution

Consider a set of $N$ objects, $r$ of which are of a special type. Suppose that we choose $n$ objects, without replacement and without regard to order. What is the probability that we get exactly $k$ of the special objects?

Let $X$ denote the number of special objects that we get. This is a discrete random variable with range $0, 1, \ldots, n$, and its pmf is easily found by combinatorial methods.

***Definition 2.16.*** If $X$ has probability mass function

$$p(k) = \frac{\binom{r}{k}\binom{N-r}{n-k}}{\binom{N}{n}}, \quad k = 0, 1, \ldots, n$$

it is said to have a *hypergeometric distribution* with parameters $N$, $r$, and $n$, written $X \sim \text{hypergeom}(N, r, n)$.

The parameters must be such that $n \leq r$ and $n - k \leq N - r$, which we assume implicitly without spelling it out. In Example 1.22, we had a hypergeometric distribution with parameters $N = 52$, $r = 13$, and $n = 5$. We next state expressions for the mean and variance and defer the proof to Section 3.6.2.

**Proposition 2.20.** If $X \sim \text{hypergeom}(N, r, n)$, then

$$E[X] = \frac{nr}{N} \quad \text{and} \quad \text{Var}[X] = n \times \frac{N-n}{N-1} \times \frac{r}{N}\left(1 - \frac{r}{N}\right)$$

Now suppose that $N$ is large and $r$ is moderate and let $p = r/N$, the proportion of special objects. If $n$ is small relative to $N$, we would expect that drawing without replacement would not be very different from drawing with replacement. For example, if the first object drawn is of the special type, the proportion changes from $r/N$ to $(r-1)/(N-1)$, which is then still approximately equal to $p$. But drawing with replacement gives a binomial distribution with success probability $p$, and we have argued that the hypergeometric distribution can be approximated by the binomial distribution for suitable parameter values. Also, compare the means and variances for the two distributions.

### 2.5.6 Describing Data Sets

The concepts of mean, variance, and standard deviation that we use to describe random variables are often used to describe data sets. For example, the mean and standard deviation may be given to describe the performance by a school class on a test or to summarize a survey of home prices in a neighborhood. Suppose that we have a data set

$$D = \{x_1, x_2, \ldots, x_n\}$$

where the $x_k$ are not necessarily different from each other. We then define the mean of this set as the arithmetic average

$$\bar{x} = \frac{1}{n}\sum_{k=1}^{n} x_k$$

and the variance is defined as[4]

$$s_x^2 = \frac{1}{n}\sum_{k=1}^{n}(x_k - \bar{x})^2$$

How does this fit into the framework of random variables? If we choose a value $X$ at random from the data set, then $X$ is a discrete random variable whose range is $S$, the set of distinct values in $D$. The pmf of $X$ is precisely the histogram of the data set:

$$p_X(x) = \frac{\#\{k : x_k = x\}}{n}, \quad x \in S$$

The expected value is

$$\mu = \sum_{x\in S} x p_X(x) = \sum_{x\in S} x \frac{\#\{k : x_k = x\}}{n} = \frac{1}{n}\sum_{k=1}^{n} x_k = \bar{x}$$

and the variance

$$\sigma^2 = \sum_{x\in S}(x - \bar{x})^2 p_X(x) = \frac{1}{n}\sum_{k=1}^{n}(x_k - \bar{x})^2 = s_x^2$$

and the standard deviation as usual is the square root of the variance. Thus, we can describe data sets in a way that is commonly done and view this as a special case of our theory of random variables and their distributions. Note that $\bar{x}$ and $s_x^2$ change if we add more data to $D$, so this corresponds to defining a new random variable. Alternatively, we could think of $X$ as an ideal random variable over all possible values that could belong to the data set, in which case $X$ could be either discrete or continuous and would have unknown mean and variance. We can then view the histogram as an approximation of the true but unknown pmf or pdf, and $\bar{x}$ and $s_x^2$ as approximations of the true mean $\mu$ and variance $\sigma^2$. We will return to this view in Chapter 6.

[4]It is also common to divide by $n - 1$ rather than $n$, and we will address the reason for this in detail in Chapter 6.

***Example 2.38.*** A survey of salaries at a company with 11 employees gave the following values: $D = \{30, 30, 30, 30, 30, 40, 40, 60, 70, 150, 220\}$ (1000 dollars per year). The mean is $\bar{x} = (30 + 30 + \cdots + 220)/11 = 66.4$ and the variance

$$s_x^2 = \frac{1}{11}\sum_{k=1}^{11}(x_k - \bar{x})^2 = 3514$$

which gives standard deviation $s \approx 59.3$. In this case, there is no more data to be collected, so the mean and variance are exact. Note that only three salaries are above the mean and eight below, and merely stating the mean as a summary of the data does not reveal this fact. We say that the data set is skewed (to the left). □

***Example 2.39.*** A die was rolled seven times and gave $D = \{2, 2, 2, 2, 4, 5, 5\}$. The mean of this data set is $\bar{x} = (2 + 2 + \cdots + 5)/7 = 3.1$, and the variance

$$s_x^2 = \frac{1}{7}\sum_{k=1}^{7}(x_k - \bar{x})^2 = 1.8$$

If we choose at random from $D$ and call the number $X$, the distribution of $X$ is

$$P(X = 2) = \frac{4}{7}, \quad P(X = 4) = \frac{1}{7}, \quad P(X = 5) = \frac{2}{7}$$

In this case, it is more natural to think of $\bar{x}$ and $s_x^2$ as approximations to the true mean and variance, which if the die is fair are 3.5 and $\frac{35}{12} \approx 2.9$, respectively. If the die is rolled repeatedly to increase the data set, the approximations become increasingly more accurate. □

## 2.6 THE EXPONENTIAL DISTRIBUTION

We will introduce several named continuous distributions. In an analogy with the discrete case, these are defined in terms of the pdf, and we may apply the name to the distribution or the random variable itself. Also as in the discrete case, there are essentially two different types of distribution: those that describe a clearly defined experiment or mathematical model from which we can derive the pdf and those that have been empirically determined to fit certain types of data sets. We have already seen one distribution of the first category; the uniform distribution describes how to choose a point at random from an interval. We continue with another distribution in this category.

Let $X$ be a continuous random variable that denotes the lifetime of something (e.g., an electronic component) that *does not age* in the sense that the probability of functioning yet another time unit does not depend on its current age. We can describe this in a formula as

$$P(X > x + y | X > y) = P(X > x) \tag{2.1}$$

for $x, y \geq 0$. This means that the probability of surviving another $x$ time units does not depend on how many time units $y$ that the component has survived so far. For the left-hand side, we get

$$P(X > x + y | X > y) = \frac{P(\{X > x + y\} \cap \{X > y\})}{P(X > y)} = \frac{P(X > x + y)}{P(X > y)}$$

where the last equality follows since $X > x + y \Rightarrow X > y$, that is, the event $\{X > x + y\}$ is included in the event $\{X > y\}$. Hence, if we let

$$G(x) = P(X > x), \quad x \geq 0$$

we get the equation

$$G(x + y) = G(x)G(y)$$

for $x, y \geq 0$. Note that the function $G$ at a point $x$ gives the probability that the component survives the age $x$ and $G$ is therefore called the *survival function*. Also note that $G(x) = 1 - F(x)$ where $F$ is the cdf of $X$. It can be shown (see Problem 75) that the only possible solutions to the equation are

**(a)** $G(x) \equiv 0$

**(b)** $G(x) \equiv 1$

**(c)** $G(x) = e^{-\lambda x}$ where $\lambda$ is any constant

From a practical point of view, (a) and (b) are uninteresting. The solution $G(x) \equiv 0$ would mean that the component has probability 0 of surviving any time, which means that it does not function to begin with. The solution $G(x) \equiv 1$, on the other hand, means that it functions forever. These two solutions also contradict the assumption that $X$ is a continuous random variable, and we are left with (c) as the only possibility. The cdf of $X$ is therefore

$$F(x) = 1 - G(x) = 1 - e^{-\lambda x}, \quad x \geq 0$$

and by differentiation we get the pdf

$$f(x) = F'(x) = \lambda e^{-\lambda x}, \quad x \geq 0$$

For this to be a possible pdf, the constant $\lambda$ must be strictly positive.

***Definition 2.17.*** If the pdf of $X$ is

$$f(x) = \lambda e^{-\lambda x}, \quad x \geq 0$$

then $X$ is said to have an *exponential distribution* with parameter $\lambda > 0$, written $X \sim \exp(\lambda)$.

In Example 2.10, we thus had an exponential distribution with $\lambda = 0.001$. Note how we started by making one seemingly innocent assumption, that there is no aging and how this forced on us only one possible form of the pdf. The property in Equation (2.1) is called the *memoryless property*. Hence, we have the following result.

**Proposition 2.21.** A continuous random variable has the memoryless property if and only if it has an exponential distribution.

It is the shape of the pdf that is connected to the memoryless property, so this property holds for any value of $\lambda$. To investigate the role of $\lambda$, we note that larger $\lambda$ means that the pdf is higher near 0, which indicates that values of $X$ tend to be smaller. By computing the expected value, we can see that this is indeed the case.

**Proposition 2.22.** If $X \sim \exp(\lambda)$, then

$$E[X] = \frac{1}{\lambda} \quad \text{and} \quad \text{Var}[X] = \frac{1}{\lambda^2}$$

*Proof.* Direct computation gives

$$E[X] = \int_0^\infty x\lambda\, e^{-\lambda x} dx = \frac{1}{\lambda}$$

by partial integration. For the variance we need $E[X^2]$:

$$E[X^2] = \int_0^\infty x^2\lambda\, e^{-\lambda x} dx = \frac{2}{\lambda^2}$$

This gives

$$\text{Var}[X] = E[X^2] - (E[X])^2 = \frac{1}{\lambda^2}$$

■

***Example 2.40.*** The technique of *carbon 14 dating* is based on decay of the isotope carbon 14 into nitrogen 14. The *half-life*, that is, the time required for half the atoms in the sample to decay, of carbon 14 is approximately 5700 years. What is the expected lifetime of a single carbon 14 atom?

Since lifetimes are random, we need to interpret half-life. If we start with a number $n_0$ of atoms, the half-life $h$ is the time when there are roughly $n_0/2$ atoms left. Thus, half the atoms have "survived" $h$, and half have not, which indicates that the probability to survive age $h$ is $\frac{1}{2}$. Let $T$ be the lifetime of an individual atom and let $h$ be the number that satisfies

$$P(T > h) = \frac{1}{2}$$

and from Example 2.10 we recognize $h$ as the *median* lifetime.

Next, we assume that $T$ has an exponential distribution. This is certainly reasonable as an atom decays spontaneously, not as a result of old age or wear and tear. Thus, $T \sim \exp(\lambda)$, and we get

$$P(T > 5700) = e^{-5700\lambda} = \frac{1}{2}$$

which gives $\lambda = \log 2/5700$ and the expected lifetime

$$E[T] = \frac{1}{\lambda} = \frac{5700}{\log 2} \approx 8200 \text{ years}$$

We will examine the problem of half-life in radioactive decay more closely in Section 3.10.1. □

The parameter $\lambda$ is often referred to as the *failure rate* or *hazard rate* and is a measure of how likely failure is at any given age. It is a natural extension to allow the failure rate to depend on age and thus define the *failure rate function*, a concept to which we return in Section 2.10. The exponential distribution is characterized by having a constant failure rate function.

The exponential distribution is also used to model times between random events such as earthquakes, customer arrivals to a store, or incoming jobs to a computer. Recall how we previously used the Poisson distribution to model the number of such events in a fixed time period. There is an interesting connection between the exponential and the Poisson distributions, which we describe briefly here and return to address in detail in Section 3.12.

Suppose that events occur in such a way that the time between two events is $\exp(\lambda)$. Suppose further that the occurrence of an event is independent of the occurrences of other events. It can then be shown that the number of occurrences in a time interval of length $t$ is a random variable that has a Poisson distribution with mean $\lambda t$. Such a stream of events is called a *Poisson process*.

***Example 2.41.*** According to the U.S. Geological Survey, earthquakes with magnitude at least 7 occur on average 18 times a year (worldwide). **(a)** What is the probability that two consecutive such earthquakes are at least 2 months apart? **(b)** What is the probability that there are no earthquakes in a 2-month period?

Suppose that earthquakes occur according to a Poisson process with rate $\lambda = 1.5$ earthquakes per month. Let $T$ be the time between two consecutive earthquakes and

$X$ the number of earthquakes in a 2-month period. Then we have $T \sim \exp(1.5)$ and $X \sim \text{Poi}(3)$ and we get for (a)

$$P(T > 2) = e^{-1.5\times 2} \approx 0.05$$

and for (b)

$$P(X = 0) = e^{-3} \approx 0.05$$

Note that the two events $\{T > 2\}$ and $\{X = 0\}$ are identical. □

## 2.7 THE NORMAL DISTRIBUTION

The next distribution falls into the second category, those who are motivated primarily by empirical observations rather than describing a particular experiment or mathematical model. Let us first define it.

***Definition 2.18.*** If $X$ has pdf

$$f(x) = \frac{1}{\sigma\sqrt{2\pi}} e^{-(x-\mu)^2/2\sigma^2}, \quad x \in R$$

it is said to have a *normal distribution* with parameters $\mu$ and $\sigma^2$, written $X \sim N(\mu, \sigma^2)$.

This is also often called the *Gaussian* distribution, after German mathematician Carl Friedrich Gauss, who used it to describe errors in astronomical measurements. The typical bell-shaped pdf is shown in Figure 2.12. The normal distribution shows up naturally in many situations where there are measurement errors or fluctuations due to randomness, noise, and other factors, and it is the most widely used of all distributions. It is used in models of signal processing, the velocities of gas molecules, and the movement of stock market prices. It also arises in the study of many biological

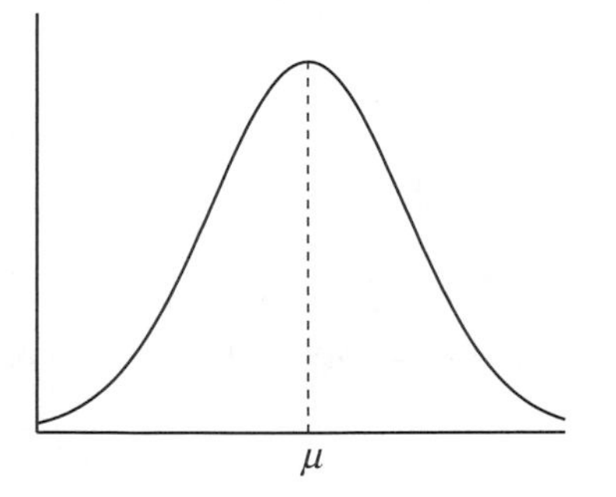


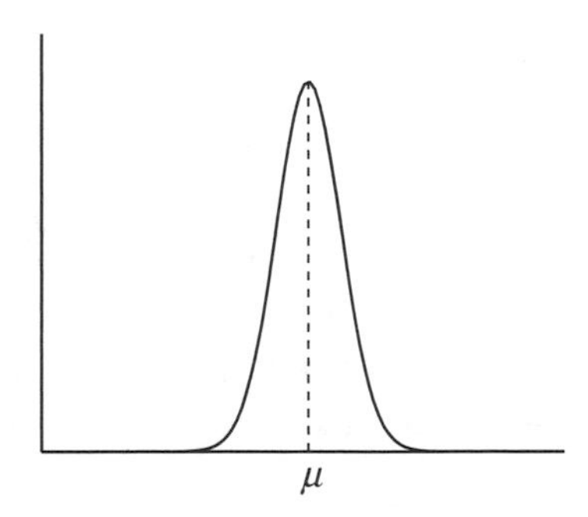


**FIGURE 2.12** The pdfs of two normal distributions with the same mean. The one on the left has higher variance.

phenomena, where the randomness is typically composed of different components. As an example, consider taking the blood pressure of a person. First, there is the variation that occurs between individuals, which gives rise to randomness due to the selection of individual. Then there is variation over time within each individual, and finally there is uncertainty in the actual measuring procedure.

Note that the pdf is symmetric around $\mu$. By computing the usual integrals, it is easily shown that

**Proposition 2.23.** If $X \sim N(\mu, \sigma^2)$, then $E[X] = \mu$ and $\mathrm{Var}[X] = \sigma^2$.

A word of caution regarding notation is due here. We have chosen to give $\sigma^2$ as the second parameter, so that if we write $N(0, 4)$, we mean that the variance is 4 (and the standard deviation 2). Some texts prefer to give the standard deviation as the second parameter, as does, for example, Matlab, so you need to pay attention.

Since we cannot find a primitive function of the pdf, there is no hope of an explicit expression for the cdf, and it must be computed numerically. The normal distribution with $\mu = 0$ and $\sigma^2 = 1$ is of particular interest and is called the *standard* normal distribution. For the standard normal distribution, we use the following special notation for the pdf and cdf:

$$\varphi(x) = \frac{1}{\sqrt{2\pi}} e^{-x^2/2}, \quad x \in R$$

$$\Phi(x) = \int_{-\infty}^{x} \varphi(t)dt, \quad x \in R$$

The significance of the standard normal distribution is spelled out in the next proposition.

**Proposition 2.24.** Suppose that $X \sim N(\mu, \sigma^2)$ and let $Z = (X - \mu)/\sigma$. Then $Z \sim N(0, 1)$.

*Proof.* Let $F_Z$ denote the cdf of $Z$. Then

$$F_Z(x) = P(Z \leq x) = P(X \leq \mu + \sigma x) = \int_{-\infty}^{\mu+\sigma x} \frac{1}{\sigma\sqrt{2\pi}} e^{-(t-\mu)^2/2\sigma^2} dt$$

Making the variable substitution $y = (t - \mu)/\sigma$ gives $dt = \sigma dy$, and the new integral limits $-\infty$ and $x$ and the expression equals

$$\int_{-\infty}^{x} \frac{1}{\sigma\sqrt{2\pi}} e^{-(\sigma y+\mu-\mu)^2/2\sigma^2} \sigma dy = \int_{-\infty}^{x} \varphi(y)dy = \Phi(x)$$

■

Hence, any normal distribution can be brought back to the standard normal distribution by subtracting the mean and dividing by the standard deviation, a procedure sometimes referred to as computing the *Z score* of $X$. We state in a corollary how this is used in computations.

**Corollary 2.3.** *Let $X \sim N(\mu, \sigma^2)$. For any $x, a, b \in R$, we have*

**(a)** $P(X \leq x) = \Phi\left(\frac{x-\mu}{\sigma}\right)$

**(b)** $P(a \leq X \leq b) = \Phi\left(\frac{b-\mu}{\sigma}\right) - \Phi\left(\frac{a-\mu}{\sigma}\right)$

*Proof.* (a) If $X \sim N(\mu, \sigma^2)$, then

$$P(X \leq x) = P\left(\frac{X-\mu}{\sigma} \leq \frac{x-\mu}{\sigma}\right) = \Phi\left(\frac{x-\mu}{\sigma}\right)$$

where the first equality holds since we subtracted and divided by the same quantities on both sides of the inequality in the probability statement, where $\sigma > 0$, thus not changing the inequality. The second equality holds by Proposition 2.24. Part (b) is true because of Proposition 2.4(a). ■

By Corollary 2.3, we have to compute only the cdf of the standard normal distribution, and Table A.1 gives numerical values of $\Phi(x)$. Note that only values for positive $x$ are given; for negative $x$, we use the following result.

**Corollary 2.4.** *For any $x \in R$*

$$\Phi(-x) = 1 - \Phi(x)$$

*Proof.* Note that $\varphi$ is symmetric around 0, which gives

$$\Phi(-x) = \int_{-\infty}^{-x} \varphi(t)dt = \int_{x}^{\infty} \varphi(t)dt = 1 - \Phi(x)$$

■

Another immediate consequence of Proposition 2.24 is that a linear transformation of a normal distribution is normal, in the following sense.

**Corollary 2.5.** *Let $X \sim N(\mu, \sigma^2)$, let $a, b \in R$, and let $Y = aX + b$. Then*

$$Y \sim N(a\mu + b, a^2\sigma^2)$$

*Proof.* Let us do the case $a > 0$. Start with the cdf of $Y$.

$$P(Y \le x) = P(aX + b \le x) \;=\; P\left(X \le \frac{x-b}{a}\right)$$

$$= \Phi\left(\frac{(x-b)/a - \mu}{\sigma}\right) \;=\; \Phi\left(\frac{x-(a\mu+b)}{a\sigma}\right)$$

which is precisely the cdf of a random variable, which is $N(a\mu + b, a^2\sigma^2)$. The case $a < 0$ is similar. ■

Let us look at a few examples.

***Example 2.42.*** The IQ of a randomly selected individual is often supposed to follow a normal distribution with mean 100 and standard deviation 15. Find the probability that an individual has an IQ **(a)** above 140 and **(b)** between 120 and 130, and **(c)** find a value $x$ such that 99% of the population has IQ at least $x$.

We have $X \sim N(100, 15^2)$ and get for (a)

$$P(X > 140) = 1 - P(X \le 140) \;=\; 1 - \Phi\left(\frac{140-100}{15}\right)$$

$$= 1 - \Phi(2.67) \;\approx\; 0.004$$

For (b), we get

$$P(120 \le X \le 130) = \Phi\left(\frac{130-100}{15}\right) - \Phi\left(\frac{120-100}{15}\right)$$

$$= \Phi(2) - \Phi(1.33) \;\approx\; 0.07.$$

For the last part, we need to find $x$ such that $P(X > x) = 0.99$. By Corollary 2.4 we get

$$P(X > x) = 1 - \Phi\left(\frac{x-100}{15}\right) = \Phi\left(\frac{100-x}{15}\right) = 0.99$$

and Table A.2 gives

$$\frac{100-x}{15} = 2.33$$

which gives $x \approx 65$. The value $x$ is called the 99th *percentile*. □

***Example 2.43.*** A power source gives an output voltage of 12 (volts). Because of random fluctuations, the true voltage at any given time is $V = 12 + X$, where $X \sim N(0, 0.1)$. The voltage is measured once an hour, and if it is outside the interval

[11.5, 12.5] the power source needs to be adjusted. What is the probability that no adjustment is needed during a 24 h period?

Let us first note that by Corollary 2.5, $V \sim N(12, 0.1)$. At any given hour, the probability that the voltage is within bounds is

$$P(11.5 \leq V \leq 12.5) = \Phi\left(\frac{11.5 - 12}{\sqrt{0.1}}\right) - \Phi\left(\frac{12.5 - 12}{\sqrt{0.1}}\right)$$

$$\approx \Phi(1.58) - \Phi(-1.58) = 2\Phi(1.58) - 1 \approx 0.89$$

and hence the probability that it is out of bounds is 0.11. Now consider the experiment to check the voltage every hour for 24 h. Assuming that hours are independent, the number of hours that has the voltage out of bounds is a binomial random variable with $n = 24$ and $p = 0.11$. Call this number $Y$ to obtain the probability that no adjustment is needed as

$$P(Y = 0) = (1 - 0.11)^{24} \approx 0.06$$

□

For values of $x$ larger than those in Table A.1, the approximation

$$\Phi(x) \approx 1 - \frac{\varphi(x)}{x}$$

can be used. See Problems 89 and 90.

## 2.8 OTHER DISTRIBUTIONS

So far we have seen the uniform, exponential, and normal distributions as examples of continuous distributions. In this section, we list a number of other continuous distributions, as well as some of their properties. We will also encounter a type of random variable that is neither discrete nor continuous.

### 2.8.1 The Lognormal Distribution

Sometimes, observations do not follow a normal distribution but their logarithms do. This may simply be an empirically observed fact but can also be argued theoretically. The normal distribution arises in situations such as those mentioned in the previous section because the randomness has an *additive* effect. It may instead be the case that randomness has a *multiplicative* effect. This is true in many applications in science and engineering, for example, reliability and material fatigue analysis, and also in models for financial markets where it is more realistic to model a price change by multiplying by a percentage than to add a dollar amount.

***Definition 2.19.*** Let $X \sim N(\mu, \sigma^2)$ and let $Y = e^X$. Then $Y$ is said to have a *lognormal distribution* with parameters $\mu$ and $\sigma^2$.

For practical reasons, this time we deviated from our usual convention of defining a continuous distribution in terms of its pdf. The reason is that we can always compute probabilities in the lognormal distribution by referring to the normal distribution. We leave it as an exercise to show the following.

**Corollary 2.6.** *Let $Y$ be lognormal with parameters $\mu$ and $\sigma^2$. Then $Y$ has pdf*

$$f_Y(x) = \frac{1}{x\sigma\sqrt{2\pi}} e^{(\log x - \mu)^2/2\sigma^2}, \quad x > 0$$

Note that the parameters $\mu$ and $\sigma^2$ are the mean and variance of the underlying normal random variable $X$, not of the lognormal random variable $Y$. By computing the usual integrals, it can be shown that the mean and variance are

$$E[Y] = e^{\mu+\sigma^2/2} \quad \text{and} \quad \text{Var}[Y] = e^{2\mu+\sigma^2}\left(e^{\sigma^2} - 1\right)$$

***Example 2.44.*** Suppose that the price $Y$ of a particular stock at closing has a lognormal distribution with $E[Y] = 20$ dollars and $\text{Var}[Y] = 4$. What is the probability that the price exceeds \$22?

First we need to find $\mu$ and $\sigma$. By the expressions above we have

$$e^{\mu+\sigma^2/2} = 20$$

and

$$e^{2\mu+\sigma^2}\left(e^{\sigma^2} - 1\right) = 4$$

where the first equation gives

$$\mu + \frac{\sigma^2}{2} = \log 20$$

which we substitute in the second equation to get

$$e^{2\log 20}\left(e^{\sigma^2} - 1\right) = 4$$

which gives

$$\sigma^2 = \log(1.01) \approx 0.01$$

and

$$\mu = \log 20 - \frac{\log 1.01}{2} \approx 3.0$$

Finally, since $\log Y$ is normal, we get

$$P(Y > 22) = 1 - P(Y \leq 22) \;=\; 1 - P(\log Y \leq \log 22)$$

$$= 1 - \Phi\left(\frac{\log 22 - 3.0}{\sqrt{0.01}}\right) \;\approx\; 0.18.$$

□

### 2.8.2 The Gamma Distribution

The gamma distribution is flexible, fits many different types of real-world data sets, and can be used as an alternative to the normal distribution when data are nonnegative and not exactly symmetric. Before we define the gamma distribution, let us introduce the *gamma function*, which is defined as

$$\Gamma(\alpha) = \int_0^\infty e^{-t} t^{\alpha-1} dt$$

If $\alpha$ is a positive integer, $n$, this reduces to

$$\Gamma(n) = (n-1)!$$

so the gamma function can be regarded as an extension of the factorial to the entire real line. We are now ready to define the gamma distribution.

***Definition 2.20.*** If $X$ has pdf

$$f(x) = e^{-\lambda x} \lambda^\alpha \frac{x^{\alpha-1}}{\Gamma(\alpha)}, \quad x \geq 0$$

it is said to have a *gamma distribution* with parameters $\alpha > 0$ and $\lambda > 0$, written $X \sim \Gamma(\alpha, \lambda)$.

Note that $\alpha = 1$ gives the exponential distribution with parameter $\lambda$. The parameter $\alpha$ is called the *shape parameter* and $\lambda$ is called the *scale parameter*. Figure 2.13 gives the pdfs for three gamma distributions with $\lambda = 1$ and different values of $\alpha$. Notice how the shape changes with the value of $\alpha$; as $\alpha$ increases, it increasingly resembles the normal distribution. Changing the parameter $\lambda$ corresponds to changing the unit

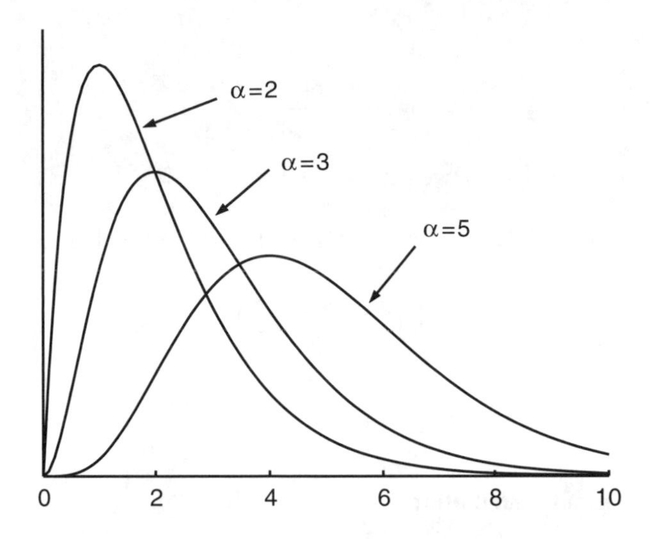


**FIGURE 2.13** Pdfs of the gamma distribution with fixed λ and different values of $\alpha$. Note the increasing resemblance to the pdf of a normal distribution.

of measure and does not affect the shape qualitatively (see Problem 95). In particular, if $\alpha$ is an integer, $n$, we get the pdf

$$f(x) = e^{-\lambda x}\lambda^n \frac{x^{n-1}}{(n-1)!}, \quad x \geq 0$$

This has an interpretation to which we will return in Section 3.10.5. In this case, we can get an explicit expression for the cdf as

$$F(x) = 1 - e^{-\lambda x} \sum_{k=0}^{n-1} \frac{\lambda^k x^k}{k!}, \quad x \geq 0$$

which can be shown by partial integration and induction. By computing the usual integrals and using some special properties of the gamma function, it can be shown that

$$E[X] = \frac{\alpha}{\lambda} \quad \text{and} \quad \mathrm{Var}[X] = \frac{\alpha}{\lambda^2}$$

In Section 3.6.2, we will show this for the integer case $\alpha = n$.

### 2.8.3 The Cauchy Distribution

We finish with a distribution that does not have much practical use. It is, however, derived from an easily described experiment and turns out to have some interesting properties, which makes it a favorite in various counterexamples in the probability and statistics literature.

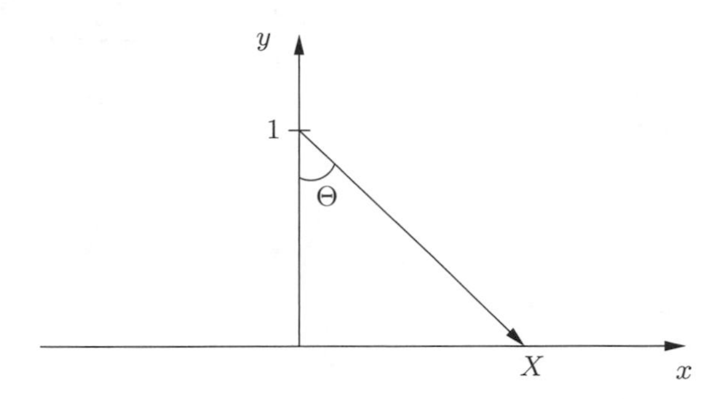


**FIGURE 2.14** If $\Theta$ is chosen at random between $-\pi/2$ and $\pi/2$, $X$ has the Cauchy distribution.

In an ordinary coordinate system, start at the point $(0, 1)$, choose an angle between $-\pi/2$ and $\pi/2$ at random and at this angle draw a line to the $x$-axis (the angle 0 gives a line to the origin). Let $X$ be the point where you hit the $x$-axis (Figure 2.14). The range of $X$ is $(-\infty, \infty)$ and we next find the pdf of $X$. Denote the angle by $\Theta$ and assume that $\Theta \sim \text{unif}(-\pi/2, \pi/2)$. Then $X = \tan \Theta$ and has cdf

$$\begin{aligned} F_X(x) = P(\tan \Theta \leq x) &= P(\Theta \leq \tan^{-1} x) \\ = F_\Theta(\tan^{-1} x) &= \frac{\tan^{-1} x + \pi/2}{\pi}, \quad x \in R \end{aligned}$$

Differentiation gives

$$f_X(x) = F_X'(x) = \frac{1}{\pi(1 + x^2)}, \quad x \in R$$

This is called the *Cauchy distribution*. Something interesting happens when we attempt to compute its expected value:

$$E[X] = \frac{1}{\pi} \int_{-\infty}^{\infty} \frac{x}{1 + x^2} dx = \frac{1}{2\pi} \left[ \log(1 + x^2) \right]_{-\infty}^{\infty}$$

This is not well defined since it is of the form "$\infty - \infty$." In this case, we say that the expected value *does not exist*. If the experiment were to be repeated over and over and consecutive averages computed, these would not converge but keep jumping around the $x$-axis forever, which is somewhat surprising since the pdf is symmetric around 0.

### 2.8.4 Mixed Distributions

The distinction between discrete and continuous random variables is fundamental but not exhaustive. There are random variables that belong to neither of the two categories, as the following example shows.

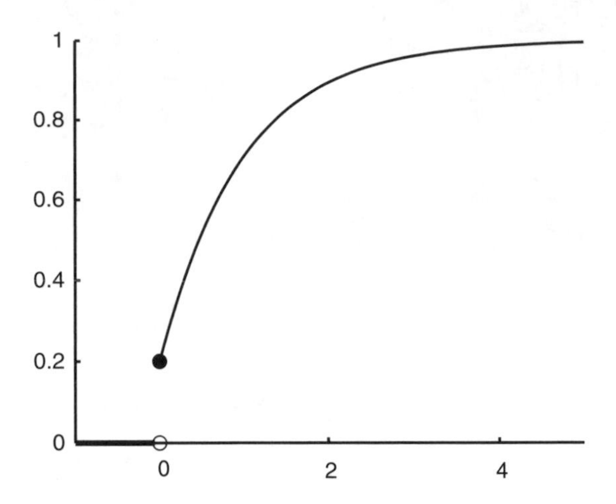


**FIGURE 2.15** The cdf of a mixed distribution with a discrete part at 0 and a continuous part on $(0, \infty)$.

***Example 2.45.*** Jobs arrive at a computer. With probability 0.8, the computer is busy and the incoming job must wait in queue for a time (in seconds), which has an exponential distribution with parameter 1. Let $X$ be the waiting time for an incoming job and find the cdf of $X$.

The crux here is that the waiting time is 0 if the computer is free and has an exponential distribution otherwise. Let $B$ be the event that the computer is free and fix an $x \geq 0$. The law of total probability gives

$$P(X \leq x) = P(X \leq x|B)P(B) + P(X \leq x|B^c)P(B^c)$$

where $P(X \leq x|B) = 1$ and $P(X \leq x|B^c) = 1 - e^{-x}$. Since $P(B) = 0.2$ and $P(B^c) = 0.8$, the cdf becomes

$$F(x) = 1 - 0.8e^{-x}, \quad x \geq 0$$

The cdf is shown in Figure 2.15. Clearly, the range is uncountable, so this is not a discrete distribution. On the other hand, since the cdf has a discontinuity at 0, it is not a continuous distribution either. This is an example of a *mixed distribution* (and $X$ is called a *mixed random variable*) that has a discrete part at 0 and a continuous part on $(0, \infty)$. □

Let us conclude this section by mentioning that there are even stranger creatures in the world of probability, namely, distributions that are neither discrete nor continuous, nor a mixture of the two. These so-called singular distributions are strange mathematical constructions and of no use in any of the applications we consider.

## 2.9 LOCATION PARAMETERS

The mean is not the only possible location parameter and in some cases may not be the best choice. For example, in Problem 40, we learn that if $X \sim \text{unif}(0, 1)$, then $E[1/X] = \infty$. Since we always get finite values of $1/X$, the mean may not be the best way to summarize the distribution. Another location parameter is the *median*, which can be considered the midpoint of the distribution in the sense that, on average, half the observations fall below and half above it. We state the formal definition.

***Definition 2.21.*** Let $X$ be a random variable. Any number $m$ that is such that

$$P(X \geq m) \geq \frac{1}{2} \quad \text{and} \quad P(X \leq m) \geq \frac{1}{2}$$

is called a *median* of $X$.

Note that the median is not necessarily unique. For example, if $X$ is the number when we roll a fair die, both 3 and 4 are medians, as are all numbers in-between. If the distribution is continuous with cdf $F$, a median always satisfies $F(m) = \frac{1}{2}$, and if the range of $X$ does not have any "holes," the median is also unique. We state the following result without proof.

**Corollary 2.7.** *Let $X$ be a continuous random variable with pdf $f$ and cdf $F$, such that $f$ is strictly positive in an entire interval $(a, b)$ and 0 otherwise, where $-\infty \leq a < b \leq \infty$. Then the median $m$ is the unique number that satisfies*

$$F(m) = \frac{1}{2}$$

The mean is thus the average and the median the midpoint of a distribution. If $X$ is continuous with a unique median and its pdf is symmetric around $\mu$, then the mean and median are equal.

***Example 2.46.*** Consider the random variable $Y = 1/X$, where $X \sim \text{unif}(0, 1)$. What is the median of $Y$?

We need to find the cdf of $Y$. Take $y > 1$ to obtain

$$\begin{aligned} F_Y(y) = P(Y \leq y) &= P\left(X \geq \frac{1}{y}\right) \\ &= 1 - F_X\left(\frac{1}{y}\right) = 1 - \frac{1}{y}, \quad y > 1. \end{aligned}$$

Solving the equation $F_Y(m) = \frac{1}{2}$ gives the median $m = 2$. On average, half the observations are less than 2 and half are larger. Values that are larger have a good chance to be much larger, in fact so much that the mean is infinite. □

For a data set, the median is defined as the value that is in the middle. If the number of observations is even, it is the average of the two middle values (note that the median for a data set is always unique). Consider the following example.

***Example 2.47.*** A survey of salaries at a small company with 11 employees gives the following values: 30, 30, 30, 30, 30, 40, 40, 60, 70, 150, 220 ($\times 1000$ dollars per year). This gives a mean salary of 66, 364 and a median of 40, 000 dollars per year. If the manager doubles his salary from 220, 000 to 440, 000, the mean goes up to over 86, 000, whereas the median stays the same. It is probably fair to say that the median better represents the data set. □

The distribution in the example is skewed, and in such a case it is common to give the median instead of the mean. Also note how the median is less sensitive to extreme values; it does not change with the manager's doubled salary, whereas the mean increases significantly. Yet another location parameter is the following.

***Definition 2.22.*** Let $X$ be a random variable with pmf or pdf $f$. Any number $x_m$ with the property

$$f(x_m) \geq f(x) \quad \text{for all } x \in R$$

is called a **mode** of $X$.

Whereas the mean is the average and the median the midpoint, the interpretation of the mode is that it is the *most likely value* of $X$. It need not be unique in either the discrete or continuous case. For example, for a uniform distribution, any number in the range is a mode. We can also define *local modes*, which have the property that the pdf or pmf has a local but not necessarily global maximum. Depending on the number of such local modes, distributions can be classified as unimodal, bimodal, and so on. For example, the pdf in Figure 2.5 is bimodal.

***Example 2.48.*** Let $X \sim \exp(\lambda)$. Find the mean, median, and mode of $X$.

We already know that the mean is $\mu = 1/\lambda$. For the median, we need to solve the equation

$$1 - e^{-\lambda m} = \frac{1}{2}$$

which gives

$$m = \frac{\log 2}{\lambda} \approx 0.69\mu$$

which means that the median is smaller than the mean for the exponential distribution. The intuition behind this is that the range above the mean is infinite and large values tend to push the average up. This distribution is said to be skewed to the right. The mode $x_m$ is the $x$ value that maximizes the pdf

$$f(x) = \lambda e^{-\lambda x}, \quad x \geq 0$$

which gives $x_m = 0$. It seems strange that 0 is the "most likely value," but since this is a continuous distribution, we need to think of the mode as the value that observations are most likely to be near, not exactly equal to. □

For a data set, the mode is simply the most frequent value, which may not be unique.

***Example 2.49.*** In the example of salaries above, the mode is 30, 000, the "typical" salary for an employee. Like the median, it does not change with the manager's doubled salary. □

## 2.10 THE FAILURE RATE FUNCTION

In this section we study nonnegative, continuous random variables, with the interpretation that they are lifetimes of some object. We use the notation $T$ for "time" rather than $X$, and the argument of cdfs and pdfs is $t$.

***Definition 2.23.*** Let $T$ be a random variable with cdf $F$ and pdf $f$. The *failure rate function* is defined as

$$r(t) = \frac{f(t)}{1 - F(t)}, \quad t \geq 0$$

Some alternative terms for $r(t)$ are *hazard rate function* and *death rate function*. The interpretation of $r(t)$ is that it is a measure of how likely failure is at time $t$ of an object that is already $t$ time units old. Compare this with the pdf $f(t)$, which is a measure of how likely failure is at $t$ of a brand new object. Hence, the failure rate is a *conditional* measure of failure and the pdf an unconditional such measure.

This is easier to understand if we instead think of a discrete case. Consider a human lifespan measured in years, and let $t = 100$ years. The pmf $p(100)$ is the probability that a newborn individual dies at age 100, which is pretty low. The failure rate $r(100)$, on the other hand, is the conditional probability that an individual who is already 100 dies before turning 101, which is much higher. With $L$ denoting lifespan, we thus have

$$p(100) = P(L = 100)$$
$$r(100) = P(L = 100 | L \geq 100)$$

Indeed, as $t$ gets large, we have $p(t)$ going to 0 but $r(t)$ going to 1. As we discussed in Example 1.3, it is convenient to model lifetimes as continuous, even though we always have a discrete timescale in practice. We will therefore treat only the continuous case, but see Problem 107 for more on the discrete failure rate function.

***Example 2.50.*** Let $T \sim \exp(\lambda)$. Find the failure rate function of $T$.

From the definition, we get

$$r(t) = \frac{f(t)}{1 - F(t)} = \frac{\lambda e^{-\lambda t}}{1 - (1 - e^{-\lambda t})} = \lambda, \quad t \geq 0$$

so the exponential function has constant failure rate. This was to be expected since we introduced the exponential distribution as the lifetime of something that does not age. □

***Example 2.51.*** Let $T \sim \Gamma(2, 1)$. Find the failure rate function of $T$.

From Section 2.8.2, we have

$$f(t) = e^{-t}t, \quad t \geq 0$$

and

$$F(t) = 1 - (1 + t)e^{-t}, \quad t \geq 0$$

which gives the failure rate function

$$r(t) = \frac{f(t)}{1 - F(t)} = \frac{t}{1 + t}, \quad t \geq 0$$

□

Note how $r(t)$ in Example 2.51 is an increasing function of $t$. This means that failure becomes more and more likely with time; that is, unlike the exponential distribution, aging is present. This is an example of a distribution with *increasing failure rate* (IFR). We could also imagine that aging is beneficial so that failure is most likely early on, and becomes less likely as time passes. This is the case of *decreasing failure rate* (DFR). Constant failure rate is appropriately labeled CFR.

In reality, most failure rate functions are combinations. Consider, for example, a human life. There is certainly an elevated risk of death both at and shortly after birth. After a couple of years, the risk of death is fairly constant, mainly due to accidents and other unpredictable causes. Then aging starts to kick in (around the time you start

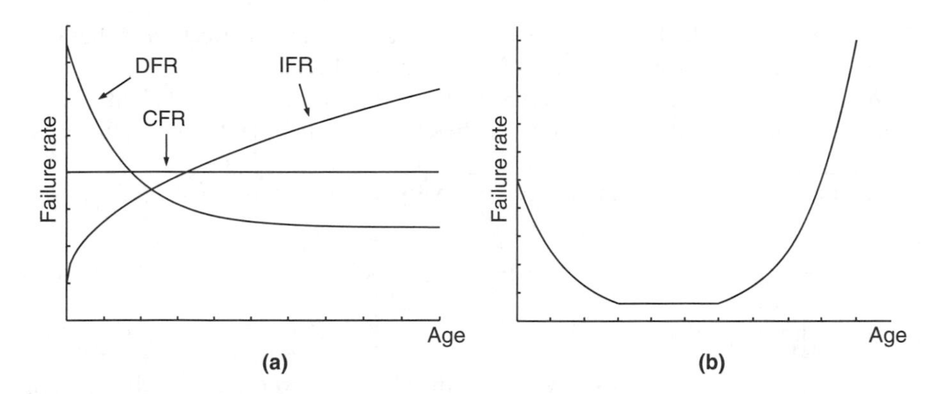


**FIGURE 2.16** Different types of failure rate functions: (a) decreasing, constant, and increasing; (b) bathtub-shaped.

college), and the risk of dying starts to increase. This is typically true for mechanical objects as well. First, there is a certain "break-in" period during which, for example, manufacturing errors may cause failure. After this is over, there is a period of fairly constant failure rate, and after a while material fatigue starts to have an impact, thus causing the failure rate to increase. Such failure rate functions are said to be "bathtub-shaped," as shown in Figure 2.16.

When the standards of living in different countries are compared, one summary statistic that is often given is the *life expectancy*. In our terms, this is the expected value of the lifetime of a newborn individual, and it varies greatly between first world and third world countries. For example, life expectancy in Japan is 81 years, and in Angola, only 37 years. What does this mean? Are there no old people in Angola? One important explanatory factor is the *infant mortality*, measured in deaths per thousand live births, or in our terminology, the probability that a newborn child survives. In Japan this number is 3.3 and in Angola 193.8! Thus, the probability that a newborn Angolan does not survive its first year is almost 0.2, and this brings the expected lifetime down significantly. Instead of just comparing the expected lifetimes, we can compare the failure rate functions, which would be more informative, and we would then notice a large gap in the beginning, corresponding to infant mortality.

If we still want a single number to describe lifetime, rather than the entire failure rate function, it might be more appropriate to use the median from Section 2.9 since this is less sensitive to a skewed distribution. We discussed human lifespans above, but the same problems are present in any kind of lifetime analysis, or, as it is more commonly termed, *survival analysis*.

### 2.10.1 Uniqueness of the Failure Rate Function

As we will soon see, the failure rate function uniquely determines the distribution. This is convenient, since when we model lifetimes, it is much easier to figure out what the failure rate function should look like, than it is to describe the cdf or pdf.

For example, the bathtub shape shown in Figure 2.16b makes perfect sense, but it is not easy to figure out what this means for the pdf or cdf.

When lifetimes are studied, it is often convenient to consider the probability that $T$ is greater than $t$ instead of less than $t$. We therefore introduce the following function.

***Definition 2.24.*** Let $T$ be a nonnegative, continuous random variable with cdf $F$. The function

$$G(t) = 1 - F(t), \quad t \geq 0$$

is called the *survival function* of $T$.

Thus, the survival function is decreasing with $G(0) = 1$ and $G(\infty) = 0$. In terms of $G$, the failure rate function is

$$r(t) = \frac{f(t)}{G(t)}$$

and we have the following relation.

**Proposition 2.25.** Let $T$ be a nonnegative, continuous random variable with failure rate function $r$. Then $T$ has survival function

$$G(t) = \exp\left(-\int_0^t r(u)du\right), \quad t \geq 0$$

*Proof.* First note that

$$r(u) = \frac{f(u)}{1 - F(u)} = -\frac{d}{du}\log(1 - F(u))$$

and integrate both sides to obtain

$$\int_0^t r(u)du = -\log(1 - F(t)) = -\log G(t)$$

which gives

$$G(t) = \exp\left(-\int_0^t r(u)du\right)$$

and we are done. We have implicitly used the boundary condition $F(0) = 0$, which always holds for a nonnegative continuous random variable. ■

***Example 2.52.*** In the late 1990s, it was reported that flu shots cut the death rate by half among elderly people in the United States and Canada. Suppose that an unvaccinated 70-year-old person has probability 0.8 to survive another 10 years. How much (in percent) does this probability increase for a vaccinated person?

Denote the lifespan for an unvaccinated person by $U$ and a vaccinated person by $V$. Denote the corresponding failure rate functions and cdfs by $r_U$, $r_V$ and $F_U$, $F_V$, respectively. Hence, $r_U(t) = 2\,r_V(t)$, and the desired probability is

$$\begin{aligned}
P(V > 80 | V > 70) &= \frac{P(V > 80)}{P(V > 70)} \\
&= \exp\left(-\int_0^{80} r_V(u)du\right) \Big/ \exp\left(-\int_0^{70} r_V(u)du\right) \\
&= \exp\left(-\int_{70}^{80} r_V(u)du\right) = \exp\left(-\frac{1}{2}\int_{70}^{80} r_U(u)du\right) \\
&= \sqrt{P(U > 80 | U > 70)} = \sqrt{0.8} \approx 0.894
\end{aligned}$$

which constitutes a 12% increase in the 10-year survival probability. □

***Example 2.53.*** Consider a certain type of ball bearing that wears down gradually at a slow but steady rate. To model this, we suppose that its failure rate function is linearly increasing, $r(t) = at$ for some $a > 0$. It has been observed that the median lifetime is 3 years. What is the probability that such a ball bearing lasts for more than 4 years?

Let $T$ be the lifetime. Then $T$ has survival function

$$G(t) = \exp\left(-\int_0^t au\,du\right) = \exp\left(-\frac{at^2}{2}\right)$$

and since the median $m = 3$ satisfies $G(3) = \frac{1}{2}$, we get $\exp(-9a/2) = \frac{1}{2}$, which gives $a = 2\log 2/9$ and

$$P(T > 4) = \exp\left(-\frac{2\log 2 \times 4^2}{18}\right) \approx 0.29$$

□

If we differentiate $F(t) = 1 - G(t)$ in the previous example, we get the pdf

$$f(t) = ate^{-at^2/2}, \quad x \geq 0$$

which is a special case of the following distribution.

***Definition 2.25.*** If $T$ has pdf

$$f(t) = \lambda\alpha\, t^{\alpha-1} e^{-\lambda t^\alpha}, \quad t \geq 0$$

it is said to have a *Weibull distribution* with parameters $\alpha > 0$ and $\lambda > 0$.

The parameter $\alpha$ is called the *shape parameter* and $\lambda$, the *scale parameter*. In the example, we had $\lambda = a/2$ and $\alpha = 2$. The cdf of the Weibull distribution is

$$F(t) = 1 - e^{-\lambda t^{\alpha}}, \quad t \geq 0$$

and the failure rate function

$$r(t) = \lambda\alpha\, t^{\alpha-1}, \quad t \geq 0$$

Thus, by choosing different values of $\alpha$ we can model increasing failure rate ($\alpha > 1$), decreasing failure rate ($\alpha < 1$), or constant failure rate ($\alpha = 1$, exponential distribution). This makes the Weibull distribution one of the most commonly used distributions for lifetime modeling. Note how $\alpha$ determines the shape of the failure rate function, and how the other parameter, $\lambda$, can be varied to change the scale, for example, to change between different units of measure.

***Example 2.54.*** Let $T \sim \text{unif}[0, 1]$. Find the failure rate function of $T$.

This is straightforward. Since $f(t) = 1$ and $F(t) = t$, we get

$$r(t) = \frac{1}{1-t}, \quad 0 \leq t \leq 1$$

where we notice that $r(t) \to \infty$ as $t$ approaches 1. If we think of this as a lifetime distribution, then failure will always occur before age 1. Thus, as the age approaches 1, failure becomes more and more likely, which is reflected in the failure rate going to infinity. Compare with the argument for human lifespans in the introduction, but remember that $r(t)$ is now a rate and not a probability. □

## PROBLEMS

### Section 2.2. Discrete Random Variables

1. The discrete random variable $X$ has cdf $F$ that is such that $F(x) = 0, x < 1$, $F(x) = \frac{1}{3}, 1 \leq x < 3$, and $F(x) = 1, x \geq 3$. Find **(a)** $F(2)$, **(b)** $P(X > 1)$, **(c)** $P(X = 2)$, and **(d)** $P(X = 3)$.
2. The discrete random variable $X$ has cdf

$$F(x) = \begin{cases} 0 & \text{for } x < 1 \\ 1/4 & \text{for } 1 \leq x < 2 \\ 3/4 & \text{for } 2 \leq x < 3 \\ 1 & \text{for } x \geq 3 \end{cases}$$

Find **(a)** $P(X = 1)$, **(b )** $P(X = 2)$, **(c)** $P(X = 2.5)$, and **(d)** $P(X \leq 2.5)$

3. Roll two dice and find the pmf of $X$ if $X$ is **(a)** the smallest number and **(b)** the difference between the largest and the smallest numbers.
4. The random variable $X$ has pmf $p(k) = ck$, $k = 1, 2, 3$. Find **(a)** the constant $c$, **(b)** the cdf $F$, **(c)** $P(X \leq 2)$, and **(d)** $P(X > 1)$.
5. The random variable $X$ has pmf $p(k) = c/2^k$, $k = 0, 1, \ldots$ Find **(a)** the constant $c$, **(b)** $P(X > 0)$, and **(c)** the probability that $X$ is even.
6. Five cards are drawn at random from a deck of cards. Let $X$ be the number of aces. Find the pmf of $X$ if the cards are drawn **(a)** with replacement and **(b)** without replacement.
7. A fair coin is flipped twice. Let $X$ be the number of heads minus the number of tails. Find the pmf and sketch the cdf of $X$.
8. Cards are drawn one by one at random from a deck of cards. Let $X$ be the number of draws needed to draw the ace of spades. Find the pmf of $X$ if we draw **(a)** with replacement and **(b)** without replacement.
9. Let the random variable $X$ have pmf $p(k) = 1/2^k$, $k = 1, 2, \ldots$, and let $Y = 1/X$. Find the cdf of $Y$.

## Section 2.3. Continuous Random Variables

10. Let $X$ and $Y$ be nonnegative random variables with the same cdf $F$. Show that, for any $x \geq 0$

$$1 - 2F(x) \leq P(X + Y > 2x) \leq 2(1 - F(x))$$

*Hint:* Problem 13, Chapter 1.

11. The concept of a random variable can be extended to allow for infinite values as was mentioned in the text following Proposition 2.3. Suppose that $X$ is a nonnegative random variable such that $P(X = \infty) = p > 0$. Show that $F(x) \to 1 - p$ as $x \to \infty$.
12. Let $f$ be the pdf of a continuous random variable $X$. Is it always true that $f(x) \leq 1$ for all $x$ in the range of $X$? Does it have to be true for some $x$ in the range of $X$?
13. The function $f$ is defined as $f(x) = cx^2$, $0 \leq x \leq 1$. **(a)** Determine the constant $c$ so that this becomes a pdf of a random variable $X$. **(b)** Find the cdf and compute $P(X > 0.5)$. **(c)** Let $Y = \sqrt{X}$ and find the pdf of $Y$.
14. Which of the following functions are possible pdfs for continuous random variables? For those that are, also find the cdf.
    **(a)** $f(x) = |x|$, $-1 \leq x \leq 1$ **(b)** $f(x) = \frac{3}{2}(x^2 - 1)$, $0 \leq x \leq 2$
    **(c)** $f(x) = 1$, $-1 \leq x \leq 0$ **(d)** $f(x) = 1/x^2$, $x \geq 1$.
15. Consider the function $f(x) = ax + b$, $0 \leq x \leq 1$. For which values of $a$ and $b$ is this a possible pdf for a continuous random variable?
16. The random variable $X$ has pdf $f(x) = 1/x^3$, $x \geq a$. Find **(a)** the value of $a$, **(b)** $P(X > 3)$, and **(c)** a value $x$ such that $P(X > x) = \frac{1}{4}$.

17. Suppose that the rate of growth of striped bass (*Morone saxatilis*) is constant between the ages of 5 and 15 years. Let $A$ be the age and $W$ the weight in pounds of a randomly sampled bass from this age interval, and suppose that $W = 2A + 3$. Let the pdf of $A$ be

$$f_A(x) = a(20 - x), \quad 5 \leq x \leq 15$$

Find **(a)** the constant $a$ and **(b)** the pdf of $W$.

18. The continuous random variable $X$ has pdf

$$f(x) = \begin{cases} 1/2 & \text{if } 0 < x \leq 1 \\ 1/(2x^2) & \text{if } x \geq 1 \end{cases}$$

**(a)** Show that this is a possible pdf of a continuous random variable. **(b)** Find the cdf of $X$ and sketch its graph. **(c)** Let $Y = 1/X$. Find the pdf of $Y$.

19. Let $X \sim \text{unif}(0, 1)$. What is the distribution of $1 - X$?
20. Let $X \sim \text{unif}(0, 1)$. Find the pdfs and sketch their graphs of the following random variables (be careful with the ranges): **(a)** $Y = 1/X$, **(b)** $Y = 1/\sqrt{X}$, and **(c)** $Y = \log X$.
21. Let $X$ have pdf $f(x) = e^{-x}, x \geq 0$. Find the pdf of $e^{-X}$. What is this distribution called?
22. Let $X$ have pdf $f(x) = e^{-x}, x \geq 0$, and let $Y = \sqrt{X}$. Find the pdf of $Y$.
23. We have seen that if $X$ is continuous and $g$ is a function, $g(X)$ can be discrete. Is it possible to have discrete $X$ and continuous $g(X)$?

## Section 2.4. Expected Value and Variance

24. Roll two dice and find $E[X]$ if $X$ is **(a)** the smallest number and **(b)** the difference between the largest and the smallest numbers.
25. **(a)** Consider a randomly chosen family with three children. What is the expected number of daughters? **(b)** Consider a randomly chosen girl who has two siblings. What is her expected number of sisters? (Recall Example 1.14.)
26. Draw three cards without replacement from a deck of cards and let $X$ be the number of spades drawn. Find $E[X]$.
27. One hundred people are to have their blood drawn to be tested for a disease. Instead of immediately analyzing each individual sample, a pooled sample of all individuals is analyzed first. If the pooled sample is negative, everybody is declared healthy; if it is positive, each individual blood sample is analyzed. Suppose that individuals have the disease with probability $p$, independent of each other. Let $X$ be the number of blood sample analyses that are done; find $E[X]$ and for which value of $p$ this is $\leq 100$.
28. In a "street bet" in roulette you bet on three numbers. If any of these come up, you win 11 times your wager, otherwise lose your wager. Let $X$ be your gain if you bet one dollar on a street bet. Find the mean and variance of $X$.

29. In a "five number bet" in roulette, you win if any of the numbers 00, 0, 1, 2, or 3 comes up. **(a)** In order to get the usual expected gain of $-2/38$, what should the payout be if you wager a dollar? **(b)** The actual payout on this bet is 6:1. What is your expected gain?
30. The game of chuck-a-luck is played with three dice, rolled independently. You bet one dollar on one of the numbers 1 through 6 and if exactly $k$ of the dice show your number, you win $k$ dollars $k = 1, 2, 3$ (and keep your wagered dollar). If no die shows your number, you lose your wagered dollar. What is your expected loss?
31. You are given two envelopes containing money and are told that one contains twice as much as the other. You choose one at random, open it, and find \$100. You are now given the options to either keep the money or switch to the other envelope and take what is inside. Since you chose at random, you figure that the other envelope with equal probabilities contains \$50 or \$200. If you switch and it contains \$50, you lose \$50 and if it contains \$200, you gain \$100. Since the average of $-50$ and 100 is 25, you figure that your expected gain is positive and that it makes sense to switch. Now you realize that you would reach the same conclusion regardless of the amount you found in the first envelope, so you did not even have to open it, just take it and immediately switch to the other. So, you might as well just pick the other envelope to start with. But then, by the same argument, you should switch to the first! This obviously does not make any sense. Where is the error?
32. In the island problem in Example 1.50, let $X$ be the number of individuals on the island with the genotype, let $p_k = P(X = k)$, $k = 0, 1, \ldots, n$, and let $\mu = E[X]$. **(a)** Suppose that we know that the murderer has the genotype. Show that the conditional distribution of $X$ given this fact is given by

$$\widehat{p}_k = \frac{kp_k}{\mu}, \quad k = 0, 1, \ldots, n$$

and explain why this is called a *size-biased* distribution. **(b)** Suppose that we have an old report stating that an islander has the genotype but that the name has been erased, and that we also find out that Mr Bloggs has it (so that we have three observations of the genotype, which could be from one, two, or three individuals). Find the conditional distribution of $X$, and show that the probability that Mr Bloggs is guilty is $E[X^2]/E[X^3]$. **(c)** Generalize (b) to $j$ independent observations of the genotype.
33. There are $n$ families living in a neighborhood. Of these, $n_k$ have $k$ children, where $k = 0, 1, \ldots, 4$. Let $p_k = n_k/n$, the probability that a randomly chosen family has $k$ children and let $\mu$ be the mean number of children. You observe a child playing in the street. Show that the probability that this child belongs to a family with $k$ children is $kp_k/\mu$. Explain how both this and Problem 32(a) can be explained by a "balls in urns" model.
34. You bid on an object at a silent auction. You know that you can sell it later for \$100 and you estimate that the maximum bid from others is uniform on

[70, 130] (for convenience, you assume that it is continuous, thus disregarding the possibility of two equal bids). How much should you bid to maximize your expected profit, and what is the maximum expected profit?

35. A stick measuring one yard in length is broken into two pieces at random. Compute the expected length of the longest piece.
36. A European roulette table has the numbers 1–36 plus 0 (but no 00), and the payout is the same as for the American table. What is the mean and variance of the gain if you **(a)** bet on odd and **(b)** make a straight bet?
37. Compute the means and variances of $A$ and $W$ in Problem 17.
38. The random variable $X$ has pdf $f(x) = 3x^2,\ 0 \le x \le 1$. **(a)** Compute $E[X]$ and $\text{Var}[X]$. **(b)** Let $Y = \sqrt{X}$ and compute $E[Y]$ and $\text{Var}[Y]$.
39. Let $X \ge 0$ be continuous. Show that $E[X] = \int_0^\infty P(X > x)dx$ and $E[X^2] = 2\int_0^\infty xP(X > x)dx$. *Hint:* $x = \int_0^t dt$ and $x^2 = 2\int_0^x t\,dt$.
40. Let $X \sim \text{unif}(0, 1)$, let $Y = 1/X$, and let $Z = 1/\sqrt{X}$. Compute the expected values and variances of $Y$ and $Z$ where possible.
41. Let $R \sim \text{unif}[0, 1]$ $V$ be the volume of a sphere with radius $R$. Compute $E[V]$ and $\text{Var}[V]$.
42. Let $X \sim \text{unif}[0,2]$ and let $V$ be the volume of a cube with side $X$. Find $E[V]$ and $\text{Var}[V]$.
43. The continuous random variable $X$ has pdf $f(x) = 2x,\ \ 0 \le x \le 1$. Find **(a)** $E[X]$ and $\text{Var}[X]$, **(b)** the expected volume of a sphere with radius $X$.
44. The random variable $X$ has pdf $f(x) = c \sin x,\ 0 \le x \le \pi$. Find **(a)** the constant $c$, **(b)** the cdf $F(x)$ of $X$, and **(c)** $E[\csc X]$.
45. Let $X$ have mean $\mu$ and variance $\sigma^2$. **(a)** What are the mean and variance of $-X$? Explain intuitively. **(b)** Find constants $a$ and $b$ such that the random variable $Y = aX + b$ has mean 0 and variance 1 (this is called *standardization*).
46. The *coefficient of variation* for a nonnegative random variable is defined as $c = \sigma/\mu$. **(a)** Let $X \sim \text{unif}[a, b]$ and find $c$. **(b)** In (a), if $a = n$ and $b = n + 1$, what happens to $c$ as $n \to \infty$?
47. Let $X$ be a nonnegative random variable with mean $\mu$ and variance $\sigma^2$. Prove the following inequalities and compute the bounds they give in Example 2.29.
**(a)** *Markov inequality:* $P(X \ge c) \le \mu/c$
**(b)** *One-sided Chebyshev inequality:* $P(X \ge \mu + a) \le \sigma^2/(\sigma^2 + a^2)$
48. Prove Proposition 2.16 by letting $c\sigma = 1/k$ in Chebyshev's inequality, and apply Proposition 1.6.

## Section 2.5. Special Discrete Distributions

49. Let $A$ and $B$ be events. Show that **(a)** $I_A^2 = I_A$, **(b)** $I_{A^c} = 1 - I_A$, **(c)** $I_{A\cap B} = I_A I_B$, and **(d)** $I_{A\cup B} = I_A + I_B - I_{A\cap B}$.
50. Suppose that the probability of a rainy day in Seattle in December is 0.8. If a day is not rainy, call it sunny. In which of the following cases is it reasonable

to assume a binomial distribution? Argue why/why not and give the parameter values where you have a binomial distribution. **(a)** You count the number of rainy days on Christmas Eve for 10 consecutive years. **(b)** You count the number of rainy days in December next year. **(c)** You count the number of rainy days on the first of each month for a year. **(d)** You count the number of sunny days on Christmas Eve for 10 consecutive years.

51. The random variable $X$ has a binomial distribution with $E[X] = 1$ and $\text{Var}[X] = 0.9$. Compute $P(X > 0)$.

52. Roll a die 10 times. What is the probability of getting **(a)** no 6s, **(b)** at least two 6s, and **(c)** at most three 6s.

53. Let $X$ be the number of 6s when a die is rolled six times, and let $Y$ be the number of 6s when a die is rolled 12 times. Find **(a)** $E[X]$ and $E[Y]$ and **(b)** $P(X > E[X])$ and $P(Y > E[Y])$.

54. A fair coin is flipped $n$ times. What is the probability of getting a total of $k$ heads if **(a)** the first flip shows heads, **(b)** the first flip shows tails, and **(c)** at least one flip shows heads?

55. Ann and Bob flip a fair coin 10 times. Each time it shows heads, Ann gets a point; otherwise Bob gets a point. **(a)** What is the most likely final result? **(b)** Which is more likely: that it ends 5–5 or that somebody wins 6–4? **(c)** If Ann wins the first three rounds, what is the probability that she ends up the winner? **(d)** If Ann wins the first four rounds, what is the probability that Bob never takes the lead? **(e)** What is the probability that the lead changes four times?

56. A multiple-choice test consists of six questions, each with four alternatives. At least four correct answers are required for a passing grade. What is the probability that you pass if you **(a)** guess at random; **(b)** know the first three answers, and guess on the rest; **(c)** for each question know the correct answer with probability $\frac{1}{2}$, otherwise guess at random? **(d)** In (c), to ensure at least 95% certainty that you will pass, how high must the probability that you know an answer be? **(e)** For (a)–(c), find the mean and variance of the number of correct answers.

57. A restaurant has 15 tables, and it is known that 70% of guests who make reservations actually show up. To compensate for this, the restaurant has a policy of taking more than 15 reservations, thus running a risk to become overbooked. How many reservations can they take to limit this risk to at most 5%?

58. Let $X \sim \text{bin}(n, p)$ and $Y \sim \text{geom}(p)$. **(a)** Show that $P(X = 0) = P(Y > n)$. Explain intuitively. **(b)** Express the probability $P(Y \leq n)$ as a probability statement about $X$.

59. You flip each of $n$ coins repeatedly until it shows heads. Let $X$ be the number of coins that require at least five flips. Find $P(X = 0)$ and $E[X]$.

60. In Example 2.35, let $S$ be the number of sons and $D$ the number of daughters. Find $P(D > S)$, $P(D = S)$, and $P(D < S)$.

61. Consider a sequence of independent trials that result in either success or failure. Fix $r \geq 1$ and let $X$ be the number of trials required until the $r$th success. Show

that the pmf of $X$ is

$$p(k) = \binom{k-1}{r-1} p^r (1-p)^{k-r}, \quad k = r, r+1, \ldots$$

This is called a *negative binomial* distribution with parameters $r$ and $p$, written $X \sim \text{negbin}(r, p)$. What is the special case $r = 1$?

62. Each workday (Monday–Friday) you catch a bus from a street corner. You have estimated that you arrive too late and miss the bus on average every 5 days. Consider a new workweek. **(a)** What is the probability that the next missed bus will be on Friday? **(b)** You decide to start biking after you have missed the bus five times. What is the probability that this happens on Friday the next week?
63. The number of customers $X$ who call a certain toll-free number in a minute has a Poisson distribution with mean 2. A minute is classified as "idle" if there are no calls and "busy" otherwise. **(a)** What is the probability that a given minute is busy? **(b)** Let $Y$ be the number of calls during a busy minute. Find the pmf of $Y$ and $E[Y]$. **(c)** If a minute is idle, what is the expected number of busy minutes before the next idle minute? What assumptions are you making?
64. Insects of a certain type lay eggs on leaves such that the number of eggs on a given leaf has a Poisson distribution with mean 1. For any given leaf, the probability is 0.1 that it will be visited by such an insect, and leaves are visited independent of each other. **(a)** What is the probability that a given leaf has no eggs? **(b)** If a leaf is inspected and has no eggs, what is the probability that it has been visited by an insect? **(c)** If 10 leaves are inspected and none have any eggs, what is the probability that at least one leaf has been visited by an insect?
65. Prove the expressions for the variances of the binomial and the Poisson distributions by first finding useful recursive expressions for $k(k-1)p(k)$ and then using the fact that $E[X^2] = E[X(X-1)] + E[X]$.
66. **(a)** Flip a coin 10 times and let $X$ be the number of heads. Compute $P(X \leq 1)$ exactly and with the Poisson approximation. **(b)** Now instead flip four coins 10 times and let $X$ be the number of times you get four heads. Compute $P(X \leq 1)$ exactly and with the Poisson approximation. **(c)** Compare (a) and (b). Where does the approximation work best and why?
67. Do Example 1.22 with the binomial approximation to the hypergeometric distribution and compare with the exact probabilities.
68. Compare the variance of the hypergeometric distribution with the variance of the binomial approximation. Which is smaller, and why?
69. Let $X$ be hypergeometric with parameters $N$, $r$, and $n$. Argue that $X$ can be approximated by a Poisson distribution. What is required of the parameters for the approximation to be good.

### Section 2.6. The Exponential Distribution

70. Jobs arrive at a computer such that the time $T$ between two consecutive jobs has an exponential distribution with mean 10 s. Find **(a)** $\text{Var}[T]$, **(b)** $P(T \leq 5)$,

(c) the probability that the next job arrives within 5 s given that the last job arrived 25 s ago, **(d)** $P(T > E[T])$.

71. A large number of lightbulbs are turned on in a new office building. A year later, 80% of them still function, and 2 years later, 30% of the original lightbulbs still function. Does it seem likely that the lifetimes follow an exponential distribution?
72. Let $X \sim \exp(\lambda)$ and let $Y = \lambda X$. Show that $Y \sim \exp(1)$.
73. The element *nobelium* has a half-life of 58 min. Let $X$ be the lifetime of an individual nobelium atom. Find **(a)** $P(X > 30)$, **(b)**, $P(X \leq 60 | X > 30)$, and **(c)** $E[X]$ and $\text{Var}[X]$.
74. Let $T \sim \exp(\lambda)$ and let $X = [T] + 1$ ("$[\cdot]$" denoting integer part). Show that $X \sim \text{geom}(1 - e^{-\lambda})$ (success probability $1 - e^{-\lambda}$). If $T$ is the lifetime of a component, what could be the interpretation of $X$?
75. Prove Proposition 2.21 by first subtracting $G(y)$ and dividing by $x$ in the equation $G(x + y) = G(x)G(y)$. Solve the resulting differential equation.
76. You are at a post office where there are two clerks, you are being served by one of them, the other clerk serves another customer, and one customer is waiting in line. If service times are independent and $\exp(\lambda)$, what is the probability that you are the last of the three customers to leave the post office?

### Section 2.7. The Normal Distribution

77. Let $X$ have a normal distribution with mean $\mu = 200$ and standard deviation $\sigma = 10$. Find **(a)** $P(X \leq 220)$, **(b)** $P(X \leq 190)$, **(c)** $P(X > 185)$, **(d)** $P(X > 205)$, **(e)** $P(190 \leq X \leq 210)$, and **(f)** $P(180 \leq X \leq 210)$.
78. Let $X$ have a normal distribution with mean $\mu$ and standard deviation $\sigma$. **(a)** Find an expression involving $\Phi$ for the probability that $X$ is within $c$ standard deviations of its mean where $c$ is a positive real number. **(b)** If you want the probability in **(a)** to be 0.99, what value of $c$ do you get?
79. Suppose that heights in a population follow a normal distribution with mean 70 and standard deviation 3 (inches). Find the 10th and the 90th percentiles.
80. Suppose that $X \sim \text{N}(\mu, \sigma^2)$. Find the $Z$-score corresponding to the $X$-value $\mu + c\sigma$ where $c$ is any real number (negative or positive).
81. Two species of fish have weights that follow normal distributions. Species A has mean 20 and standard deviation 2; species B has mean 40 and standard deviation 8. Which is more extreme: a 24 pound A-fish or a 48 pound B-fish?
82. Jane takes a test and scores 80 points. The test results in her class follow a normal distribution with mean 70 and standard deviation 10. On the second test, the mean is instead 150 and the standard deviation is 30. How much must she score to do as well as on the first test?
83. The number of free electrons in a metal fragment is measured with a measurement error that is normally distributed. More specifically, if there are $n$ free electrons, the number $Y = n + X$ is recorded where $X$ has a normal

distribution with mean 0 and standard deviation $\sigma = 0.43$. If $Y$ is rounded to the nearest integer, what is the probability that we get the correct answer?

84. A manually operated stopwatch is used to clock times in a 100 m track race. If the true time of a runner is $t$ seconds, the watch will show the time $T = t + 0.1 + X$, where $X$ has a normal distribution with mean 0 and variance 0.01. **(a)** For a given $t$, what is the distribution of $T$? **(b)** If the true time $t$ is 11.5 s, what is $P(T \leq 11.5)$? **(c)** What is the probability that $T$ is within 0.05 s of the true time $t$?
85. Let $X \sim N(0, 1)$. Express the pdfs of the following random variables in terms of the standard normal pdf $\varphi(x)$: **(a)** $-X$, **(b)** $|X|$, **(c)** $X^2$, **(d)** $e^X$.
86. Let $H$ and $W$ be the height and weight of a randomly chosen individual. In the light of the previous problem, is it reasonable to assume that they are both normal?
87. A type of metal rod is desired to have a length of 100 cm. It is first cut crudely by a machine and then finished by hand. It is known that the machine gives lengths that are normal with mean $\mu$ and variance 2, where $\mu$ can be set by the operator. If a machine-cut rod is shorter than 100, it is wasted and if it is longer than 100, the excess material is wasted. **(a)** If the length of a rod is $X$, which function of $X$ describes the amount of waste? **(b)** How should $\mu$ be chosen in order to minimize the expected waste?
88. The random variable $X$ has cdf $F(x) = 1 - e^{-x^2}$, $x \geq 0$. Use properties of the normal distribution together with Proposition 2.10 to find $E[X]$.
89. If $X$ has a normal distribution, what is the probability that it deviates more than $k$ standard deviations from its mean for $k = 1, 2, \ldots$? Compare with the bounds given by Chebyshev's inequality.
90. Use the approximation formula given in Section 2.7 to find the probability that a randomly selected person has an IQ at least as high as Marilyn's (see Example 2.29).
91. Let $X \sim N(0, 1)$. Show that $P(X > x + \epsilon | X > x) \approx e^{-x\epsilon}$ for large $x$ and small $\epsilon$.

### Section 2.8. Other Distributions

92. Let $X$ be lognormal with parameters $\mu$ and $\sigma^2$. Find the distributions of $X^2$ and $X^3$.
93. Let $W$ and $H$ be the weight and height of a randomly chosen individual. If $H$ is lognormal, is it reasonable to assume that $W$ is also lognormal?
94. Let $X$ be lognormal with parameters $\mu = 0$ and $\sigma^2 = 1$. Find **(a)** $P(X \leq 2)$, **(b)** $P(X^2 \leq 2)$, **(c)** $P(X > E[X])$, and **(d)** the median $m$ of $X$.
95. Let $X \sim \Gamma(\alpha, \lambda)$, and let $Y = \lambda X$. Show that $Y \sim \Gamma(\alpha, 1)$.
96. Let $X$ have the Cauchy distribution. **(a)** Find the cdf of $X$. **(b)** Let $Y = 1/X$. Show that $Y$ also has the Cauchy distribution.

97. On my drive to work from home, I have a left turn where there is a stoplight. About 80% of the time it is red, in which case my waiting time until it turns green is uniform on (0,30) seconds. Let $X$ be my waiting time when I arrive at the stoplight. Find the cdf of $X$ and sketch its graph. Is $X$ discrete? Continuous? Why or why not?

98. A box contains equally many electronic components of two types, I and II. A type I component has a lifetime that is exp(1); a type II component has a lifetime that is exp(2). Consider a randomly chosen component and let $X$ be its lifetime. **(a)** Find the cdf of $X$. Is $X$ exponential? **(b)** Given that a component works after $t$ hours, what is the probability that it is of type I?

## Section 2.9. Location Parameters

99. For the following pdfs, find means, medians, and modes where possible. Compare them and argue which best represents the distribution. **(a)** $f(x) = 1/x^2,\ x \geq 1$, **(b)** $f(x) = \frac{1}{2},\ x \in [0,1] \cup [2,3]$, **(c)** $f(x) = 2(1-x),\ x \in [0,1]$, **(d)** $f(x) = -x,\ -1 \leq x \leq 0$ and $f(x) = x,\ 0 \leq x \leq 1$, and **(e)** $f(x) = 1/(\pi(1+x^2)),\ x \in R$.

100. For the following pairs of expressions, explain the difference in terms of location parameters: **(a)** "taller than average" and "taller than most people," **(b)** "typical new home price" and "average new home price," and **(c)** "a majority of the salaries are between \$50, 000 and \$60, 000" and "a plurality of the salaries are between \$50, 000 and \$60, 000."

101. Let $X$ be a random variable. If there exists a constant $a$ such that $P(X \geq a + x) = P(X \leq a - x)$ for all $x \geq 0$, $X$ is said to have a *symmetric* distribution. **(a)** Which of the following distributions are symmetric? (i) $N(0, 1)$, (ii) $\exp(\lambda)$, (iii) unif[0, 1], (iv) Poi($\lambda$), (v) bin($n, \frac{1}{2}$), and (vi) bin($n, p$), where $p \neq \frac{1}{2}$. **(b)** What does the pmf or pdf look like for a symmetric distribution? **(c)** Suppose that $X$ is continuous and has a symmetric distribution with mean $\mu$. Show that $a = \mu$, and that the median equals the mean (under the assumptions of Corollary 2.7).

102. Let $X$ have mean $\mu$ and variance $\sigma^2$. The *skewness* of $X$ is defined as $\text{skw}[X] = E[(X-\mu)^3]/\sigma^3$ and is a measure of the asymmetry of the distribution. Show that $\text{skw}[X] = (E[X^3] - 3\mu E[X^2] + 2\mu^3)/\sigma^3$ (compare Corollary 2.2).

103. Find skw[$X$] if **(a)** $X \sim \exp(\lambda)$ (*Note:* $E[X^3] = 6/\lambda^3$), **(b)** $X \sim \text{Poi}(\lambda)$ (*Note:* $E[X^3] = \lambda^3 + 3\lambda^2 + \lambda$). **(c)** What happens in (a) and (b) as $\lambda$ increases? Explain!

104. Let $X$ have mean $\mu$ and variance $\sigma^2$. The *kurtosis* of $X$ is defined as $\text{kur}[X] = E[(X-\mu)^4]/\sigma^4$ and is a measure of the peakedness of the pdf. Show that $\text{kur}[X] = (E[X^4] - 4E[X^3]\mu + 6E[X^2]\mu^2 - 3\mu^4)/\sigma^4$ (compare with Corollary 2.2).

105. Find kur[$X$] if **(a)** $X \sim N(\mu, \sigma^2)$ (*Note:* $E[Z^4] = 3$ if $Z \sim N(0, 1)$), **(b)** $S \sim \exp(\lambda)$ (*Note:* $E[X^4] = 24/\lambda^4$), **(c)** $X \sim$ unif[0, 1], and **(d)** $X \sim \text{Poi}(\lambda)$ (*Note:* $E[X^4] = \lambda^4 + 6\lambda^3 + 7\lambda^2 + \lambda$).

106. The kurtosis can be used to check for deviations from the normal distribution. For this purpose, the *excess kurtosis* can be defined as $\text{xkur}[X] = \text{kur}[X] - 3$. Find $\text{xkur}[X]$ for the distributions in the previous problem and interpret in terms of deviations from normality.

## Section 2.10. The Failure Rate Function

107. Let $X$ be a discrete random variable with range $\{0, 1, 2, \ldots\}$. The (discrete) failure rate function is then defined as

$$r(k) = \frac{P(X = k)}{P(X \geq k)}$$

**(a)** Show that $r(k) = P(X = k | X \geq k)$. **(b)** Let $X$ be the number when you roll a fair die. Find the failure rate function of $X$. Sketch the pmf and the failure rate function of $X$ and explain the difference between the two graphs. Also suggest an interpretation in terms of lifetimes.

108. Let $T$ be a nonnegative, continuous random variable. Express $E[T]$ as an integral that includes the failure rate function $r(t)$ but not the pdf, cdf, or survival function.

109. Find the failure rate function of the random variable $T$ if it has pdf **(a)** $f(t) = 2t, 0 \leq t \leq 1$, **(b)** $f(t) = 1/t^2, t \geq 1$, **(c)** $f(t) = 2t\exp(-t^2), t \geq 0$.

110. Find the cdfs of the nonnegative continuous random variables with the following failure rate functions: **(a)** $r(t) = 1/(1 + t)$, **(b)** $r(t) = 2t$, **(c)** $r(t) = t^3$, **(d)** $r(t) = e^{-t}$.

111. A certain type of lightbulb has failure rate function $r(t)$. The probability is 0.2 that such a lightbulb functions for more than 5000 h. Suppose that we want to double this probability by decreasing the failure rate function by a factor $c < 1$, to $cr(t)$. How should we choose $c$?

112. The lifetime $L$ of a certain insect has median 2 months and failure rate function

$$r(t) = at^2, \quad t \geq 0$$

Find **(a)** the constant $a$ and **(b)** the probability that a newborn insect lives at least 3 months.

113. The time $T$ in seconds between consecutive jobs arriving to a computer has failure rate function

$$r(t) = \begin{cases} 1 - t & \text{if } 0 \leq t \leq 1 \\ t - 1 & \text{if } t \geq 1 \end{cases}$$

Find **(a)** $P(0.5 \leq T \leq 1.5)$ and **(b)** the pdf $f$ and the median $m$ of $T$.

114. The lifetime $T$ in years of a lawn mower has failure rate function

$$r(t) = \begin{cases} 1/2 & \text{if } 0 \leq t \leq 2 \\ t/4 & \text{if } 2 \leq t \leq 6 \\ 9 & \text{if } t \geq 6 \end{cases}$$

Find **(a)** the probability that a new mower breaks down within 6 months, **(b)** the probability that a 2-year-old mower works for yet another year, **(c)** the probability that a 7-year-old mower breaks down within a month, **(d)** the median lifelength of a new mower, and **(e)** the median remaining lifelength of a 2-year-old mower.

115. The lifetime $T$ in hours of a type of electronic component is a continuous random variable with failure rate function $r(t)$ that equals the constant $c$ for $t \leq 100$. For $t \geq 100$, $r(t)$ is such that the failure rate doubles every 2 h and $r(t)$ is continuous at 100. Find **(a)** the expression for $r(t)$, **(b)** the value of $c$ if it is known that the median lifetime is 110 h, and **(c)** the probability that a lifetime is at most 100 h.

# 3

# JOINT DISTRIBUTIONS

## 3.1 INTRODUCTION

In the previous chapter, we introduced random variables to describe random experiments with numerical outcomes. We restricted our attention to cases where the outcome is a single number, but there are many cases where the outcome is a vector of numbers. We have already seen one such experiment, in Example 1.5, where a dart is thrown at random on a dartboard of radius $r$. The outcome is a pair $(X, Y)$ of random variables that are such that $X^2 + Y^2 \leq r^2$. For another example, suppose that we measure voltage and current in an electric circuit with known resistance. Owing to random fluctuations and measurement error, we can view this as an outcome $(V, I)$ of a pair of random variables.

These examples have in common that there is a relation between the random variables that we measure, and by describing them only one by one, we do not get all the possible information. The dart coordinates are restricted by the board, and voltage and current are related by Ohm's law. In this chapter, we extend the notion of random variables to random vectors.

## 3.2 THE JOINT DISTRIBUTION FUNCTION

We will focus primarily on random vectors in two dimensions. Let us begin by formally defining our fundamental object of study.

*Probability, Statistics, and Stochastic Processes*, Second Edition. Peter Olofsson and Mikael Andersson.

***Definition 3.1.*** Let $X$ and $Y$ be random variables. The pair $(X, Y)$ is then called a (two-dimensional) *random vector*.

Our aim is to describe random vectors in much the same way as we described random variables in Chapter 2, and we will thus define analogs of the cdf, pmf, and pdf. For a given random vector $(X, Y)$, we will often be interested in events of the type $\{X \in A\} \cap \{Y \in B\}$, that is, the event that $X$ belongs to the set $A$ and $Y$ at the same time belongs to the set $B$. To ease the notation, we write $P(X \in A, Y \in B)$ instead of $P(\{X \in A\} \cap \{Y \in B\})$. With this in mind, we can state the following definition.

***Definition 3.2.*** The *joint distribution function* (joint cdf) of $(X, Y)$ is defined as

$$F(x, y) = P(X \leq x, Y \leq y)$$

for $x, y \in R$.

The joint cdf is a function of two variables. It has properties similar to those of the one-dimensional cdf, but since it requires three dimensions to plot, it is more difficult to visualize. It also turns out to be somewhat less central for computation of probabilities than its one-dimensional analog. The reason for this is that in one dimension, virtually any event involving a random variable is of the form $\{X = x\}$ or $\{a < X \leq b\}$, and the probabilities of these events can be directly expressed in terms of the cdf according to Propositions 2.2 and 2.4.

In two dimensions, things become more difficult. The cdf in a point $(x, y)$ is the probability that $X \leq x$ and $Y \leq y$, that is, the probability that $(X, Y)$ is in the set $(-\infty, x] \times (-\infty, y]$, the infinite rectangle "southwest" of the point $(x, y)$. For a finite rectangle $B = (a, b] \times (c, d]$, it is easy to show (Problem 1) that

$$P(a < X \leq b, c < Y \leq d) = F(b, d) - F(b, c) - F(a, d) + F(a, c)$$

and we can find probabilities of other types of rectangular sets in a similar way. However, in two dimensions, there are many other types of sets. For example, if $C$ is the unit circle disk, $C = \{(x, y) : x^2 + y^2 \leq 1\}$, then we cannot express $P((X, Y) \in C)$ directly by the joint cdf. (Try it!) The same problems arise for events such as $\{X \leq Y\}$ or $\{|X - Y| > 1\}$. The plane $R^2$ is much more complicated than the nicely ordered real line.

If we are given the joint cdf $F$ of $(X, Y)$, we can obtain the individual cdfs of $X$ and $Y$ by noticing that the event $\{X \leq x\}$ is the same as the event $\{X \leq x, Y < \infty\}$, which puts no restrictions on $Y$. Since this has probability

$$P(X \leq x, Y < \infty) = F(x, \infty)$$

we get the following result.

**Proposition 3.1.** If $(X, Y)$ has joint cdf $F$, then $X$ and $Y$ have cdfs

$$F_X(x) = F(x, \infty) \quad \text{and} \quad F_Y(y) = F(\infty, y)$$

for $x, y \in R$.

The cdfs of $X$ and $Y$ are called the *marginal* cdfs of $(X, Y)$.

## 3.3 DISCRETE RANDOM VECTORS

Just as in the one-dimensional case, we make the distinction between discrete and continuous random vectors.

***Definition 3.3.*** If $X$ and $Y$ are discrete random variables, then $(X, Y)$ is called a *discrete random vector*.

If the range of $X$ is $\{x_1, x_2, \ldots\}$ and the range of $Y$ is $\{y_1, y_2, \ldots\}$, then the range of $(X, Y)$ is $\{(x_j, y_k), j, k = 1, 2, \ldots\}$, which is also a countable set. Hence, a discrete random vector $(X, Y)$ is characterized by a countable range, in the same way as is a discrete random variable. In an analogy with the one-dimensional case, we state the following definition.

***Definition 3.4.*** If $(X, Y)$ is discrete with range $\{(x_j, y_k) : j, k = 1, 2, \ldots\}$, the function

$$p(x_j, y_k) = P(X = x_j, Y = y_k)$$

is called the *joint pmf* of $(X, Y)$.

***Example 3.1.*** Roll a die twice. Let $X$ be the first and $Y$ the second number. Find the joint pmf of $(X, Y)$.

The range is $\{(j, k) : j = 1, 2, \ldots, 6, k = 1, 2, \ldots, 6\}$, and for any pair $(j, k)$, we get

$$\begin{aligned} p(j, k) &= P(X = j, Y = k) \\ &= P(X = j)P(Y = k) = \frac{1}{36} \end{aligned}$$

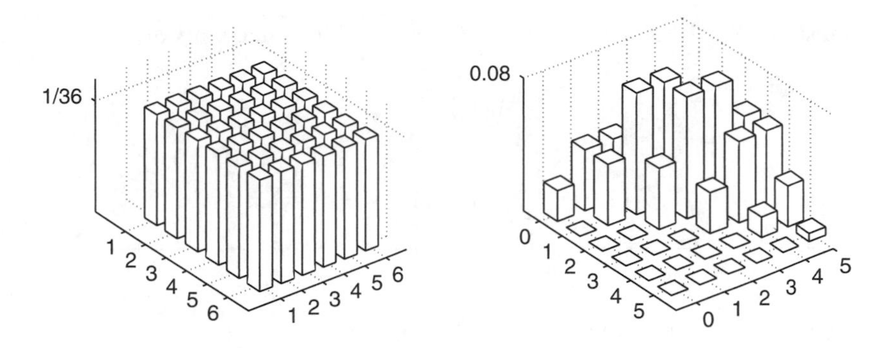


**FIGURE 3.1** Joint pmfs for Examples 3.1 and 3.2, respectively.

since the events $\{X = j\}$ and $\{Y = k\}$ are independent. We say that $(X, Y)$ has a uniform distribution on $\{(j, k) : j = 1, 2, \ldots, 6, k = 1, 2, \ldots, 6\}$. □

***Example 3.2.*** Phone calls arrive to a mail order company such that the number of phone calls in a minute has a Poisson distribution with mean 4. A given caller is female with probability 0.5, independent of other callers. In a given minute, let $X$ be the number of female callers and $Y$ the total number of callers. Find the joint pmf of $(X, Y)$.

Unlike the previous example, events pertaining to $X$ and $Y$ are not independent, which is clear since $X$ is always less than or equal to $Y$. However, we can use conditional probabilities by noting that if $Y = k$, then the number of female callers is binomial with parameters $k$ and 0.5, and we get

$$\begin{aligned} p(j, k) &= P(X = j | Y = k) P(Y = k) \\ &= \binom{k}{j} 0.5^j 0.5^{k-j} e^{-4} \frac{4^k}{k!} \\ &= e^{-4} \frac{2^k}{j!(k-j)!}, \quad 0 \leq j \leq k, \ k = 0, 1, 2, \ldots \end{aligned}$$

□

The joint pmfs for the two examples are depicted in Figure 3.1.

If we are given the pmf of $(X, Y)$, we might be interested in finding the marginal (one-dimensional) pmfs of the random variables $X$ and $Y$. Since the event $\{X = x_j\}$ can be described in terms of $(X, Y)$ as $\{X = x_j, \text{“}Y = \text{anything”}\}$, we get

$$p_X(x_j) = \sum_{k=1}^{\infty} P(X = x_j, Y = y_k) = \sum_{k=1}^{\infty} p(x_j, y_k)$$

with the obvious analog for $p_Y$. We have the following result.

**Proposition 3.2.** If $(X, Y)$ has joint pmf $p$, then the marginal pmfs of $X$ and $Y$ are

$$p_X(x_j) = \sum_{k=1}^{\infty} p(x_j, y_k), \quad j = 1, 2, \ldots$$

$$p_Y(y_k) = \sum_{j=1}^{\infty} p(x_j, y_k), \quad k = 1, 2, \ldots.$$

***Example 3.3.*** Find the marginal pmfs in Example 3.2.

From the problem it is given that $Y \sim \text{Poi}(4)$. To find the pmf of $X$, start from the joint pmf

$$p(j, k) = e^{-4} \frac{2^k}{j!(k-j)!}, \quad 0 \leq j \leq k, \ k = 0, 1, 2, \ldots$$

and note that for fixed $j$, $k$ ranges from $j$ to $\infty$. This gives

$$\begin{aligned} p_X(j) &= e^{-4} \sum_{k=j}^{\infty} \frac{2^k}{j!(k-j)!} \\ &= e^{-4} \frac{2^j}{j!} \sum_{k=j}^{\infty} \frac{2^{k-j}}{(k-j)!} = e^{-2} \frac{2^j}{j!}, \quad j = 0, 1, \ldots \end{aligned}$$

which means that $X \sim \text{Poi}(2)$. □

## 3.4 JOINTLY CONTINUOUS RANDOM VECTORS

***Definition 3.5.*** If there exists a function $f$ such that

$$P((X, Y) \in B) = \iint_B f(x, y) dx\, dy$$

for all subsets $B \subseteq R^2$, then $X$ and $Y$ are said to be *jointly continuous*. The function $f$ is called the *joint pdf*.

We may also say simply that the vector $(X, Y)$ is continuous. The notation $\iint_B$ means that we integrate over the two-dimensional region $B$.[1] In particular, the choice $B = \{(s, t) : s \leq x, t \leq y\} = (-\infty, x] \times (-\infty, y]$ gives

$$F(x, y) = \int_{-\infty}^{y} \int_{-\infty}^{x} f(s, t) ds\, dt$$

which also gives the following proposition.

**Proposition 3.3.** If $X$ and $Y$ are jointly continuous with joint cdf $F$ and joint pdf $f$, then

$$f(x, y) = \frac{\partial^2}{\partial x \partial y} F(x, y), \quad x, y \in R$$

The intuition is the same as in the one-dimensional case; $f(x, y)$ is a measure of how likely it is that $(X, Y)$ is in the neighborhood of the point $(x, y)$, not exactly equal to $(x, y)$. Again, the probability that $(X, Y)$ equals $(x, y)$ is 0 for all $x$ and $y$. Even more holds in the jointly continuous case. Since lines and curves in $R^2$ have area 0, integrating the joint pdf over them also gives 0 and hence the probability that $(X, Y)$ takes values on any fixed line or curve is 0. In particular, $P(X = Y) = 0$.

Since probabilities are always nonnegative and $P((X, Y) \in R^2) = 1$, we get the following analog of Proposition 2.6.

**Proposition 3.4.** A function $f$ is a possible joint pdf for the random variables $X$ and $Y$ if and only if

(a) $f(x, y) \geq 0$ for all $x, y \in R$

(b) $\displaystyle\int_{-\infty}^{\infty} \int_{-\infty}^{\infty} f(x, y) dx\, dy = 1$

***Example 3.4.*** The random vector $(X, Y)$ has joint pdf

$$f(x, y) = c(x + 2y), \quad 0 \leq x \leq 1, \ 0 \leq y \leq 1$$

Find **(a)** the constant $c$ and **(b)** the joint cdf $F(x, y)$.

[1] As in the one-dimensional case, some restrictions must in general be imposed on the class of possible subsets $B$, but we can again safely disregard this problem.

For part (a), we use Proposition 3.4. Clearly, $f \geq 0$ and since

$$\int_0^1 \int_0^1 f(x, y)dx\,dy = c \int_0^1 \int_0^1 (x + 2y)dx\,dy$$
$$= c \int_0^1 \left(\frac{1}{2} + 2y\right) dy = \frac{3c}{2}$$

we get $c = \frac{2}{3}$. For part (b), we get

$$F(x, y) = \int_0^y \int_0^x \frac{2}{3}(s + 2t)ds\,dt = \int_0^y \left(\frac{x^2}{3} + \frac{4xt}{3}\right) dt$$
$$= \left[\frac{tx^2}{3} + \frac{4xt^2}{6}\right]_{t=0}^{y} = \frac{1}{3}\left(x^2 y + 2xy^2\right), \quad 0 \leq x \leq 1,\ 0 \leq y \leq 1$$

□

In an analogy with the result regarding marginal pmfs, we have the following result in the jointly continuous case.

**Proposition 3.5.** Suppose that $X$ and $Y$ are jointly continuous with joint pdf $f$. Then $X$ and $Y$ are continuous random variables with marginal pdfs

$$f_X(x) = \int_{-\infty}^{\infty} f(x, y)dy, \quad x \in R$$

$$f_Y(y) = \int_{-\infty}^{\infty} f(x, y)dx, \quad y \in R$$

***Example 3.5.*** Find the marginal pdfs in Example 3.4.

The joint pdf is

$$f(x, y) = \frac{2}{3}(x + 2y), \quad 0 \leq x \leq 1,\ 0 \leq y \leq 1$$

which gives

$$f_X(x) = \frac{2}{3} \int_0^1 (x + 2y)dy$$
$$= \frac{2}{3}\left[xy + y^2\right]_{y=0}^{1} = \frac{2}{3}(x + 1), \quad 0 \leq x \leq 1$$

and

$$f_Y(y) = \frac{2}{3}\int_0^1 (x + 2y)dx$$

$$= \frac{2}{3}\left[\frac{x^2}{2} + 2xy\right]_{x=0}^1 = \frac{1}{3}(1 + 4y), \quad 0 \le y \le 1$$

□

***Example 3.6.*** Choose a point $(X, Y)$ at random in the unit disk $\{(x, y) : x^2 + y^2 \le 1\}$. Find the joint pdf and the marginal pdfs.

In an analogy with the one-dimensional case, "at random" means that the joint pdf must be constant on the unit disk (and 0 outside). Since the joint pdf must also integrate to 1 and the area of the unit disk is $\pi$, we realize that $(X, Y)$ has joint pdf

$$f(x, y) = \frac{1}{\pi}, \quad x^2 + y^2 \le 1$$

We call this a uniform distribution on the unit disk. Since the boundary has probability 0, we can replace "$\le$" by "$<$" if we wish, and thus it makes no difference whether we consider the closed or open disk (compare the one-dimensional uniform distribution).

What is the marginal distribution of $X$? First note that the range of $X$ is $[-1, 1]$. Is $X$ uniform on this range? To find the marginal pdf of $X$, fix $x \in [-1, 1]$ and compute

$$f_X(x) = \int_{-\infty}^{\infty} f(x, y)dy$$

The integral limits are determined by where the joint pdf $f(x, y)$ is strictly positive, which is where the point $(x, y)$ is inside the unit disk. For fixed $x$, this happens when $y$ is between $-\sqrt{1 - x^2}$ and $\sqrt{1 - x^2}$. Hence, we get

$$f_X(x) = \frac{1}{\pi}\int_{-\sqrt{1-x^2}}^{\sqrt{1-x^2}} dy = \frac{2}{\pi}\sqrt{1 - x^2}, \quad -1 \le x \le 1$$

and by symmetry, $Y$ has the same marginal distribution as $X$. Now, this is certainly not the pdf of a uniform distribution on $[-1, 1]$. To understand why $X$ is not uniform, consider an interval $I_h$ of a fixed length $h$ inside $[-1, 1]$. For a uniform distribution, the probability $P(X \in I_h)$ is the same regardless of where the interval is located, but in our case this probability is larger the closer the interval is to the origin. Remember that we are choosing a point uniformly in the unit disk, and getting $X$ in a specific interval corresponds to $(X, Y)$ being in a specific region. For a fixed interval length, this region is larger the closer the interval is to 0 (draw a figure). □

## 3.5 CONDITIONAL DISTRIBUTIONS AND INDEPENDENCE

Recall the concept of conditional probability from Section 1.5. The intuition behind $P(A|B)$ is that this is the probability of $A$ if we know that the event $B$ has occurred. Now suppose that the event $A$ is related to a discrete random variable $Y$, for example, $A = \{Y = y\}$. The conditional probability of the event $\{Y = y\}$ given $B$ can then be regarded as the *conditional pmf* of $Y$, given the event $B$, evaluated in the point $y$. If the event $B$ is also related to a discrete random variable, say, $X$, we can state the following definition.

***Definition 3.6.*** Let $X$ and $Y$ be discrete random variables with ranges $\{x_1, x_2, \ldots\}$ and $\{y_1, y_2, \ldots\}$, respectively, and joint pmf $p$. The *conditional pmf* of $Y$ given $X = x_j$ is defined as

$$p_Y(y_k|x_j) = \frac{p(x_j, y_k)}{p_X(x_j)} \quad \text{for } y_k \text{ in the range of } Y$$

We view this as a function of $y_k$ where the value $x_j$ is fixed. An application of the law of total probability gives

$$p_Y(y_k) = \sum_{j=1}^{\infty} p_Y(y_k|x_j)p_X(x_j)$$

for $y_k$ in the range of $Y$, which is the same formula as in Proposition 3.2. To get the conditional cdf, we compute the sum

$$F_Y(y|x_j) = \sum_{k:y_k \le y} p_Y(y_k|x_j)$$

for $y \in R$.

***Example 3.7.*** Consider Example 3.2, where $X$ is the number of female callers and $Y$ the total number. Find the conditional pmf of $Y$ given $X = j$.

We computed the joint pdf and found it to be

$$p(j, k) = e^{-4}\frac{2^k}{j!(k-j)!}, \quad 0 \le j \le k, \ k = 0, 1, 2, \ldots$$

and in Example 3.3, we found that $X \sim \text{Poi}(2)$. Conditioned on $X = j$, the range of $Y$ is $j, j+1, \ldots$ and we get the conditional pmf

$$p_Y(k|j) = \frac{e^{-4}\,2^k/(j!(k-j)!)}{e^{-2}\,2^j/j!} = e^{-2}\frac{2^{k-j}}{(k-j)!}, \quad k = j, j+1, \ldots$$

which means that, given that there are $j$ female callers, the total number of callers is $j$ plus a number that has a Poisson distribution with mean 2, which is of course the number of male callers. □

The continuous case is similar, but we have to be careful since pdfs are not probabilities. Let us first consider what we should mean by the conditional pdf of $Y$ given an event $B$. Intuitively, it is clear that this should be a pdf that is computed given the information that the event $B$ has occurred. If $Y$ were discrete, we would simply go ahead according to the definition above, but for continuous $Y$, the event $\{Y = y\}$ has probability 0 for all $y$, and we need to proceed differently. Since the pdf is not a probability but the cdf is, let us first define the *conditional cdf* as

$$F(y|B) = P(Y \le y|B) = \frac{P(\{Y \le y\} \cap B)}{P(B)}$$

and then the conditional pdf of $Y$ given $B$ as

$$f(y|B) = \frac{d}{dy}F(y|B)$$

This is fine as long as $B$ is an event with $P(B) > 0$, but what if $B$ is also related to a continuous random variable of the form $B = \{X = x\}$ so that $B$ has probability 0? Clearly, it still makes intuitive sense to talk about the "distribution of $Y$ given that $X = x$," but if $X$ is continuous, we cannot use the definition of conditional probability as we did above. Our definition of conditional pdf instead mimics its discrete counterpart, and we state the following.

***Definition 3.7.*** Let $(X, Y)$ be jointly continuous with joint pdf $f$. The *conditional pdf of $Y$ given $X = x$* is defined as

$$f_Y(y|x) = \frac{f(x, y)}{f_X(x)}, \quad y \in R$$

We interpret this as the pdf of $Y$ if we know that $X = x$. Note that $x$ is a fixed value and the argument of the pdf is $y$. To define the conditional cdf, we integrate the conditional pdf and get

$$F_Y(y|x) = P(Y \le y|X = x) = \int_{-\infty}^{y} f_Y(t|x)dt$$

for $y \in R$. More generally, for any set $B \subseteq R$, we have

$$P(Y \in B|X = x) = \int_B f_Y(y|x)dy$$

***Example 3.8.*** Consider again Example 3.6, where a point $(X, Y)$ is chosen uniformly in the unit disk. What is the conditional pdf of $Y$ given $X = x$?

The joint pdf is

$$f(x, y) = \frac{1}{\pi}, \quad x^2 + y^2 < 1$$

and the marginal pdf of $X$ is

$$f_X(x) = \frac{2}{\pi}\sqrt{1 - x^2}, \quad -1 \le x \le 1$$

For a fixed $x$, the range of $Y$ is $(-\sqrt{1-x^2}, \sqrt{1-x^2})$, and the conditional pdf is

$$f_Y(y|x) = \frac{f(x, y)}{f_X(x)} = \frac{1}{2\sqrt{1-x^2}}, \quad -\sqrt{1-x^2} \le y \le \sqrt{1-x^2}$$

But this functional expression does not depend on $y$, so we conclude that $f_Y(y|x)$ is constant as a function of $y$ (remember that $x$ is a fixed value). Hence, the conditional distribution of $Y$ given that $X = x$ is uniform on $(-\sqrt{1-x^2}, \sqrt{1-x^2})$, which we may write as

$$Y|X = x \sim \text{unif}\left(-\sqrt{1-x^2}, \sqrt{1-x^2}\right)$$

and by symmetry we also get

$$X|Y = y \sim \text{unif}\left(-\sqrt{1-y^2}, \sqrt{1-y^2}\right)$$

□

In the above example, we started by describing $(X, Y)$ and found the marginals and conditional distributions. It is also possible to go the other way, that is, to first define the marginal pdf of $X$ and then the conditional pdf of $Y$ given $X = x$. This is natural if the experiment is performed in stages, as the following example shows.

***Example 3.9.*** Choose a point $X$ uniformly on $[0, 1]$ and given $X = x$, choose $Y$ uniformly on $[0, x]$. Find the joint pdf of $(X, Y)$.

First note that the range of $(X, Y)$ is the triangle $\{(x, y) : 0 \le x \le 1, 0 \le y \le x\}$. The pdf of $X$ is

$$f_X(x) = 1, \quad 0 \le x \le 1$$

and the conditional pdf of $Y$ given $X = x$ is

$$f_Y(y|x) = \frac{1}{x}, \quad 0 \le y \le x$$

which gives joint pdf

$$f(x, y) = f_Y(y|x)f_X(x) = \frac{1}{x}, \quad 0 \le x \le 1, \ 0 \le y \le x$$

Note that this is not a uniform distribution on the triangle since values tend to be more likely near the origin. Think about why this is the case. □

The following proposition is a continuous version of the law of total probability.

**Proposition 3.6.** Let $X$ and $Y$ be jointly continuous. Then

(a) $f_Y(y) = \int_{-\infty}^{\infty} f_Y(y|x) f_X(x)dx, \quad y \in R$

(b) $P(Y \in B) = \int_{-\infty}^{\infty} P(Y \in B|X = x) f_X(x)dx, \quad B \subseteq R$

*Proof.* For (a), just combine Proposition 3.5 with the definition of conditional pdf and for (b), part (a) gives

$$
\begin{aligned}
P(Y \in B) &= \int_B f_Y(y)dy = \int_B \int_{-\infty}^{\infty} f_Y(y|x) f_X(x)dx\,dy \\
&= \int_{-\infty}^{\infty} \int_B f_Y(y|x) f_X(x)dy\,dx \\
&= \int_{-\infty}^{\infty} P(Y \in B|X = x) f_X(x)dx
\end{aligned}
$$

as desired. ■

Note again how intuition is quite clear. To find, for example, $P(Y \in B)$, we condition on a particular value, $X = x$, and compute $P(Y \in B|X = x)$. Then we compute a weighted average over all possible values of $X$ and use the pdf of $X$ to find the weights. Since $X$ is continuous, the averaging is done by an integral instead of a sum. In particular, with $B = (-\infty, y]$, we get

$$F_Y(y) = \int_{-\infty}^{\infty} F_Y(y|x) f_X(x)dx$$

Another version of the law of total probability that we state without proof can be helpful for computing probabilities of events involving both $X$ and $Y$.

**Proposition 3.7.** Let $X$ and $Y$ be jointly continuous. Then, for $B \subseteq R^2$

$$P((X, Y) \in B) = \int_{-\infty}^{\infty} P((x, Y) \in B | X = x) f_X(x) dx$$

Note how $P((x, Y) \in B | X = x)$ is in fact the probability of a statement about $Y$ alone, for fixed $x$. The usefulness of the proposition is best illustrated by an example.

***Example 3.10.*** Let $X \sim$ unif[0, 1], and given $X = x$, let $Y \sim$ unif$[0, x]$. Find $P(Y \leq X^2)$.

The region $B$ is the set $\{(x, y) : 0 \leq x \leq 1, 0 \leq y \leq x^2\}$, and we have

$$P((x, Y) \in B | X = x) = P(Y \leq x^2 | X = x) = x$$

since given $X = x$, $Y$ has cdf $P(Y \leq t | X = x) = t/x$. We get

$$P(Y \leq X^2) = P((X, Y) \in B) = \int_0^1 x\, dx = \frac{1}{2}$$

□

Part (b) in Proposition 3.6 is also valid if $Y$ is discrete and $X$ continuous. In this case, the pair $(X, Y)$ is neither jointly discrete nor jointly continuous, but it still makes sense to talk about conditional distributions. We illustrate this in an example.

***Example 3.11.*** Let $X \sim$ unif[0, 1] and given $X = x$, let $Y \sim$ geom$(x)$. Thus, we consider a geometric distribution where the success probability is chosen at random from [0, 1]. Find the distribution of $Y$.

The quickest way is to note that $P(Y > k | X = x) = (1 - x)^k$, which gives

$$P(Y > k) = \int_0^1 (1 - x)^k dx = \frac{1}{k + 1}, \quad k = 0, 1, 2, \ldots$$

which in turn gives

$$P(Y = k) = P(Y > k - 1) - P(Y > k) = \frac{1}{k(k + 1)}, \quad k = 1, 2, \ldots$$

and it is interesting to note that this is not a geometric distribution. □

### 3.5.1 Independent Random Variables

Recall Example 3.6, where a point $(X, Y)$ is chosen uniformly in the unit disk. When we computed the marginal pdf of $X$, we saw that $X$ tends more likely to be near the midpoint 0 than the endpoints $-1$ and 1. By symmetry, $Y$ has the same marginal distribution as $X$ on the $y$ axis and hence also $Y$ tends to be more likely to be near 0. But this seems to indicate that the pair $(X, Y)$ is more likely to be near the origin $(0, 0)$, which would contradict the assumption that $(X, Y)$ is uniform on the unit disk. How is this possible?

The solution to this apparent paradox is that $X$ and $Y$ do tend to be concentrated near 0 when considered one by one but *not at the same time*. Recall the conditional distributions that state that if, for example, $X = x$, $Y$ is uniform on the interval $(-\sqrt{1-x^2}, \sqrt{1-x^2})$. Hence, if $X$ is near 0, there is a lot of room for variability in $Y$, but if $X$ is far from 0, there is less room and $Y$ must be near 0.

It would be natural to say that $X$ and $Y$ are *dependent*, since knowing the value of one of them gives us information on the possible values and hence the distribution of the other. We next state the formal definition of independence of random variables.

***Definition 3.8.*** The random variables $X$ and $Y$ are said to be *independent* if

$$P(X \in A, Y \in B) = P(X \in A)P(Y \in B)$$

for all $A, B \subseteq R$.

Note that this means that $\{X \in A\}$ and $\{Y \in B\}$ are independent events. It can be difficult to check this for all possible sets $A$ and $B$, but it turns out that there is a simpler characterization of independence.

**Proposition 3.8.** The random variables $X$ and $Y$ are independent if and only if

$$F(x, y) = F_X(x)F_Y(y)$$

for all $x, y \in R$.

The proof of the proposition is beyond the scope of this book. If $X$ and $Y$ are independent and $g$ and $h$ are any functions, it can be shown that $g(X)$ and $h(Y)$ are also independent. While intuitively reasonable, the proof is quite advanced and we will not give it. For some special cases, see Problem 37.

In the case when $X$ and $Y$ are discrete, independence can be easily characterized directly in terms of the probability mass functions, as the following proposition shows.

**Proposition 3.9.** Suppose that $(X, Y)$ is discrete with joint pmf $p$. Then $X$ and $Y$ are independent if and only if

$$p(x, y) = p_X(x)p_Y(y)$$

for all $x, y \in R$.

*Proof.* Suppose that $X$ and $Y$ are independent. In the definition of independence, choose $A = \{x\}$ and $B = \{y\}$ to obtain

$$p(x, y) = P(X = x, Y = y) = P(X = x)P(Y = y) = p_X(x)p_Y(y)$$

Conversely, suppose that $p(x, y) = p_X(x)p_Y(y)$ and take two subsets of $R$, $A$ and $B$. Then

$$P(X \in A, Y \in B) = \sum_{x \in A} \sum_{y \in B} p(x, y) = \sum_{x \in A} p_X(x) \sum_{y \in B} p_Y(y)$$

which equals $P(X \in A)P(Y \in B)$, and hence $X$ and $Y$ are independent. ■

Not surprisingly, an analogous relation characterizes independence in the jointly continuous case. The proof, which relies on Proposition 3.8, is left as an exercise.

**Proposition 3.10.** Suppose that $X$ and $Y$ are jointly continuous with joint pdf $f$. Then $X$ and $Y$ are independent if and only if

$$f(x, y) = f_X(x)f_Y(y)$$

for all $x, y \in R$.

Thus, there are several equivalent characterizations of independence, depending on what type of random variables we are dealing with. Regardless of whether we are talking about cdfs, pmfs, or pdfs, we can keep the following informal description in mind.

**Corollary 3.1.** *The random variables X and Y are independent if and only if "the joint is the product of the marginals."*

It also follows from the definitions that independence is equivalent to equality between conditional and unconditional distributions. Thus, all our intuition from independent events carries over to independent random variables.

***Example 3.12.*** Recall Example 3.6, where a point $(X, Y)$ is chosen uniformly in the unit disk. The joint pdf is

$$f(x, y) = \frac{1}{\pi}, \quad x^2 + y^2 \leq 1$$

and the marginals

$$f_X(x) = \frac{2}{\pi}\sqrt{1 - x^2}, \quad -1 \leq x \leq 1$$

and

$$f_Y(y) = \frac{2}{\pi}\sqrt{1-y^2}, \quad -1 \leq y \leq 1$$

and since $f(x, y) \neq f_X(x)f_Y(y)$, $X$ and $Y$ are not independent, as was to be expected from the discussion at the beginning of this section. □

In the last example, it is immediately clear that $X$ and $Y$ cannot be independent since if we know the value of one of them, this changes the range of possible values of the other. This leads us to realize that two random variables $X$ and $Y$ can be independent only if the range in two dimensions of the pair $(X, Y)$ has a shape such that the ranges of $X$ and $Y$ individually do not change in this way. Thus, only ranges that are shaped as rectangles (possibly infinite) parallel to the axes are possible to get independent random variables. Note that this is a necessary but not sufficient condition. Even if the shape is rectangular, the joint pdf may still not equal the product of the marginals (as in Example 3.4).

When we compute probabilities pertaining to independent random variables, the following corollary to Proposition 3.7 can be very useful.

**Corollary 3.2.**

$$P((X, Y) \in B) = \int_{-\infty}^{\infty} P((x, Y) \in B) f_X(x)dx$$

*Proof.* By independence

$$P((x, Y) \in B | X = x) = P((x, Y) \in B)$$

and everything else follows from Proposition 3.7. ■

***Example 3.13 (Buffon's Needle).*** A table is ruled by equidistant parallel lines, 1 in. apart. A needle of length 1 in. is tossed at random on the table. What is the probability that it intersects a line?

We describe the needle's position by distance $D$ from the center of the needle to the nearest line and the (smallest) angle $A$ between the needle and that line. In Figure 3.2, we see how the left needle and the line form a triangle that has a hypotenuse of length $D/\sin A$. The right needle does not intersect the line but would also form such a triangle if extended toward the line. Thus, we realize that the needle intersects the line if and only if

$$\frac{D}{\sin A} \leq \frac{1}{2}$$

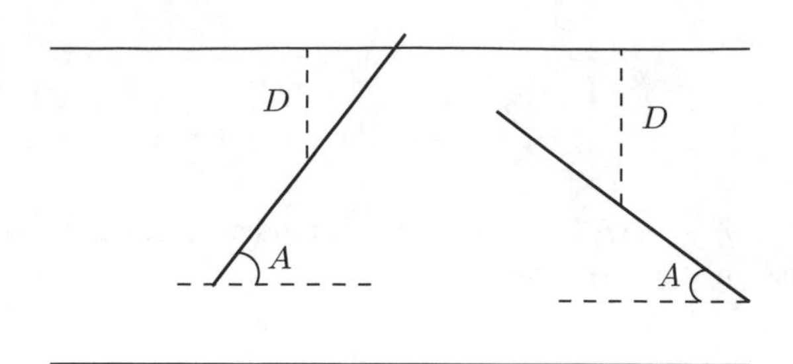


**FIGURE 3.2** Buffon's needle. The needle to the left has $D/\sin A \le \frac{1}{2}$ and thus intersects a line.

where the range of $D$ is $[0, 1/2]$ and that of $A$ is $[0, \pi/2]$. If we interpret "tossed at random" as $D$ and $A$ being independent and uniform on their respective ranges, we can apply Corollary 3.2 to obtain the probability of an intersection as

$$\begin{aligned} P\left(\frac{D}{\sin A} \le \frac{1}{2}\right) &= \int_0^{\pi/2} P\left(\frac{D}{\sin a} \le \frac{1}{2}\right) f_A(a)da \\ &= \frac{2}{\pi}\int_0^{\pi/2} P\left(D \le \frac{\sin a}{2}\right) da \\ &= \frac{2}{\pi}\int_0^{\pi/2} \sin a\, da = \frac{2}{\pi} \end{aligned}$$

yet another of the many appearances of the number $\pi$ in mathematics.[2]

□

## 3.6 FUNCTIONS OF RANDOM VECTORS

Just as in the one-dimensional case, we may be interested in a function $g$ of $(X, Y)$. There are two principal cases of interest: $g: R^2 \to R$, resulting in a random variable, and $g: R^2 \to R^2$, resulting in another random vector. We start with the first of these.

### 3.6.1 Real-Valued Functions of Random Vectors

The typical situation here is that we have a random vector $(X, Y)$ and apply a real-valued function $g$ to get the random variable $g(X, Y)$. To compute probabilities of the type $P(g(X, Y) \in B)$ for $B \subseteq R$, we need to identify the outcomes of $(X, Y)$ that are mapped to $B$. Let us illustrate this with a couple of examples.

[2]In the late eighteenth century, Count de Buffon, French naturalist and avid needle tosser, designed this experiment to estimate a numerical value of $\pi$. We will return to this aspect of the problem in Chapter 4.

***Example 3.14.*** Let $X$ and $Y$ be independent and unif[0, 1], and let $A$ be the area of a rectangle with sides $X$ and $Y$. Find the pdf of $A$.

Here, $A = XY$ and the range of $A$ is $[0, 1]$. Take $a$ in this range and start with the cdf to obtain

$$F_A(a) = P(XY \leq a) = \iint_B f(x, y)dx\,dy$$

where $B$ is the two-dimensional region

$$B = \{(x, y) : 0 \leq x \leq 1, 0 \leq y \leq 1, xy \leq a\}$$

and $f(x, y) = f_X(x)f_Y(y) = 1$ by independence. Draw a picture to realize that

$$\begin{aligned} F_A(a) &= a + \int_a^1 \int_0^{a/x} dy\,dx \\ &= a + a\int_a^1 \frac{1}{x}dx = a - a\log a, \quad 0 \leq a \leq 1 \end{aligned}$$

which we differentiate to get the pdf

$$f_A(a) = F_A'(a) = -\log a, \quad 0 \leq a \leq 1$$

□

***Example 3.15.*** Let $X$ and $Y$ be independent and $\exp(\lambda)$, and let $Z = X/(X + Y)$. Find the pdf of $Z$.

First note that the range of $Z$ is $[0, 1]$. Take $z$ in this range and consider the cdf

$$F_Z(z) = P\left(\frac{X}{X+Y} \leq z\right) = P\left(Y \geq \left(\frac{1}{z} - 1\right)X\right)$$

This means that we integrate the joint pdf over the region

$$B = \left\{(x, y) : y \geq \left(\frac{1}{z} - 1\right)x\right\}$$

By independence, the joint pdf is

$$f(x, y) = \lambda e^{-\lambda x}\lambda e^{-\lambda y}, \quad x \geq 0,\ y \geq 0$$

and with $a = 1/z - 1$, we get

$$F_Z(z) = \int_0^\infty \int_{ax}^\infty \lambda e^{-\lambda y} \lambda e^{-\lambda x} dy\, dx$$

$$= \int_0^\infty \lambda e^{-\lambda a x} e^{-\lambda x} dx = \frac{1}{a+1}$$

in which we substitute $a$ by $1/z - 1$ to obtain

$$F_Z(z) = \frac{1}{1/z - 1 + 1} = z, \quad 0 \le z \le 1$$

which we recognize as the uniform distribution on [0, 1]. It is tempting to believe that this has nothing to do with the exponential distribution and that $X/(X+Y)$ is uniform on [0, 1] as long as $X$ and $Y$ are independent and have the same distribution. However, this is not the case; see, for example, Problem 43(b). In fact, it is a special property of the exponential distribution that gives this result, and we return to this in Section 3.12. □

Sometimes, two-dimensional methods can be used even if there is seemingly nothing two-dimensional in the problem. Here is one typical example.

***Example 3.16.*** Choose two points at random on a yardstick. What is the probability that they are at most half a yard apart?

Let us first formulate this strictly as a probability problem. Thus, we let $X$ and $Y$ be independent unif[0, 1] and wish to find $P(|X - Y| \le \frac{1}{2})$. Although there is nothing two-dimensional in this problem from a physical point of view, we can solve it by viewing $(X, Y)$ as a random vector in two dimensions. By independence, the joint pdf is

$$f(x, y) = f_X(x) f_Y(y) = 1, \quad 0 \le x \le 1,\ 0 \le y \le 1$$

and the region of interest $B = \{(x, y) : |x - y| \le \frac{1}{2}\}$, which is illustrated in Figure 3.3. Since $f(x, y) \equiv 1$, the integral of $f$ over $B$ is equal to the area of $B$, which gives

$$P\left(|X - Y| \le \frac{1}{2}\right) = \frac{3}{4}$$

□

We have seen examples of how to find the distribution of a function $g(X, Y)$ of a random vector $(X, Y)$. If all we are interested in is the expected value $E[g(X, Y)]$, we do not need to find the distribution, according to the following two-dimensional analog of Proposition 2.12, which we state without proof.

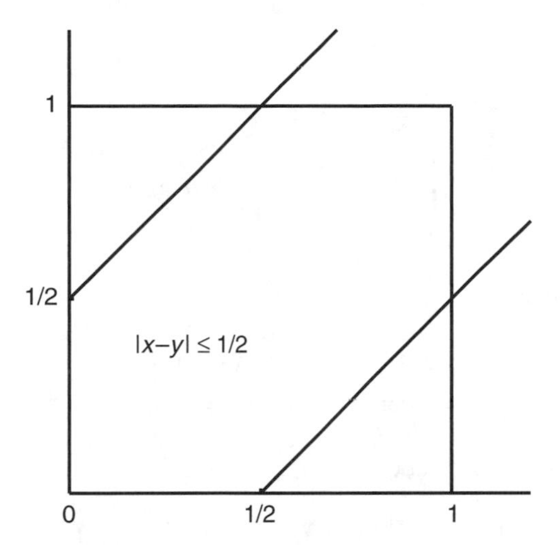


**FIGURE 3.3** Illustration of Example 3.16. Every outcome of $(X, Y)$ in the region bounded by the axes and the lines has $|X - Y|$ at most half a yard apart.

**Proposition 3.11.** Let $(X, Y)$ be a random vector with joint pmf $p$ or joint pdf $f$ and let $g: R \times R \to R$ be any function. Then

$$E[g(X, Y)] = \begin{cases} \displaystyle\sum_{j=1}^{\infty}\sum_{k=1}^{\infty} g(x_j, y_k)p(x_j, y_k) & \text{if } (X, Y) \text{ is discrete} \\ \displaystyle\int_{-\infty}^{\infty}\int_{-\infty}^{\infty} g(x, y)f(x, y)dx\,dy & \text{if } (X, Y) \text{ is continuous} \end{cases}$$

***Example 3.17.*** Choose a point at random in the unit disk. What is its expected distance to the origin?

If the point is $(X, Y)$, the distance is $R = \sqrt{X^2 + Y^2}$. We thus have $g(x, y) = \sqrt{x^2 + y^2}$ in the proposition above, and the joint pdf of $(X, Y)$ is

$$f(x, y) = \frac{1}{\pi}, \quad x^2 + y^2 \leq 1$$

which by Proposition 3.11 gives

$$E[R] = E\left[\sqrt{X^2 + Y^2}\right] = \frac{1}{\pi}\int\int_{x^2+y^2\leq 1} \sqrt{x^2 + y^2}dx\,dy$$

which we solve by changing to polar coordinates. Thus, let $x = r\cos\theta$, $y = r\sin\theta$, which gives region of integration $0 \leq r \leq 1$ and $0 \leq \theta \leq 2\pi$. The Jacobian matrix for the transformation $(r, \theta) \to (x, y)$ is

$$J = \begin{pmatrix} \dfrac{\partial x}{\partial r} & \dfrac{\partial x}{\partial \theta} \\ \\ \dfrac{\partial y}{\partial r} & \dfrac{\partial y}{\partial \theta} \end{pmatrix} = \begin{pmatrix} \cos\theta & -r\sin\theta \\ \sin\theta & r\cos\theta \end{pmatrix}$$

which has determinant

$$\begin{aligned} |J| &= \cos\theta \times r\cos\theta - (-r\sin\theta) \times \sin\theta \\ &= r(\cos^2\theta + \sin^2\theta) \ = \ r \end{aligned}$$

which gives $dx\, dy = r\, dr\, d\theta$ and

$$E[R] = \frac{1}{\pi}\int_0^{2\pi}\int_0^1 \sqrt{r^2}\, r\, dr\, d\theta = \frac{1}{\pi}\int_0^{2\pi} d\theta \int_0^1 r^2 dr = \frac{2}{3}$$

□

### 3.6.2 The Expected Value and Variance of a Sum

We can use Proposition 3.11 to show that expected values are linear in the sense that the expected value of a sum is the sum of the expected values.

**Proposition 3.12.** Let $X$ and $Y$ be any random variables. Then

$$E[X + Y] = E[X] + E[Y]$$

*Proof.* Let us do the jointly continuous case. By Proposition 3.11, with the function $g(x, y) = x + y$, we get

$$\begin{aligned} E[X + Y] &= \int_{-\infty}^{\infty}\int_{-\infty}^{\infty} (x + y) f(x, y) dx\, dy \\ &= \int_{-\infty}^{\infty}\int_{-\infty}^{\infty} x f(x, y) dx\, dy + \int_{-\infty}^{\infty}\int_{-\infty}^{\infty} y f(x, y) dx\, dy \\ &= \int_{-\infty}^{\infty} x f_X(x) dx + \int_{-\infty}^{\infty} y f_Y(y) dy \ = \ E[X] + E[Y] \end{aligned}$$

where the second to last equality follows from Proposition 3.5. The discrete case is similar, replacing integrals by sums. ■

Combining Propositions 3.12 and 2.11 gives the following result.

**Corollary 3.3.** *Let $X$ and $Y$ be any random variables, and let $a$ and $b$ be real numbers. Then*

$$E[aX + bY] = aE[X] + bE[Y]$$

From Proposition 2.15, we already know that the variance is not linear since $\text{Var}[aX] = a^2\text{Var}[X]$, but the question remains whether the variance is additive, that is, if $\text{Var}[X + Y] = \text{Var}[X] + \text{Var}[Y]$. The following simple example shows that there might be a problem.

***Example 3.18.*** Let $X \sim \text{unif}[0, 1]$ and let $Y = -X$. By Propositions 2.13 and 2.15, we have

$$\text{Var}[X] = \text{Var}[Y] = \frac{1}{12}$$

and hence

$$\text{Var}[X] + \text{Var}[Y] = \frac{1}{6}$$

However, $X + Y \equiv 0$, so by Proposition 2.16, $\text{Var}[X + Y] = 0$ and

$$\text{Var}[X + Y] \neq \text{Var}[X] + \text{Var}[Y]$$

in this case. □

The problem in the example is that there is no variability at all in the sum $X + Y$, even though $X$ and $Y$ have variability individually. Intuitively, this is because variation in $X$ is canceled by variation in $Y$ in the opposite direction. Note that $X$ and $Y$ are dependent (and very strongly so), and it turns out that independence is an assumption that allows us to add variances. We state this next, together with a result about the expected value of a product.

**Proposition 3.13.** Let $X$ and $Y$ be independent random variables. Then

(a) $E[XY] = E[X]E[Y]$
(b) $\text{Var}[X + Y] = \text{Var}[X] + \text{Var}[Y]$

*Proof.* Let us again consider the continuous case only. First note that when $X$ and $Y$ are independent, then $f(x, y) = f_X(x)f_Y(y)$ by Proposition 3.10. By

Proposition 3.11, we get

$$E[XY] = \int_{-\infty}^{\infty} \int_{-\infty}^{\infty} xyf(x, y)dx\, dy$$

$$= \int_{-\infty}^{\infty} xf_X(x)dx \int_{-\infty}^{\infty} yf_Y(y)dy = E[X]E[Y]$$

which proves part (a). For part (b), use Corollary 2.2 together with linearity of expected values to obtain

$$\text{Var}[X+Y] = E\left[(X+Y)^2\right] - \left(E[X+Y]\right)^2$$

$$= E[X^2] + 2E[XY] + E[Y^2] - E[X]^2 - 2E[X]E[Y] - E[Y]^2$$

$$= E[X^2] - E[X]^2 + E[Y^2] - E[Y]^2 = \text{Var}[X] + \text{Var}[Y]$$

where we used part (a) for the second to last equality. ■

Finally, combining Propositions 2.15 and 3.13(b), we get the following corollary.

**Corollary 3.4.** *Let X and Y be independent random variables, and let a and b be real numbers. Then*

$$Var[aX + bY] = a^2 Var[X] + b^2 Var[Y]$$

***Example 3.19.*** You have an instrument to measure length, which gives a small measurement error. If the true length is $l$, the instrument gives the estimated length $L = l + \epsilon$, where $\epsilon$ is a random variable with mean 0 and variance $\sigma^2$. You have two rods of different lengths and are allowed a total of two measurements to determine their lengths. Can you do better than one measurement of each?

Yes, you can. Let the true lengths be $a$ and $b$, where $a > b$. If you take one measure of the longer rod, you get the measurement $A = a + \epsilon$, which has mean $E[A] = a + E[\epsilon] = a$ and variance $\text{Var}[A] = \text{Var}[\epsilon] = \sigma^2$. Similarly, the shorter rod is measured to be $B$, with mean $b$ and variance $\sigma^2$.

Instead, put the rods side by side and measure the difference $D$, and then put them end to end and measure the sum $S$. To estimate $a$ and $b$, let $A = (S + D)/2$ and $B = (S - D)/2$. Since $S = a + b + \epsilon_1$ and $D = a - b + \epsilon_2$, where $\epsilon_1$ and $\epsilon_2$ are the two errors that we assume are independent, Corollary 3.3 gives

$$E[A] = \frac{1}{2}E[S + D] = \frac{1}{2}(a + b + a - b) = a$$

and similarly $E[B] = b$, so your estimates are correct on average. The precision of $A$ and $B$ are measured by their variances, and Corollary 3.4 gives

$$\begin{aligned}\text{Var}[A] &= \frac{1}{2^2}(\text{Var}[S] + \text{Var}[D]) = \frac{\sigma^2}{2}\\ \text{Var}[B] &= \frac{1}{2^2}(\text{Var}[S] + (-1)^2\text{Var}[D]) = \frac{\sigma^2}{2}\end{aligned}$$

and the precision is better than that of two individual measurements. The reason is that in effect we get two measurements on each rod instead of one. We are assuming that there is no error in the alignment of the rods side by side; see Problem 49 for a variant of this. □

***Example 3.20.*** Let $X$ and $Y$ be independent and uniform on $[0, 1]$, and let $C$ be the circumference of a rectangle with sides $X$ and $Y$. Find the mean and variance of $C$.

Since $C = 2X + 2Y$, Propositions 3.3 and 3.4 give

$$\begin{aligned}E[C] &= 2E[X] + 2E[Y] = 2\\ \text{Var}[C] &= 4\text{Var}[X] + 4\text{Var}[Y] = \frac{2}{3}\end{aligned}$$

□

If we have a sequence $X_1, \ldots, X_n$ of random variables, Corollary 3.3 and induction give the general result about linearity of expected values:

**Proposition 3.14.** Let $X_1, X_2, \ldots, X_n$ be random variables and let $a_1, a_2, \ldots, a_n$ be real numbers. Then

$$E\left[\sum_{k=1}^{n} a_k X_k\right] = \sum_{k=1}^{n} a_k E[X_k]$$

We will look at several examples. The first two examples concern repeated rolls of a die and are special cases of more general problems. The first is a *coupon collecting problem* and the second an *occupancy problem.*

***Example 3.21.*** Roll a die repeatedly, and let $X$ be the number of rolls it takes to get all the numbers. Find $E[X]$.

The first number comes in the first roll. Then we wait for any number that is different from the first. Since the probability of this in each roll is $\frac{5}{6}$, the time it takes has a geometric distribution with success probability $\frac{5}{6}$. Once the second number has appeared, we start waiting for the third one, and now the waiting time is geometric with success probability $\frac{4}{6}$ and so on. Hence,

$$X = 1 + \sum_{k=2}^{6} X_k$$

where $X_k \sim \text{geom}\left(\frac{7-k}{6}\right)$ and

$$E[X] = 1 + \sum_{k=2}^{6} E[X_k] = 1 + \left(\frac{6}{5} + \frac{6}{4} + \cdots + \frac{6}{1}\right) \approx 14.7$$

To explain the term "coupon collecting problem," replace the die with coupons numbered 1, 2, ..., 6, which are collected until we have all of them. For a general version, see Problem 54. □

***Example 3.22.*** Roll a die six times and let $X$ be the number of different numbers that are represented. For example, if you get 1, 6, 3, 5, 3, 5, then $X = 4$ since 1, 3, 5, and 6 are represented. What is $E[X]$?

Note that the range of $X$ is 1, 2, ..., 6. It is possible to find the pmf of $X$ and compute $E[X]$ according to the definition, but it is much quicker to use indicators. Let

$$I_k = \begin{cases} 1 & \text{if } k \text{ is represented} \\ 0 & \text{otherwise} \end{cases}$$

so that

$$X = \sum_{k=1}^{6} I_k$$

and

$$E[X] = \sum_{k=1}^{6} E[I_k]$$

Now, $I_k$ equals 1 unless all six rolls gave numbers different from $k$. The probability in one roll to get a number different from $k$ is $\frac{5}{6}$ and hence

$$E[I_k] = P(I_k = 1) = 1 - \left(\frac{5}{6}\right)^6$$

which gives

$$E[X] = 6\left(1 - \left(\frac{5}{6}\right)^6\right) \approx 3.99$$

If we call a number "occupied" once it has been rolled, we understand the term "occupancy problem" (see also Problem 55). □

***Example 3.23.*** Recall the matching problem from Example 1.27. What is the expected number of matches?

Again, let us use indicators. Thus, let

$$I_k = \begin{cases} 1 & \text{if there is a match at } k \\ 0 & \text{otherwise} \end{cases}$$

so that

$$X = \sum_{k=1}^{n} I_k$$

Following the calculations in Example 1.27, we get

$$E[I_k] = P(I_k = 1) = \frac{1}{n}$$

for all $I_k$. Hence,

$$E[X] = \sum_{k=1}^{n} E[I_k] = n \times \frac{1}{n} = 1$$

so on average one match is expected, regardless of the value of $n$.

□

To deal with more than two random variables, let us first state the obvious generalization of the independence concept.

***Definition 3.9.*** The random variables $X_1, X_2, \ldots$ are said to be *independent* if

$$P(X_{i_1} \in B_1, X_{i_2} \in B_2, \ldots, X_{i_k} \in B_k) = \prod_{i=1}^{k} P(X_{i_k} \in B_k)$$

for all choices of $i_1 < \cdots < i_k$ and sets $B_1, \ldots, B_k \subseteq R$, $k = 2, 3, \ldots$

Note that we have defined independence for an infinite sequence, exactly as we did for events in Definition 1.7. As in the two-dimensional case, independence can be characterized in terms of joint and marginal distributions, and we return to this in Section 3.10. For now, let us concentrate on the general variance formula. By Proposition 3.13, Corollary 3.4, and induction, we get the following result:

**Proposition 3.15.** If $X_1, \ldots, X_n$ are independent, then

(a) $E[X_1 X_2 \cdots X_n] = E[X_1]E[X_2] \cdots E[X_n]$

(b) $\mathrm{Var}\left[\sum_{k=1}^{n} a_k X_k\right] = \sum_{k=1}^{n} a_k^2 \mathrm{Var}[X_k]$

***Example 3.24.*** Let $X \sim \mathrm{bin}(n, p)$. Find the mean and variance of $X$.

Recall that a binomial distribution counts the number of successes in $n$ independent trials. By introducing the indicators

$$I_k = \begin{cases} 1 & \text{if the } k\text{th trial gives a success} \\ 0 & \text{otherwise} \end{cases}$$

we can write

$$X = \sum_{k=1}^{n} I_k$$

where $I_k$ are independent and have mean $E[I_k] = p$ and variance $\mathrm{Var}[I_k] = p(1-p)$ (see Section 2.5.1). Hence,

$$E[X] = \sum_{k=1}^{n} E[I_k] = np$$

and

$$\mathrm{Var}[X] = \sum_{k=1}^{n} \mathrm{Var}[I_k] = np(1-p)$$

which is in agreement with Proposition 2.17, but note how much simpler it was to use indicators. □

### 3.6.3 Vector-Valued Functions of Random Vectors

Previously, we saw how we can apply a real-valued function $g$ to map $(X, Y)$ to a random variable $g(X, Y)$. The following example illustrates a situation where $(X, Y)$ is mapped to another two-dimensional random vector.

***Example 3.25.*** Choose a point $(X, Y)$ at random in the unit disk, and let its polar coordinates be $(R, \Theta)$. What is the joint distribution of $R$ and $\Theta$?

First note that the random radius $R$ and the random angle $\Theta$ relate to $X$ and $Y$ through

$$X = R\cos\Theta \quad \text{and} \quad Y = R\sin\Theta$$

where the range of $R$ is $[0, 1]$ and the range of $\Theta$ is $[0, 2\pi]$. Are they uniform on these ranges? Let us try to find the joint cdf of $(R, \Theta)$, that is,

$$F(r, \theta) = P(R \leq r, \Theta \leq \theta)$$

for fixed $r$ in $[0, 1]$ and $\theta$ in $[0, 2\pi]$. Now let

$$C = \left\{(x, y) : 0 \leq \sqrt{x^2 + y^2} \leq r,\ \ 0 \leq \tan^{-1}\left(\frac{y}{x}\right) \leq \theta\right\}$$

so that we have

$$F(r, \theta) = P((X, Y) \in C)$$

(see Figure 3.4). Since $(X, Y)$ is uniform, the probability that it belongs to the sector $C$ is simply the area of $C$ divided by the area of the unit disk, and we get

$$F(r, \theta) = P((X, Y) \in C) = \frac{\pi r^2 \theta / 2\pi}{\pi} = \frac{\theta r^2}{2\pi}$$

By differentiating, we get the joint pdf

$$f(r, \theta) = \frac{\partial^2}{dr\, d\theta} F(r, \theta) = \frac{r}{\pi}, \quad 0 \leq r \leq 1, \quad 0 \leq \theta \leq 2\pi$$

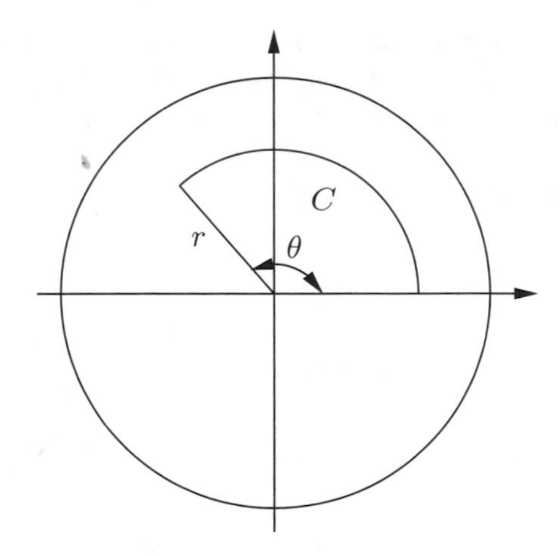


**FIGURE 3.4** A point in the disk has angle less than $\theta$ and radius less than $r$ if it is chosen in the region $C$.

which has marginals

$$f_R(r) = \frac{r}{\pi}\int_0^{2\pi} d\theta = 2r, \quad 0 \le r \le 1$$

and

$$f_\Theta(\theta) = \frac{1}{\pi}\int_0^1 r\,dr = \frac{1}{2\pi}, \quad 0 \le \theta \le 2\pi$$

Hence, $\Theta$ is uniform but $R$ is not, which should not be too surprising. It is easier to get a radius near 1 than near 0 since a ring of fixed width has larger area near 1. There is no reason why any particular angle would be more likely than any other, though. Note also that $f(r, \theta) = f_R(r)f_\Theta(\theta)$, which means that the radius and the angle are independent. Think about why this makes sense. □

The following is a two-dimensional analog of Proposition 2.8, which is stated without proof.

**Proposition 3.16.** Let $(X, Y)$ be jointly continuous with joint pdf $f_{(X,Y)}$. Furthermore, let $(U, V) = (u(X, Y), v(X, Y))$, where the map $(x, y) \to (u, v)$ is invertible. The pair $(U, V)$ then has joint pdf

$$f_{(U,V)}(u, v) = f_{(X,Y)}(x(u, v), y(u, v))\,|J(x(u, v), y(u, v))|$$

where $|J(x(u, v), y(u, v))|$ is the absolute value of the Jacobian determinant of the inverse map $(u, v) \to (x(u, v), y(u, v))$.

***Example 3.26.*** Let $X$ and $Y$ be independent standard normal random variables and consider the pair $(X, Y)$ in its polar representation $(R, \Theta)$. Find the joint pdf of $(R, \Theta)$ and the marginal distributions of $R$ and $\Theta$.

By independence, the joint pdf of $(X, Y)$ is

$$f_{(X,Y)}(x, y) = f_X(x)f_Y(y) = \frac{1}{2\pi}e^{-(x^2+y^2)/2}$$

and $(X, Y)$ is mapped to $(R, \Theta)$ according to

$$R = \sqrt{X^2 + Y^2} \quad \text{and} \quad \Theta = \tan^{-1}\left(\frac{Y}{X}\right)$$

The inverse map is

$$X = R\cos\Theta \quad \text{and} \quad Y = R\sin\Theta$$

and we apply Proposition 3.16. We know from before that $|J| = r$ and get

$$\begin{aligned} f_{(R,\Theta)}(r,\theta) &= f_{(X,Y)}(r\cos\theta, r\sin\theta) \times r \\ &= \frac{1}{2\pi} r e^{-r^2/2}, \quad r \geq 0,\ 0 \leq \theta \leq 2\pi \end{aligned}$$

The marginals are easily found to be

$$f_R(r) = re^{-r^2/2}, \quad r \geq 0$$

and

$$f_\Theta(\theta) = \frac{1}{2\pi}, \quad 0 \leq \theta \leq 2\pi$$

so that $R$ and $\Theta$ are independent and $\Theta$ is uniform on $[0, 2\pi]$. If we consider $R^2$, the squared distance to the origin, Proposition 2.8 with $g(r) = r^2$ gives

$$f_{R^2}(x) = \frac{1}{2\sqrt{x}} f_R(\sqrt{x}) = \frac{1}{2} e^{-x/2}, \quad x \geq 0$$

which we recognize as an exponential distribution with parameter $\lambda = \frac{1}{2}$. We will later see that this observation is useful to simulate observations from a normal distribution. □

## 3.7 CONDITIONAL EXPECTATION

Once we have defined conditional distributions, it is a logical step to also define *conditional expectations*. The intuition is clear; these are simply the expected values in the conditional distributions, and the definitions are straightforward. We start with conditioning on an event.

***Definition 3.10.*** Let $Y$ be a random variable and $B$ an event with $P(B) > 0$. The *conditional expectation* of $Y$ given $B$ is defined as

$$E[Y|B] = \begin{cases} \displaystyle\sum_{k=1}^{\infty} y_k P(Y = y_k|B) & \text{if } Y \text{ is discrete with range } \{y_1, y_2, \ldots\} \\ \displaystyle\int_{-\infty}^{\infty} y f_Y(y|B)dy & \text{if } Y \text{ is continuous} \end{cases}$$

Now recall the law of total probability that allows us to compute probabilities by finding suitable conditioning events. There is an analog for expected values, a "law of total expectation," which states that

$$E[Y] = \sum_{k=1}^{\infty} E[Y|B_k]P(B_k)$$

under the same conditions as the law of total probability. As usual, the sum may be finite or infinite. We omit the proof and illustrate with an example.

***Example 3.27.*** Consider the computer from Example 2.45, which is busy with probability 0.8, in which case an incoming job must wait for a time that is exp(1). Find the expected waiting time of an incoming job.

The two cases pertain to whether the computer is busy, so let $B$ be the event that it is busy to obtain

$$E[Y] = E[Y|B]P(B) + E[Y|B^c]P(B^c)$$

where $E[Y|B^c] = 0$ since there is no wait if the computer is free and $E[Y|B] = 1$ since the expected wait is 1 s if it is busy. Hence, the expected waiting time is

$$E[Y] = 0 \times 0.2 + 1 \times 0.8 = 0.8 \text{ s}$$

□

Next, we condition on the outcome of a random variable. If this random variable is discrete, the situation is the same as above, with events of the type $B_j = \{X = x_j\}$.

***Definition 3.11.*** Suppose that $X$ and $Y$ are discrete. We define

$$E[Y|X = x_j] = \sum_{k=1}^{\infty} y_k p_Y(y_k|x_j)$$

The law of total expectation now takes the form

$$E[Y] = \sum_{j=1}^{\infty} E[Y|X = x_j]p_X(x_j)$$

where the sum is over the range of $X$. Finally, we consider the continuous case where, as usual, sums are replaced by integrals.

***Definition 3.12.*** Suppose that $X$ and $Y$ are jointly continuous. We define

$$E[Y|X = x] = \int_{-\infty}^{\infty} y f_Y(y|x) dy$$

Following the usual intuitive interpretation, this is the expected value of $Y$ if we know that $X = x$. The law of total expectation now takes the following form.

**Proposition 3.17.** Suppose that $X$ and $Y$ are jointly continuous. Then

$$E[Y] = \int_{-\infty}^{\infty} E[Y|X = x] f_X(x)dx$$

*Proof.* By definition of expected value and Proposition 3.6

$$E[Y] = \int_{-\infty}^{\infty} y f_Y(y)dy = \int_{-\infty}^{\infty}\int_{-\infty}^{\infty} y f_Y(y|x) f_X(x)dx\, dy$$

where we change the order of integration to obtain

$$E[Y] = \int_{-\infty}^{\infty}\int_{-\infty}^{\infty} y f_Y(y|x)dy f_X(x)dx$$

where the inner integral equals $E[Y|X = x]$ by definition, and we are done. ∎

***Example 3.28.*** Consider Example 3.9 where $X \sim \text{unif}[0, 1]$ and the conditional distribution of $Y$ is $Y|X = x \sim \text{unif}[0, x]$. What is $E[Y]$?

There are two possible ways to solve this: using the definition of expected value or using Proposition 3.17.

The first solution is as follows. By definition

$$E[Y] = \int_0^1 y f_Y(y)dy$$

where

$$f_Y(y) = \int_{-\infty}^{\infty} f(x, y)dx = \int_y^1 \frac{1}{x}dx$$

since for fixed $y$, the joint pdf is strictly positive when $x$ is between $y$ and 1. Hence,

$$f_Y(y) = \left[\log x\right]_y^1 = -\log y, \quad 0 < y \leq 1$$

The expected value is

$$E[Y] = -\int_0^1 y \log y\, dy$$

which can be shown to equal $\frac{1}{4}$.

The second solution is as follows. Since $Y|X = x \sim \text{unif}[0, x]$, we have

$$E[Y|X = x] = \frac{x}{2}$$

and by Proposition 3.17, we obtain

$$E[Y] = \int_0^1 E[Y|X = x] f_X(x) dx = \int_0^1 \frac{x}{2} dx = \frac{1}{4}$$

□

***Example 3.29.*** A *Geiger counter* is an instrument used to detect radiation, such as emission of alpha particles. When a particle is detected, the so-called dead period follows, during which the counter cannot register anything. Dead periods are often given as single numbers, but it is more realistic to assume that they are random. Suppose that alpha particles are emitted at a rate of 10,000 per second and that a dead period lasts for a random time that has pdf

$$f(t) = 6t(1 - t), \quad 0 \leq t \leq 1$$

where we take the basic time unit to be 100 μs. What is the expected number of particles that go undetected during a dead period?

Let us denote the number of particles in an interval of fixed length $t \times 100$ μs by $X(t)$ and assume that $X(t)$ has a Poisson distribution with mean $t$ (emissions per 100 μs). Since the length of a dead period is not fixed but random, say, $T$, the number of emissions during this period is $X(T)$, and we condition to obtain

$$E[X(T)] = \int_0^1 E[X(T)|T = t] f_T(t) dt$$
$$= 6 \int_0^1 t \times t(1 - t) dt = \frac{1}{2}$$

so, on average, we miss one particle for every two that are detected. □

We have considered both jointly discrete and jointly continuous random vectors. It is also possible to have a mixture in the sense that, for example, $X$ is discrete and $Y$ is continuous. In that case, neither a joint pmf nor a joint pdf exists, but we can still use conditional distributions and expectations.

***Example 3.30.*** Let $Y \sim \text{unif}[0, 1]$ and independently choose one of the endpoints 0 and 1 such that 0 is chosen with probability $p$. What is the expected distance between $Y^2$ and the chosen point?

Let $X$ be the point so that $P(X = 0) = p$ and $P(X = 1) = 1 - p$, and we are looking for $E[|Y^2 - X|]$. Now, $X$ is discrete and $Y$ continuous, so we solve this by conditioning on $X$. First note that

$$|Y^2 - X| = \begin{cases} Y^2 & \text{if } X = 0 \\ 1 - Y^2 & \text{if } X = 1 \end{cases}$$

This means that

$$E\left[|Y^2 - X| \,\Big|\, X = 0\right] = E[Y^2|X = 0] = E[Y^2]$$

since $X$ and $Y$ are independent. Similarly,

$$E\left[|Y^2 - X| \,\Big|\, X = 1\right] = 1 - E[Y^2]$$

and since $E[Y^2] = \frac{1}{3}$, we get

$$\begin{aligned} E\left[|Y^2 - X|\right] &= E\left[|Y^2 - X| \,\Big|\, X = 0\right] P(X = 0) \\ &\quad + E\left[|Y^2 - X| \,\Big|\, X = 1\right] P(X = 1) \\ &= E[Y^2]p + (1 - E[Y^2])(1 - p) \; = \; \frac{2 - p}{3} \end{aligned}$$

□

### 3.7.1 Conditional Expectation as a Random Variable

In the previous section, we learned that the conditional expectation $E[Y|X = x]$ is the expected value of $Y$ if we know that $X = x$. We can therefore view $E[Y|X = x]$ as a function of $x$, $g(x)$, and it is natural to define a random variable $g(X)$ that equals $E[Y|X = x]$ whenever $X = x$. We use the notation $E[Y|X]$ for this random variable and get the following definition.

***Definition 3.13.*** The *conditional expectation* of $Y$ given $X$, $E[Y|X]$, is a random variable that equals $E[Y|X = x]$ whenever $X = x$.

To get a better understanding, let us reconsider Example 3.28. There we had $E[Y|X = x] = x/2$ and hence

$$E[Y|X] = \frac{X}{2}$$

the conditional expectation of $Y$ given $X$. We also showed that

$$E[Y] = \int_0^1 \frac{x}{2} dx$$

and since the integral is the expected value of the random variable $X/2$, we see that we have

$$E[Y] = E\left[\frac{X}{2}\right]$$

In fact, it is a simple consequence of the laws of total expectation that we always have this type of relation.

**Corollary 3.5.**

$$E[Y] = E\left[E[Y|X]\right]$$

We can now restate Example 3.28 in a more compact way. Let $Y \sim \text{unif}[0, X]$, where $X \sim \text{unif}[0, 1]$. Then it is clear that $E[Y|X] = X/2$, the midpoint, and by Corollary 3.5 we get

$$E[Y] = E\left[\frac{X}{2}\right] = \frac{1}{4}$$

Make sure that you understand the difference between conditioning on the *event* $\{X = x\}$ and the *random variable* $X$. In the first case, the conditional expectation is a number (dependent on $x$) and in the second case it is a random variable (dependent on $X$). When we condition on $X$, we can think of $X$ as a known quantity, and the following result, stated without proof, is easy to believe.

**Corollary 3.6.**

(a) $E[XY|X] = XE[Y|X]$
(b) *If $X$ and $Y$ are independent, then* $E[Y|X] = E[Y]$
(c) *For any function $g$,* $E[g(X)|X] = g(X)$

Note that (b) and (c) are the two extremes of independence and total dependence. The conditional expectation changes from being equal to $E[Y]$ (no information from $X$) to $g(X)$ (complete information from $X$).

***Example 3.31.*** Let $X \sim \text{unif}[0, 1]$ and let $Y|X \sim \text{unif}[0, X^2]$. Let $A$ be the area of a rectangle with sidelengths $X$ and $Y$ and find $E[A]$.

By Corollary 3.5, we get $E[A] = E[E[A|X]]$, and since $A = XY$, Corollary 3.6(a) gives

$$E[A|X] = E[XY|X] = XE[Y|X]$$

where $E[Y|X] = X^2/2$, and we get

$$E[A] = \frac{1}{2}E[X^3] = \frac{1}{2}\int_0^1 x^3 dx = \frac{1}{8}$$

□

### 3.7.2 Conditional Expectation and Prediction

A common problem in many applications is that we are interested in one random variable but observe another. For example, if a signal is received and we know that there is noise that distorts the transmission, we wish to predict the most likely signal that was sent. Another application might be that we are interested in the concentration of a chemical compound in a solution but can measure only the concentration of a by-product of the reaction that creates the compound. A third application could be a company predicting sales figures for next year based on this year's sales.

In all these cases, we are interested in a random variable $Y$ but observe another $X$ and want to predict $Y$ by a function of $X$, $g(X)$, called a *predictor* of $Y$. Clearly, we want $g(X)$ to be as close to $Y$ as possible and to be able to quantify this idea, we need a "measure of closeness." The following is the most common.

***Definition 3.14.*** Let $g(X)$ be a predictor of $Y$. The *mean square error* is defined as

$$E\left[(Y - g(X))^2\right]$$

It turns out that the best predictor of $Y$ is the conditional expectation $E[Y|X]$ in the following sense.

**Proposition 3.18.** Among all predictors $g(X)$ of $Y$, the mean square error is minimized by $E[Y|X]$.

We omit the proof and instead refer to an intuitive argument. Suppose that we want to predict $Y$ as much as possible by a constant value $c$. Then, we want to minimize

$E[(Y-c)^2]$, and with $\mu = E[Y]$ we get

$$\begin{aligned} E\left[(Y-c)^2\right] &= E\left[(Y-\mu+\mu-c)^2\right] \\ &= E\left[(Y-\mu)^2\right] + 2(\mu-c)E[(Y-\mu)] + (\mu-c)^2 \\ &= \text{Var}[Y] + (\mu-c)^2 \end{aligned}$$

since $E[Y-\mu]=0$. But the last expression is minimized when $c=\mu$ and hence $\mu$ is the best predictor of $Y$ among all constants. This is not too surprising; if we do not know anything about $Y$, the best guess should be the expected value $E[Y]$. Now, if we observe another random variable $X$, the same ought to be true: $Y$ is best predicted by its expected value given the random variable $X$, that is, $E[Y|X]$.

Note here how we view the conditional expectation as a random variable. If we observe a particular value $x$, the best predicted value of $Y$ is $E[Y|X=x]$. Thus, if we plot observations on $(X, Y)$, the curve $y = E[Y|X=x]$ gives the best fit to the data.

### 3.7.3 Conditional Variance

Once we have introduced conditional expectation, the next logical definition is that of conditional variance.

***Definition 3.15.*** The *conditional variance* of $Y$ given $X$ is defined as

$$\text{Var}[Y|X] = E\left[(Y - E[Y|X])^2|X\right]$$

Note that the conditional variance is also a random variable and we think of it as the variance of $Y$ given the value $X$. In particular, if we have observed $X=x$, then we can denote and define

$$\text{Var}[Y|X=x] = E\left[(Y - E[Y|X=x])^2 \,\middle|\, X=x\right]$$

Also note that if $X$ and $Y$ are independent, $E[Y|X]=E[Y]$, and the definition boils down to the regular variance. There is an analog of Corollary 2.2, which we leave to the reader to prove.

**Corollary 3.7.**

$$Var[Y|X] = E\left[Y^2|X\right] - (E[Y|X])^2$$

There is also a "law of total variance," which looks slightly more complicated than that of total expectation.

**Proposition 3.19.**

$$\text{Var}[Y] = \text{Var}\left[E[Y|X]\right] + E\left[\text{Var}[Y|X]\right]$$

*Proof.* Take expected values in Corollary 3.7 to obtain

$$E\left[\text{Var}[Y|X]\right] = E[Y^2] - E\left[(E[Y|X])^2\right] \tag{3.1}$$

and since $E[E[Y|X]] = E[Y]$, we have

$$\text{Var}\left[E[Y|X]\right] = E\left[(E[Y|X])^2\right] - (E[Y])^2 \tag{3.2}$$

and the result follows by adding Equations (3.1) and (3.2). ■

The first term in the formula in Proposition 3.19 accounts for how much of the variability in $Y$ is explained by $Y$'s dependence on $X$; the second, how much is explained by other sources. To understand this, note that if $X$ and $Y$ are highly dependent, then it should make little difference whether we consider $Y$ or $E[Y|X]$, so their variances should be about the same. On the other hand, if the dependence is weak, then $E[Y|X]$ should not change much as a result of variation in $X$, and thus have a small variance. The extreme cases are if $Y$ is a function of $X$, in which case the second term is 0, and if $X$ and $Y$ are independent, in which case the first term is 0 (see Problem 76). For all cases in between, the size of the first term relative to Var[$Y$] measures how good $X$ is at predicting $Y$. We will formalize the idea of strength of dependence in Section 3.8.

***Example 3.32.*** Let us revisit Example 3.28 again and use the proposition to find Var[$Y$]. Since $Y \sim \text{unif}[0, X]$, Proposition 2.13 gives

$$E[Y|X] = \frac{X}{2} \quad \text{and} \quad \text{Var}[Y|X] = \frac{X^2}{12}$$

and we get

$$\text{Var}[Y] = \text{Var}\left[\frac{X}{2}\right] + E\left[\frac{X^2}{12}\right] = \frac{1}{4} \times \frac{1}{12} + \frac{1}{12}\int_0^1 x^2 dx = \frac{7}{144}$$

□

### 3.7.4 Recursive Methods

In Section 1.6.3, we saw several examples of how the law of total probability can be used to compute probabilities with recursive methods. The same type of technique can be applied to compute expected values and variances, and we start by a simple illustration of this.

***Example 3.33.*** Let $X \sim \text{geom}(p)$. Show that $E[X] = 1/p$.

Recall that the geometric distribution counts the number of independent trials needed to get the first success. We condition on the first trial, which can be either success ($S$) or failure ($F$). Let $\mu = E[X]$ to obtain

$$\mu = E[X] = E[X|S]P(S) + E[X|F]P(F)$$

where $P(S) = p$ and $P(F) = 1 - p$. For the conditional expectations, note that if the first trial is a success, then we know that $X = 1$ and hence $E[X|S] = 1$. If the first trial is a failure, one trial has gone by and we start over to wait for the first success and hence $E[X|F] = 1 + \mu$. We get

$$\mu = 1 \times p + (1 + \mu)(1 - p) = 1 + \mu - \mu p$$

which gives $\mu = 1/p$ as desired. There was a little bit of handwaving in the argument. To make it strict, we should let $I_S$ and $I_F$ be the indicators of success or failure in the first trial and let $Y$ be a random variable with the same distribution as $X$. Then, we have the relation

$$X = I_S + I_F(1 + Y) \tag{3.3}$$

and by linearity of expected values and independence of $I_F$ and $Y$, we get

$$\mu = E[I_S] + E[I_F](1 + E[Y]) = p + (1 - p)(1 + \mu)$$

which is the same equation as above. We will generally not be this picky since it is intuitively clear what we are doing. In Problem 77, you are asked to compute the variance with a recursive argument, and then Equation (3.3) comes in handy. □

***Example 3.34.*** A fair coin is flipped repeatedly and we count the number of flips required until the first occurrence of the pattern *HH*. Let $X$ be this number and find the expected value of $X$.

To compute $E[X]$ directly, we would need to find the probabilities $P(X = k)$, which leads to a challenging combinatorial problem. Instead, let us use a recursive approach. There are three different possibilities: (1) the first flip gives $T$ and we start over, (2) the first two flips give *HT* and we start over, and (3) the first two flips give *HH* and we are done (see Figure 3.5). With $\mu = E[X]$, we get

$$\mu = E[X|T] \times \frac{1}{2} + E[X|HT] \times \frac{1}{4} + E[X|HH] \times \frac{1}{4}$$

Clearly, $E[X|HH] = 2$, and for the other conditional expectations, note that in each case we have already spent a number of flips and then start over to wait for *HH*, which takes an additional number of flips with mean $\mu$. Hence,

$$\mu = (1 + \mu)\frac{1}{2} + (2 + \mu)\frac{1}{4} + 2 \times \frac{1}{4} = \frac{6}{4} + \frac{3}{4}\mu$$

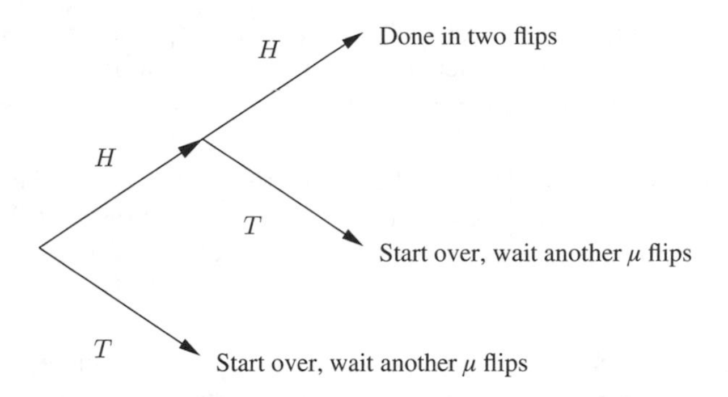


**FIGURE 3.5** Waiting for *HH*.

which gives $\mu = 6$. Hence, on average, we have to wait six flips until the first occurrence of *HH* (see also Problem 80).

Now let us instead consider the same problem for the pattern *TH*. Since the probability to get *TH* in any two flips is $\frac{1}{4}$, the same as the probability to get *HH*, we might guess that the mean number of flips is still 6. Let us compute it by conditioning on the first flip. If this is $H$, we have spent one flip and start over. If it is $T$, we have spent one flip and wait for the first $H$, which takes a number of flips with mean 2 (the mean of a geometric distribution with success probability $\frac{1}{2}$). We get

$$\mu = (1 + \mu)\frac{1}{2} + (1 + 2) \times \frac{1}{2} = \frac{\mu}{2} + 2$$

which gives $\mu = 4$. Although the two patterns *HH* and *TH* are equally likely to occur in any particular two flips, *TH* tends to show up before *HH* after repeated flipping. To compensate for its tardiness, *HH* has the possibility of repeating itself immediately in the next flip (in *HHH* there are two occurrences of *HH*), whereas we must wait at least an additional two flips before we can see *TH* repeat itself. Hence, in a large number of flips, we see on average equally many occurrences of *HH* and *HT*, but *HH* tends to come in bursts and *TH* more evenly spread out.

This "pattern of patterns" becomes more pronounced the longer our pattern sequence is. Recall the game of Penney-ante from Example 1.61, which is constructed so that the inherent asymmetries work to your advantage. It can be shown that the expected number of flips until *HHH* is 14, whereas the same number for *HTH* is 10 and for *HTT* and *HHT* only 8 (see Problem 82).

It turns out that the key in determining the length of the waiting time for a particular pattern is how much *overlap* it has (with itself). For example, the pattern *HTH* has an overlap of length 1 since the last letter can be the first letter of another occurrence of the pattern. For the same reason, the pattern *HHH* has one overlap of length 1 but also one of length 2 since the last two letters can be the first two in another occurrence of the pattern. Finally, the patterns *HHT* and *HTT* have no overlap at all. The higher the amount of overlap, the longer the waiting time and there is a nice and simple formula that expresses the expected waiting time in terms of the amount of overlap.

In Problem 83, you are asked to find this formula in a special case. For a general overview of waiting times for patterns and plenty of other good stuff, see *Problems and Snapshots from the World of Probability* by Blom et al. [2]. □

***Example 3.35.*** Recall Example 1.59, the gambler's ruin problem when Ann starts with a dollar and plays against the infinitely wealthy Bob. We saw that she eventually goes broke, and we now address the question of how long the game can be expected to last.

Let $N$ be the time when Ann goes broke and let $\mu = E[N]$. Again, we condition on the first flip. If this is heads, Ann goes broke immediately, and hence $E[N|H] = 1$. If the first flip is tails, one round of the game has passed and Ann's fortune is \$2. In order for her to go broke eventually, she must first arrive at a fortune of \$1, which takes a time that has mean $\mu$. Then she starts over, and the remaining time until ruin has mean $\mu$. Hence, the total time until ruin, given that the first flip gives tails, is $E[N|T] = 1 + \mu + \mu$ and we get

$$\begin{aligned}\mu &= E[N|H]P(H) + E[N|T]P(T)\\ &= 1 \times \frac{1}{2} + (1 + 2\mu)\frac{1}{2} \;=\; 1 + \mu\end{aligned}$$

and since the only possible solution to the equation $\mu = 1 + \mu$ is $\mu = \infty$, we have shown that the expected time until ruin is infinite! Note that it is the mean of $N$ that is infinite, not the random variable $N$ itself (compare with the St. Petersburg paradox in Example 2.23). Ann will eventually go broke, but it may take a long time. □

## 3.8 COVARIANCE AND CORRELATION

So far we have distinguished between random variables that are independent and those that are not. This is a crude distinction, and in the case of dependent random variables, we would like to be able to quantify the strength of the dependence. Another consideration is that dependence can go in different directions. For example, consider the ideal gas law from chemistry. This states that in 1 mol of gas, the pressure $P$, volume $V$, and temperature $T$ are related through the formula

$$P = \frac{RT}{V}$$

where $R$ is the universal gas constant. Hence, pressure increases with increasing temperature but decreases with increasing volume. We could express this as having a "positive" dependence between $P$ and $T$ but a "negative" dependence between $P$ and $V$.

It would be convenient to have a quantitative measure of the degree and direction of dependence. If $X$ and $Y$ are two random variables, it is reasonable to require such a measure to

- be 0, if $X$ and $Y$ are independent
- be $> 0$, if larger $X$ values on average correspond to larger $Y$ values
- be $< 0$, if larger $X$ values on average correspond to smaller $Y$ values
- be higher (in absolute value) the "stronger" the dependence between $X$ and $Y$.

Only the first requirement is mathematically precise, but hopefully there is a clear intuition for what we want to achieve. Our first attempt at a measure of dependence is the following.

***Definition 3.16.*** The *covariance* of $X$ and $Y$ is defined as

$$\text{Cov}[X, Y] = E\left[(X - E[X])(Y - E[Y])\right]$$

To compute the covariance, we can use Proposition 3.11, but it is often easier to use the following formula, which is easily proved by the repeated use of Corollary 3.3.

**Proposition 3.20.**

$$\text{Cov}[X, Y] = E[XY] - E[X]E[Y]$$

The covariance satisfies the requirements we stated above. For example, by Proposition 3.13(a), we get

**Corollary 3.8.** *If $X$ and $Y$ are independent, then $Cov[X, Y] = 0$.*

In the next section, we will see that the converse is not true; the covariance can be 0 even if $X$ and $Y$ are dependent. The covariance satisfies the other requirements as well. Suppose, for example, that $X$ and $Y$ are "positively dependent," that is, that larger $X$ values tend to correspond to larger $Y$ values. This means that when the factor $X - E[X]$ is positive, the factor $Y - E[Y]$ also tends to be positive and so is the product. On the other hand, if larger $X$ values correspond to smaller $Y$ values, then positive values of $X - E[X]$ correspond to negative values of $Y - E[Y]$, and the product is negative. Hence, we have $\text{Cov}[X, Y] > 0$ in the first case and Cov[X, Y] $< 0$ in the second.

In the same way, we can argue that "strong positive dependence" means that large values of $X - E[X]$ correspond closely to large values of $Y - E[Y]$ and the product becomes large as well. A weak positive dependence means that the correspondence is less precise and the product, while still positive, may not be as large.

***Example 3.36.*** Let $X$ and $Y$ be independent and uniform on $[0, 1]$. Let $C$ be the circumference and $A$ the area of a rectangle with sides $X$ and $Y$. Find the covariance of $A$ and $C$.

Since $A = XY$ and $C = 2X + 2Y$, we get $AC = 2X^2Y + 2XY^2$ and hence

$$E[AC] = 2E[X^2]E[Y] + 2E[X]E[Y^2] = 2 \times \frac{1}{3} \times \frac{1}{2} + 2 \times \frac{1}{2} \times \frac{1}{3} = \frac{2}{3}$$

which gives

$$\text{Cov}[A, C] = E[AC] - E[A]E[C] = \frac{2}{3} - \frac{1}{4} \times 2 = \frac{1}{6}$$

□

One important use of the covariance is that it allows us to state a general formula for the variance of a sum of random variables.

**Proposition 3.21.**

$$\text{Var}[X + Y] = \text{Var}[X] + \text{Var}[Y] + 2\,\text{Cov}[X, Y]$$

*Proof.* By the definitions of variance and covariance and repeated use of properties of expected values, we get

$$\begin{aligned}
\text{Var}[X + Y] &= E\left[(X + Y - E[X + Y])^2\right] \\
&= E\left[(X - E[X] + Y - E[Y])^2\right] \\
&= E\left[(X - E[X])^2 + (Y - E[Y])^2 + 2(X - E[X])(Y - E[Y])\right] \\
&= \text{Var}[X] + \text{Var}[Y] + 2\,\text{Cov}[X, Y]
\end{aligned}$$

and we are done. ■

Note that Proposition 3.13(b) is a special case of this with $\text{Cov}[X, Y] = 0$.

***Example 3.37.*** In Example 3.18, we had $X \sim \text{unif}[0, 1]$ and $Y = -X$. Verify that $\text{Var}[X + Y] = 0$.

We have $\text{Var}[X] = \text{Var}[Y] = \frac{1}{12}$ and covariance

$$\text{Cov}[X, Y] = \text{Cov}[X, -X] = E[X(-X)] - E[X]E[-X] = -\text{Var}[X]$$

and Proposition 3.21 gives $\text{Var}[X + Y] = 0$. □

Let us investigate some of the properties of the covariance. We start with the case of two random variables and proceed to a more general result. The proof of the following proposition is a straightforward application of the definition and the properties of expected values.

**Proposition 3.22.** Let $X$, $Y$, and $Z$ be random variables and let $a$ and $b$ be real numbers. Then

(a) $\text{Cov}[X, X] = \text{Var}[X]$
(b) $\text{Cov}[aX, bY] = ab\,\text{Cov}[X, Y]$
(c) $\text{Cov}[X + Y, Z] = \text{Cov}[X, Z] + \text{Cov}[Y, Z]$

By Proposition 3.22 and induction, we can now show that covariances are *bilinear* in the following sense.

**Proposition 3.23.** Let $X_1, \ldots, X_n$ and $Y_1, \ldots, Y_m$ be random variables and let $a_1, \ldots, a_n$ and $b_1, \ldots, b_m$ be real numbers. Then

$$\text{Cov}\left[\sum_{j=1}^{n} a_j X_j, \sum_{k=1}^{m} b_k Y_k\right] = \sum_{j=1}^{n}\sum_{k=1}^{m} a_j b_k \text{Cov}[X_j, Y_k]$$

We also get the following general variance formula, proved by combining Propositions 3.22(a) and 3.23.

**Proposition 3.24.** Let $X_1, \ldots, X_n$ be random variables and let $a_1, \ldots, a_n$ be real numbers. Then

$$\text{Var}\left[\sum_{k=1}^{n} a_k X_k\right] = \sum_{k=1}^{n} a_k^2 \text{Var}[X_k] + \sum_{i \neq j} a_i a_j \text{Cov}[X_i, X_j]$$

***Example 3.38.*** Let $X$ have a hypergeometric distribution with parameters $N$, $r$, and $n$. Prove Proposition 2.20 about the mean and variance of $X$.

Recall that $X$ is the number of special objects, when we draw $n$ out of $N$ objects, $r$ of which are special. Just as for the binomial distribution, we can solve this by introducing indicators. Thus, let

$$I_k = \begin{cases} 1 & \text{if the } k\text{th draw gives a special object} \\ 0 & \text{otherwise} \end{cases}$$

to obtain

$$X = \sum_{k=1}^{n} I_k$$

It is easily realized that $I_k$ have the same distribution and that

$$P(I_k = 1) = \frac{r}{N}$$

which is the proportion of special objects. This gives the mean of $X$ as

$$E[X] = \sum_{k=1}^{n} E[I_k] = \sum_{k=1}^{n} \frac{r}{N} = \frac{nr}{N}$$

which proves the first part of Proposition 2.20. The $I_k$ are not independent this time since, for example,

$$P(I_2 = 1) = \frac{r}{N}$$

but

$$P(I_2 = 1 | I_1 = 1) = \frac{r-1}{N-1}$$

since repeated drawing without replacement changes the proportion of special objects. To find the variance of $X$, we thus need to use the general formula from Proposition 3.24. The variance of $I_k$ is

$$\text{Var}[I_k] = \frac{r}{N}\left(1 - \frac{r}{N}\right)$$

and we also need to find the covariances $\text{Cov}[I_j, I_k]$. By symmetry, we realize that these are the same for all $j \neq k$, so let us find $\text{Cov}[I_1, I_2]$. We have

$$\text{Cov}[I_1, I_2] = E[I_1 I_2] - E[I_1]E[I_2]$$

where

$$E[I_1 I_2] = P(I_1 = 1, I_2 = 1)$$

since this is the only case when $I_1 I_2 \neq 0$. But

$$P(I_1 = 1, I_2 = 1) = P(I_2 = 1 | I_1 = 1)P(I_1 = 1) = \frac{r-1}{N-1} \times \frac{r}{N}$$

and we get

$$\text{Cov}[I_1, I_2] = \frac{r-1}{N-1} \times \frac{r}{N} - \frac{r}{N}\frac{r}{N} = -\frac{r(N-r)}{N^2(N-1)}$$

which finally gives

$$\begin{aligned}\mathrm{Var}[X] &= \sum_{k=1}^{n} \mathrm{Var}[I_k] + \sum_{i \neq j} \mathrm{Cov}[I_j, I_k] \\ &= n\frac{r}{N}\left(1 - \frac{r}{N}\right) - n(n-1)\frac{r(N-r)}{N^2(N-1)} \\ &= n \times \frac{N-n}{N-1} \times \frac{r}{N}\left(1 - \frac{r}{N}\right)\end{aligned}$$

after doing some algebra. Note that if we draw *with* replacement, we get a binomial distribution with parameters $n$ and $p = r/N$. The mean is the same, but the variance is changed by a factor $(N-n)/(N-1)$. If $N$ is large relative to $n$, this factor is approximately 1, which indicates that the hypergeometric distribution in that case can be approximated by a binomial distribution. □

Note how the general variance formula shows that the assumptions of independence in Propositions 3.13 and 3.15 can be replaced by $\mathrm{Cov}[X_j, X_k] = 0$, $j \neq k$, a weaker assumption. However, since independence is a central concept, we stick to the formulations that we have.

### 3.8.1 The Correlation Coefficient

The covariance is an important and widely used concept in probability theory, but it turns out that it has shortcomings as a measure of dependence. To see where the problem lies, consider the experiment of measuring the weight and height of a randomly chosen individual. Let $X$ be the weight in pounds and $Y$ the height in inches. The degree of dependence between $X$ and $Y$ is then $\mathrm{Cov}[X, Y]$. Now let us instead use the metric system, and let $S$ denote the weight in kilograms and $T$ the height in centimeters. Then $S = 0.45X$ and $T = 2.5Y$ and Proposition 3.22(b) gives

$$\mathrm{Cov}[S, T] = \mathrm{Cov}[0.45X, 2.5Y] = 1.12\,\mathrm{Cov}[X, Y]$$

so if covariance measures the degree of dependence, we would thus have "12% stronger dependence" if we use kilograms and centimeters than if we use pounds and inches. If we further change from centimeters to millimeters, the covariance increases by another factor of 100! Clearly this makes no sense, and we realize that any reasonable measure of dependence must be *dimensionless*, that is, not be affected by such changes in units of measure. It turns out that we can achieve this by scaling the covariance by the standard deviations, and we state the following definition.

***Definition 3.17.*** The *correlation coefficient* of $X$ and $Y$ is defined as

$$\rho(X, Y) = \frac{\mathrm{Cov}[X, Y]}{\sqrt{\mathrm{Var}[X]\mathrm{Var}[Y]}}$$

The correlation coefficient is dimensionless. To demonstrate this, take $a, b > 0$ and note that

$$\rho(aX, bY) = \frac{\text{Cov}[aX, bY]}{\sqrt{\text{Var}[aX]\text{Var}[bY]}}$$

$$= \frac{ab\text{Cov}[X, Y]}{\sqrt{a^2\text{Var}[X]b^2\text{Var}[Y]}} = \rho(X, Y)$$

We also call $\rho(X, Y)$ simply the *correlation*[3] between $X$ and $Y$. Here are some more good properties of the correlation coefficient.

**Proposition 3.25.** The correlation coefficient of any pair of random variables $X$ and $Y$ satisfies

(a) $-1 \leq \rho(X, Y) \leq 1$
(b) If $X$ and $Y$ are independent, then $\rho(X, Y) = 0$
(c) $\rho(X, Y) = 1$ if and only if $Y = aX + b$, where $a > 0$
(d) $\rho(X, Y) = -1$ if and only if $Y = aX + b$, where $a < 0$

*Proof.* Let $\text{Var}[X] = \sigma_1^2$ and $\text{Var}[Y] = \sigma_2^2$. For (a), first apply Proposition 3.21 to the random variables $X/\sigma_1$ and $Y/\sigma_2$ and use properties of the variance and covariance to obtain

$$0 \leq \text{Var}\left[\frac{X}{\sigma_1} + \frac{Y}{\sigma_2}\right] = \frac{\text{Var}[X]}{\sigma_1^2} + \frac{\text{Var}[Y]}{\sigma_2^2} + \frac{2\text{Cov}[X, Y]}{\sigma_1\sigma_2} = 2 + 2\rho$$

which gives $\rho \geq -1$. To show that $\rho \leq 1$, instead use $X/\sigma_1$ and $-Y/\sigma_2$. Part (b) follows from Corollary 3.8, and parts (c) and (d) follow from Proposition 2.16, applied to the random variables $X/\sigma_1 - Y/\sigma_2$ and $X/\sigma_1 + Y/\sigma_2$, respectively. Note that this also gives $a$ and $b$ expressed in terms of the means, variances, and correlation coefficient (see Problem 90). ■

The correlation coefficient is thus a number between $-1$ and 1, where $-1$ and 1 denote the maximum degrees of dependence in the sense that $Y$ can be computed from $X$ with certainty. Part (b) states that independent random variables have correlation 0, but it turns out that the converse of this is not true; the correlation coefficient can be 0 even if the random variables are dependent, as we will see in an example.

[3]In the electrical engineering literature, the term "correlation" is used in a slightly different meaning, namely, to denote $E[XY]$. Statisticians refer to this quantity as the *product moment*.

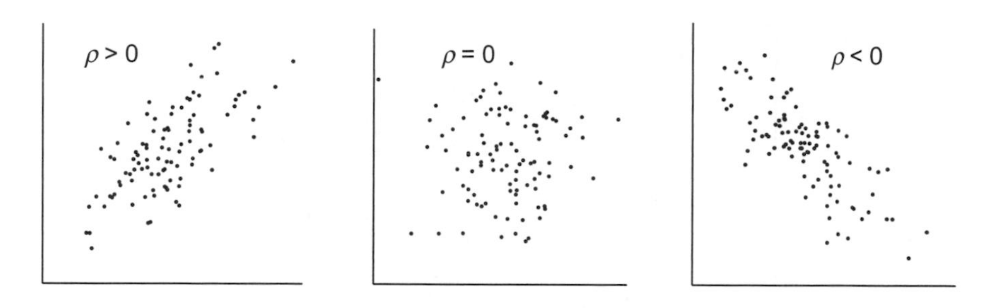


**FIGURE 3.6** Simulated observations on pairs $(X, Y)$ that are positively correlated, uncorrelated, and negatively correlated, respectively.

***Example 3.39.*** Choose a point $(X, Y)$ uniformly in the unit disk. What is $\rho(X, Y)$?

The joint pdf is $f(x, y) = 1/\pi$. Denote the unit disk by $D$ to obtain

$$E[XY] = \frac{1}{\pi} \iint_D xy\,dx\,dy$$

which we realize must equal 0 by symmetry. Also, by symmetry, $E[X] = E[Y] = 0$, so we get $\text{Cov}[X, Y] = 0$ and hence also $\rho(X, Y) = 0$. Thus, the correlation is 0, but we already know that $X$ and $Y$ are dependent random variables. □

The explanation is that the correlation coefficient measures only the degree of *linear dependence*. Thus, if we have a number of observations on $(X, Y)$, the correlation coefficient measures how well the observations fit a straight line. If $\rho(X, Y) > 0$, $X$ and $Y$ are said to be *positively correlated*; if $\rho(X, Y) < 0$, they are *negatively correlated*; and if $\rho(X, Y) = 0$, they are *uncorrelated* (see Figure 3.6 for an illustration with simulated values). We have seen that

$$X, Y \text{ independent} \Rightarrow X, Y \text{ uncorrelated}$$

but not the converse. In the case of the uniform points in the disk, although $X$ and $Y$ are dependent, there is no linear dependence (see Figure 3.7 for an illustration with simulated values). Note how the boundaries of the disk are clear but how there is no particular straight line that would fit better than any other. Thus, there is no linear dependence and, in fact, no functional dependence at all.

The following example shows how there can be functional dependence (in this case quadratic) and how the correlation coefficient can fail to detect it but how the dependence structure can still be revealed by an appropriate change of random variables.

***Example 3.40.*** Let $X \sim \text{unif}[0, 1]$ and let $A$ be the area of a disk with radius $X$. What is $\rho(X, A)$?

**FIGURE 3.7** One hundred simulated points, uniform in a disk. Note how the boundary is clearly discernible, indicating dependence, but how there is no obvious way to fit a straight line, indicating uncorrelatedness.

We have $E[X] = \frac{1}{2}$, $\text{Var}[X] = \frac{1}{12}$, and since $A = \pi X^2$, we get

$$E[A] = \pi \int_0^1 x^2 dx = \frac{\pi}{3}, \quad E[A^2] = \pi^2 \int_0^1 x^4 dx = \frac{\pi^2}{5}$$

which gives $\text{Var}[A] = 4\pi^2/45$. We also obtain

$$E[XA] = \pi E[X^3] = \pi \int_0^1 x^3 dx = \frac{\pi}{4}$$

which gives $\text{Cov}[X, A] = E[XA] - E[X]E[A] = \pi/12$. We get

$$\rho(X, A) = \frac{\pi/12}{\sqrt{(1/12) \times (4\pi^2/45)}} \approx 0.97$$

which is certainly high but does not reveal the fact that $A$ is completely determined by $X$. However, if we instead compute the correlation coefficient of $A$ and the random variable $Y = X^2$, we have $A = \pi Y$, and by Proposition 3.25(c), $\rho(A, Y) = 1$. Hence, by changing the random variable, we can view the quadratic dependence on $X$ as linear dependence on $X^2$ and reveal the completely deterministic relation between $X$ and $A$. □

***Example 3.41.*** If $X \sim \text{unif}[0, 1]$ and given $X = x$, $Y \sim \text{unif}[0, x]$, what is $\rho(X, Y)$?

From previous treatment of this example, we know that the means are

$$E[X] = \frac{1}{2} \quad \text{and} \quad E[Y] = \frac{1}{4}$$

and the variances

$$\text{Var}[X] = \frac{1}{12} \quad \text{and} \quad \text{Var}[Y] = \frac{7}{144}$$

so all we need to find is the covariance $\text{Cov}[X, Y]$ and since

$$\text{Cov}[X, Y] = E[XY] - E[X]E[Y]$$

all that remains is to compute $E[XY]$. A direct application of Proposition 3.11 gives

$$E[XY] = \int_0^1 \int_0^x xy\frac{1}{x}dy\,dx = \frac{1}{6}$$

which gives the correlation

$$\rho(X, Y) = \frac{1/6 - (1/2) \times (1/4)}{\sqrt{(1/12) \times (7/144)}} \approx 0.65$$

□

The logical next question is "What is a high correlation?" We can understand the extremes of correlations of 0, −1, or 1 and also values close to them, as in Example 3.40. But how about intermediate values? For example, is the 0.65 we got in the last example a high correlation? What does it tell us about the strength of the relationship between $X$ and $Y$? To come up with an interpretation of the value of $\rho$, we revisit the concept of prediction from Section 3.7.2.

Recall how we argued that $E[Y|X]$ is the best predictor of $Y$, based on observing $X$. To compute the conditional expectation, we need to know the joint distribution of $(X, Y)$. This is not necessarily the case, and even if we do, the calculations may be intractable. If we restrict our goal to finding the best *linear* predictor, that is, of the form $l(X) = aX + b$, we need to know only the means, variances, and the correlation coefficient, as the following result shows.

**Proposition 3.26.** Let $X$ and $Y$ be random variables with means $\mu_X$ and $\mu_Y$, variances $\sigma_X^2$ and $\sigma_Y^2$, and correlation coefficient $\rho$. The *best linear predictor* of $Y$ based on $X$ is

$$l(X) = \mu_Y + \rho\frac{\sigma_Y}{\sigma_X}(X - \mu_X)$$

*Proof.* Suppose first that $\mu_X = \mu_Y = 0$ and $\sigma_X^2 = \sigma_Y^2 = 1$. The mean square error is then

$$E[(Y - (aX + b))^2] = E[Y^2] + a^2 E[X^2] + b^2 - 2aE[XY] = 1 + b^2 + a(a - 2\rho)$$

which is minimized for $b = 0$ and $a = \rho$. For the general case, consider the random variables $(X - \mu_X)/\sigma_X$ and $(Y - \mu_Y)/\sigma_Y$, which have means 0 and variances 1, and apply the result we just proved. ■

Thus, for any two random variables $X$ and $Y$, the correlation coefficient can be regarded as a parameter that is chosen so that the relation between $X$ and $Y$ becomes "as linear as possible." If we have a number of observations on $(X, Y)$, we can regard the line $y = l(x)$ as the best linear fit, whereas the curve $y = E[Y|X = x]$ is the best possible fit among all functions.

***Example 3.42.*** Let $X \sim \text{unif}[0, 1]$ and given $X = x$, let $Y \sim \text{unif}[0, x^2]$. Find $E[Y|X = x]$ and $l(x)$.

It is clear from the description that $E[Y|X = x] = x^2/2$. For $l(x)$, we need to find all the parameters involved in its definition. We have $E[X] = \frac{1}{2}$ and $\text{Var}[X] = \frac{1}{12}$, and to find the mean of $Y$, as well as the correlation coefficient, we use conditional expectations and variances. Since

$$\rho \frac{\sigma_Y}{\sigma_X} = \frac{\text{Cov}[X, Y]}{\sigma_X^2}$$

we do not need to compute the variance of $Y$. We have $E[Y|X] = X^2/2$, which gives

$$\mu_Y = E[Y] = E\left[\frac{X^2}{2}\right] = \frac{1}{2}\int_0^1 x^2 dx = \frac{1}{6}$$

and for the covariance we also need $E[XY]$, which equals $\frac{1}{8}$, by Example 3.31. We get

$$\text{Cov}[X, Y] = E[XY] - E[X]E[Y] = \frac{1}{24}$$

which finally gives

$$l(x) = \frac{1}{6} + \frac{1/24}{1/12}\left(x - \frac{1}{2}\right) = \frac{x}{2} - \frac{1}{12}$$

which is depicted in Figure 3.8 together with $E[Y|X = x]$ and 100 simulated observations. Overall, the fits are both pretty good, but the detail in Figure 3.8b shows that the linear fit is poor near the origin. □

In the previous example, the line $l(x)$ is computed from the values of the parameters $\mu_X$, $\mu_Y$, and so on, and we can see how well the line fits the observations. In practice,

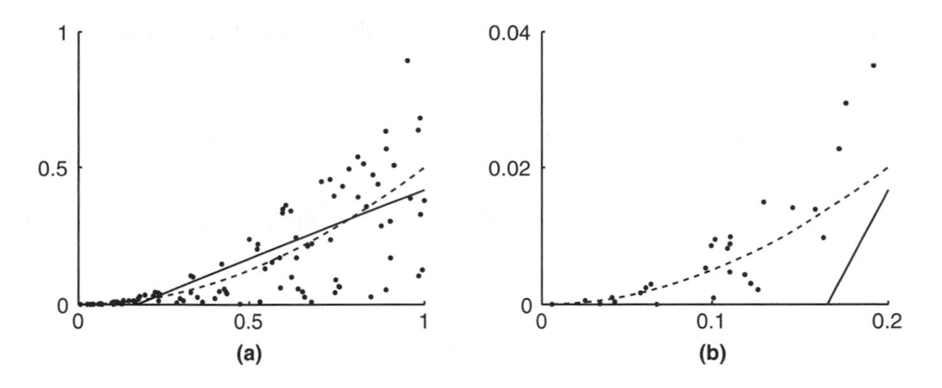


**FIGURE 3.8** (a) One hundred simulated values of $(X, Y)$ in Example 3.42 together with the curve $y = E[Y|X = x]$ (dashed) and the line $y = l(x)$ (solid). (b) A detail of the region near the origin.

the situation is usually the reverse; we start from a set of observations, do not know the values of the parameters, and wish to find the straight line that best fits the data. This means that we are trying to find the ideal line $l(x)$ by finding approximations of the parameters on the basis of the data. We will return to this relationship between theory and data in Chapter 6.

We still do not know what a particular value of $\rho$ means, so let us now finally address this question. One way to evaluate the linear fit is to compare the variances of $Y$ and $l(X)$. The random variable $Y$ has a certain variance $\sigma_Y^2$. If $Y$ is strongly correlated with $X$, then $Y$ is closely approximated by the straight line $l(X)$, and most of the variation in $Y$ can be attributed to variation in $X$. By computing the variance of $l(X)$, we get

$$\text{Var}[l(X)] = \rho^2 \frac{\sigma_Y^2}{\sigma_X^2} \text{Var}[X] = \rho^2 \sigma_Y^2$$

and see that $\text{Var}[l(X)] \leq \text{Var}[Y]$ always holds. Moreover, we see that

$$\frac{\text{Var}[l(X)]}{\text{Var}[Y]} = \rho^2$$

which gives an interpretation of $\rho^2$; it measures how much of the variation in $Y$ can be explained by a linear relationship to $X$. The number $\rho^2$ is called the *coefficient of determination*. Hence, a value of $\rho = 0.65$, as in Example 3.41, gives $\rho^2 = 0.42$ with the interpretation that $X$ explains about 42% of the variation in $Y$, whereas the rest is due to other sources of randomness. The following example is illustrative and good to remember.

***Example 3.43.*** Roll two fair dice, letting $X$ and $Y$ be the numbers and letting $Z = X + Y$, the total number. What is $\rho(X, Z)$?

First, we find the covariance of $X$ and $Z$. By Proposition 3.22 and Example 2.25, we have

$$\mathrm{Cov}[X, Z] = \mathrm{Cov}[X, X+Y] = \mathrm{Var}[X] + \mathrm{Cov}[X, Y] = \frac{35}{12}$$

since $X$ and $Y$ are independent and hence $\mathrm{Cov}[X, Y] = 0$. Furthermore, by Proposition 3.13(b)

$$\mathrm{Var}[Z] = \mathrm{Var}[X] + \mathrm{Var}[Y] = \frac{35}{6}$$

and we get

$$\rho(X, Z) = \frac{\mathrm{Cov}[X, Z]}{\sqrt{\mathrm{Var}[X]\mathrm{Var}[Z]}} = \frac{1}{\sqrt{2}}$$

and if we square this, we get the coefficient of determination $\rho^2 = \frac{1}{2}$. Thus, half the variation in $Z$ is explained by variation in $X$, which is intuitively clear since $Z = X + Y$, where $X$ and $Y$ are independent and on average contribute equally to the sum (see also Problem 99). □

Recall from Section 2.5.6 how we can summarize a data set by stating its mean and variance and how this can be viewed in the context of random variables. The same is true for the correlation. Given a data set $D = \{(x_1, y_1), \ldots, (x_n, y_n)\}$, we define its correlation as

$$r = \frac{\sum_{k=1}^{n}(x_k - \bar{x})(y_k - \bar{y})}{\sqrt{\sum_{k=1}^{n}(x_k - \bar{x})^2 \sum_{k=1}^{n}(y_k - \bar{y})^2}}$$

Again, we can think of this as an approximation of the true but unknown correlation, computed from a set of observations.

***Example 3.44.*** The following data are eight of astronomer Edwin Hubble's 1929 measurements on a galaxy's distance from Earth (megaparsecs) and recession velocity (km/s).

Distance: 0.032, 0.263, 0.45, 0.63, 0.90, 1.00, 1.4, 2.0
Velocity: 170, −70, 200, 200, 650, 920, 500, 850

Computing the sums in the expression above gives $r = 0.78$. We view this as an approximation of the true value, an approximation that would get better the more data we have. We will return to Hubble's data in Chapter 6. □

It is important to realize that the correlation coefficient does not measure *causation*, only *association* between random variables. This means that even if there is a high correlation, it does not necessarily follow that large values of one random variable *cause* large values of the other, only that such large values tend to appear together. One amusing example is to let $X$ be the number of firefighters sent to a fire and $Y$ the economic damage in dollars of the fire. It can then be observed that there is a positive correlation between $X$ and $Y$. Conclusion? To keep your economic loss down in case of a fire, don't call the fire department? In this case, there is a third variable that explains the correlation: the size of the fire. The correlation is thus caused by an underlying factor that affects both $X$ and $Y$.

A similar effect can also be caused by an intermediate factor, for example, in the Berkeley admissions data in Example 1.46, where a correlation between gender and admission rate was demonstrated, but the real cause was that female students applied for majors that were more difficult to be admitted to. Thus, gender affected the choice of major, which in turn affected the chance of being admitted.

## 3.9 THE BIVARIATE NORMAL DISTRIBUTION

In Section 2.7, the importance of the normal distribution was pointed out. In this section, its two-dimensional analog is introduced.

***Definition 3.18.*** If $(X, Y)$ has joint pdf

$$f(x, y) = \frac{1}{2\pi\sigma_1\sigma_2\sqrt{1-\rho^2}}$$

$$\times \exp\left\{-\frac{1}{2(1-\rho^2)}\left(\frac{(x-\mu_1)^2}{\sigma_1^2} + \frac{(y-\mu_2)^2}{\sigma_2^2} - \frac{2\rho(x-\mu_1)(y-\mu_2)}{\sigma_1\sigma_2}\right)\right\}$$

for $x, y \in R$, then $(X, Y)$ is said to have a *bivariate normal distribution*.

The formula is ugly, but the pdf itself is beautiful (see Figure 3.9). The bivariate normal distribution has five parameters: $\mu_1, \mu_2 \in R, \sigma_1, \sigma_2 > 0$, and $-1 < \rho < 1$. The center of the bell in Figure 3.9 is above the point $(\mu_1, \mu_2)$, and its shape is determined by the other three parameters. The notation suggests that these are means, variances, and the correlation coefficient of $X$ and $Y$, but we need to prove that this is actually the case.

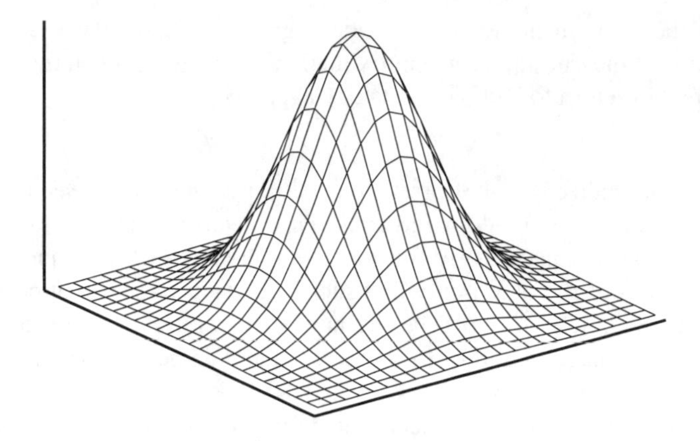

**FIGURE 3.9** The joint pdf of a bivariate normal distribution.

**Proposition 3.27.** Let $(X, Y)$ have a bivariate normal distribution with parameters $\mu_1, \mu_2, \sigma_1, \sigma_2, \rho$. Then

(a) $X \sim N(\mu_1, \sigma_1^2)$ and $Y \sim N(\mu_2, \sigma_2^2)$
(b) $\rho$ is the correlation coefficient of $X$ and $Y$

*Proof.* (a) Let $f(x, y)$ be the joint pdf of $(X, Y)$. To find the pdf of $X$, we need to integrate the joint pdf with respect to $y$. We first make the change of variables $u = (x - \mu_1)/\sigma_1$ and $v = (y - \mu_2)/\sigma_2$, so that $dy = \sigma_2\, dv$ and the integral limits remain $-\infty$ and $\infty$. We get

$$
\begin{aligned}
f_X(x) &= \int_{-\infty}^{\infty} f(x, y)dy \\
&= \frac{1}{2\pi\sigma_1\sqrt{1-\rho^2}} \int_{-\infty}^{\infty} \exp\left(-\frac{1}{2(1-\rho^2)}(u^2 + v^2 - 2\rho u v)\right) dv
\end{aligned}
$$

Now note that $u^2 + v^2 - 2\rho u v = (v - \rho u)^2 + u^2(1 - \rho^2)$ and hence

$$
\begin{aligned}
f_X(x) = &\frac{1}{\sigma_1\sqrt{2\pi}} \exp\left(-\frac{u^2}{2}\right) \\
&\times \int_{-\infty}^{\infty} \frac{1}{\sqrt{1-\rho^2}\sqrt{2\pi}} \exp\left(-\frac{1}{2(1-\rho^2)}(v - \rho u)^2\right) dv
\end{aligned}
$$

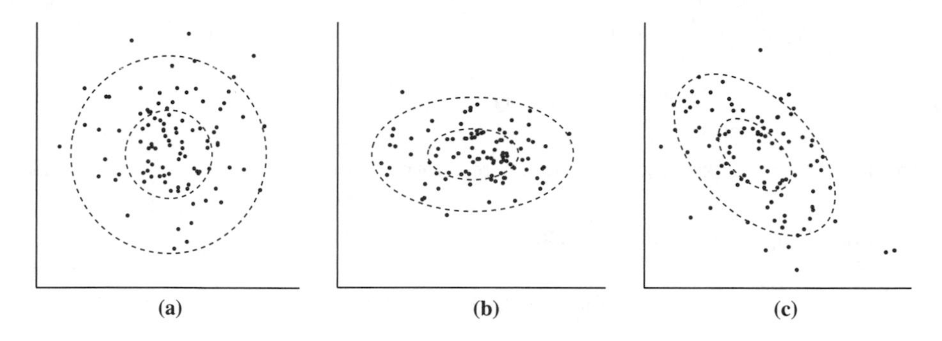


**FIGURE 3.10** Simulated observations from bivariate normal distributions: (a) $\rho = 0$, $\sigma_1^2 = \sigma_2^2$; (b) $\rho = 0, \sigma_2^2 > \sigma_1^2$; (c) $\rho < 0$.

Now note that the integrand in the last integral is the pdf of a normal distribution with mean $\rho u$ and variance $1 - \rho^2$, and hence this integral equals 1. Finally, substituting back $u = (x - \mu_1)/\sigma_1$ gives

$$f_X(x) = \frac{1}{\sigma_1\sqrt{2\pi}} \exp\left(-\frac{(x - \mu_1)^2}{2\sigma_1^2}\right)$$

which we recognize as the pdf of a normal distribution with mean $\mu_1$ and variance $\sigma_1^2$. We prove (b) in Example 3.45. ■

It is instructive to consider the *contours* of the joint pdf, that is, the sets of points $(x, y)$ for which $f(x, y)$ equals some fixed constant. These contours are ellipses that reflect how observations are typically spread in the plane. Such an ellipse is centered in $(\mu_1, \mu_2)$, and its shape and direction depend on the other parameters. If $\rho = 0$, the ellipse is parallel to the axes, and if, in addition, $\sigma_1^2 = \sigma_2^2$, it is in fact a circle (see Figure 3.10 for an illustration of some different cases). Each plot shows 100 simulated observations from a bivariate normal distribution, displayed together with two contours.

We have seen that a bivariate normal distribution has normal marginals. Is the converse also true; that is, if $X$ and $Y$ are normal, then the pair $(X, Y)$ is bivariate normal? The answer is "almost," but if $Y$ is a linear function of $X$, the pair $(X, Y)$ is concentrated on a straight line and there is no joint pdf. We can also see this in the expression for $f(x, y)$ above since $\rho$ would then be $-1$ or 1. It is also possible to construct a nonnormal joint pdf such that both marginals are normal, but this is not of much practical interest and we will not address it further. A nice property of the bivariate normal distribution is that not only the marginals but also the conditional distributions are normal.

**Proposition 3.28.** Let $(X, Y)$ be bivariate normal. Then, for fixed $x \in R$

$$Y|X = x \sim N\left(\mu_2 + \rho\frac{\sigma_2}{\sigma_1}(x - \mu_1), \sigma_2^2(1 - \rho^2)\right)$$

*Proof.* Carry out the division in

$$f_Y(y|x) = \frac{f(x, y)}{f_X(x)}$$

and identify the expression with the desired normal pdf. ■

Note that Proposition 3.28 states that

$$E[Y|X] = \mu_2 + \rho\frac{\sigma_2}{\sigma_1}(X - \mu_1)$$

which we note is a linear function of $X$. Hence, in the case of a bivariate normal distribution, the best predictor of $Y$ given $X$ is linear (see also Proposition 3.26). This indicates that the dependence that exists between $X$ and $Y$ can be attributed to linear dependence, and the following result shows that this is indeed so.

**Proposition 3.29.** Let $(X, Y)$ be bivariate normal. Then $X$ and $Y$ are independent if and only if they are uncorrelated.

*Proof.* Simply note that the joint pdf equals the product of the marginal pdfs if and only if $\rho = 0$. ■

This is a convenient result since it says that if we cannot find any linear dependence structure in observations that are bivariate normal, then there is no dependence structure at all. Another ramification of Proposition 3.28 is that we can easily prove Proposition 3.27(b) that $\rho$ is the correlation coefficient of $X$ and $Y$.

***Example 3.45.*** Let $(X, Y)$ be bivariate normal. Show that $\rho(X, Y) = \rho$.

First, let $Z$ and $W$ be standard normal. By Corollaries 3.5 and 3.6, we obtain

$$\begin{aligned}\text{Cov}[Z, W] = E[ZW] &= E\left[E[ZW|Z]\right] \\ &= E\left[ZE[W|Z]\right] = \rho E[Z^2] = \rho\end{aligned}$$

and since both variances are 1, this is also the correlation coefficient. Next, consider the random variables $X = \mu_1 + \sigma_1 Z$ and $Y = \mu_2 + \sigma_2 W$. Then $X \sim N(\mu_1, \sigma_1^2)$ and $Y \sim N(\mu_2, \sigma_2^2)$, and since

$$\text{Cov}[X, Y] = \text{Cov}[X - \mu_1, Y - \mu_2]$$

we get

$$\rho(X, Y) = \rho(\sigma_1 Z, \sigma_2 W) = \rho(Z, W) = \rho$$

□

***Example 3.46.*** At a measuring station for air pollutants, the amounts of ozone and carbon particles are recorded at noon every day. Let $X$ be the concentration of carbon ($\mu$g/m$^3$) and $Y$ the concentration of ozone (ppm), and suppose that $(X, Y)$ has a bivariate normal distribution such that $X \sim N(10.7, 29.0)$, $Y \sim N(0.1, 0.02)$, and $\rho = 0.72$. The ozone level is considered unhealthy if it exceeds 0.30. Suppose that the equipment used to measure ozone fails, so we can measure only the carbon level. If this turns out to be 20.0 $\mu$g/m$^3$, what is (a) the predicted ozone level and (b) the probability that the ozone level is unhealthy?
We have observed $X = 20.0$. For (a), we need to find $E[Y|X = 20]$. From Proposition 3.28, we know that $Y$ is normal with mean

$$E[Y|X = 20] = 0.1 + 0.72\frac{\sqrt{0.02}}{\sqrt{29.0}}(20 - 10.7) \approx 0.28$$

so this is the predicted ozone level asked for in part (a). For (b), we wish to find the probability

$$P(Y > 0.30|X = 20.0)$$

and also need the conditional variance, which is

$$\text{Var}[Y|X = 20] = 0.02(1 - 0.72^2) \approx 0.01$$

which gives

$$\begin{aligned} P(Y > 0.30|X = 20.0) &= 1 - P(Y \leq 0.30|X = 20) \\ &= 1 - \Phi\left(\frac{0.30 - 0.28}{\sqrt{0.01}}\right) \approx 0.42 \end{aligned}$$

□

The line

$$y = E[Y|X = x] = \mu_2 + \rho\frac{\sigma_2}{\sigma_1}(x - \mu_1)$$

is called the *regression line*. For an observed value $X = x$, it is the best predicted value of $y$, in the sense discussed in Section 3.7.2. It might be tempting to believe that the regression line is an axis of the ellipses describing the pdf, but this is not the case. The reason is probably best understood by considering a plot of observed values. Figure 3.11 shows 500 simulated values from a bivariate normal distribution with means 0, variances 1, and correlation $\rho = 0.5$. The major axis is the dashed line $y = x$ and the regression line, the solid line $y = 0.5x$. Since the regression line in a point $x$ is the conditional mean of $Y$, given $X = x$, it has roughly the same number of points above and below for any given $x$. Look, for example, at $x$ in the vicinity of 2. Note how the regression line cuts through the points roughly in the middle, and compare with the ellipse axis, which is entirely above all the points.

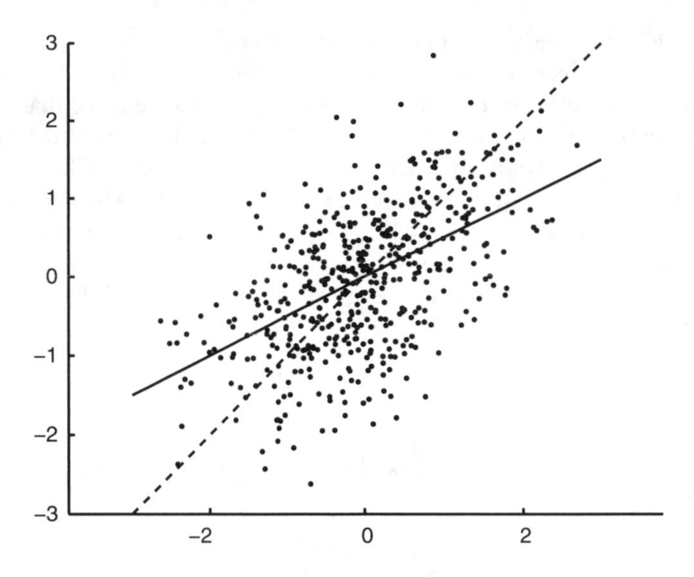


**FIGURE 3.11** Simulated observations on $(X, Y)$, with the corresponding regression line (solid) and major ellipse axis (dashed).

There is an interesting observation to be made here. Suppose, for example, that the points represent standardized test scores ($\mu = 0, \sigma = 1$) for students where $X$ is the score on the first test and $Y$ the score on the second. The positive correlation reflects that students with high scores on one test are expected to have high scores on the other as well. Now consider students with good scores on the first test, say, above 1. On the second test, most of them did worse! What happened? Well, this is completely normal. If your test score is $X = 1$ on the first test, your predicted score on the second test is $E[Y|X = 1] = 0.5$. In the same way, those who did poorly on the first test are expected to do better on the second. It is important to understand that such changes are natural and do not necessarily indicate anything in particular. The phenomenon is called *regression to the mean*[4] and is frequently misinterpreted.

By working through more tedious calculations, it can be shown that linear combinations are also normal. More specifically, consider the following proposition.

**Proposition 3.30.** Let $(X, Y)$ be bivariate normal and let $a$ and $b$ be real numbers. Then

$$aX + bY \sim N\left(a\mu_1 + b\mu_2, a^2\sigma_1^2 + b^2\sigma_2^2 + 2ab\rho\sigma_1\sigma_2\right)$$

In particular, if $X$ and $Y$ are independent, this leads to the following corollary.

[4]The British aristocrat and scientist Sir Francis Galton, who was one of the first to use linear regression methods, used the term "regression to mediocrity."

**Corollary 3.9.** *If $X \sim N(\mu_1, \sigma_1^2)$, $Y \sim N(\mu_2, \sigma_2^2)$, and $X$ and $Y$ are independent, then*

$$aX + bY \sim N\left(a\mu_1 + b\mu_2, a^2\sigma_1^2 + b^2\sigma_2^2\right)$$

The noteworthy property of the bivariate normal distribution is not that the means and variances add the way as they do; we know that this holds generally. What is noteworthy is that the normal distribution is retained by taking a linear combination of $X$ and $Y$, independent or not. This is not true for most distributions; for example, the uniform and exponential distributions do not have this property. With induction, we get the following immediate extension.

**Corollary 3.10.** *If $X_1, \ldots, X_n$ are independent with $X_k \sim N(\mu_k, \sigma_k^2)$ and $a_1, \ldots, a_n$ are constants, then*

$$\sum_{k=1}^{n} a_k X_k \sim N\left(\sum_{k=1}^{n} a_k \mu_k, \sum_{k=1}^{n} a_k^2 \sigma_k^2\right)$$

***Example 3.47.*** Two different types of diets are used on a salmon farm. Suppose that one group of fish is given diet A and the other diet B. To check if there is a difference in weights, $n$ salmon from each group are weighed, and the average weight is computed for each group. Suppose that the weights are $N(20, 5)$ and $N(18, 4)$ (kilograms) in the two groups. How large must $n$ be in order for us to be at least 99% certain to determine that diet A yields bigger salmon?

Call the weights $X_1, \ldots, X_n$ and $Y_1, \ldots, Y_n$, and let $\bar{X}$ and $\bar{Y}$ be the two averages. Corollary 3.10 with $a_1 = \cdots = a_n = \frac{1}{n}$ gives

$$\bar{X} \sim N\left(20, \frac{5}{n}\right) \quad \text{and} \quad \bar{Y} \sim N\left(18, \frac{4}{n}\right)$$

and Corollary 3.9 with $a = 1, b = -1$ gives

$$\bar{X} - \bar{Y} \sim N\left(2, \frac{9}{n}\right)$$

We get

$$0.99 \leq P(\bar{X} \geq \bar{Y}) = P(\bar{X} - \bar{Y} \geq 0) = 1 - \Phi\left(\frac{0-2}{3/\sqrt{n}}\right) = \Phi\left(\frac{2}{3/\sqrt{n}}\right)$$

which gives $2\sqrt{n}/3 \geq 2.33$, so $n$ must be at least 13. □

## 3.10 MULTIDIMENSIONAL RANDOM VECTORS

Everything we have done for two-dimensional random variables in the previous sections generalizes in an obvious way to $n$ dimensions as well. Thus, if $X_1, \ldots, X_n$ are random variables, we call $(X_1, \ldots, X_n)$ an $n$-dimensional random vector. If the $X_k$ are discrete, we define the joint pmf as $p(x_1, \ldots, x_n) = P(X_1 = x_1, \ldots, X_n = x_n)$, and if there is a function $f : R^n \to R$ such that

$$P((X_1, \ldots, X_n) \in B) = \int \cdots \int_B f(x_1, \ldots, x_n) dx_1 \cdots dx_n$$

for sets $B \in R^n$, then $X_1, \ldots, X_n$ are said to be jointly continuous with joint pdf $f$. The joint cdf is defined as

$$F(x_1, \ldots, x_n) = P(X_1 \leq x_1, \ldots, X_n \leq x_n)$$

and we have the relations

$$F(x_1, \ldots, x_n) = \int_{-\infty}^{x_n} \cdots \int_{-\infty}^{x_1} f(t_1, \ldots, t_n) dt_1 \cdots dt_n$$

and

$$f(x_1, \ldots, x_n) = \frac{\partial^n}{\partial x_1 \cdots \partial x_n} F(x_1, \ldots, x_n)$$

for $x_1, \ldots, x_n \in R$, in the jointly continuous case. The marginals for the $X_k$ are obtained by summing the joint pmf or integrating the joint pdf over other variables. Note, however, that when $n \geq 3$, we also get "multidimensional marginals," for example, the joint distribution of $(X, Y)$ when we start from $(X, Y, Z)$. Independence between random variables now means that all possible joint distributions are products of the possible marginals.

Rather than stating formal multidimensional analogs of the two-dimensional results, we illustrate these in an example that deals with a uniform distribution in three dimensions.

***Example 3.48.*** Suppose that we choose a point $(X, Y, Z)$ uniformly in the unit sphere $\{(x, y, z) : x^2 + y^2 + z^2 \leq 1\}$. By an argument similar to that for the uniform distribution in one or two dimensions, we must have the joint pdf

$$f(x, y, z) = \frac{3}{4\pi}, \quad x^2 + y^2 + z^2 \leq 1$$

since the volume of the sphere is $4\pi/3$. We might now wonder what the joint distribution of $X$ and $Y$ is. Is it the uniform distribution on the unit disk in the $(x, y)$ plane? We need to integrate over $z$, so let us first note that for fixed $x$ and $y$, $z$ ranges from

$-\sqrt{1-x^2-y^2}$ to $\sqrt{1-x^2-y^2}$ and we get

$$f_{(X,Y)}(x,y) = \frac{3}{4\pi}\int_{-\sqrt{1-x^2-y^2}}^{\sqrt{1-x^2-y^2}} dz = \frac{3}{2\pi}\sqrt{1-x^2-y^2}, \quad x^2+y^2 \le 1$$

which is certainly not a uniform distribution. The intuition is that a region of fixed size in the $(x, y)$ plane is more probable if it is near the origin $(0, 0)$ because it then corresponds to a larger volume to pick $(X, Y, Z)$ from. With the experience from Example 3.6, this was to be expected. We also expect that the conditional distribution of $(X, Y)$ given $Z = z$ is uniform on the disk with radius $\sqrt{1-z^2}$. Let us verify this. The conditional pdf of $(X, Y)$ given $Z = z$ is defined in the obvious way:

$$f_{(X,Y)}(x,y|z) = \frac{f(x,y,z)}{f_Z(z)}, \quad x^2+y^2 \le 1-z^2$$

and we must thus find the marginal pdf $f_Z(z)$. We get

$$f_Z(z) = \int\int_{x^2+y^2\le 1-z^2} f(x,y,z)dx\,dy = \frac{3}{4\pi}\int\int_{x^2+y^2\le 1-z^2} dx\,dy$$

where we change to polar coordinates to obtain

$$f_Z(z) = \frac{3}{4\pi}\int_0^{2\pi}\int_0^{\sqrt{1-z^2}} r\,dr\,d\theta$$

$$= \frac{3}{4\pi} \times 2\pi\left[\frac{r^2}{2}\right]_0^{\sqrt{1-z^2}} = \frac{3}{4}(1-z^2), \quad -1 \le z \le 1$$

We now get

$$f_{(X,Y)}(x,y|z) = \frac{1}{\pi(1-z^2)}, \quad x^2+y^2 \le 1-z^2$$

and it is reassuring to note that this is indeed a uniform distribution on the disk with center 0 and radius $\sqrt{1-z^2}$. □

If the random variables $X_1, X_2, \ldots$ can be regarded as repeated measurements on some quantity, they can be assumed to be independent, and they also all have the same distribution. If this is the case, we use the following terminology.

***Definition 3.19.*** If $X_1, X_2, \ldots$ are independent and have the same distribution, we say that they are *independent and identically distributed* or *i.i.d.* for short.

***Example 3.49 (Waiting Time Paradox).*** Bob often catches a bus from a street corner to go downtown. He arrives at different timepoints during the day but has estimated that his average wait is 17 min. His friend Joe, who can see the bus stop from his living room window, assures Bob that the buses run on average every half hour. But then Bob's average wait ought to be 15 min, not 17. Explain!

This is an example of the *waiting time paradox* and the explanation is simple. The times between consecutive arrivals are i.i.d. random variables with mean 30 min. If Bob arrives at an arbitrary timepoint, he is more likely to arrive in a time interval between buses that is longer than usual. For example, suppose that three consecutive buses arrive at 1:00, 1:40, and 2:00. The time intervals are 40 and 20 min, which have the average 30 min. However, if Bob arrives at a uniform time between 1:00 and 2:00, the probability that he arrives in the first interval is $\frac{2}{3}$, in which case his expected wait is 20 min and with probability $\frac{1}{3}$, his expected wait is 10 min. This gives the expected wait $20 \times \frac{2}{3} + 10 \times \frac{1}{3} \approx 17$ min. Hence, it is perfectly normal that buses run on average every half hour, but Bob's expected wait is more than 15 min. □

Call the times between successive arrivals in the example $T_1, T_2, \ldots$, which are i.i.d. random variables. This is an example of a *renewal process*. Now let $\mu = E[T]$, the mean *interarrival time*, and consider the time $T'$ until the next arrival at some arbitrary timepoint $t$. We argued above that $E[T']$ does not equal $\mu/2$. It can be shown that for large $t$ (the process has been going on for a while), this expected time is

$$E[T'] = \frac{E[T^2]}{2\mu}$$

For example, if the $T_k$ are uniform [0, 1], the mean is $\mu = \frac{1}{2}$ and $E[T^2] = \frac{1}{3}$, which gives $E[T'] = \frac{1}{3}$. Note that this means that the expected length of the interval that contains $t$ is $E[T^2]/\mu$, which is reminiscent of the size-biased distributions studied in Problems 32 and 33 in Chapter 2. If you sample a person at random, this person is more likely to belong to a large family. If you arrive at random at a bus stop, you are more likely to arrive in a long time interval and we might call this a *length-biased* distribution.

### 3.10.1 Order Statistics

As we know from many applications and real-life situations, it is often of practical interest to order observations by size. In terms of random variables, we have the following definition.

***Definition 3.20.*** Let $X_1, \ldots, X_n$ be independent and denote by $X_{(j)}$ the $j$th smallest of the $X_k$. The random variables $X_{(1)} \leq X_{(2)} \leq \cdots \leq X_{(n)}$ are called the *order statistics* of the random variables $X_1, \ldots, X_n$.

Note that $X_k$ are independent but $X_{(k)}$ are obviously not. Of particular interest are $X_{(1)}$ and $X_{(n)}$, the smallest and largest of $X_k$.

***Example 3.50.*** Let $X_1, \ldots, X_n$ be independent where $X_k \sim \exp(\lambda_k)$, $k = 1, \ldots, n$. What is the distribution of $X_{(1)}$?

First note that $X_{(1)} > x$ if and only if $X_1 > x, X_2 > x, \ldots, X_n > x$. Hence, using independence, we obtain

$$\begin{aligned} P(X_{(1)} > x) &= P(X_1 > x, \ldots, X_n > x) \\ &= P(X_1 > x) \cdots P(X_n > x) \\ &= e^{-\lambda_1 x} \cdots e^{-\lambda_n x} = e^{-(\lambda_1 + \cdots + \lambda_n)x}, \quad x \geq 0 \end{aligned}$$

which gives the cdf

$$F_{X_{(1)}}(x) = 1 - e^{-(\lambda_1 + \cdots + \lambda_n)x}, \quad x \geq 0$$

which we recognize from Section 2.8.2 as the cdf of an exponential distribution with parameter $\lambda_1 + \cdots + \lambda_n$. Hence,

$$X_{(1)} \sim \exp(\lambda_1 + \cdots + \lambda_n)$$

□

***Example 3.51.*** Recall Example 2.40, where the concept of half-life of a radioactive material was first introduced informally as the time required until half the atoms have decayed and then defined as the median lifetime of an individual atom. Let us now examine this more closely.

Suppose that we start from a number $2n$ of atoms. The half-life is then the time it takes until we have $n$ atoms left. If we assume that lifetimes are i.i.d. $\exp(\lambda)$ for some $\lambda$, the time until the first decay is $T_{(1)}$, the minimum of $2n$ exponentials. We then start over with $2n - 1$ atoms and wait for the next decay. The time until this is $T_{(2)}$, which by the memoryless property is the minimum of $2n - 1$ exponentials (note that this is also the second smallest among the original $2n$ lifetimes if $T_{(1)}$ is subtracted from all of them, and hence the subscript). The time until half the atoms have decayed is then

$$T = \sum_{k=1}^{n} T_{(k)}$$

which is a random time. To describe this by a number, we define half-life as the *expected* time until half the atoms have decayed:

$$h = E[T]$$

Now, by Example 3.50, $T_{(1)} \sim \exp(2n\lambda), \ldots, T_{(n)} \sim \exp((n+1)\lambda)$, which gives

$$h = \sum_{k=1}^{n} E[T_{(k)}] = \frac{1}{\lambda} \sum_{k=n+1}^{2n} \frac{1}{k} = \frac{1}{\lambda} \left( \sum_{k=1}^{2n} \frac{1}{k} - \sum_{k=1}^{n} \frac{1}{k} \right)$$

It is a well-known result that the harmonic series $H_n = 1 + \frac{1}{2} + \cdots + \frac{1}{n}$ is approximately equal to $\log n$ for large $n$, and we get[5]

$$h \approx \frac{1}{\lambda}(\log(2n) - \log n) = \frac{1}{\lambda}\log 2$$

which is precisely the median of $T$. Thus, if the radioactive sample is large, the interpretation of half-life as the median in the individual lifetime distribution works fine. However, as the sample gets smaller, this becomes less and less accurate and the decay tends to speed up. Consider the extreme case when only two atoms remain. The time until only one is left is then the minimum of two exponentials, which gives $h = 1/(2\lambda)$, which is smaller than the median $\log 2/\lambda$. □

The definition of half-life as "the time until half the atoms have decayed" is a *deterministic* description, and our approach with exponential lifetimes is *probabilistic* (or *stochastic*). We have seen that they agree when the number of atoms is large, but only the probabilistic description is accurate for small numbers.

If the random variables are not just independent but also have the same distribution, there are some nice and simple formulas for the distribution of individual order statistics, as well as their joint distribution. We start by a result that gives the cdfs of the minimum and the maximum.

**Proposition 3.31.** Let $X_1, \ldots, X_n$ be i.i.d. with cdf $F$. Then $X_{(1)}$ and $X_{(n)}$ have cdfs

$$F_{(1)}(x) = 1 - (1 - F(x))^n, \quad x \in R$$
$$F_{(n)}(x) = F(x)^n, \quad x \in R$$

*Proof.* We do the maximum and leave the minimum as an exercise. First note that

$$\{X_{(n)} \leq x\} = \{X_1 \leq x, \ldots, X_n \leq x\}$$

that is, the maximum is less than or equal to $x$ if and only if all the $X_k$ are less than or equal to $x$. We get

$$\begin{aligned} F_{(n)}(x) = P(X_{(n)} \leq x) &= P(X_1 \leq x, \ldots, X_n \leq x) \\ = P(X_1 \leq x) \cdots P(X_n \leq x) &= F(x)^n, \quad x \in R \end{aligned}$$

where the second to last equality is due to independence of the $X_k$. For the minimum, instead start with the event $\{X_{(1)} > x\}$. ■

[5]The exact result is that the difference between the harmonic series $H_n$ and $\log n$ approaches a number $\gamma$ as $n \to \infty$. The number $\gamma$ is known as *Euler's constant* and is approximately equal to 0.58.

This result is valid for both discrete and continuous random variables, but there are some differences between the two cases when it comes to more properties of order statistics. In the discrete case, we may get duplicate values, and the order statistics are not necessarily unique. Thus, we can, for example, have maximum values attained at two or more different $X_k$, and computing the joint pmf of the order statistics is complicated by this fact. In the continuous case, however, duplicate values are impossible (recall from Section 3.4 that $P(X_j = X_k) = 0$ for all $j \neq k$), so there is only one possible set of order statistics.

***Example 3.52.*** You are selling your car and receive consecutive bids. The bidders do not know each other's bids and for each bid you need to decide immediately whether or not to take it. If you decline, you cannot accept the offer later. Your strategy is to decline the very first bid, and then accept the first bid that is larger. How long can you expect to wait?

Let us assume that the consecutive bids, $X_1, X_2, \ldots$, are i.i.d. continuous random variables. You are thus waiting for the first time $n$ such that

$$X_1 > X_2, \ldots, X_1 > X_{n-1}, \; X_1 < X_n$$

which means that $X_n$ is the largest and $X_1$ the second largest among the $X_k$. We thus define

$$N = \min\{n : X_{(n)} = X_n, X_{(n-1)} = X_1\}$$

and wish to find $E[N]$. The assumption of continuous $X_k$ means that all $X_k$ are different from each other, and since they are i.i.d., each $X_k$ is equally likely to be the maximum. Once a maximum is decided, each of the remaining $X_k$ is equally likely to be the second largest. Hence, the probability that $N$ equals a particular $n$ is

$$\begin{aligned} P(N = n) &= P(X_{(n)} = X_n, X_{(n-1)} = X_1) \\ &= P(X_{(n-1)} = X_1 | X_{(n)} = X_n) P(X_{(n)} = X_n) \\ &= \frac{1}{n-1} \times \frac{1}{n} = \frac{1}{n(n-1)}, \quad n = 2, 3, \ldots \end{aligned}$$

and the expected value of $N$ is

$$E[N] = \sum_{n=2}^{\infty} n P(N = n) = \sum_{n=2}^{\infty} \frac{1}{n-1} = \infty$$

so you can definitely expect to wait for a long time! See Problem 109 for a strategy that enables you to sell your car faster. □

In the continuous case, it turns out that it is easy to find the marginal pdf of the order statistics as well as their joint pdf. We start by the following corollary, which follows immediately by differentiating the expressions in Proposition 3.31.

**Corollary 3.11.** *Let $X_1, \ldots, X_n$ be i.i.d. and continuous with pdf $f$ and cdf $F$. Then $X_{(1)}$ and $X_{(n)}$ have pdf's*

$$f_{(1)}(x) = nf(x)(1 - F(x))^{n-1} \quad \text{and} \quad f_{(n)}(x) = nf(x)F(x)^{n-1}$$

*for $x \in R$.*

In Problem 113, you are asked to find the pdfs for the remaining order statistics. In the case of i.i.d. continuous random variables, we can easily find not only the marginals but also the entire joint distribution of the order statistics.

**Proposition 3.32.** Let $X_1, X_2, \ldots, X_n$ be i.i.d. and continuous with the common pdf $f_X$. Then the vector of order statistics $(X_{(1)}, X_{(2)}, \ldots, X_{(n)})$ has joint pdf

$$f(x_1, x_2, \ldots, x_n) = n!\, f_X(x_1) f_X(x_2) \cdots f_X(x_n)$$

for $-\infty < x_1 \le x_2 \le \cdots \le x_n < \infty$

The intuition behind the expression in the proposition is quite clear if we skip the strict definition of the joint pdf for a moment and instead think of it as a probability. Then, the "probability" that the ordered $X_k$ values equal the particular vector $(x_1, x_2, \ldots, x_n)$ is simply the "probability" that the unordered random variables $X_1, X_2, \ldots, X_n$ equal the numbers $x_1, x_2, \ldots, x_n$ in any order. Since there are $n!$ different ways to order these numbers and the "probability" of any particular order is $f_X(x_1) f_X(x_2) \cdots f_X(x_n)$, the expression follows.

*Proof.* We will do the proof only in the case $n = 2$. Thus, fix $x \le y$ and consider the joint cdf of $(X_{(1)}, X_{(2)})$:

$$F(x, y) = P(X_{(1)} \le x, X_{(2)} \le y), \quad x \le y$$

In terms of the unordered $X_1$ and $X_2$, this means that both must be less than or equal to $y$ but that at most one can be between $x$ and $y$. Denoting the cdf of $X_k$ by $F_X$, we thus get

$$\begin{aligned} F(x, y) &= P(X_1 \le y, X_2 \le y) - P(x < X_1 \le y, x < X_2 \le y) \\ &= F_X(y)^2 - (F_X(y) - F_X(x))^2 \end{aligned}$$

which we differentiate twice to get the joint pdf of $(X_{(1)}, X_{(2)})$:

$$\begin{aligned} f(x, y) &= \frac{\partial^2}{\partial x\, \partial y}\left(F_X(y)^2 - (F_X(y) - F_X(x))^2\right) \\ &= \frac{d}{dy} 2(F_X(y) - F_X(x)) f_X(x) \;=\; 2 f_X(y) f_X(x) \end{aligned}$$

as desired. The general case is proved in the same way, only keeping track of more different cases. ■

### 3.10.2 Reliability Theory

Recall Example 1.42, where we defined the reliability of a system as the probability that it functions, given that its components function independent of each other. We will now instead consider this as a process in time. Thus, suppose that each component has a random lifetime that has some continuous distribution and consider the time $T$ until failure of the system. Let us first consider the case of two components.

***Example 3.53.*** Suppose that the individual lifetimes are $T_1$ and $T_2$, which are i.i.d. continuous and nonnegative random variables with pdf $f$ and cdf $F$. What is the pdf of the lifetime $T$ for a series system and for a parallel system?

Since a series system functions as long as both components function, the time until failure is the minimum of $T_1$ and $T_2$. Hence, by Corollary 3.11

$$f_T(x) = 2f(x)(1 - F(x)), \quad x \geq 0$$

The parallel system functions as long as at least one component functions, and hence $T = T_{(2)}$, the largest of $T_1$ and $T_2$. We get

$$f_T(x) = 2f(x)F(x), \quad x \geq 0$$

(see Figure 3.12). In particular, if we assume that lifetimes are exp($\lambda$), we get

$$f_T(x) = 2\lambda e^{-2\lambda x}, \quad x \geq 0$$

for the series system and

$$f_T(x) = 2\lambda(e^{-\lambda x} - e^{-2\lambda x}), \quad x \geq 0$$

for the parallel system. Note how the lifetime for the series system is exp($2\lambda$) (see also Problem 112). □

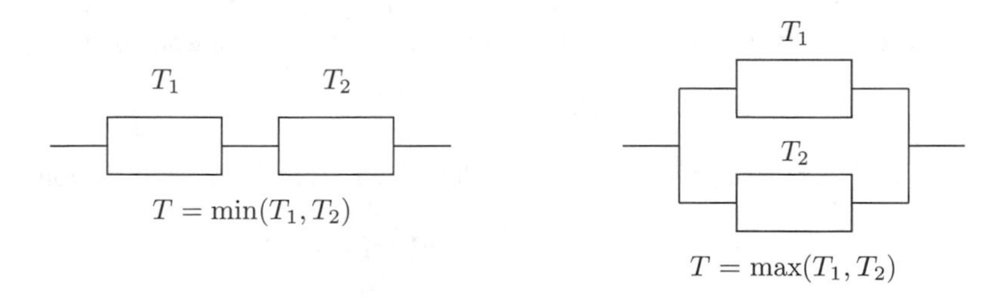


**FIGURE 3.12** Systems in series and parallel.

***Example 3.54.*** Recall the failure rate function from Section 2.10. Consider a series system where the $k$th component has failure rate function $r_k$ and components function independent of each other. What is the failure rate function of the system?

Denote the lifetimes by $X_1, \ldots, X_n$, and let the corresponding survival functions be $G_1, \ldots, G_n$. Then the lifetime of the system is the minimum $X_{(1)}$, whose survival function and failure rate function we denote by $G_{(1)}$ and $r_{(1)}$, respectively. We get

$$G_{(1)}(t) = P(X_1 > t, \ldots, X_n > t) = G_1(t) \cdots G_n(t)$$

by independence. By Proposition 2.25, we obtain

$$G_{(1)}(t) = \prod_{k=1}^{n} \exp\left(-\int_0^t r_k(u)du\right) = \exp\left(-\int_0^t \sum_{k=1}^{n} r_k(u)du\right)$$

and also since

$$G_{(1)}(t) = \exp\left(-\int_0^t r_{(1)}(u)du\right)$$

we identify the exponents to conclude that

$$r_{(1)}(u) = \sum_{k=1}^{n} r_k(u), \quad u \geq 0$$

Note how Example 3.50 was a special case of this for exponential lifetimes. □

***Example 3.55.*** In Example 2.53, we considered a type of ball bearing whose failure rate function $r(t) = at$ was chosen to describe an increasing risk of failure due to continuing wear and tear. Now suppose that in addition to this, a ball bearing may also suddenly break as a result of external factors such as explosions or other accidents, which occur at random. What failure rate function does this give?

We need to interpret the expression "at random." If we assume that this means that the time until the next accident is independent of the age of the ball bearing, then the time until failure because of an accident is exponential with some parameter $b > 0$. Thus, if we denote the time until failure due to wear by $T_1$ and the time until failure due to an accident by $T_2$, the time $T$ until failure of the ball bearing satisfies

$$T = \min(T_1, T_2)$$

where $T_1$ and $T_2$ are independent. By Example 3.54, $T$ has failure rate function

$$r_T(t) = r_{T_1}(t) + r_{T_2}(t) = at + b, \quad t \geq 0$$

Adding a constant to a failure rate function thus corresponds to introducing failure due to random incidents. □

### 3.10.3 The Multinomial Distribution

Recall how the binomial distribution counts the number of successes in $n$ independent trials, where each trial can result in success or failure. An immediate extension of this is to consider the case where each trial has more than two different outcomes. For example, consider a soccer match and the three outcomes "win," "loss," and "tie" for the home team. Suppose that we consider $n$ matches and that we let the vector $(W, L, T)$ be the numbers of wins, losses, and ties, respectively. What is the joint distribution of $(W, L, T)$, expressed in terms of $n$ and the probabilities $p = P(W)$, $q = P(L)$, and $r = P(T)$? Let us consider the joint pmf in a point $(w, l, t)$, that is,

$$p(w, l, t) = P(W = w, L = l, T = t)$$

where $w + l + t = n$ and the probabilities $p$, $q$ and $r$ add to one. First note that each configuration of $w$ wins, $l$ losses, and $t$ ties has probability $p^w q^l r^t$, so the issue is how many such configurations are there. First, there are $\binom{n}{w}$ ways to choose positions for the $w$ wins. For each such choice, there are then $\binom{n-w}{l}$ ways to choose positions for the $l$ losses. Once this is done, the $t$ ties are put in the remaining positions. Hence, the probability is

$$p(w, l, t) = \binom{n}{w}\binom{n-w}{l} p^w q^l r^t$$

for $w \geq 0, l \geq 0, t \geq 0$ and $w + l + t = n$. Now note that

$$\binom{n}{w}\binom{n-w}{l} = \frac{n!}{w!(n-w)!} \times \frac{(n-w)!}{l!\,(n-w-l)!} = \frac{n!}{w!\,l!\,t!}$$

a quantity that is called the *multinomial coefficient*. We generalize to $r$ dimensions.

***Definition 3.21.*** If $(X_1, \ldots, X_r)$ has joint pmf

$$p(n_1, \ldots, n_r) = \frac{n!}{n_1!\,n_2! \cdots n_r!} p_1^{n_1} p_2^{n_2} \cdots p_r^{n_r}$$

for $n_1 + \cdots + n_r = n$ and $p_1 + \cdots + p_r = 1$, then $(X_1, \ldots, X_r)$ is said to have a *multinomial distribution* with parameters $(n, p_1, \ldots, p_r)$.

Note that the binomial distribution is a special case with $r = 2$, $p_1 = p$, and $p_2 = 1 - p$.

***Example 3.56.*** Roll a die 12 times. What is the probability that you get exactly two of each number $1, \ldots, 6$?

With $X_k$ = the number of $k$ values, we have a multinomial distribution with $r = 6$ and parameters $(12, \frac{1}{6}, \frac{1}{6}, \ldots, \frac{1}{6})$ and get

$$p(2, 2, \ldots, 2) = \frac{12!}{2!\,2!\cdots 2!}\left(\frac{1}{6}\right)^2\left(\frac{1}{6}\right)^2\cdots\left(\frac{1}{6}\right)^2 \approx 0.003$$

□

What are the one-dimensional marginal distributions in a multinomial distribution? If we consider $X_k$, this is the number of times we get the $k$th outcome, and it is easy to realize that the distribution of $X_k$ is binomial with parameters $n$ and $p_k$. To prove this formally, we would have to sum over all the other variables in the joint pmf, but we will not do this. Thus, each $X_k$ is binomial when considered individually, but clearly $X_k$ are not independent; for example, $P(X_1 = n) = p_1^n$, but $P(X_1 = n | X_2 > 0) = 0$. In Problem 124, you are asked to find the covariance and correlation.

### 3.10.4 The Multivariate Normal Distribution

In Section 3.9, we introduced the bivariate normal distribution. Notice how the exponent in Definition 3.18 is a quadratic form, which we will now rewrite in convenient matrix notation. Introduce the column vectors

$$\mathbf{X} = \begin{pmatrix} X \\ Y \end{pmatrix} \quad \text{and} \quad \mathbf{x} = \begin{pmatrix} x \\ y \end{pmatrix} \quad \text{for } x, y \in R$$

and define the *mean vector* $\boldsymbol{\mu}$ and *covariance matrix* $\Sigma$ as

$$\boldsymbol{\mu} = \begin{pmatrix} \mu_1 \\ \mu_2 \end{pmatrix} \quad \text{and} \quad \Sigma = \begin{pmatrix} \sigma_1^2 & \rho\sigma_1\sigma_2 \\ \rho\sigma_1\sigma_2 & \sigma_2^2 \end{pmatrix}$$

Note that all entries in $\Sigma$ are covariances, since $\text{Cov}[X, Y] = \text{Cov}[Y, X] = \rho\sigma_1\sigma_2$, $\text{Cov}[X, X] = \text{Var}[X] = \sigma_1^2$, and $\text{Cov}[Y, Y] = \sigma_2^2$. The covariance matrix has determinant

$$|\Sigma| = \sigma_1^2\sigma_2^2 - \rho^2\sigma_1^2\sigma_2^2 = \sigma_1^2\sigma_2^2(1 - \rho^2)$$

and inverse

$$\Sigma^{-1} = \frac{1}{\sigma_1^2\sigma_2^2(1 - \rho^2)}\begin{pmatrix} \sigma_2^2 & -\rho\sigma_1\sigma_2 \\ -\rho\sigma_1\sigma_2 & \sigma_1^2 \end{pmatrix}$$

and we get, with the superscript $T$ denoting matrix transposition,

$$(\mathbf{x}-\boldsymbol{\mu})^T\Sigma^{-1}(\mathbf{x}-\boldsymbol{\mu})$$

$$=\frac{1}{1-\rho^2}\left(\frac{(x-\mu_1)^2}{\sigma_1^2}+\frac{(y-\mu_2)^2}{\sigma_2^2}-\frac{2\rho(x-\mu_1)(y-\mu_2)}{\sigma_1\sigma_2}\right)$$

and taken altogether, we can write the joint pdf of the bivariate normal distribution as

$$f_{\mathbf{X}}(\mathbf{x})=\frac{1}{2\pi\sqrt{|\Sigma|}}\exp\left(-\frac{1}{2}(\mathbf{x}-\boldsymbol{\mu})^T\Sigma^{-1}(\mathbf{x}-\boldsymbol{\mu})\right),\quad \mathbf{x}\in R^2$$

We can now generalize and define an $n$-dimensional multivariate normal distribution. Let $X_1, X_2, \ldots, X_n$ be random variables with means $\mu_1, \mu_2, \ldots, \mu_n$, and let

$$\mathbf{X}=\begin{pmatrix}X_1\\X_2\\\vdots\\X_n\end{pmatrix}\quad\text{and}\quad\boldsymbol{\mu}=\begin{pmatrix}\mu_1\\\mu_2\\\vdots\\\mu_n\end{pmatrix}$$

Furthermore, let $\Sigma$ be a $n\times n$ matrix where the $(i, j)$th entry is $\Sigma_{ij}=\mathrm{Cov}[X_i, X_j]$. This means that the diagonal elements are the variances of the $X_i$ and the off-diagonal elements can be written as $\Sigma_{ij}=\rho\sigma_i\sigma_j$.

***Definition 3.22.*** If $\mathbf{X}$ has $n$-dimensional joint pdf

$$f(\mathbf{x})=\frac{1}{\sqrt{(2\pi)^n|\Sigma|}}\exp\left(-\frac{1}{2}(\mathbf{x}-\boldsymbol{\mu})^T\Sigma^{-1}(\mathbf{x}-\boldsymbol{\mu})\right)$$

for $\mathbf{x}=(x_1, x_2, \ldots, x_n)^T\in R^n$, it is said to have a *multivariate normal distribution* with parameters $\boldsymbol{\mu}$ and $\Sigma$, written $\mathbf{X}\sim\mathbf{N_n}(\boldsymbol{\mu}, \Sigma)$.

As in the bivariate case, marginals are normal, keeping in mind that marginals can now themselves be multidimensional. Linear combinations and conditional distributions are also normal.

### 3.10.5 Convolution

We have seen how to find the expected value and variance of a sum of random variables. Sometimes, we want to know more, and in the case of independent random variables, it turns out that we can establish nice formulas for the entire distribution of their sum. We will start with the discrete case, for which we already possess the necessary tools.

Let us start by examining random variables with finite ranges. Suppose that $X$ and $Y$ are independent discrete random variables, both with range $\{1, 2, \ldots, n\}$ and pmfs $p_X$ and $p_Y$, respectively. We wish to find the pmf of the sum $X + Y$. First, note that the range of $X + Y$ is $\{2, 3, \ldots, 2n\}$, take a value $k$ in this range, and apply the law of total probability to obtain

$$
\begin{aligned}
P(X + Y = k) &= \sum_{j=1}^{n} P(X + Y = k | X = j) P(X = j) \\
&= \sum_{j=1}^{n} P(Y = k - j | X = j) P(X = j) \\
&= \sum_{j=1}^{n} P(Y = k - j) P(X = j)
\end{aligned}
$$

where we used the independence of $X$ and $Y$ in the last equality. In terms of the pmfs, we have thus shown that

$$
p_{X+Y}(k) = \sum_{j=1}^{n} p_Y(k - j) p_X(j), \quad k = 2, 3, \ldots, 2n
$$

This is called the *convolution* of the two pmfs $p_X$ and $p_Y$. The assumption of finite range was made only to simplify things a little, but it does not change anything vital, and we have the following general result.

**Proposition 3.33.** Let $X$ and $Y$ be independent discrete random variables, with ranges $\{x_1, x_2, \ldots\}$ and $\{y_1, y_2, \ldots\}$, respectively, and pmfs $p_X$ and $p_Y$. The sum $X + Y$ then has pmf

$$
p_{X+Y}(x) = \sum_{j=1}^{\infty} p_Y(x - x_j) p_X(x_j) \quad \text{for } x \text{ in the range of } X + Y
$$

Once we have established the formula in the discrete case, we can guess that the corresponding formula in the continuous case is similar, with pdfs instead of pmfs and that the sum is replaced by an integral.

**Proposition 3.34.** Let $X$ and $Y$ be independent continuous random variables with pdfs $f_X$ and $f_Y$, respectively. The pdf of the sum $X + Y$ is then

$$
f_{X+Y}(x) = \int_{-\infty}^{\infty} f_Y(x - u) f_X(u) du, \quad x \in R
$$

For a formal proof, start with the cdf of $X + Y$ and use Corollary 3.2 to express this as an integral involving $F_Y$ and $f_X$. It can be shown that it is allowed to interchange differentiation and integration; use this fact to obtain the pdf of $X + Y$ and finish the proof.

Just as in the discrete case, this is called the *convolution* of $f_X$ and $f_Y$. The actual limits of the integral are determined according to where the integrand is positive, which means that we must have both $f_X(u) > 0$ and $f_Y(x - u) > 0$.

***Example 3.57.*** Let $X$ and $Y$ be independent and uniform on $[0, 1]$. Find the pdf of the sum $X + Y$.

By Proposition 3.34,

$$f_{X+Y}(x) = \int_{-\infty}^{\infty} f_Y(x - u) f_X(u) du$$

where we need to determine what the actual integral limits are. First, we note that the range of $X + Y$ is $[0, 2]$, so we need to pick our $x$ from that interval. Now, $f_X(u) = 1$ if $u$ is between 0 and 1 and 0 otherwise. For the other factor in the integrand, note that $f_Y(x - u) = 1$ if its argument $x - u$ is between 0 and 1, that is, $0 \leq x - u \leq 1$ and 0 otherwise. Since $x$ is fixed and $u$ is the variable of integration, we rewrite this as $x - 1 \leq u \leq 1$. Hence, for any $x \in [0, 2]$, the two inequalities

$$0 \leq u \leq 1 \quad \text{and} \quad x - 1 \leq u \leq x \tag{3.4}$$

must be satisfied and if they are, the integrand equals 1. We need to distinguish between two cases:

*Case 1*: $x \in [0, 1]$. In this case, the two conditions in Equation (3.4) are satisfied if $0 \leq u \leq x$. Hence, for $0 \leq x \leq 1$

$$f_{X+Y}(x) = \int_0^x du = x, \quad 0 \leq x \leq 1$$

*Case 2*: $x \in [1, 2]$. Here, the conditions are satisfied if $x - 1 \leq u \leq 1$, and we get

$$f_{X+Y}(x) = \int_{x-1}^1 du = 2 - x, \quad 1 \leq x \leq 2$$

and the pdf of $X + Y$ is therefore

$$f_{X+Y}(x) = \begin{cases} x & \text{if } 0 \leq x \leq 1 \\ 2 - x & \text{if } 1 \leq x \leq 2 \end{cases}$$

This is for obvious reasons called a *triangular distribution*. □

***Example 3.58.*** Consider radioactive decay that is such that times between consecutive emissions are independent and $\exp(\lambda)$. Find the pdf of the time between first and third emissions.

This example asks for the pdf of $X + Y$, where $X$ and $Y$ are independent and $\exp(\lambda)$. We have

$$f_X(u) = \lambda e^{-\lambda u} \quad \text{if } u \geq 0$$

and

$$f_Y(x-u) = \lambda e^{-\lambda(x-u)} \quad \text{if } x - u \geq 0$$

Recalling that $x$ is fixed and $u$ the variable of integration, we get

$$\begin{aligned} f_{X+Y}(x) &= \int_0^x \lambda e^{-\lambda u} \lambda e^{-\lambda(x-u)} du \\ &= e^{-\lambda x} \lambda^2 \int_0^x du \;=\; e^{-\lambda x} \lambda^2 x, \quad x \geq 0 \end{aligned}$$

which we recognize from Section 2.8.2 as the gamma distribution with $n = 2$, that is, $X + Y \sim \Gamma(2, \lambda)$. □

The last example can be extended to sums of more than one random variable. For example, if $X$, $Y$, and $Z$ are independent and exponential with parameter $\lambda$, then $X + Y \sim \Gamma(2, \lambda)$, and we can apply Proposition 3.34 to the random variables $X + Y$ and $Z$ to obtain

$$\begin{aligned} f_{X+Y+Z}(x) &= \int_{-\infty}^{\infty} f_Z(x-u) f_{X+Y}(u) du \\ &= \int_0^x \lambda e^{-\lambda(x-u)} e^{-\lambda u} \lambda^2 u \, du \\ &= e^{-\lambda x} \lambda^3 \int_0^x u \, du \;=\; e^{-\lambda x} \lambda^3 \frac{x^2}{2}, \quad x \geq 0 \end{aligned}$$

by integration by parts of the last integral. Again, we recognize a gamma distribution, this time with parameters $n = 3$ and $\lambda$. By induction it is easily shown that this holds generally, and we get the following nice result.

**Proposition 3.35.** Let $X_1, \ldots, X_n$ be i.i.d. variables that have the exponential distribution with parameter $\lambda$, and let $S = \sum_{k=1}^n X_k$. Then $S \sim \Gamma(n, \lambda)$.

This means that in the case of integer values of the parameter $\alpha$, we have a direct interpretation of the gamma distribution as the sum of independent exponentially distributed random variables.

## 3.11 GENERATING FUNCTIONS

*Generating functions*, or *transforms*, are very useful in probability theory as in other fields of mathematics. Several different generating functions are used, depending on the type of random variable. We will discuss two, one that is useful for discrete random variables and the other for continuous random variables.

### 3.11.1 The Probability Generating Function

When we study nonnegative, integer-valued random variables, the following function proves to be a very useful tool.

***Definition 3.23.*** Let $X$ be nonnegative and integer valued. The function

$$G_X(s) = E[s^X], \quad 0 \leq s \leq 1$$

is called the *probability generating function* (pgf) of $X$.

Note that by Proposition 2.12 we compute $G_X$ as the power series

$$G_X(s) = \sum_{k=0}^{\infty} s^k p_X(k), \quad 0 \leq s \leq 1$$

where $p_X$ is the pmf of $X$. If the range of $X$ is finite, the sum is finite. Note that

$$G_X(s) = p_X(0) + \sum_{k=1}^{\infty} s^k p_X(k)$$

which immediately gives the following corollary.

**Corollary 3.12.** *Let $X$ be nonnegative and integer valued with pgf $G_X$. Then*

$$G_X(0) = p_X(0) \quad \text{and} \quad G_X(1) = 1$$

Before we examine further properties of the pgf, let us find the pgf for two of our common distributions.

***Example 3.59.*** $X \sim \text{bin}(n, p)$. The pgf is

$$G_X(s) = \sum_{k=0}^{n} \binom{n}{k} (ps)^k (1-p)^{n-k} = (1 - p + ps)^n, \quad 0 \leq s \leq 1$$

by the binomial theorem. □

***Example 3.60.*** $X \sim \text{Poi}(\lambda)$. The pgf is

$$G_X(s) = e^{-\lambda} \sum_{k=0}^{\infty} \frac{(\lambda s)^k}{k!} = e^{\lambda(s-1)}, \quad 0 \leq s \leq 1$$

by Taylor's theorem. □

One important property of the pgf is that it uniquely determines the distribution. In other words, if we are given a pgf, exactly one pmf corresponds to it and we can compute this pmf explicitly. Above, we saw that $p_X(0) = G_X(0)$, and to obtain $p_X(1)$, we first differentiate the pgf

$$\begin{aligned} G'_X(s) &= \frac{d}{ds}\left(p_X(0) + \sum_{k=1}^{\infty} s^k p_X(k)\right) \\ &= \sum_{k=1}^{\infty} k s^{k-1} p_X(k) \;=\; p_X(1) + \sum_{k=2}^{\infty} k s^{k-1} p_X(k) \end{aligned}$$

where we have used a result from calculus that says that this power series may be differentiated termwise. We now see that if we are given the pgf, we have

$$p_X(1) = G'_X(0)$$

To see the general pattern, let us differentiate once more:

$$\begin{aligned} G''_X(s) &= \frac{d}{ds}\left(p_X(1) + \sum_{k=2}^{\infty} k s^{k-1} p_X(k)\right) \\ &= \sum_{k=2}^{\infty} k(k-1) s^{k-2} p_X(k) \\ &= 2p_X(2) + \sum_{k=3}^{\infty} k(k-1) s^{k-2} p_X(k) \end{aligned}$$

which gives

$$p_X(2) = \frac{G''_X(0)}{2}$$

If we continue to differentiate, we get the general formula that we state as a proposition.

**Proposition 3.36.** Let $X$ be nonnegative and integer valued with pgf $G_X$. Then

$$p_X(k) = \frac{G_X^{(k)}(0)}{k!}, \quad k = 0, 1, \ldots$$

where $G_X^{(k)}$ denotes the $k$th derivative of $G_X$.

Another property of the pgf is that we can obtain the mean and variance, also by differentiating. From the formulas for the derivatives above, we get

$$G'_X(1) = E[X]$$
$$G''_X(1) = E[X(X-1)]$$

and we get the following result.

**Proposition 3.37.** If $X$ has pgf $G_X$, then

$$E[X] = G'_X(1) \quad \text{and} \quad \text{Var}[X] = G''_X(1) + G'_X(1) - G'_X(1)^2$$

Let us check the formulas for the Poisson distribution. If $X \sim \text{Poi}(\lambda)$, then it has pgf

$$G(s) = e^{\lambda(s-1)}$$

which has first and second derivatives

$$G'(s) = \lambda e^{\lambda(s-1)} \quad \text{and} \quad G''(s) = \lambda^2 e^{\lambda(s-1)}$$

which gives

$$G'(1) = \lambda \quad \text{and} \quad G''(1) = \lambda^2$$

and

$$E[X] = \lambda \quad \text{and} \quad \text{Var}[X] = \lambda^2 + \lambda - \lambda^2 = \lambda$$

which is in accordance with Proposition 2.19 (or, if you wish, a proof of the proposition). Note how it is simpler to find the mean and variance using pgfs than to work directly with the definition. Another central property of the pgf is that it turns sums into products, in the following sense.

**Proposition 3.38.** Let $X_1, X_2, \ldots, X_n$ be independent random variables with pgfs $G_1, G_2, \ldots, G_n$, respectively and let $S_n = X_1 + X_2 + \cdots + X_n$. Then $S_n$ has pgf

$$G_{S_n}(s) = G_1(s)G_2(s)\cdots G_n(s), \quad 0 \leq s \leq 1$$

*Proof.* Since $X_1, \ldots, X_n$ are independent, the random variables $s^{X_1}, \ldots, s^{X_n}$ are also independent for each $s$ in $[0, 1]$, and we get

$$\begin{aligned} G_{S_n}(s) &= E[s^{X_1+X_2+\cdots+X_n}] \\ &= E[s^{X_1}]E[s^{X_2}]\cdots E[s^{X_n}] \\ &= G_1(s)G_2(s)\cdots G_n(s) \end{aligned}$$

and we are done. ■

This result is useful for proving results for sums of random variables. Let us look at one example.

***Example 3.61.*** Let $X_1, \ldots, X_n$ be independent such that $X_k \sim \text{Poi}(\lambda_k)$. What is the distribution of $S_n = X_1 + \cdots + X_n$?

By Proposition 3.38 and Example 3.60, $S_n$ has pgf

$$G_{S_n}(s) = e^{\lambda_1(s-1)} \cdots e^{\lambda_n(s-1)} = e^{(\lambda_1+\cdots+\lambda_n)(s-1)}$$

which we recognize as the pgf of a Poisson distribution with parameter $\lambda_1 + \cdots + \lambda_n$. We have shown that the sum of independent Poisson is again Poisson. We could have done this with the convolution methods from Section 3.34, but note how much quicker it is to use generating functions. □

In particular, if $X_k$ are i.i.d. with pgf $G(s)$, then the sum has pgf $G(s)^n$. We look at one example of this.

***Example 3.62.*** Let $X \sim \text{bin}(n, p)$. Then $X$ can be written as $\sum_{k=1}^{n} I_k$, where $I_k$ are i.i.d. indicators with $P(I_k = 1) = p$. The pgf of each $I_k$ is

$$G(s) = s^0(1-p) + s^1 p = 1 - p + ps$$

and hence $X$ has pgf $(1 - p + ps)^n$, in accordance with Example 3.59. □

A different situation arises if the number of summands is not fixed but random, which is a very natural assumption in many applications. Suppose, for example, that customers arrive at a convenience store and each customer buys a number of lottery tickets. How many tickets are sold in a day? We can suppose that the number of customers is a random variable $N$ and that the $k$th customer buys $X_k$ tickets, where $X_1, X_2, \ldots$ are i.i.d. nonnegative integer-valued random variables, independent of $N$.

The total number of tickets sold is then

$$S_N = \sum_{k=1}^{N} X_k$$

which is the sum of a random number of random variables, where we interpret $S_N$ as 0 if $N = 0$. If we are given distributions of $N$ and $X_k$, we can find the distribution of $S_N$ via probability generating functions according to the following result.

**Proposition 3.39.** Let $X_1, X_2, \ldots$ be i.i.d. nonnegative and integer valued with common pgf $G_X$, and let $N$ be nonnegative and integer valued, and independent of the $X_k$, with pgf $G_N$. Then $S_N = X_1 + \cdots + X_N$ has pgf

$$G_{S_N}(s) = G_N(G_X(s))$$

the composition of $G_N$ and $G_X$.

*Proof.* We condition on $N$ to obtain

$$G_{S_N}(s) = \sum_{n=0}^{\infty} E\left[s^{S_N} | N = n\right] P(N = n) = \sum_{n=0}^{\infty} E\left[s^{S_n}\right] P(N = n)$$

since $N$ and $S_n$ are independent. Now note that

$$E\left[s^{S_n}\right] = G_X(s)^n$$

by Proposition 3.38 and we get

$$G_{S_N}(s) = \sum_{n=0}^{\infty} G_X(s)^n P(N = n) = G_N(G_X(s))$$

the pgf of $N$ evaluated at the point $G_X(s)$. ■

Note the similarity between the cases of fixed and random numbers of summands:

$$G_{S_n}(s) = G_X(s)^n$$
$$G_{S_N}(s) = E\left[G_X(s)^N\right]$$

***Example 3.63.*** Suppose that customers arrive at a rural convenience store such that the number of customers in a not so busy hour has a Poisson distribution with mean 5. Each customer buys a number of lottery tickets, independent of other customers, and this number has a Poisson distribution with mean 2. For this hour, find (a) the pgf of the total number of tickets sold, (b) the probability that no tickets are sold, and (c) the expected number of tickets sold.

We have the situation in Proposition 3.39 with the pgfs

$$G_X(s) = e^{2(s-1)}$$
$$G_N(s) = e^{5(s-1)}$$

and get

$$G_{S_N}(s) = \exp\left(5(e^{2(s-1)} - 1)\right)$$

which answers (a). For (b), we let $s = 0$ to obtain

$$P(S_N = 0) = \exp\left(5(e^{-2} - 1)\right) \approx 0.01$$

and for (c), we differentiate and get

$$G'_{S_N}(s) = 10e^{2(s-1)} \exp\left(5(e^{2(s-1)} - 1)\right)$$

and plugging in $s = 1$ gives

$$E[S_N] = 10$$

□

There are two interesting observations to be made in the last example: (1) $S_N$ does not have a Poisson distribution, so "the sum of independent Poisson is Poisson" from Example 3.61 is not true for random sums, and (2) we can note that $E[S_N] = 10$ equals the product of $E[N] = 5$ and $E[X] = 2$. This is no coincidence, and we next state a general result for the mean and variance of a random sum.

**Corollary 3.13.** *Under the assumptions of Proposition 3.39, it holds that*

$$E[S_N] = E[N]\mu$$
$$Var[S_N] = E[N]\sigma^2 + Var[N]\mu^2$$

*where $\mu = E[X_k]$ and $\sigma^2 = Var[X_k]$.*

*Proof.* First note that

$$G'_{S_N}(s) = \frac{d}{ds} G_N(G_X(s)) = G'_N(G_X(s))G'_X(s)$$

by the chain rule. Combining this with Propositions 3.37 and 3.39 now gives

$$E[S_N] = G'_{S_N}(1) = G'_N(G_X(1))G'_X(1) = G'_N(1)G'_X(1) = E[N]\mu$$

where we also used the fact that $G_X(1) = 1$. We leave the proof of the variance formula as an exercise. ■

Recall from Section 3.6.2 that for a sum of $n$ i.i.d. random variables, we have $E[S_n] = n\mu$ and $\text{Var}[S_n] = n\sigma^2$. In the case of a random number of summands $N$, we can thus replace $n$ by $E[N]$ for the mean, but things are a little more complicated for the variance. The first term in the variance formula accounts for the variability of the $X_k$ and the second for the variability of $N$.

### 3.11.2 The Moment Generating Function

The probability generating function is an excellent tool for nonnegative and integer-valued random variables. For other random variables, we can instead use the following more general generating function.

***Definition 3.24.*** Let $X$ be a random variable. The function

$$M_X(t) = E[e^{tX}], \quad t \in R$$

is called the *moment generating function* (mgf) of $X$.

If $X$ is continuous with pdf $f_X$, we compute the mgf by

$$M_X(t) = \int_{-\infty}^{\infty} e^{tx} f_X(x)dx$$

and for discrete $X$, we get a sum instead. Note that if $X$ is nonnegative integer valued with pgf $G_X$, then

$$M_X(t) = G_X(e^t)$$

which immediately gives the mgf for the distributions for which we computed the pgf above. Let us look at some continuous distributions.

***Example 3.64.*** Let $X \sim \text{unif}[0, 1]$. Find the mgf of $X$.

We get

$$M_X(t) = \int_0^1 e^{tx}dx = \frac{e^t - 1}{t}, \quad t \in R$$

where we interpret $M_X(0)$ as the limit $\lim_{t \to 0} M_X(t) = 1$. □

***Example 3.65.*** Let $X \sim \exp(\lambda)$. Find the mgf of $X$.

We get

$$M_X(t) = \int_0^{\infty} e^{tx}\lambda e^{-\lambda x}dx = \lambda \int_0^{\infty} e^{(t-\lambda)x}dx = \frac{\lambda}{\lambda - t}, \quad t < \lambda$$

□

Note that the integral in the calculation above is infinite if $t \geq \lambda$, and hence the mgf of the exponential distribution is defined only for $t < \lambda$. In general, the mgf is not necessarily defined on all of $R$, and for some random variables it turns out that the mgf does not exist at all (Problem 146). Before the next example, we state a useful property of the mgf. We leave the proof as an exercise.

**Corollary 3.14.** *Let $X$ be a random variable that has mgf $M_X$ and let $Y = aX + b$. Then $Y$ has mgf*

$$M_Y(t) = e^{bt} M_X(at)$$

***Example 3.66.*** Let $X \sim N(\mu, \sigma)$. Find $M_X(t)$.

Let us start with the standard normal distribution. Thus, let $Z \sim N(0, 1)$ and find $M_Z(t)$. We get

$$\begin{aligned} M_Z(t) = \int_{-\infty}^{\infty} e^{tx}\varphi(x)dx &= \frac{1}{\sqrt{2\pi}} \int_{-\infty}^{\infty} e^{(t^2/2-(x-t)^2/2)}dx \\ = e^{t^2/2} \int_{-\infty}^{\infty} \frac{1}{\sqrt{2\pi}} e^{-(x-t)^2/2}dx &= e^{t^2/2}, \quad t \in R \end{aligned}$$

where the last equality follows since the integrand is the pdf of a normal distribution with mean $t$ and variance 1, and hence the integral equals 1. Now let $X = \mu + \sigma Z$ so that $X \sim N(\mu, \sigma^2)$, to obtain

$$M_X(t) = e^{t\mu} M_Z(\sigma t) = e^{t\mu + \sigma^2 t^2/2}, \quad t \in R$$

by an application of Corollary 3.14. □

The moment generating function has many properties that are analogous to those of the probability generating function. Thus, the mgf uniquely determines the distribution of $X$, a fact that we will not prove. We can also obtain the mean and variance from the mgf, according to the following formula.

**Corollary 3.15.** *If $X$ has mgf $M_X(t)$, then*

$$E[X] = M'_X(0) \quad \text{and} \quad Var[X] = M''_X(0) - (M'_X(0))^2$$

*Proof.* By differentiating $M_X(t)$ with respect to $t$, we get

$$M'_X(t) = \frac{d}{dt} E[e^{tX}] = E[Xe^{tX}]$$

where we assumed that we can interchange differentiation and expectation. Note how $t$ is the variable of differentiation and we view $X$ as fixed. In the case of a discrete $X$, this amounts to differentiating a sum termwise, and in the case of a continuous $X$, it means that we can differentiate under the integral sign. This is by no means obvious but can be verified. We will not address this issue further. Differentiating once more gives

$$M_X''(t) = \frac{d}{dt}E[Xe^{tX}] = E[X^2e^{tX}]$$

which gives

$$\text{Var}[X] = E[X^2] - (E[X])^2 = M_X''(0) - (M_X'(0))^2$$

as desired. ■

Note that we get the general result

$$E[X^n] = M_X^{(n)}(0), \quad n = 1, 2, \ldots$$

where $M_X^{(n)}$ is the $n$th derivative of $M_X$. The number $E[X^n]$ is called the $n$th *moment* of $X$, hence the term moment generating function. Compare it with the probability generating function that generates probabilities by differentiating and setting $s = 0$. The moment generating function also turns sums into products, according to the following proposition, which you may prove as an exercise.

**Proposition 3.40.** Let $X_1, X_2, \ldots, X_n$ be independent random variables with mgfs $M_1, M_2, \ldots, M_n$, respectively, and let $S_n = X_1 + \cdots + X_n$. Then $S_n$ has mgf

$$M_{S_n}(t) = M_1(t)M_2(t)\cdots M_n(t), \quad t \in R$$

***Example 3.67.*** Let $X \sim N(\mu_1, \sigma_1^2)$ and $Y \sim N(\mu_2, \sigma_2^2)$ and suppose that $X$ and $Y$ are independent. Show that $X + Y \sim N(\mu_1 + \mu_2, \sigma_1^2 + \sigma_2^2)$.

By Proposition 3.40 and Example 3.66, the mgf of $X + Y$ is

$$\begin{aligned} M_{X+Y}(t) &= \exp\left(t\mu_1 + \frac{\sigma_1^2t^2}{2}\right)\exp\left(t\mu_2 + \frac{\sigma_2^2t^2}{2}\right) \\ &= \exp\left(t(\mu_1 + \mu_2) + \frac{(\sigma_1^2 + \sigma_2^2)t^2}{2}\right) \end{aligned}$$

which we recognize as the mgf of a normal distribution with parameters $\mu_1 + \mu_2$ and $\sigma_1^2 + \sigma_2^2$, and by uniqueness of the mgf we are done. □

We have seen how summation of independent random variables corresponds to multiplication of their pgfs or mgfs. Also, we have previously learned that the distribution of the sum of independent random variables can be obtained as the convolution of their distributions in both discrete and continuous cases. We summarize this observation in an informal statement that can be useful to remember.

The following are equivalent:

(a) Summation of independent random variables
(b) Convolution of distributions
(c) Multiplication of generating functions

## 3.12 THE POISSON PROCESS

Suppose that we are observing some system where changes occur randomly in time. For example, we could be observing incoming jobs to a computer, customers arriving at a store, accidents occurring on a certain highway, earthquakes, or emission of radioactive particles. Each of these examples has in common that we can represent them as points on a timeline; see Figure 3.13, where each "×" marks the time when something was observed.

We call this a *point process*. To be able to investigate the behavior of this point process, we need to make assumptions on how the points are distributed. We therefore assume that the times between consecutive points are independent random variables $T_1, T_2, \ldots$, which have exponential distributions with the same parameter $\lambda$. Assuming an exponential distribution means that we assume that the process is very unpredictable. $T_k$ are called the *interarrival times*. By the memoryless property, this means that at any fixed time, the time until the next point is $\exp(\lambda)$ regardless of when the previous point came. Such a process is sometimes referred to as *completely random*, but we use another term, already mentioned in Section 2.6.

***Definition 3.25.*** A point process where times between consecutive points are i.i.d. random variables that are $\exp(\lambda)$ is called a *Poisson process* with rate $\lambda$.

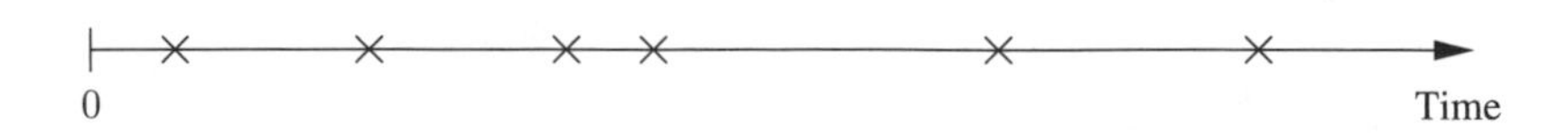


**FIGURE 3.13** A point process in time. Each "×" marks the occurrence of some event of interest.

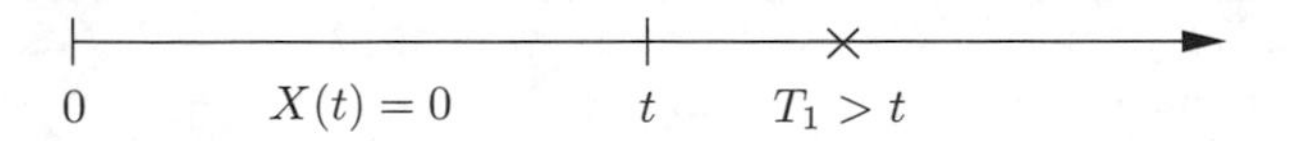


**FIGURE 3.14** The relation between the random variables $X(t)$ and $T_1$.

The rate $\lambda$ has unit mean number of points per time unit. Hence, larger values of $\lambda$ correspond to more points per time unit, which is also clear since the expected time between points is $1/\lambda$.

One random variable of interest is the number of points in a interval. Fix $t \geq 0$ and define the random variable

$$X(t) = \text{ the number of points in a time interval of length } t$$

and note that $X(t)$ is a discrete random variable with range $\{0, 1, 2, \ldots\}$. Also note that by the memoryless property, it does not matter where the interval is located since at any given time, the time until the next point has the same distribution. Thus, we can without loss of generality consider the interval $[0, t]$. Let us first investigate $P(X(t) = 0)$. If we denote the time for the first point by $T_1$, we realize that $X(t) = 0$ if and only if $T_1 > t$ (see Figure 3.14). Hence, we get

$$P(X(t) = 0) = P(T_1 > t) = e^{-\lambda t}$$

For general $k$, note that $X(t) = k$ if and only if the $k$th point came before $t$ and the $(k+1)$th point after $t$. With $S_k = T_1 + \cdots + T_k$, the time for the $k$th point, we can express this in terms of events as

$$\{X(t) = k\} = \{S_k \leq t\} \setminus \{S_{k+1} \leq t\}$$

the event that the $k$th but not the $(k+1)$th point came before $t$. Also, since $S_{k+1} \leq t$ implies that $S_k \leq t$, we have $\{S_{k+1} \leq t\} \subseteq \{S_k \leq t\}$, and we get

$$P(X(t) = k) = P(S_k \leq t) - P(S_{k+1} \leq t)$$

and by Proposition 3.35, $S_k \sim \Gamma(k, \lambda)$ for all $k$. From Section 2.8.2, we now get

$$\begin{aligned} P(X(t) = k) &= 1 - e^{-\lambda t} \sum_{j=0}^{k-1} \frac{\lambda^j t^j}{j!} - \left(1 - e^{-\lambda t} \sum_{j=0}^{k} \frac{\lambda^j t^j}{j!}\right) \\ &= e^{-\lambda t} \frac{(\lambda t)^k}{k!}, \quad k = 0, 1, 2, \ldots \end{aligned}$$

which we recognize as the Poisson distribution from Section 2.5.4. Hence, we arrive at the following proposition.

**Proposition 3.41.** Consider a Poisson process with rate $\lambda$, where $X(t)$ is the number of points in an interval of length $t$. Then

$$X(t) \sim \text{Poi}(\lambda t)$$

Recall that the parameter in the Poisson distribution is also the expected value. Hence, we have

$$E[X(t)] = \lambda t$$

which makes sense since $\lambda$ is the mean number of points per time unit and $t$ is the length of the time interval. In practical applications, we need to be careful to use the same time units for $\lambda$ and $t$.

Now consider two disjoint time intervals. By the memoryless property, we realize that the number of points in one of the intervals is independent of the number of points in the other. Thus, in a Poisson process with intensity $\lambda$, the numbers of points in disjoint time intervals are independent and have Poisson distributions with means $\lambda$ times the interval lengths. Formally, let $I_1, I_2, \ldots$ be any sequence of disjoint time intervals of lengths $t_1, t_2, \ldots$, and let $X(t_j)$ denote the number of points in $I_j$. Furthermore, let $T_1, T_2, \ldots$ be the times between points defined above. We have the following proposition.

**Proposition 3.42.** In a Poisson process with rate $\lambda$

(a) $T_1, T_2, \ldots$ are independent and $\exp(\lambda)$

(a) $X(t_1), X(t_2), \ldots$ are independent and $X(t_j) \sim \text{Poi}(\lambda t_j), \quad j = 1, 2, \ldots$

We have used (a) as the definition of the Poisson process and argued that (b) follows. In fact, we could as well use (b) as the definition and argue that (a) follows. The independence assumption in (b) implies the memoryless property for times between points and the relation above between numbers of points and times between consecutive points gives the exponential distribution. Hence,

The properties (a) and (b) in Proposition 3.42 are equivalent.

We know that the number of points in the interval $[0, t]$ in a Poisson process with rate $\lambda$ has a Poisson distribution with mean $\lambda t$. Now suppose that we know there is exactly one point in the interval. What is the distribution of its position?

By the memoryless property, it does not matter whether there is also a point exactly at time $t$, so we can assume that the second point arrived at $t$. What is the distribution for the position of the first point? In terms of the interarrival times, we ask for the conditional distribution of $T_1$ given $T_1 + T_2 = t$. To find this, consider the transformation

$$S_1 = T_1, \quad S_2 = T_1 + T_2$$

that is, $S_1$ and $S_2$ are simply the arrival times of the first and second points, respectively. We wish to find the conditional distribution of $S_1$ given $S_2 = t$. The joint distribution of $(T_1, T_2)$ is, by independence,

$$f_{(T_1,T_2)}(t_1, t_2) = \lambda^2 e^{-\lambda(t_1+t_2)}, \quad t_1, t_2 \in R$$

The joint pdf of $(S_1, S_2)$ is obtained by noting that the Jacobian matrix for the transformation $(t_1, t_2) \to (t_1, t_1 + t_2)$ has determinant 1 (and so does its inverse), and Proposition 3.16 gives

$$f_{(S_1,S_2)}(s, t) = f_{(T_1,T_2)}(s, t - s) = \lambda^2 e^{-\lambda t}, \quad 0 \le s \le t$$

We already know that $S_2 \sim \Gamma(2, \lambda)$, so $S_2$ has pdf

$$f_{S_2}(t) = \lambda^2 t e^{-\lambda t}$$

which gives the conditional pdf

$$f_{S_1}(s_1|t) = \frac{f_{(S_1,S_2)}(s, t)}{f_{S_2}(t)} = \frac{1}{t}, \quad 0 \le s \le t$$

which is a uniform distribution on $[0, t]$. Thus, given that the second point arrives at time $t$, the position for the first is uniform between 0 and $t$. Considering the memorylessness of the exponential interarrival times, this comes as no big surprise. Also, by the memoryless property, it is equivalent to condition on the fact that the second point arrived at $t$ and that there is exactly one point in the interval $[0, t]$. Whether there is in fact a point at time $t$ does not matter. More generally, if we condition on $n$ points in $[0, t]$ or, equivalently, that the $(n + 1)$th point arrived at time $t$, the $n$ points are distributed as independent uniforms on the interval. Let us state this formally.

**Proposition 3.43.** Consider a Poisson process with rate $\lambda$. If there are $n$ points in the time interval $[0, t]$, their joint distribution is the same as that of $n$ i.i.d. random variables that are unif$[0, t]$.

The property in the proposition is called the *order statistic property* of the Poisson process since the points are distributed as $n$ order statistics from a uniform distribution on $[0, t]$. The proof is similar to what we did previously in the case of $n = 1$, invoking Proposition 3.32.

***Example 3.68.*** Suppose that earthquakes in a region occur according to a Poisson process with an average of 104 earthquakes per year. Find the probability that (a) a given week has at most one earthquake, (b) three consecutive earthquakes are at

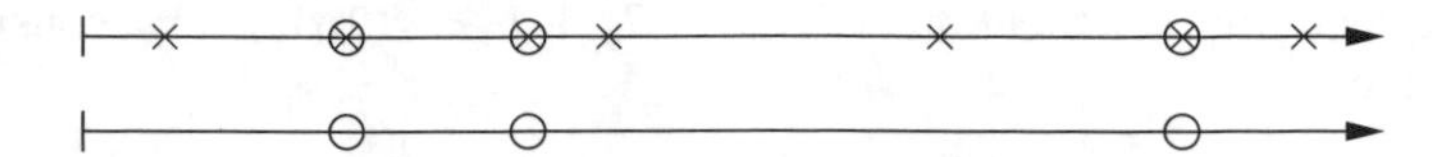

**FIGURE 3.15** The original Poisson process with observed points encircled and the resulting thinned process.

least 1 week apart from each other, and (c) a week with two earthquakes has them on different days.

The rate is 104 earthquakes per year. (a) With $X$ as the number of earthquakes in a week, we have $\lambda = 104$ and $t = \frac{1}{52}$ and hence $X \sim \text{Poi}(2)$ and get

$$P(X \leq 1) = P(X = 0) + P(X = 1) = e^{-2}(1 + 2) \approx 0.41$$

For (b), the times $T_1$ and $T_2$ between consecutive earthquakes are independent and exp(2), which gives

$$P(T_1 > 1, T_2 > 1) = P(T_1 > 1)P(T_2 > 1) = e^{-4} \approx 0.02$$

For (c), by the order statistic property, the two earthquakes are distributed as two independent uniform random variables over the week. Hence, any given day has probability $\frac{1}{7}$, and the probability that the other falls on another day is $\frac{6}{7}$ (formally, condition on the day of one of the quakes). □

### 3.12.1 Thinning and Superposition

Suppose now that we have a Poisson process with rate $\lambda$, where we do not observe every point, either by accident or on purpose. For example, phone calls arrive as a Poisson process, but some calls are lost when the line is busy. Hurricanes are formed according to a Poisson process, but we are interested only in those that make landfall. These are examples of *thinning* of a Poisson process (see Figure 3.15).

We assume that each point is observed with probability $p$ and that different points are observed independent of each other. Then, it turns out that the thinned process is also a Poisson process.

**Proposition 3.44.** The thinned process is a Poisson process with rate $\lambda p$.

*Proof.* We work with characterization (b) in Proposition 3.42. Clearly, the numbers of observed points in disjoint intervals are independent. To show the Poisson distribution, we use probability generating functions. Consider an interval of length $t$, letting $X(t)$ be the total number of points and $X_p(t)$ be the number of observed points in this interval. Then,

$$X_p(t) = \sum_{k=1}^{X(t)} I_k$$

where $I_k$ is 1 if the $k$th point was observed and 0 otherwise. By Proposition 3.39, $X_p(t)$ has pgf

$$G_{X_p}(s) = G_{X(t)}(G_I(s))$$

where

$$G_{X(t)}(s) = e^{\lambda t(s-1)}$$

and

$$G_I(s) = 1 - p + ps$$

and we get

$$G_{X_p}(s) = e^{\lambda t(1-p+ps-1)} = e^{\lambda pt(s-1)}$$

which we recognize as the pgf of a Poisson distribution with parameter $\lambda pt$. ■

It is definitely reasonable that the expected number of observed points in the thinned process is $\lambda pt$. The more interesting aspect of Proposition 3.44 is that the Poisson process properties are retained. It is important, though, that the thinning is done in the independent way described above; otherwise, the Poisson process properties are ruined. The following result is not only interesting but also surprising.

**Proposition 3.45.** The processes of observed and unobserved points are independent.

*Proof.* Fix an interval of length $t$, let $X(t)$ be the total number of points, and $X_p(t)$ and $X_{1-p}(t)$ the number of observed and unobserved points, respectively. Hence, $X(t) = X_p(t) + X_{1-p}(t)$ and by Proposition 3.44, we obtain

$$X_p(t) \sim \text{Poi}(\lambda pt) \quad \text{and} \quad X_{1-p}(t) \sim \text{Poi}(\lambda(1-p)t)$$

Also, given that $X(t) = n$, the number of observed points has a binomial distribution with parameters $n$ and $p$. We get

$$\begin{aligned}
P(X_p(t) = j, X_{1-p}(t) = k) &= P(X_p(t) = j, X(t) = k + j) \\
&= P(X_p(t) = j | X(t) = k + j) P(X(t) = k + j) \\
&= \binom{k+j}{j} p^j (1-p)^k e^{-\lambda t} \frac{(\lambda t)^{k+j}}{(k+j)!} \\
&= \frac{(k+j)!}{k!j!} p^j (1-p)^k e^{-\lambda t} \frac{(\lambda t)^{k+j}}{(k+j)!} \\
&= e^{-\lambda p t} \frac{(\lambda p t)^j}{j!} e^{-\lambda(1-p)t} \frac{(\lambda(1-p)t)^k}{k!} \\
&= P(X_p(t) = j) P(X_{1-p}(t) = k)
\end{aligned}$$

■

This result should end with an exclamation mark instead of a mere period. For example, suppose that we have a Poisson process such that we expect 10 points per hour. If in 1 h we observe 20 points and in another no points at all, it seems that the second hour ought to have more unobserved points. However, the last result tells us that this is not so, but the distribution of unobserved points is the same in both cases.

***Example 3.69.*** Consider the earthquakes in Example 3.68, occurring on average twice a week. Each time there is an earthquake, it is of category "major" with probability 0.01; otherwise, "minor." (a) What is the probability that there are no major earthquakes in a given year? (b) What is the probability that there are no major earthquakes and at least one minor quake in a given week?

For (a), note that the number $X$ of major earthquakes in a year has a Poisson distribution with mean $\lambda p = 104 \times 0.01 = 1.04$ and hence

$$P(X = 0) = e^{-1.04} \approx 0.35$$

For (b), the mean of $X$ is instead 0.02 (quakes per week) and if we let $Y$ be the number of minor quakes, then $Y$ has a Poisson distribution with mean $2 \times 0.99 = 1.98$ and by independence of $X$ and $Y$, we obtain

$$P(X = 0, Y > 0) = P(X = 0)(1 - P(Y = 0)) = e^{-0.02}(1 - e^{-1.98}) \approx 0.84$$

□

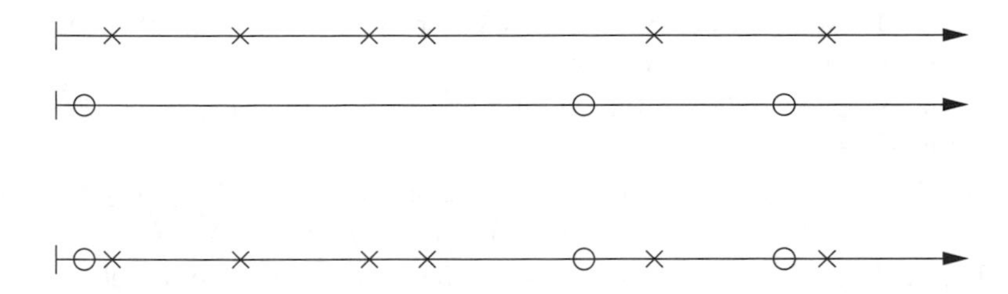

**FIGURE 3.16** Two independent Poisson processes and their superposition.

Let us finally consider a situation that is in a sense the opposite of thinning. Instead of removing points from a Poisson process, we add points according to another Poisson process. Hence, we are considering a situation where we have two Poisson processes and observe the total number of points in both of them; that is, we observe the *superposition* of the two. For example, jobs may arrive at a computer according to Poisson processes from different sources or cars pass a point on a road according to two Poisson processes, one in each direction (see Figure 3.16). Just as in the case of thinning, the Poisson process properties are retained, provided the two processes are independent.

**Proposition 3.46.** Consider two independent Poisson processes with rates $\lambda_1$ and $\lambda_2$, respectively. The superposition of the two processes is a Poisson process with rate $\lambda_1 + \lambda_2$.

*Proof.* For this we use characterization (a) in Proposition 3.42. Fix a time where there is a point in either of the two processes. Denote the time until the next point in the first process by $S$ and the corresponding time in the second by $T$. Since the two processes are independent, $S$ and $T$ are independent random variables. Also, by the memoryless property, the distributions are $S \sim \exp(\lambda_1)$ and $T \sim \exp(\lambda_2)$, regardless of which of the two processes the last point came from. Hence, the time until the next point in the superposition process is the minimum of $S$ and $T$, and by Example 3.50, this is $\exp(\lambda_1 + \lambda_2)$. Thus, the superposition process is a Poisson process with rate $\lambda_1 + \lambda_2$ and we are done. ■

## PROBLEMS

### Section 3.2. The Joint Distribution Function

1. Let $(X, Y)$ have joint cdf $F$. Show that

$$P(a < X \leq b, c < Y \leq d) = F(b, d) + F(a, c) - F(a, d) - F(b, c)$$

2. Let $(X, Y)$ have joint cdf $F$ and let $x \leq y$. For which set do we get the probability if we compute $F(y, y) - F(x, x)$?
3. Let $F$ be a joint cdf. Find the limit as $x \to \infty$ of **(a)** $F(x, x)$, **(b)** $F(-x, x)$, **(c)** $F(-x, -x)$.
4. Are the following statements true or false in general? **(a)** $F(x, y) \leq F_X(x)$ for all $x, y \in R$, **(b)** $P(X > x, Y > y) = 1 - F(x, y)$ for all $x, y \in R$.
5. Let $(X, Y)$ have joint cdf $F$ and let $G$ be the cdf of the random variable $X + Y$. Show that $F(x, x) \leq G(2x)$ for all $x \in R$.
6. Find the marginal cdf's of $X$ and $Y$ for the following joint cdfs: **(a)** $F(x, y) = 1 - e^{-x} - e^{-y} + e^{-(x+y)}, x, y \geq 0$, **(b)** $F(x, y) = x^2\sqrt{y}$, $0 \leq x \leq 1, 0 \leq y \leq 1$, **(c)** $F(x, y) = 2xy, 0 \leq x \leq \frac{1}{2}, 0 \leq y \leq 1$, and **(d)** $F(x, y) = \frac{1}{3}(x^2 y + 2xy^2), 0 \leq x \leq 1, 0 \leq y \leq 1$.

## Section 3.3. Discrete Random Vectors

7. Let $(X, Y)$ be uniform on the four points $(0, 0), (1, 0), (1, 1), (2, 1)$. **(a)** Find the marginal pmfs of $X$ and $Y$. **(b)** For which joint pmf of $(X, Y)$ are $X$ and $Y$ uniform on their respective ranges?
8. Is it true in general that $p(x, y) \leq p_X(x)$ for all $x, y$?
9. Consider a family with three children. Let $X$ be the number of daughters and $Y$ the number of sons. Find the joint pmf of $(X, Y)$.
10. Roll a die twice and let $Y$ be the sum of the two rolls. Find the joint pmf of $(X, Y)$ if $X$ is **(a)** the number on the first roll, and **(b)** the smallest number.
11. Draw three cards without replacement from a deck of cards. Let $H$ be the number of hearts and $S$ the number of spades drawn. **(a)** Find the joint pmf of $(H, S)$. **(b)** Find $P(H = S)$.
12. Draw one card from a deck of cards. Let $H$ be the number of hearts and $A$ the number of aces drawn. **(a)** Find the joint pmf of $(H, A)$. **(b)** Find $P(H \leq A)$.
13. Consider a population of individuals that reproduce in such a way that the number of children of an individual has probability distribution $(\frac{1}{4}, \frac{1}{2}, \frac{1}{4})$ on the set $\{0, 1, 2\}$ and that individuals reproduce independently. Consider a randomly chosen individual, let $X$ be the number of children and $Y$ the number of grandchildren of this individual, and let $p$ be the joint pmf of $(X, Y)$. Find **(a)** $p(0, 0)$, **(b)** $p(2, 2)$, and **(c)** $P(Y = 0)$.

## Section 3.4. Jointly Continuous Random Vectors

14. Show that the following functions are possible joint pdfs: **(a)** $f(x, y) = e^{-y}, 0 \leq x \leq y < \infty$, **(b)** $f(x, y) = xe^{-x(y+1)}, x, y \geq 0$, **(c)** $f(x, y) = 1/(x^2 y^2), x, y \geq 1$, and **(d)** $f(x, y) = x/\sqrt{y}, 0 \leq x \leq 1, 0 \leq y \leq 1$.
15. Which of the following functions are possible joint pdfs on $[0, 1] \times [0, 1]$: **(a)** $f(x, y) = xy$, **(b)** $f(x, y) = 4y - 2x$, **(c)** $f(x, y) = 3|y - x|$, and **(d)** $f(x, y) = 2x$?

16. The random variables $X$ and $Y$ have joint pdf

$$f(x, y) = c(x - y), \quad 0 \leq y \leq x \leq 1$$

Find **(a)** the constant $c$, **(b)** $P(X > \frac{1}{2}, Y \leq \frac{1}{2})$, **(c)** $P(X \leq 2Y)$, and **(d)** the marginal pdfs of $X$ and $Y$.

17. The random variables $X$ and $Y$ have joint cdf $F(x, y) = \frac{1}{2}(x^2y + xy^2), 0 \leq x \leq 1, 0 \leq y \leq 1$. Find the joint pdf and the marginal pdfs of $X$ and $Y$.

18. Let $X \sim \text{unif}[0, 1]$ and let $A$ be the area of a square with sidelength $X$. Show that $X$ and $A$ are continuous when viewed one by one but that $(X, A)$ is not jointly continuous.

19. Let $(X, Y)$ be uniform on the set $([0, \frac{1}{2}] \times [0, \frac{1}{2}]) \cup ([\frac{1}{2}, 1] \times [\frac{1}{2}, 1])$ (draw a figure). Find the marginals of $X$ and $Y$.

20. Consider a dart thrown at a dartboard with radius 1. Suppose that the player aims at the center and that this is reflected in the joint pdf of $(X, Y)$ being $f(x, y) = c(1 - (x^2 + y^2)), x^2 + y^2 \leq 1$. Find **(a)** the constant $c$, and **(b)** the probability that the dart hits within distance $d$ of the center.

21. Are the following statements true in general for a joint pdf: **(a)** $f(x, y) \leq f_X(x)$ for all $x, y \in R$, **(b)** $f(x, y) \leq f_X(x)$ for some $x, y \in R$, **(c)** $f(x, y) \leq 1$ for all $x, y \in R$, and **(d)** $f(x, y) \leq 1$ for some $x, y \in R$?

## Section 3.5. Conditional Distributions and Independence

22. Eggs are delivered to a restaurant by the gross (1 gross = 12 dozen). From each gross, a dozen of eggs are chosen at random. If none are cracked, the gross is accepted, and if more than one egg is cracked, the gross is rejected. If exactly one egg is cracked, an additional dozen eggs from the same gross are inspected. If this lot has no cracked eggs, the entire gross is accepted, otherwise it is rejected. Suppose that a gross has eight cracked eggs. What is the probability that it is accepted?

23. Customers arrive to a store such that the number of arriving customers in an hour has a Poisson distribution with mean 4. A customer is male or female with equal probabilities. Let $X$ be the number of female customers in an hour and find $P(X = 0)$.

24. Let $X$ be a random variable and $c$ a constant. Show that $X$ and $c$ are independent.

25. Let $X$ and $Y$ be independent and have Poisson distributions with means $\lambda_1$ and $\lambda_2$, respectively. Show that the conditional distribution of $X$ given $X + Y = n$ is binomial and identify the parameters.

26. Let $X$ and $Y$ be independent and have the same binomial distribution with parameters $n$ and $p$. Show that the distribution of $X$ given $X + Y = n$ is hypergeometric and does not depend on $p$. Explain intuitively.

27. Let $X$ and $Y$ be independent and have the same geometric distribution with success probability $p$. Find the conditional distribution of $X$ given $X + Y = n$. Explain intuitively.

28. A salesman has a weekly net income that is uniform on $[-1, 2]$ (\$1000; a negative number means a net loss). Let $X$ be his income in a week when he makes a profit. Find the pdf of $X$.
29. Consider Example 2.22 where the amount $X$ of the compound has pdf $f(x) = 2x$, $0 \leq x \leq 1$ and is kept if $X \geq 0.5$. What is the conditional pdf of the compound in a test tube that is kept?
30. Let $(X, Y)$ be jointly continuous. Is it then always true that $F_Y(y|x) = F(x, y)/F_X(x)$? If not, give a condition for when it is true.
31. For the joint pdfs in Problem 14, find the conditional pdf of $Y$ given $X = x$.
32. Let $X$ have a uniform distribution on $(0, 1)$, and given that $X = x$, let the conditional distribution of $Y$ be uniform on $(0, 1/x)$. **(a)** Find the joint pdf $f(x, y)$ and sketch the region where it is positive. **(b)** Find $f_Y(y)$, the marginal pdf of $Y$, and sketch its graph. **(c)** Compute $P(X > Y)$.
33. Let $X \sim \text{unif}[0, 1]$. Find the pmf of $Y$ if the conditional distribution of $Y$ given $X = x$ is $\text{bin}(n, x)$. Does $Y$ have a binomial distribution? *Note:* $\int_0^1 x^a(1-x)^b dx = a!b!/(a+b+1)!$
34. Adam and Billy Bob have agreed to meet at 12:30. Assume that their arrival times are independent random variables, Adam's uniformly distributed between 12:30 and 1:00 and Billy Bob's uniformly distributed between 12:30 and 1:15. **(a)** Compute the probability that Billy Bob arrives first. **(b)** Compute the probability that the one who arrives first must wait more than 10 min.
35. Let $X$ and $Y$ be nonnegative, independent continuous random variables. **(a)** Show that

$$P(X < Y) = \int_0^\infty F_X(x) f_Y(x) dx$$

**(b)** What does this become if $X \sim \exp(\lambda_1)$ and $Y \sim \exp(\lambda_2)$?
36. Decide whether $X$ and $Y$ are independent if $(X, Y)$ has the following joint pdfs on $[0, 1] \times [0, 1]$: **(a)** $f(x, y) = 4xy$, **(b)** $f(x, y) = x + y$, **(c)** $f(x, y) = x(2y + 1)$, **(d)** $f(x, y) = 6xy^2$, **(e)** $f(x, y) = 2y$.
37. **(a)** Let $X$ and $Y$ be independent random variables. Show that $X^2$ and $Y^2$ are also independent. **(b)** The converse of (a) is not true. Let $X$, $Y$, and $U$ be independent such that $X$ and $Y$ are $\exp(1)$ and $U$ is $-1$ or $1$ with equal probabilities. Let $S = UX$ and $T = UY$. Show that $S^2$ and $T^2$ are independent but that $S$ and $T$ are not independent. **(c)** Let $X$ and $Y$ be independent random variables and $g$ and $h$ be two functions that are strictly monotone. Show that $g(X)$ and $h(Y)$ are independent.
38. Let $R_1$ and $R_2$ be independent and $\text{unif}[0, 1]$, and let $V_1$ be the volume of a sphere with radius $R_1$ and $V_2$ the volume of a cube with side $2R_2$. Find $P(V_1 > V_2)$.
39. Consider the quadratic equation $x^2 + Bx + C = 0$ where $B$ and $C$ are independent and have uniform distributions on $[-n, n]$. Find the probability that the equation has real roots. What happens as $n \to \infty$?

40. Consider the bass from Problem 17 in Chapter 2. A more realistic model is that the weight at age $A$ is $W = 2A + 3 + X$, where $X$ is a random variable, independent of $A$, accounting for the fact that not all fish of a particular age have exactly the same weight. Suppose that $X \sim \text{unif}[-2, 2]$ and find the probability that **(a)** a 10-year-old bass weighs less than 22 pounds, **(b)** a randomly chosen bass is older than 8 years and weighs more than 20 pounds, and **(c)** two randomly chosen bass of the same age differ in weight by at least 2 pounds.
41. Let $f$ and $g$ be two pdfs, let $p \in (0, 1)$, and let $h(x) = pf(x) + (1 - p)g(x)$, $x \in R$. **(a)** Show that $h$ is a pdf. **(b)** If $X$ has pdf $f$ and $Y$ has pdf $g$, what does the pdf $h$ describe?

## Section 3.6. Functions of Random Vectors

42. Let $X$ and $Y$ have the same distribution (but not necessarily be independent) and let $Z = X - Y$. Show that $Z$ has a symmetric distribution (see Problem 101 in Chapter 2).
43. Let $X$ and $Y$ be independent and unif[0, 1]. Find the cdf and pdf of the random variables **(a)** $|X - Y|$, **(b)** $X/(X + Y)$.
44. A current of $I$ flows through a resistance of $R$, thus generating the power $W = I^2 R$. Suppose that $I$ has pdf $f(x) = 2x, 0 \leq x \leq 1$, $R \sim \text{unif}[0, 1]$ and that $I$ and $R$ are independent. Find the pdf of $W$.
45. In Problem 34, compute the expected time that **(a)** Billy Bob must wait for Adam, **(b)** the first person to arrive must wait for the second.
46. Water flows in and out of a dam such that the daily inflow is uniform on [0, 2] (megaliters) and the daily outflow is uniform on [0, 1], independent of the inflow. Each day the surplus water (if there is any) is collected for an irrigation project. Compute the expected amount of surplus water in a given day.
47. Let $X$ and $Y$ be independent and exp(1). Find $E[e^{-(X+Y)/2}]$.
48. Let $X$ and $Y$ be independent and unif[0, 1]. Find **(a)** $E[XY]$, **(b)** $E[X/Y]$, **(c)** $E[\log(XY)]$, and **(d)** $E[|Y - X|]$.
49. Consider the two rods in Example 3.19. Suppose that there is an error $X$ when we lay the rods side by side that is independent of the measurements and such that $E[X] = 0$ and $\text{Var}[X] = \tau^2$. Now measure the sum and difference $S$ and $D$ and estimate $a$ by $A = (S + D)/2$. **(a)** Find the mean and variance of $A$. **(b)** When is the precision with this method better than taking one single measurement of the longer rod?
50. Let $X_1, X_2, \ldots, X_n$ be i.i.d. random variables with mean $\mu$ and variance $\sigma^2$, let $S_n = X_1 + \cdots + X_n$, and let $\bar{X} = S_n/n$ (called the *sample mean*). Find $E[\bar{X}]$ and $\text{Var}[\bar{X}]$.
51. Let $X_1, \ldots, X_n$ be independent with the same mean $\mu$ and the same variance $\sigma^2$. Express the following expected values in terms of $n$, $\mu$, and $\sigma^2$: **(a)** $E[X_1^2 + \cdots + X_n^2]$, **(b)** $E[(X_1 + \cdots + X_n)^2]$.

52. Let $X$ have a negative binomial distribution with parameters $r$ and $p$ (see Problem 61, Chapter 2). Find $E[X]$ and $\text{Var}[X]$ without using the definition; instead, consider how $X$ can be written as a sum of independent random variables.
53. Recall the concepts of skewness and kurtosis from Problems 102 and 104 in Chapter 2. Let $X_1, \ldots, X_n$ be independent and have the same distribution with skewness $s$ and kurtosis $c$, and let $S_n = \sum_{k=1}^{n} X_k$. **(a)** Show that $\text{skw}[S_n] = s/\sqrt{n}$. **(b)** Show that $\text{kur}[S_n] = 3 + (c-3)/n$. **(c)** Find $\text{skw}[X]$ and $\text{kur}[X]$ if $X \sim \text{bin}(n, p)$.
54. Consider coupons numbered $1, 2, \ldots, n$ that you collect until you have all numbers. Each time you get a coupon, its number is independent of previous numbers and all numbers are equally likely. Let $N$ be the number of coupons you need and show that **(a)** $E[N] = n \sum_{k=1}^{n}(1/k) \approx n \log n$ for large $n$, and **(b)** there is a constant $c$ such that $\text{Var}[N] \approx cn^2$ for large $n$. What is $c$?
55. In the previous problem, let $X$ be the number of different coupons when you have collected $k$ coupons. **(a)** Find $E[X]$. *Hint:* Let $I_j$ be the indicator that there is at least one coupon with number $j$, among the $k$ coupons. **(b)** Show that, for large $n$ and $k = cn$, $E[X] \approx n(1 - e^{-c})$, $c = 1, 2, \ldots$ .
56. You roll a die repeatedly until all numbers have shown up. What is the expected number of 6s? *Hint:* Let $N_k$ be the expected number of $k$'s and $N$ as in Problem 54 and argue that $E[N_1] + \cdots + E[N_6] = E[N]$.
57. You throw darts at random on a chessboard (64 squares). After $n$ throws, let $X$ be the number of squares that are not hit. Find $E[X]$ and show that $E[X] \approx ne^{-1/64}$ for large $n$. For which values of $n$ is the approximation useful?
58. Consider $N$ married couples. If there are $n$ deaths, what is the expected number of married couples remaining?
59. Consider a sample containing $2n$ carbon-14 atoms (see Example 2.40). Show that the expected time until all have decayed is approximately $5700 \log n / \log 2$ years. How can you come up with this number in the deterministic description?
60. Consider the inclusion–exclusion formula in Proposition 1.5. Prove this by first expressing the indicator of the union in terms of indicators of intersections, and then take expected values.
61. Let $X$ and $Y$ be nonnegative and have joint pdf $f$ and let $Z = Y/X$. **(a)** Express the joint pdf of $(X, Z)$ in terms of $f$. **(b)** If $X$ and $Y$ are independent exp(1), find the joint pdf of $(X, Z)$ and the marginal pdf of $Z$.
62. Let $X$ and $Y$ be independent and unif[0, 1]. Find the joint pdf of $U = X + Y$ and $V = X/(X + Y)$.
63. Let $X$ and $Y$ be independent $N(0, 1)$ and let $U = X - Y$, $V = X + Y$. Find the joint pdf of $(U, V)$.
64. A point $(X, Y)$ is chosen in the unit disk by letting its polar coordinates $R$ and $\Theta$ be independent and uniform on their respective ranges. Find the joint pdf of $(X, Y)$.

65. A battery with voltage $U$ is connected to a resistance of $R$, thus creating a current of $I = U/R$ and a power of $W = U^2/R$. Suppose that $U$ has pdf $f(u) = 2u, 0 \leq u \leq 1$, $R$ has pdf $f(r) = 1/r^2, r \geq 1$ and that $U$ and $R$ are independent. Find the joint pdf of $(I, W)$ and the marginal pdfs of $I$ and $W$ (be careful with the ranges).
66. Let $(X, Y)$ have joint pdf $f(x, y) = x + y, 0 \leq x \leq 1, 0 \leq y \leq 1$ and let $U = 2X$, $V = X + Y$. Find the joint pdf of $(U, V)$ (be careful with the range).
67. Let $(X, Y)$ have joint pdf

$$f(x, y) = \frac{1}{\pi\sqrt{3}} e^{-2(x^2+y^2-xy)/3}, \quad x, y \in R$$

and let $U = X + Y$, $V = X - Y$. Find the joint pdf of $(U, V)$.

### Section 3.7. Conditional Expectation

68. Let $X$ be a continuous random variable with pdf $f$ and let $b$ be a real number. Show that

$$E[X|X > b] = \frac{\int_b^\infty xf(x)dx}{P(X > b)}$$

69. A saleswoman working for a company sells goods worth $X$ \$1000 per week, where $X$ is unif[0, 2]. Of this, she must pay the company back up to \$800 and gets to keep the rest. Compute her expected profit **(a)** in a given week, and **(b)** in a week when she makes a profit.
70. In Problem 10(a) and (b), compute $E[Y|X = k]$ for $k = 1, \ldots, 6$.
71. Compute the conditional expectations $E[Y|X = x]$ in Problem 14.
72. Let $X \sim$ unif[0, 1]. Compute $E[Y]$ if the conditional distribution of $Y$ given $X = x$ is **(a)** unif$[0, x^2]$, **(b)** unif$[0, \sin(\pi x)]$, **(c)** unif$[0, 1/x]$, and **(d)** exp$(1/x)$.
73. Let $X \sim$ exp(1). Compute $E[Y]$ if the conditional distribution of $Y$ given $X = x$ is **(a)** unif$[0, x]$, **(b)** unif$[x, x + 2]$, and **(c)** exp$(x)$.
74. Let $X \sim$ unif[0, 1]. Compute $E[Y]$ if the conditional distribution of $Y$ given $X = x$ is **(a)** geom$(x)$, and **(b)** bin$(n, x)$ (recall Example 3.11 and Problem 33).
75. Suppose that $X$ and $Y$ are such that $E[Y|X] = 2X$. Thus, if we know $X$, we would expect $Y$ to be twice as much. Does this imply that $E[X|Y] = Y/2$?
76. Consider the variance formula $\text{Var}[Y] = \text{Var}[E[Y|X]] + E[\text{Var}[Y|X]]$. **(a)** Show that $\text{Var}[E[Y|X]] = 0$ if $X$ and $Y$ are independent. **(b)** Show that $E[\text{Var}[Y|X]] = 0$ if $X$ and $Y$ are totally dependent in the sense that $Y = g(X)$ for some function $g$ (recall Corollary 3.6).
77. Let $X \sim$ geom$(p)$. Use a recursive argument to show that $\text{Var}[X] = (1 - p)/p^2$. *Hint:* First, let $\tau = E[X^2]$ and obtain an equation for $\tau$ by using Equation (3.3) and properties of indicators.

78. Recall Example 1.56, where Ann and Bob play tennis and Ann wins a point as server with probability $p$. Suppose that the players are at deuce and that Ann serves. What is the expected number of played points until the game is over?
79. Recall Example 1.57, where Ann and Bob play badminton and Ann wins a rally with probability $p$. If Ann is about to serve, what is the expected number of rallies until the next point is scored?
80. In Example 3.34, we arrived at the equation $\mu = \frac{6}{4} + \frac{3}{4}\mu$, which has solution $\mu = 6$. However, $\mu = \infty$ also is a solution, so we need to argue why this can be ruled out. Show that $\mu$ can be at most 8 (think about coins flipped two by two).
81. In Example 3.34, compute the variances of the waiting times for the patterns $HH$ and $TH$.
82. Flip a fair coin repeatedly and verify the expected number of flips until the first occurrence of $HHH$, $HHT$, $HTH$, and $HTT$, as given in Example 3.34.
83. Flip a fair coin repeatedly and wait for the first occurrence of $n$ consecutive heads, $HH \cdots H$. Find an expression for the expected number of flips until this occurs.
84. Roll a die repeatedly. What is the expected number of rolls until the first occurrence of the pattern **(a)** 36, **(b)** 666?

## Section 3.8. Covariance and Correlation

85. Let $U$ and $V$ be independent and unif[0, 1] and let $X = \min(U, V)$ and $Y = \max(U, V)$. Find $\text{Cov}[X, Y]$ and comment on its sign.
86. Find the variance of the number of matches in Example 3.23. First argue that $I_k$ are not independent but that $\text{Cov}[I_j, I_k]$ is the same for all $j \neq k$, and then compute it.
87. Draw three cards without replacement from a deck of cards. Let $H$ be the number of hearts and $S$ the number of spades drawn. Find $\rho(H, S)$ and comment on its sign.
88. In Definition 3.17, it is necessary that $\text{Var}[X] > 0$ and $\text{Var}[Y] > 0$. If this is not the case, what value is reasonable to assign to the correlation?
89. Let $\rho$ be the correlation between $X$ and $Y$. What is the correlation between $X + a$ and $Y + b$, where $a$ and $b$ are constants?
90. Prove Proposition 3.25(c) using the method suggested in the proof and also express $a$ and $b$ in terms of $\mu_1, \mu_2, \sigma_1^2, \sigma_2^2$, and $\rho$.
91. Compute $\rho(I_j, I_k)$ in Example 3.38. What is its sign and how does it depend on the parameters of the hypergeometric distribution? Explain intuitively.
92. Let $(X, Y)$ be uniformly distributed on the triangle with corners in (0, 0), (0, 1), and (1, 0). **(a)** Compute the correlation coefficient $\rho(X, Y)$. **(b)** If you have done (a) correctly, the value of $\rho(X, Y)$ is negative. Explain intuitively.

93. Let $A$ and $B$ be two events. The degree of dependence between $A$ and $B$ can then be measured by the correlation between their indicators $I_A$ and $I_B$. Suppose $P(A) = P(B) = \frac{1}{2}$ and express the correlation coefficient $\rho(I_A, I_B)$ as a function of $P(A|B)$. Give an intuitive interpretation.

94. Let $A$ and $B$ be independent events. Show that $I_A + I_B$ and $|I_A - I_B|$ are uncorrelated. Are they independent?

95. For the following random variables, $X$ and $Y$, determine whether they are uncorrelated and independent. **(a)** $X \sim \text{unif}[0, 1]$ and $Y = X^2$, **(b)** $X \sim \text{unif}[-1, 1]$ and $Y = X^2$, **(c)** $X = \cos(U)$ and $Y = \sin(U)$, where $U \sim \text{unif}[0, 2\pi]$, and **(d)** $X \sim N(0, 1)$ and $Y = X^2$.

96. Let $X$ and $Y$ be independent and uniform on $[0, 1]$. Let $A$ be the area and $C$ the circumference of a rectangle with sides $X$ and $Y$. Find the correlation coefficient of $A$ and $C$.

97. Let $X$ and $Y$ be independent with the same variance $\sigma^2$, and let $S = X + Y$ and $T = XY$. Under what conditions are $S$ and $T$ uncorrelated?

98. Let $(X, Y)$ be uniform on the region $\{(x, y) : 0 \leq x \leq 1, 0 \leq y \leq x^2\}$. Find $E[Y|X]$ and the best linear predictor $l(X)$.

99. Let $X$ and $Y$ be independent, with the same mean $\mu$ and the same variance $\sigma^2$, and let $Z = X + Y$. Find $\rho(X, Z)$. Explain intuitively in terms of the coefficient of determination.

100. You are dealt a bridge hand (13 cards). Let $X$ be the number of aces and $Y$ the number of hearts. Show that $X$ and $Y$ are uncorrelated but not independent. *Hint:* Write both $X$ and $Y$ as sums of indicators and show that indicators in one sum are independent of indicators in the other.

101. Recall the sample mean $\bar{X}$ from Problem 50. **(a)** Show that $\text{Cov}[X_1, \bar{X}] = \sigma^2/n$ and find $\rho(X_1, \bar{X})$. Interpret the value of $\rho$ in terms of the coefficient of determination. **(b)** Show that $\bar{X}$ and $X_k - \bar{X}$ are uncorrelated for each $k = 1, 2, \ldots, n$.

### Section 3.9. The Bivariate Normal Distribution

102. Which of the following are plots from a bivariate normal distribution? For those that are, what can you say about variances and correlation? For those that are not, what looks wrong?

103. A company manufactures rectangular metal plates of size $5 \times 10$ (inches). Owing to random fluctuations, a randomly selected plate has a size of $X \times Y$ in. where $(X, Y)$ follows a bivariate normal distribution with means 5 and 10, variances 0.01 and 0.04, and correlation coefficient 0.8. Let $C$ be the circumference and $A$ the area of a plate. **(a)** Find $E[C]$ and $E[C|X = x]$. **(b)** Find $E[A]$ and $E[A|X = x]$. **(c)** A plate is useless if $C$ is less than 29 or more than 31 in. What is the probability that this happens? **(d)** If the sidelength $X$ of a plate is measured to be 5.1 in., what is the probability that the plate is useless? **(e)** One way to improve the process is to reduce the variances of $X$ and $Y$. Suppose that

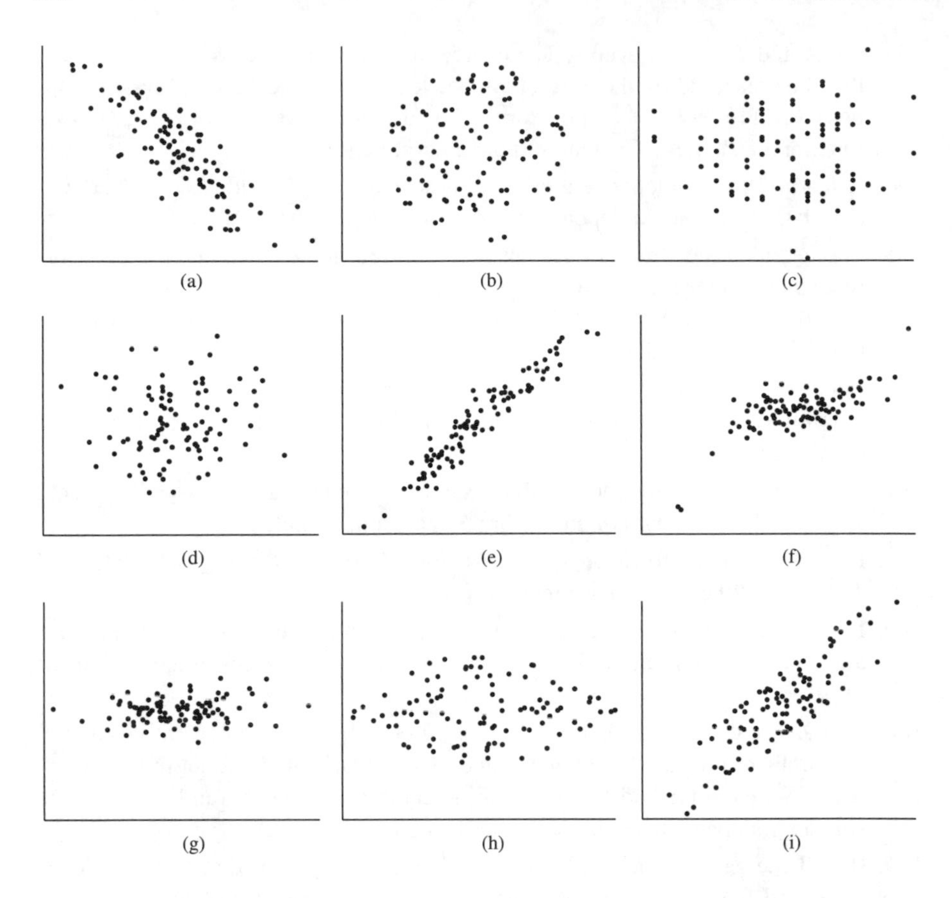

we can calibrate the process so that both variances are reduced by a factor $c$ (so that $X$ has variance $0.01c$ and $Y$ $0.04c$). To get the probability in (c) down below 0.01, how small must $c$ be?

104. Suppose that the weights (in kilograms) of Norwegian salmon of a certain age follow a normal distribution with mean 20 and variance 15 and that those of Canadian salmon of the same age are normal with mean 22 and variance 21. What is the probability that a randomly chosen Norwegian salmon weighs more than a randomly chosen Canadian one?

105. Lead fishing sinkers of two types are being manufactured. One type has weights $N(5, 1)$ and the other $N(7, 1)$ (ounces). Choose one of each, and let $X$ and $Y$ be their weights. Find **(a)** $P(X \geq Y)$, and **(b)** $P(Y \geq 2X)$. **(c)** Let $\bar{X}$ and $\bar{Y}$ be the average weights of $n$ sinkers of the two types. How large must $n$ be for $P(\bar{Y} \geq \bar{X})$ to be at least 0.99?

106. Let $X \sim N(0, 1)$ and let $Y$ be a random variable such that

$$Y = \begin{cases} X & \text{if } |X| \leq 1 \\ -X & \text{if } |X| > 1 \end{cases}$$

**(a)** Show that $Y \sim N(0, 1)$. **(b)** Is $(X, Y)$ bivariate normal?

107. Let $(X, Y)$ be bivariate normal with means 0, variances 1, and correlation coefficient $\rho > 0$, and let $U = X + Y$, $V = X - Y$. What is the joint distribution of $(U, V)$?

108. Two types of lightbulbs are being manufactured, one whose lifetime is $N(1000, 100)$ and the other whose lifetime is $N(800, 81)$ (hours). **(a)** Choose one of each type and let $Z$ be the average lifetime. Express the pdf of $Z$ in terms of the standard normal pdf $\varphi$. **(b)** Choose a lightbulb at random, so that it is equally likely to be of either type, and let $W$ be its lifetime. Express the pdf of $X$ in terms of the standard normal pdf $\varphi$. Does $W$ have a normal distribution? What does its pdf look like? *Hint:* Write $W$ as $IX + (1 - I)Y$, where $I$ is a suitably chosen indicator. **(c)** Find the means and variances of $Z$ and $W$. Compare.

## Section 3.10. Multidimensional Random Vectors

109. Consider the following variation of Example 3.52. Instead of accepting the first bid that is higher than the first, you decide to accept the first bid that is higher than the one immediately preceding it. Let $N$ be the number of the bid you accept. Define $N$ formally and find $E[N]$.

110. In the previous problem, suppose that you instead decide to take the first offer that exceeds $c$ dollars. If the independent bids have cdf $F$, what is $E[N]$?

111. The Old Faithful geyser in Yellowstone National Park is so named for its regular eruptions, about once an hour on average. However, most tourists have to wait for more than half an hour, which they think indicates that the geyser is slowing down. **(a)** Explain why their longer waiting times do not contradict hourly eruptions. **(b)** Suppose that times between eruptions are i.i.d. random variables that are uniform between 30 and 90 min. What is the expected wait for an arriving tourist (recall Example 3.49)?

112. Consider a parallel system of $n$ components whose lifetimes are i.i.d. $\exp(\lambda)$. Find the pdf of the lifetime of the system.

113. Let $X_1, \ldots, X_n$ be i.i.d. and continuous with cdf $F$ and pdf $f$. Find the pdf of $X_{(j)}$, the $j$th smallest order statistic. *Hint:* Let $N(x)$ be the number of $X_k$ that are less than or equal to $x$ and express the event $\{X_{(j)} \leq x\}$ in terms of $N(x)$. What is the distribution of $N(x)$?

114. Consider the sinkers from Problem 105 with weights that are $N(5, 1)$ and $N(7, 1)$, respectively. Choose one of each and let $M$ be the maximum weight of the two. **(a)** Express the pdf of $M$ in terms of the standard normal cdf $\Phi$ and pdf $\varphi$. **(b)** Find $P(M \leq 6)$.

115. Let $X_1, \ldots, X_n$ be i.i.d. exponential with parameter $\lambda$. Show that the expected value of the maximum $X_{(n)}$ is approximately $\log n/\lambda$ for large $n$. *Hint:* Example 3.51.

116. Let $X_1, \ldots, X_n$ be independent such that $X_k$ has cdf $F_k(x)$ and pdf $f_k(x)$. Find the cdf and pdf of $X_{(1)}$ and $X_{(n)}$.

117. Let $X$, $Y$, $Z$ and $W$ be i.i.d. continuous random variables. Find the probability that **(a)** $X > Y$ and $Z < W$, **(b)** $X < Y < Z < W$, **(c)** $X < Y < Z > W$, and **(d)** $X < Y > Z > W$.

118. You and four other people are bidding at a silent auction. The bids are independent and you assume that the other bids are uniform on [100, 200] dollars. How much must you bid to be at least 90% sure to win the bidding?

119. An electronic component has failure rate function $r(t) = a + r_1(t) + r_2(t)$, where $a$ is a constant, $r_1(t)$ is an increasing function, and $r_2(t)$ is decreasing. What does the failure rate function $r(t)$ describe?

120. Consider a parallel system of two components that have lifetimes that are exp(1). When one of them breaks down, the other has to take more stress and has a remaining lifetime that is exp(2). Find the failure rate function of the system.

121. A machine manufactures two types of components with exponential lifetimes, one with mean 1 and one with mean $\frac{1}{2}$ year. Find the failure rate function of the lifetime of a randomly chosen component.

122. Let $X_1$ and $X_2$ be jointly continuous with joint pdf $f$ (not necessarily independent). Find an expression for the joint pdf of $(X_{(1)}, X_{(2)})$ in terms of $f$.

123. Consider the ABO classification of blood types from Problem 52 in Chapter 1. If you choose eight people at random, what is the probability that you get two of each type?

124. Let $(X_1, \ldots, X_r)$ have a multinomial distribution with parameters $n$ and $(p_1, \ldots, p_r)$. Use indicators to show that

$$\mathrm{Cov}[X_j, X_k] = -np_jp_k$$

for $j \neq k$. Why is it intuitively clear that the covariance is negative? What is the correlation coefficient?

125. Let $X$, $Y$, and $Z$ be independent and $N(1, 1)$. Find the probability that $X + Y \geq 3Z$.

126. Jobs arrive at a computer. When the computer is free, the waiting time $W$ until the next incoming job has an exponential distribution with mean 1 ms. The time $C$ it takes to complete a job has a uniform distribution with mean 0.5 ms. Consider a time when the computer is free, and let $T$ be the time until the next job is completed. Find the pdf of $T$ and sketch its graph.

127. Recall the Geiger counter from Example 3.29. Consider a time when the counter registers a particle and let $T$ be the time until the next registration. Find the pdf of $T$.

128. Let $X$ and $Y$ be independent, both with pdf $f(x) = \exp(1 - x)$, $x \geq 1$. Find the pdf of $X + Y$.

129. Let $X \sim \exp(\lambda_1)$ and $Y \sim \exp(\lambda_2)$ be independent. Find the pdf of $X + Y$.

130. Let $(X, Y)$ have joint pdf $f(x, y)$. Show that the pdf of the sum $X + Y$ is

$$f_{X+Y}(x) = \int_{-\infty}^{\infty} f(u, x - u)du$$

131. Let $(X, Y)$ have joint pdf

$$f(x, y) = \frac{1}{2}(x + y)e^{-(x+y)}, \quad x, y \geq 0$$

Find the pdf of $X + Y$.

## Section 3.11. Generating Functions

132. Describe the random variables that have the following pgfs: **(a)** $G(s) \equiv 1$, **(b)** $G(s) = s$, and **(c)** $G(s) = \frac{1}{2}(s + s^2)$.

133. Let $G$ be the pgf of a random variable $X$. Show that $G$ is increasing and convex. When is it strictly increasing? Strictly convex?

134. **(a)** Let $X$ be the number when a fair die is rolled. Find the pgf of $X$. **(b)** Let $S$ be the sum when five fair dice are rolled. Find the pgf of $S$ and describe how you can use it to find the probability $P(S = 20)$.

135. **(a)** Let $X \sim \text{geom}(p)$. Find the pgf of $X$. **(b)** Let $Y \sim \text{negbin}(r, p)$ (see Problem 61, Chapter 2). Use (a) and Proposition 3.38 to find the pgf of $Y$.

136. Let $X$ and $Y$ be independent random variables with range $\{1, 2, \ldots\}$ and pgfs $G_X$ and $G_Y$. Show that

$$E\left[\frac{X}{X+Y}\right] = \int_0^1 G'_X(s)G_Y(s)\, ds$$

137. Let $X_1, X_2, \ldots$ be i.i.d. nonnegative and integer valued, and let $N \sim \text{Poi}(\lambda)$, independent of $X_k$. If the random sum $S_N = X_1 + \cdots + X_N$ has a Poisson distribution with mean $\mu$, what is the distribution of $X_k$?

138. In the department I am working in, there is a self-service coffee bar downstairs from my office. Each week I go down for coffee a number of times $N$, which has a Poisson distribution with mean 20. Each time, I am equally likely to choose the small size for 75 cents or the medium for one dollar, and I am equally likely to pay directly or to record my purchase to pay later. At the end of a week, what is my expected debt?

139. Jobs arrive at a computer at a rate of 1.8 jobs/day. Each job requires an execution time (in milliseconds) that has a binomial distribution with $n = 10$ and $p = 0.5$ (where a zero execution time means that the job is rejected). Find **(a)** the probability that the total execution time in a day for this type of job is at least 2 ms, and **(b)** the expected total execution time in a day.

140. Cars arrive at a park entrance such that the number of cars in 1 h has a Poisson distribution with mean 4. Each car is equally likely to contain one, two, three, or four people. Let $Y$ be the number of people who arrive at the park during an hour.

Find (after making appropriate assumptions) **(a)** the pgf of $Y$, **(b)** $P(Y = 0)$, and **(c)** $E[Y]$ and $\text{Var}[Y]$.

141. Roll a fair die repeatedly and keep track of the numbers. When the first 6 shows up, you add what you have thus far, not including the 6. Hence, we have $X_1, X_2, \ldots$ i.i.d. die rolls, $N \sim \text{geom}(1/6)$, and observe

$$Y = \sum_{k=1}^{N-1} X_k$$

where $Y$ is 0 if we get 6 in the first roll. We wish to find the expected value of $Y$. By Proposition 3.13, we get $E[Y] = E[N-1]E[X]$ that equals $5 \times 3.5 = 17.5$. We could also argue that the values we add cannot contain any 6s and are thus uniform on $1, 2, \ldots, 5$, which would instead give $E[Y] = 5 \times 3 = 15$. Which of the answers is correct? Does Proposition 3.13 apply?

142. Let $X_1, X_2, \ldots$ be i.i.d. nonnegative with cdf $F$ and let $N$ be nonnegative and integer valued with pgf $G$, independent of $X_k$. Let $M_N = \max(X_1, \ldots, X_N)$ with $M_N = 0$ if $N = 0$. **(a)** Show that $P(M_N \leq x) = G(F(x))$. If $X_k$ are continuous, what type of distribution does $M_N$ have? **(b)** You are bidding on a silent auction where the number of other bidders has a Poisson distribution with mean 2. Bids are independent and uniform on [0, 100] dollars. What is the minimum amount that you would need to bet to be at least 90% sure to win the bidding?

143. If we allow random variables to take on infinite values, the pgf can still be defined in the same way, but many of the results are no longer valid. Let $X$ take values in $\{0, 1, \ldots, \infty\}$ and have pgf $G$, and let $p = P(X = \infty) > 0$. **(a)** Show that $G(1) = 1 - p$ (recall Problem 11 in Chapter 2). **(b)** Give an example to show that we may have $E[X] \neq G'(1)$.

144. Let $X$ have mgf $M(t)$. Show that

$$M(t) = \sum_{k=0}^{\infty} \frac{t^k}{k!} E[X^k]$$

145. Let $X \sim N(0, 1)$. Use the mgf to find $E[X^3]$ and $E[X^4]$.

146. Let $X$ have the Cauchy distribution from Section 2.8.3. Show that the mgf of $X$ does not exist (in the sense that the integral defining it is of the form "$\infty - \infty$").

147. There is an analog to Proposition 3.39 that deals with mgfs. Thus, let $X_1, X_2, \ldots$ be i.i.d. random variables with common mgf $M_X$ and common mean $\mu$, and let $N$ be nonnegative and integer valued, and independent of the $X_k$, with pgf $G_N$ (note that this is the pgf). **(a)** Show that $S_N$ has mgf

$$M_{S_N}(t) = G_N(M_X(t)), \quad t \in R$$

**(b)** Let $X_k \sim \exp(\lambda)$ and let $N \sim \text{geom}(p)$. Show that $S_N \sim \exp(\lambda p)$. **(c)** Show that $E[S_N] = E[N]\mu$.

148. Let $X_1, X_2, \ldots$ be i.i.d. $N(0, 1)$, and let $S_n = X_1 + \cdots + X_n$, in which case we know that $S_n \sim N(0, n)$. Now roll a die to decide how many $X_k$ to sum. Find the mgf of the resulting sum. Is it normal, and if so, what are the parameters?
149. If $X$ and $Y$ are nonnegative and integer valued, the function $G(s, t) = E[s^X t^Y]$, $0 \le s \le 1, 0 \le t \le 1$ is called the *joint pgf* of $X$ and $Y$. **(a)** Express the pgfs of $X$ and $Y$ in terms of $G$. **(b)** Use differentiation to find formulas for expected values, variances, and the covariance of $X$ and $Y$. **(c)** Show that the joint pgf $G$ uniquely determines the joint pmf $p$ of $(X, Y)$ and that $X$ and $Y$ are independent if and only if $G = G_X G_Y$. **(d)** Suggest how to define the joint mgf and do (a) and (b) in this case.

**Section 3.12. The Poisson Process**

150. Alaska has over half of all earthquakes in the United States. In particular, earthquakes with magnitude >8 on the Richter scale occur in Alaska on average every 13 years. Suppose that these occur according to a Poisson process. Compute the probability that **(a)** a decade has no such earthquakes, **(b)** two consecutive such earthquakes are at least 5 years apart, **(c)** a 13-year period has exactly one such earthquake, and **(d)** three consecutive decades have exactly one such earthquake each.
151. Traffic accidents in a town occur according to a Poisson process at a rate of two accidents per week. **(a)** What is the probability that a given day has no accidents? **(b)** Whenever there is an accident, the risk is 1 in 10 that it will cause personal injury. What is the probability that a given month has at least one such accident? **(c)** Let $N$ be the number of accident-free weeks in a year. What is the distribution of $N$?
152. In any given hurricane season (6 months from June to November), there is about a 50–50 chance that the Gulf Coast will be hit by a hurricane. Assume a Poisson process with rate $\lambda$ and use time unit "months." **(a)** Find the value of $\lambda$. **(b)** Let $X$ be the number of hurricanes that hit the Gulf Coast during the months of August and September. What is the distribution of $X$ (name and parameter(s))? **(c)** Let $Y$ be the number of months that have no hits during one season. What is the distribution of $Y$ (name and parameter(s))?
153. Large meteorites hit Earth on average once every 1000 years. Find **(a)** the probability that Earth gets hit within the next 1000 years, **(b)** the probability that Earth gets hit more than once within the next 1000 years, and **(c)** the probability that there are no hits within the next 100 years.
154. Consider a radioactive sample that decays at rate $\lambda$. You turn on the Geiger counter, wait for the first registered emission, and wonder how long this takes on average. Since the timepoint at which you turned it on is completely arbitrary and independent of the decay process, you figure out that you ought to be on average in the middle between two consecutive emissions, and since the average time between two emissions is $1/\lambda$, you decide that the average wait must be

$1/(2\lambda)$. However, experience tells you that the average wait is longer. Explain! What is the average wait?

155. Explain Problem 147(b) in terms of thinning of Poisson processes.

156. Consider two independent Poisson processes with rates $\lambda_1$ and $\lambda_2$. Given that a total of $n$ points are observed in the interval $[0, t]$, what is the probability that $k$ of those came from the first process? *Hint:* Problem 25.

157. Southbound and northbound cars pass a point on a rural highway according to two independent Poisson processes at rates two and three cars per minute, respectively. **(a)** What is the probability that exactly two cars pass in a given minute? **(b)** What is the probability that exactly one northbound car and one southbound car pass in a given minute? **(c)** If two cars pass in a minute, what is the probability that they are both northbound? **(d)** If four cars pass between 11:55 am and 12:05 pm, what is the probability that two pass before and two after noon.

158. Accidents in a town occur according to a Poisson process at a rate of two accidents per week. Two towing companies, Adam's Towing and Barry's Wrecker, have agreed to take turns in dealing with the accidents on a weekly basis. Adam, who has been in town longer, takes care of the first accident of the week, Barry the second, and so on. **(a)** Consider the process of accidents that Adam takes care of. Is this a Poisson process? If so, what is the rate? **(b)** What is the probability that Barry gets no business in a given week? **(c)** What is the probability that Adam and Barry get an equal amount of business in a given week?

159. Consider the superposition of two independent Poisson processes with rates $\mu$ and $\lambda$, respectively. Let $X$ be the number of points in the first between two consecutive points in the second. Show that $X$ has a geometric distribution including 0 (see Section 2.5.3) and identify the success probability. Explain intuitively. You may need to utilize the fact that $\int_0^\infty e^{-ax}x^k dx = k!/a^{k+1}$ or you can use generating functions.

# 4

# LIMIT THEOREMS

## 4.1 INTRODUCTION

In advanced studies of probability theory, limit theorems form the most important class of results. A limit theorem typically starts with a sequence of random variables, $X_1, X_2, \ldots$ and investigates properties of some function of $X_1, X_2, \ldots, X_n$ as $n \to \infty$. From a practical point of view, this allows us to use the limit as an approximation to an exact quantity that may be difficult to compute.

When we introduced expected values, we argued that these could be considered averages of a large number of observations. Thus, if we have observations $X_1, X_2, \ldots, X_n$ and we do not know the mean $\mu$, a reasonable approximation ought to be the *sample mean*

$$\bar{X} = \frac{1}{n} \sum_{k=1}^{n} X_k$$

in other words, the average of $X_1, \ldots, X_n$. Suppose now that the $X_k$ are i.i.d. with mean $\mu$ and variance $\sigma^2$. By the formulas for the mean and variance of sums of independent random variables, we get

$$E[\bar{X}] = E\left[\frac{1}{n} \sum_{k=1}^{n} X_k\right] = \sum_{k=1}^{n} \frac{1}{n} E[X_k] = \mu$$

*Probability, Statistics, and Stochastic Processes*, Second Edition. Peter Olofsson and Mikael Andersson.

and

$$\text{Var}[\bar{X}] = \text{Var}\left[\frac{1}{n}\sum_{k=1}^{n} X_k\right] = \sum_{k=1}^{n} \frac{1}{n^2}\text{Var}[X_k] = \frac{\sigma^2}{n}$$

that is, $\bar{X}$ has the same expected value as each individual $X_k$ and a variance that becomes smaller the larger the value of $n$. This indicates that $\bar{X}$ is likely to be close to $\mu$ for large values of $n$. Since the variance goes to 0, we might want to say that "$\bar{X}$ converges to $\mu$ as $n \to \infty$," but exactly what does this mean? For a sequence of real numbers, $a_n \to a$ means that for any $\epsilon > 0$, we can always make $|a_n - a| < \epsilon$ if $n$ is large enough. This cannot be true for $\bar{X}$ and $\mu$, though. Since $\bar{X}$ is random, we can never say for certain that we will have $|\bar{X} - \mu| < \epsilon$ from some $n$ on. We need to come up with a definition of what convergence means for random quantities such as $\bar{X}$, and in the next section we address this issue.

## 4.2 THE LAW OF LARGE NUMBERS

Although we can never guarantee that $|\bar{X} - \mu|$ is smaller than a given $\epsilon$, we can say that it is very likely that $|\bar{X} - \mu|$ is small if $n$ is large. That is the idea behind the following result.

***Theorem 4.1. (The Law of Large Numbers).*** *Let $X_1, X_2, \ldots$ be a sequence of i.i.d. random variables with mean $\mu$, and let $\bar{X}$ be their sample mean. Then, for every $\epsilon > 0$*

$$P(|\bar{X} - \mu| > \epsilon) \to 0 \quad as \quad n \to \infty$$

*Proof.* Assume that the $X_k$ have finite variance, $\sigma^2 < \infty$. Apply Chebyshev's inequality to $\bar{X}$ and let $c = \epsilon\sqrt{n}/\sigma$. Since $E[\bar{X}] = \mu$ and $\text{Var}[\bar{X}] = \sigma^2/n$, we get

$$P(|\bar{X} - \mu| > \epsilon) \leq \frac{\sigma^2}{n\epsilon^2} \to 0 \quad \text{as} \quad n \to \infty$$

The assumption of finite variance is necessary for this proof to work. However, the law of large numbers is true also if the variance is infinite, but the proof in that case is more involved and we will not give it. ■

We say that $\bar{X}$ *converges in probability* to $\mu$ and write

$$\bar{X} \xrightarrow{P} \mu \quad \text{as} \quad n \to \infty$$

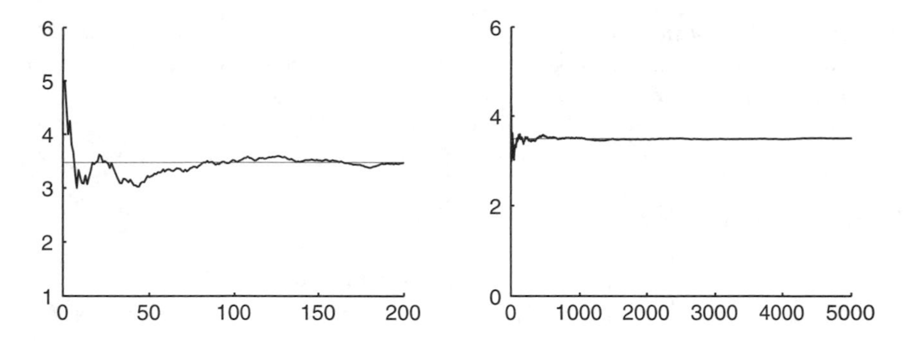


**FIGURE 4.1** Consecutive averages of 200 (left) and 5000 (right) simulated rolls of a fair die.

The law of large numbers thus states that although we can never be certain that $\bar{X}$ is within $\pm\epsilon$ of $\mu$, the probability that this happens approaches one as $n$ increases.[1]

In general, if $Y_1, Y_2, \ldots$ is any sequence of random variables and $a$ a constant, $Y_n \stackrel{P}{\rightarrow} a$ means that $P(|Y_n - a| > \epsilon) \rightarrow 0$ as $n \rightarrow \infty$ for all $\epsilon > 0$. It is also possible to have a random variable $Y$ in the limit instead of the constant $a$ but we will not consider any such examples.

To illustrate the law of large numbers, let us consider the example of rolling a die. In the left plot in Figure 4.1, consecutive averages in a total of 200 rolls are plotted. Note the typical wild fluctuations in the beginning, followed by a fairly rapid stabilization around the mean 3.5. After about 130 rolls, the average appears to have settled, but then there is a sudden decline after roll 160. By pure chance, between rolls 160 and 180, there were unusually large numbers of ones and twos which dragged the average down. In the right plot in Figure 4.1, the sequence is continued to 5000 rolls. The dip around 180 is visible, as are a few subsequent excursions above and below 3.5, but after that, all deviations are significantly smaller in size. If another sequence of 20 rolls similar to that between 160 and 180 would occur after 5000 rolls, it would have no visible impact.

A consequence of the law of large numbers is that we can now prove that the probability of an event is the limit of the relative frequencies. Recall how we mentioned this as a source of inspiration for the axioms of probability and how we have often argued that we can think of a probability as a relative frequency in a large number of trials.

[1]More precisely, for fixed $\epsilon > 0$ we can never be certain that $\bar{X}$ is in $\mu \pm \epsilon$ for any *fixed n*. However, it can be shown that the probability is one that $\bar{X}$ will eventually be in $\mu \pm \epsilon$ but this occurs at a *random N* so in practice we never know whether it has actually occurred. The exact formulation is $P(\bar{X} \rightarrow \mu) = 1$, a type of convergence called *convergence almost surely*, which is stronger than convergence in probability. The distinction between these two convergence concepts is crucial in a more advanced treatment of probability theory; we will consider convergence only in probability.

**Corollary 4.1.** *Consider an experiment where the event A occurs with probability p. Repeat the experiment independently, let $S_n$ be the number of times we get the event A in n trials, and let $f_n = S_n/n$, the relative frequency. Then*

$$f_n \xrightarrow{P} p \quad as \quad n \to \infty$$

*Proof.* Define the indicators

$$I_k = \begin{cases} 1 & \text{if we get } A \text{ in the } k\text{th trial} \\ 0 & \text{otherwise} \end{cases}$$

for $k = 1, 2, \ldots, n$. Then the $I_k$ are i.i.d. and we know from Section 2.5.1 that they have mean $\mu = p$. Since $f_n$ is the sample mean of the $I_k$, the law of large numbers gives $f_n \xrightarrow{P} p$ as $n \to \infty$. ■

The law of large numbers is also popularly known as the "law of averages" and is frequently misunderstood. Let us illustrate some of the common misperceptions by considering the experiment of flipping a fair coin.

The first mistake we examine is the confusion between *relative* and *absolute* frequencies. If we flip the coin $n$ times and let $S_n$ be the number of heads, this is the absolute frequency and $f_n = S_n/n$ is the relative frequency. The expected value of $S_n$ is $\frac{n}{2}$, and the expected value of $f_n$ is $\frac{1}{2}$. The law of large numbers states that $f_n$ gets close to $\frac{1}{2}$, *not* that $S_n$ gets close to $\frac{n}{2}$. In fact, the difference $|S_n - \frac{n}{2}|$ tends to increase with $n$, which is illustrated in Figure 4.2. In the next section, we will see that the rate of increase is on the order of $\sqrt{n}$.

The second mistake is the incorrect notion that extreme occurrences in one direction are compensated by extreme occurrences in the other. As an illustration, look at Figure 4.3, where the consecutive relative frequencies in a total of 200 coin flips are plotted. This is a sequence of coin flips that had an unusually large number of tails

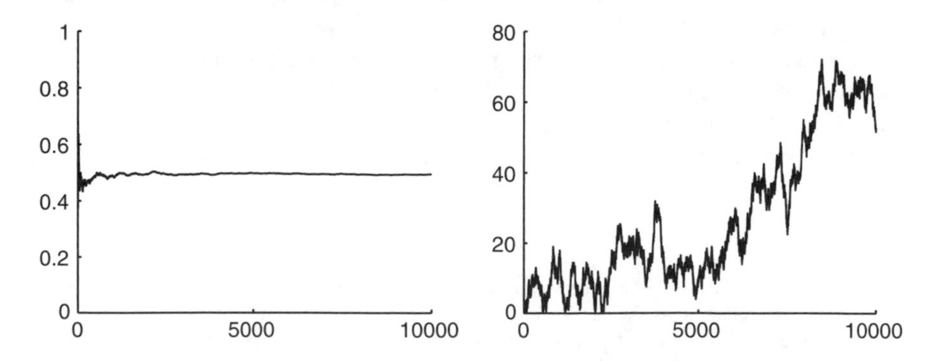


**FIGURE 4.2** Successive relative frequencies $f_n$ (left) and absolute differences $|S_n - \frac{n}{2}|$ (right) of repeated coin flips. Note the stabilization to the left and the increase to the right.

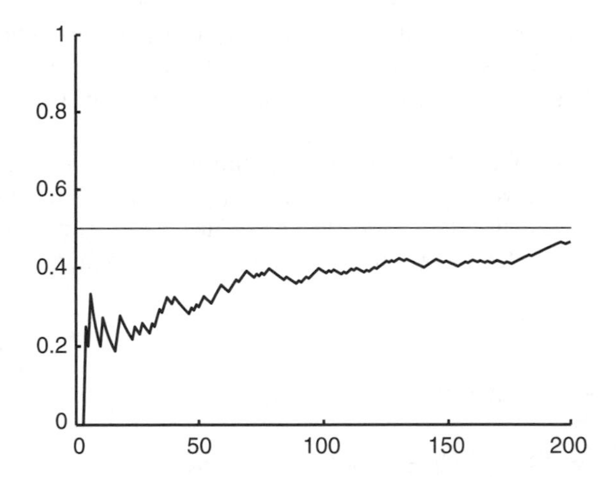


**FIGURE 4.3** Successive relative frequencies of repeated coin flips.

in the beginning. After 100 flips, there had only been 39 heads, quite far from the expected number of 50, and thus the relative frequency $f_{100}$ is only 0.39. The relative frequencies slowly but steadily work their way up toward 0.5, and after 200 flips we get quite close. Now, this must mean that the unusually low number of heads in the first 100 flips was compensated for by an unusually high number of heads in the last 100 flips, right?

Wrong! Actually, there were 54 heads in the last 100 flips, not far from the expected number of 50. If we had had exactly 50 heads in the last 100 flips, the relative frequency $f_{200}$ would have been $\frac{89}{200} = 0.445$, significantly higher than 0.39. Even if we got a little bit *less* than the expected number of heads, say, 45, the relative frequency would still go up, to 0.42. As we keep flipping, the unusual number of heads in the first 100 flips will have less and less impact on the relative frequency. Thus, there is no trend to compensate, only the persistence of completely normal behavior that in the long run erodes the effects of any abnormalities.[2]

Sometimes, we are interested in a function of $\bar{X}$ rather than $\bar{X}$ itself. The following result is useful.

**Corollary 4.2.** *Let g be a continuous function. Under the conditions of the law of large numbers*

$$g(\bar{X}) \xrightarrow{P} g(\mu) \quad as \quad n \to \infty$$

[2]My good friend, the eminent probabilist Jeff Steif, once pointed out to me over a roulette table that it is sometimes better to be completely ignorant than to have only a little bit of knowledge. After seven straight occurrences of red, the ignorant gambler does not blink (nor does Jeff), whereas he who has heard of, but not understood, the law of large numbers believes that more occurrences of black must follow.

It certainly feels reasonable that if $\bar{X}$ is close to $\mu$, then $g(\bar{X})$ by continuity must be close to $g(\mu)$. However, since "close" now refers to convergence in probability, it must be shown that $P(|g(\bar{X}) - g(\mu)| > \epsilon) \to 0$. We omit the technical proof.

***Example 4.1.*** In Example 3.13, we examined Buffon's needle tossing scheme that can be used as a primitive way to estimate $\pi$, based on the law of large numbers. We saw that the probability that the needle intersects a line is $2/\pi$, and if we let $f_n$ be the relative frequency of intersections in $n$ tosses, the law of large numbers tells us that

$$f_n \xrightarrow{P} \frac{2}{\pi} \quad \text{as} \quad n \to \infty$$

and by Corollary 4.2 with $g(x) = 2/x$ we get

$$\frac{2}{f_n} \xrightarrow{P} \pi \quad \text{as} \quad n \to \infty$$

which suggests the estimate $2/f_n$ of $\pi$. We will later investigate how good an estimate this is. □

We finish this section with an example of convergence in probability that does not involve the sample mean.

***Example 4.2.*** Let $X_1, X_2, \ldots$ be i.i.d. $\exp(\lambda)$, and as usual let $X_{(1)}$ denote the minimum of $X_1, \ldots, X_n$. Show that $X_{(1)} \xrightarrow{P} 0$ as $n \to \infty$.

By Example 3.50 we have $X_{(1)} \sim \exp(n\lambda)$ and hence for fixed $\epsilon > 0$

$$P(|X_{(1)} - 0| > \epsilon) = P(X_{(1)} > \epsilon) = e^{-n\lambda\epsilon} \to 0$$

as $n \to \infty$. □

## 4.3 THE CENTRAL LIMIT THEOREM

In the previous section, we saw that $\bar{X} \xrightarrow{P} \mu$ as $n \to \infty$ or, in other words, that $\bar{X} \approx \mu$ if $n$ is large. It would be good to also have an idea of how accurate this approximation is, that is, have an idea of the extent to which $\bar{X}$ tends to deviate from $\mu$. The next result gives the answer.

***Theorem 4.2 (The Central Limit Theorem).*** *Let $X_1, X_2, \ldots$ be i.i.d. random variables with mean $\mu$ and variance $\sigma^2 < \infty$ and let $S_n = \sum_{k=1}^{n} X_k$. Then, for each $x \in R$, we have*

$$P\left(\frac{S_n - n\mu}{\sigma\sqrt{n}} \leq x\right) \to \Phi(x)$$

*as $n \to \infty$, where $\Phi$ is the cdf of the standard normal distribution.*

Before the proof, let us point out that this is a truly remarkable result. It states that the cdfs of the random variables $(S_n - n\mu)/\sigma\sqrt{n}$ converge to the cdf of a standard normal distribution. In other words, for large $n$

$$P\left(\frac{S_n - n\mu}{\sigma\sqrt{n}} \leq x\right) \approx \Phi(x)$$

which we can also write as

$$\frac{S_n - n\mu}{\sigma\sqrt{n}} \stackrel{d}{\approx} N(0, 1)$$

where "$\stackrel{d}{\approx}$" is notation for approximate distribution. Since $S_n$ has mean $n\mu$ and variance $n\sigma^2$, we can also write this as

$$S_n \stackrel{d}{\approx} N(n\mu, n\sigma^2)$$

Thus, the central limit theorem states that the sum of i.i.d. random variables has an approximate normal distribution *regardless of the distribution of the* $X_k$! The $X_k$ do not even have to be continuous random variables; if we just add enough random variables, the sum becomes approximately normal anyway.

The central limit theorem gives theoretical justification for why the normal distribution tends to show up so often in practice. If a quantity is the result of many small independent contributions, it is likely to be approximately normal. For example, the weight of a bag of potato chips is the sum of the weights of many individual chips. Whatever the distribution of the weight of a single chip, the sum has an approximate normal distribution. For another example, consider the change in location of a dust particle in the air. This is due to the bombardment of a huge number of air molecules from different directions, and when added up, the changes in coordinates are likely to follow normal distributions.

*Proof.* We will only outline the proof and skip the details. The main idea is to work with moment generating functions instead of working directly with the distribution functions. Thus, we will show that the moment generating function of $(S_n - n\mu)/\sigma\sqrt{n}$ converges to the moment generating function of a standard normal distribution.

Let us first assume that $\mu = 0$ and $\sigma^2 = 1$. Let $Y_n = S_n/\sqrt{n}$, and let $M(t)$ be the mgf of the $X_k$. Combining Corollary 3.14 and Proposition 3.40, we see that $Y_n$ has mgf

$$M_{Y_n}(t) = \left(M\left(\frac{t}{\sqrt{n}}\right)\right)^n \tag{4.1}$$

Now do a Taylor expansion of $M(s)$ around $s = 0$ to obtain

$$M(s) = M(0) + sM'(0) + s^2\frac{M''(0)}{2}$$

where we have neglected the error term. By Corollary 3.15, we get

$$M(s) = 1 + \frac{s^2}{2}$$

(remember that $\mu = 0$ and $\sigma^2 = 1$). By inserting this into Equation (4.1), we obtain

$$M_{Y_n}(t) = \left(1 + \frac{t^2}{2n}\right)^n \rightarrow e^{t^2/2} \quad \text{as} \quad n \rightarrow \infty$$

which we recognize from Example 3.66 as the mgf of the standard normal distribution. The result for general $\mu$ and $\sigma^2$ follows by considering the random variables $X_k^* = (X_k - \mu)/\sigma$, which have mean 0 and variance 1, and we leave it to the reader to finish the proof.

The details that we have omitted include an explanation of why convergence of the mgfs is the same as convergence of the distribution functions (which is a fairly deep result and not exactly a detail) and why the error term in the Taylor expansion can be neglected as $n \rightarrow \infty$. ■

***Example 4.3.*** You play 1000 rounds of roulette, each time betting \$1. What is the probability that you end up with a gain if you **(a)** bet on odd and **(b)** play straight bets?

Denote your gain in round $k$ by $X_k$ so that your total gain in $n$ rounds is

$$S_n = \sum_{k=1}^{n} X_k$$

By the central limit theorem, $S_n$ has an approximate normal distribution with mean $n\mu$ and variance $n\sigma^2$. We computed the means and variances in Example 2.26, and for (a) we have

$$\mu = -\frac{1}{19} \quad \text{and} \quad \sigma^2 = \frac{360}{361}$$

and with $n = 1000$ we get

$$\begin{aligned} P(S_n > 0) &= 1 - P(S_n \leq 0) \\ &\approx 1 - \Phi\left(\frac{0 - 1000 \times (-1/19)}{\sqrt{1000 \times 360/361}}\right) \\ &= 1 - \Phi(1.67) \approx 0.05 \end{aligned}$$

For (b) we have

$$\mu = -\frac{1}{19} \quad \text{and} \quad \sigma^2 = \frac{11988}{361}$$

which gives

$$P(S_n > 0) \approx 1 - \Phi\left(\frac{0 - 1000 \times (-1/19)}{\sqrt{1000 \times 11988/361}}\right) \approx 0.39$$

so you are much more likely to be ahead with the straight bet strategy. Now, the expected gains are the same for the two strategies, so the strategy to bet on odd must have some other advantage instead. To illustrate this, let us compute the probability that you lose more than \$100. With the straight bet strategy this is

$$P(S_n \leq -100) \approx \Phi\left(\frac{-100 - 1000 \times (-1/19)}{\sqrt{1000 \times 11988/361}}\right) \approx 0.40$$

and with the strategy to bet on odd

$$P(S_n \leq -100) \approx \Phi\left(\frac{-100 - 1000 \times (-1/19)}{\sqrt{1000 \times 360/361}}\right) \approx 0.07$$

so you are more likely to lose more money with the straight bet strategy. The smaller variance of the strategy to bet on odd means that your loss is likely to be fairly close to the expected loss of $1000 \times \frac{1}{19} \approx 53$ dollars. With the straight bet strategy, there is more fluctuation and you take a greater risk for the chance to gain more. □

***Example 4.4.*** We have previously seen that the binomial distribution can be represented as a sum of indicators. Thus, if $X \sim \text{bin}(n, p)$, the central limit theorem states that

$$\frac{X - np}{\sqrt{np(1-p)}} \overset{d}{\approx} N(0, 1)$$

or equivalently

$$X \overset{d}{\approx} N(np, np(1-p))$$

Historically, this was among the first versions of the central limit theorem that was proved. It is often called the *de Moivre–Laplace theorem*. Figure 4.4 shows the pmfs of three binomial distributions with $p = 0.8$ and $n = 5, 10$, and 100, respectively. The pdfs of the corresponding approximating normal distributions are plotted for comparison. Note how the binomial pmf for $n = 5$ is quite asymmetric but how the pmfs become more symmetric as $n$ increases. This suggests that the accuracy of the normal approximation depends not only on $n$ but also on $p$; the more symmetric the binomial pmf is to start with, the better the approximation. This is indeed true; for fixed $n$, the approximation works best if $p = \frac{1}{2}$ and gets worse as $p$ approaches 0 or 1. A general rule of thumb for a decent approximation is that both $np$ and $n(1-p)$ should be at least 5. □

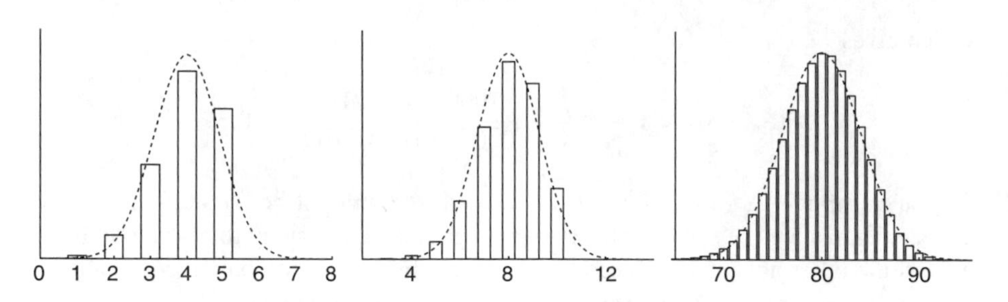


**FIGURE 4.4** The central limit theorem in action; pmfs for three binomial distributions with parameters $p = 0.8$ and $n = 5$, 10, and 100, respectively. The dashed curve shows the corresponding normal pdf.

If we divide $S_n$ by $n$, we get

$$\bar{X} \stackrel{d}{\approx} N\left(\mu, \frac{\sigma^2}{n}\right)$$

or in the case of relative frequencies,

$$f_n \stackrel{d}{\approx} N\left(p, \frac{p(1-p)}{n}\right) \tag{4.2}$$

Hence, we have two results about the sample mean $\bar{X}$: the law of large numbers, which states that

$$\bar{X} \approx \mu$$

and the central limit theorem, which states that

$$\bar{X} \stackrel{d}{\approx} N\left(\mu, \frac{\sigma^2}{n}\right)$$

Note how the first result talks about the *value* of $\bar{X}$ whereas the second talks about its *distribution*. The type of convergence in the central limit theorem is therefore called *convergence in distribution*, and we will return to this in Section 4.4.

We can now get an idea of how close $\bar{X}$ tends to be to $\mu$. The central limit theorem gives

$$P(|\bar{X} - \mu| > \epsilon) \approx 2\left(1 - \Phi\left(\frac{\epsilon\sqrt{n}}{\sigma}\right)\right)$$

regardless of the distribution of the $X_k$.

***Example 4.5.*** Consider again Buffon's needle problem. Recall that the probability that the randomly tossed needle intersects a line is $2/\pi$ and how we argued in the previous section that $2/f_n \stackrel{P}{\to} \pi$ as $n \to \infty$. Let $\widehat{\pi}$ denote our estimate of $\pi$ after $n$ tosses:

$$\widehat{\pi} = \frac{2}{f_n}$$

As mentioned previously, Buffon himself actually intended that this experiment be used to estimate $\pi$. Let us say that on some occasion he tossed the needle 1000 times. What is the probability that he got the estimate correct to two decimals?

We wish to find

$$P(|\widehat{\pi} - 3.14| \leq 0.005) = P\left(\frac{2}{3.145} \leq f_n \leq \frac{2}{3.135}\right)$$

where, by Equation (4.2) with $p = 2/\pi$

$$f_n \overset{d}{\approx} N\left(\frac{2}{\pi}, \frac{2(\pi - 2)}{n\pi^2}\right)$$

With $n = 1000$ we now get

$$\begin{aligned} P(|\widehat{\pi} - 3.14| \leq 0.005) &= P\left(\frac{2}{3.145} \leq f_{1000} \leq \frac{2}{3.135}\right) \\ &= \Phi\left(\frac{2/3.135 - 2/\pi}{\sqrt{2(\pi - 2)/1000\pi^2}}\right) - \Phi\left(\frac{2/3.145 - 2/\pi}{\sqrt{2(\pi - 2)/1000\pi^2}}\right) \\ &= \Phi(0.09) - \Phi(-0.05) \approx 0.06 \end{aligned}$$

which is not a very good reward for all that needle tossing. □

### 4.3.1 The Delta Method

We have learned that the sample mean $\bar{X}$ has an approximate normal distribution. It is often the case that we are interested in some function of the sample mean, $g(\bar{X})$, and we already know from Corollary 4.2 that $g(\bar{X})$ converges in probability to $g(\mu)$. It would be useful to also supplement this with a result about the approximate distribution of $g(\bar{X})$. The following proposition, usually referred to as the *delta method*, gives the answer. Although it should formally be stated as a limit result, we state it as an approximation, which is how it is useful to us.

**Proposition 4.1 (The Delta Method).** Let $X_1, X_2, \ldots$ be i.i.d. with mean $\mu$ and variance $\sigma^2$, and let $\bar{X}$ be the sample mean. Let further $g$ be a function such that $g'(\mu) \neq 0$. Then

$$g(\bar{X}) \overset{d}{\approx} N\left(g(\mu), \frac{\sigma^2(g'(\mu))^2}{n}\right)$$

for large $n$.

*Proof.* A first-order Taylor expansion of $g$ around $\mu$ gives

$$g(\bar{X}) \approx g(\mu) + (\bar{X} - \mu)g'(\mu)$$

where we know that

$$\bar{X} \stackrel{d}{\approx} N\left(\mu, \frac{\sigma^2}{n}\right)$$

and since $g(\bar{X})$ is (approximately) a linear function of $\bar{X}$, we know that it has (approximately) a normal distribution. The mean is

$$E[g(\bar{X})] \approx g(\mu) + g'(\mu)E[\bar{X} - \mu] = g(\mu)$$

since $E[\bar{X} - \mu] = E[\bar{X}] - \mu = 0$. The variance is

$$\text{Var}[g(\bar{X})] \approx (g'(\mu))^2 \text{Var}[\bar{X}] = \frac{\sigma^2 (g'(\mu))^2}{n}$$

as was to be shown. ■

Note how there are two factors in the variance of $g(\bar{X})$: $\sigma^2/n$, which measures how much $\bar{X}$ deviates from $\mu$, and $(g'(\mu))^2$, which measures how sensitive the function $g$ is to deviations from $\mu$. The smaller these quantities are, the better the approximation.

***Example 4.6.*** Let $X_1, X_2, \ldots, X_n$ be i.i.d. random variables. The *geometric mean* $G_n$ is defined as

$$G_n = (X_1 X_2 \cdots X_n)^{1/n}$$

Suppose that the $X_k$ are unif[0, 1]. Find the limit of $G_n$ and its approximate distribution for large $n$.

To be able to apply the law of large numbers, we need to transform the product into a sum. Thus, let $Y_k = \log X_k$ to obtain

$$\log G_n = \frac{1}{n} \sum_{k=1}^{n} Y_k$$

that is

$$G_n = e^{\bar{Y}}$$

By the law of large numbers, $\bar{Y} \stackrel{P}{\rightarrow} \mu$, where

$$\mu = E[Y_k] = E[\log X_k] = \int_0^1 \log x \, dx = -1$$

Hence, $\bar{Y} \stackrel{P}{\to} -1$, and since the function $g(x) = e^x$ is continuous, Proposition 4.1 gives

$$G_n \stackrel{P}{\to} e^{-1}$$

as $n \to \infty$. To get the approximate distribution, we apply the delta method to $\bar{Y}$ and the function $g(x) = e^x$. We have $g'(x) = e^x$ that gives $(g'(\mu))^2 = e^{-2}$, and since $\sigma^2 = \text{Var}[Y_k] = 1$, we get the approximate distribution

$$G_n \stackrel{d}{\approx} N\left(e^{-1}, \frac{e^{-2}}{n}\right)$$

□

## 4.4 CONVERGENCE IN DISTRIBUTION

In the previous sections, we looked at two limit results: the law of large numbers and the central limit theorem. As we pointed out, these are different in the sense that the first deals with convergence of $\bar{X}$ itself and the second, with its distribution. In this section, we take a closer look at this second type of convergence.

### 4.4.1 Discrete Limits

We first consider the case of discrete random variables and state the following definition.

***Definition 4.1.*** Let $X_1, X_2, \ldots$ be a sequence of discrete random variables such that $X_n$ has pmf $p_{X_n}$. If $X$ is a discrete random variable with pmf $p_X$ and

$$p_{X_n}(x) \to p_X(x) \quad \text{as } n \to \infty \text{ for all } x$$

then we say that $X_n$ *converges in distribution* to $X$, written $X_n \stackrel{d}{\to} X$.

This limit result is used much in the same way as we used our previous limit results, as the approximation $P(X_n = x) \approx P(X = x)$. The largest and most useful class of limit theorems for discrete random variables is when the limiting random variable has a Poisson distribution. Let us look at the simplest example of this: the Poisson approximation of the binomial distribution. In Section 2.5.4, we learned that if $X$ is $\text{bin}(n, p)$, where $n$ is large and $p$ small, then $X$ is approximately $\text{Poi}(np)$. We will now state a limit result that motivates the approximation.

**Proposition 4.2.** Let $X_1, X_2, \ldots$ be a sequence of random variables such that $X_n \sim \text{bin}(n, p_n)$, where $np_n \to \lambda > 0$ as $n \to \infty$, and let $X \sim \text{Poi}(\lambda)$. Then $X_n \stackrel{d}{\to} X$.

*Proof.* We sketch a proof based on probability generating functions. Let $G_n(s)$ be the pgf of $X_n$ so that

$$G_n(s) = (1 - p_n + p_n s)^n$$

of which we take the logarithm to obtain

$$\log G_n(s) = n \log(1 + p_n(s - 1))$$

Now, by a Taylor expansion around 0, we obtain

$$\log(1 + x) \approx x \tag{4.3}$$

for small $x$. Since $np_n \to \lambda > 0$, we must have $p_n \to 0$, and the approximation in Equation (4.3) can be used to get

$$\begin{aligned} \log G_n(s) &= n \log(1 + p_n(s - 1)) \\ &\approx np_n(s - 1) \to \lambda(s - 1) \quad \text{as} \quad n \to \infty \end{aligned}$$

Hence

$$G_n(s) \to e^{\lambda(s-1)} \quad \text{as } n \to \infty$$

where we recognize the limit as the pgf of a Poisson distribution with parameter $\lambda$. It is true but not trivial that convergence of the pmfs is equivalent to convergence of the pgfs. We will not prove this. ■

Hence, the informal statement "large $n$ and small $p$" can be formalized as $np_n \to \lambda$, and the limit result legitimizes the use of the approximation.

Another approximation was suggested in Section 2.5.5, where the hypergeometric distribution was considered. We argued that under some circumstances, sampling with or without replacement ought to give approximately the same result. Thus, if $X \sim$ hypergeom$(N, r, n)$, then

$$P(X = k) = \frac{\binom{r}{k}\binom{N-r}{n-k}}{\binom{N}{n}} \approx \binom{n}{k}\left(\frac{r}{N}\right)^k\left(1 - \frac{r}{N}\right)^{n-k}$$

The limit result that justifies the approximation is stated in the following proposition. It is proved by a direct computation according to Definition 4.1 and we leave the proof as an exercise.

**Proposition 4.3.** Let $X_N \sim$ hypergeom$(N, r, n)$, where $N \to \infty$ and $r/N \to p > 0$, and let $X \sim$ bin$(n, p)$. Then

$$X_N \xrightarrow{d} X \quad \text{as } N \to \infty$$

In words, the approximation works well if $n$ is small and $r$ is moderate relative to $N$. We leave it to the reader to contemplate what may go wrong if $n$ is too large or if $r$ is either too large or too small.

### 4.4.2 Continuous Limits

Let us next consider the case when the limiting random variable is continuous. As we already know from the de Moivre–Laplace theorem, the limit can be continuous even if the random variables themselves are not.

***Definition 4.2.*** Let $X_1, X_2, \ldots$ be a sequence of random variables such that $X_n$ has cdf $F_n$. If $X$ is a continuous random variable with cdf $F$ and

$$F_n(x) \to F(x) \quad \text{as } n \to \infty \text{ for all } x \in R$$

we say that $X_n$ *converges in distribution* to $X$, written $X_n \stackrel{d}{\to} X$.

The most important result of this type is the central limit theorem. Another class of important results regarding convergence in distribution deals with the so-called *extreme values*, for example, the minimum or maximum in a sequence of random variables.

***Example 4.7.*** Let $X_1, X_2, \ldots$ be i.i.d. random variables that are unif[0, 1], and let $X_{(1)} = \min(X_1, \ldots, X_n)$, the minimum of the $n$ first $X_k$. As $n$ increases, $X_{(1)}$ can only get smaller so as $n \to \infty$, we expect $X_{(1)}$ to go to 0. However, if we adjust for this by multiplying $X_{(1)}$ by a suitable factor, we can get something interesting in the limit. Thus, let $Y_n = nX_{(1)}$ and let $Y \sim \exp(1)$. Then $Y_n \stackrel{d}{\to} Y$ as $n \to \infty$.

Note that the range of $Y_n$ is $[0, n]$ and that of $Y$ is $[0, \infty)$. We need to find the cdf of $Y_n$ and get

$$F_{Y_n}(x) = P(nX_{(1)} \le x) = P\left(X_{(1)} \le \frac{x}{n}\right)$$

In Section 3.10.1, we studied order statistics, and by Proposition 3.31 we have

$$P(X_{(1)} \le t) = 1 - (1-t)^n, \quad 0 \le t \le 1$$

which gives

$$F_{Y_{(n)}}(x) = 1 - \left(1 - \frac{x}{n}\right)^n \to 1 - e^{-x} \quad \text{as} \quad n \to \infty$$

which is the cdf of $Y \sim \exp(1)$. Note that this holds for any $x$ since eventually $n$ will be large enough so that $x$ is in the range of $Y_n$. Thus, we have that $nX_{(1)} \stackrel{d}{\approx} \exp(1)$, which can be used to approximate probabilities for $X_{(1)}$, for example, as

$$P(X_{(1)} \le x) = P(nX_{(1)} \le nx) \approx 1 - e^{-nx}, \quad 0 \le x \le 1$$

□

## PROBLEMS

### Section 4.2. The Law of Large Numbers

1. Let $X_1, X_2, \ldots$ be a sequence of random variables with the same mean $\mu$ and variance $\sigma^2$, which are such that $\text{Cov}[X_j, X_k] < 0$ for all $j \neq k$. Show that $\bar{X} \xrightarrow{P} \mu$ as $n \to \infty$.
2. Let $X_1, X_2, \ldots$ and $Y_1, Y_2, \ldots$ be two sequences of random variables and $a$ and $b$ two constants such that $X_n \xrightarrow{P} a$ and $Y_n \xrightarrow{P} b$. Show that $X_n + Y_n \xrightarrow{P} a + b$. *Hint:* Problem 10 in Chapter 2.
3. Let $X_1, X_2, \ldots$ be i.i.d. unif[0, 1] and let $g : [0, 1] \to R$ be a function. What is the limit of $\sum_{k=1}^{n} g(X_k)/n$ as $n \to \infty$? How can this result be used?
4. Let $X_1, X_2, \ldots, X_n$ be i.i.d. random variables. The *harmonic mean* is defined as

$$H_n = \left( \frac{1}{n} \sum_{k=1}^{n} \frac{1}{X_k} \right)^{-1}$$

   Suppose that the pdf of the $X_k$ is $f(x) = 3x^2, \quad 0 \leq x \leq 1$, and find the limit of $H_n$ as $n \to \infty$.
5. Let $X_1, X_2, \ldots$ be i.i.d. continuous with a pdf that is strictly positive in some interval $[0, a]$. Show that $X_{(1)} \xrightarrow{P} 0$ as $n \to \infty$.
6. Let $X_1, X_2, \ldots$ be i.i.d. exp(1). Show that $X_{(n)}/\log n \xrightarrow{P} 1$ as $n \to \infty$.

### Section 4.3. The Central Limit Theorem

7. In Problem 53(a) and (b), Chapter 3, what happens as $n \to \infty$? Explain this in the light of the central limit theorem.
8. Use the central limit theorem to argue that the following random variables are approximately normal; also give the parameters: **(a)** $X \sim \Gamma(n, \lambda)$ for large $n$ and **(b)** $X \sim \text{Poi}(\lambda)$ for large $\lambda$.
9. Radioactive decay of an element occurs according to a Poisson process with rate 10,000 per second. What is the approximate probability that the millionth decay occurs within 100.2 s?
10. In any given day, a certain email account gets a number of spam emails that has a Poisson distribution with mean 200. What is the approximate probability that it receives less than 190 spam emails in a day?
11. How many times do you need to roll a die to be at least $\approx 99\%$ certain that the sample mean is between 3 and 4?
12. Let $X \sim \text{bin}(5, 0.8)$. Compute $P(X \leq k)$ for $k = 4, 5$, and 6, both exactly and with the approximate normal distribution. Compare and comment.

13. A multiple-choice test has 100 questions, each with four alternatives. At least 80 correct answers are required for a passing grade. On each question, you know the correct answer with probability $\frac{3}{4}$, otherwise you guess at random. What is the (approximate) probability that you pass?
14. In Buffon's needle problem, what is the probability that the value of $\pi$ is correct to one decimal?
15. A parking lot is planned for a new apartment complex with 200 apartments. For each apartment it is assumed that the number of cars is 0, 1, or 2, with probabilities 0.1, 0.6, and 0.3, respectively. In order to be approximately 95% certain that there is room for all cars, how many spaces must the parking lot have?
16. Let $X$ be a random variable with mean $\mu$ and variance $\sigma^2$, and let $g$ be a differentiable function. Use the idea in the proof of the delta method to deduce the two approximations

$$E[g(X)] \approx g(\mu) \quad \text{and} \quad E[g(X)] \approx g(\mu) + \frac{\sigma^2 g''(\mu)}{2}$$

Compare these if $X \sim \text{unif}[0, 1]$ for **(a)** $g(x) = x$, **(b)** $g(x) = x^2$, **(c)** $g(x) = x^4$, **(d)** $g(x) = e^x$, **(e)** $g(x) = e^{-x}$, **(f)** $g(x) = \sin(\pi x)$, and **(g)** $g(x) = \sin(2\pi x)$.

17. Consider Buffon's needle problem. Use the delta method to compute how many times we must toss the needle in order to be at least 95% certain to be within $\pm 0.01$ of $\pi$?
18. Consider Problem 4. Find the approximate distribution of $H_n$ for large $n$.

## Section 4.4. Convergence in Distribution

19. Let $X \sim \text{hypergeom}(N, r, n)$. Argue that $X$ has an approximate Poisson distribution and what is required of the parameters for the approximation to be justified. Try to both state a formal limit result and give an intuitive argument.
20. Let $X_1, X_2, \ldots$ be a sequence of random variables such that $X_n$ is uniform on the set $\{0, 1, \ldots, n\}$, and let $X \sim \text{unif}[0, 1]$. Show that $X_n/n \stackrel{d}{\to} X$ as $n \to \infty$.
21. Let $X_1, X_2, \ldots$ be a sequence of random variables such that $X_n \sim \text{geom}(1/n)$, and let $X \sim \exp(1)$. Show that $X_n/n \stackrel{d}{\to} X$ as $n \to \infty$.
22. Let $X_1, X_2, \ldots$ be i.i.d. unif[0, 1], let $Y_n = n(1 - X_{(n)})$, where $X_{(n)}$ is the maximum, and let $Y \sim \exp(1)$. Show that $Y_n \stackrel{d}{\to} Y$ as $n \to \infty$.
23. Let $X_1, X_2, \ldots$ be i.i.d. $\exp(\lambda)$, and let $Y \sim \exp(1)$. Show that $X_1/\bar{X} \stackrel{d}{\to} Y$ as $n \to \infty$. *Hint:* Let the $X_k$ be interarrival times in a Poisson process, and use the order statistic property from Proposition 3.43.

24. Let $X_1, X_2, \ldots$ be i.i.d. random variables with pdf $f(x) = 2x, 0 \leq x \leq 1$. Show that the sequence of random variables $Y_n = \sqrt{n}X_{(1)}$ converges in distribution to a random variable $Y$ that has a Weibull distribution. What are the parameters?
25. Let $X_1, X_2, \ldots$ be i.i.d. random variables with pdf $f(x) = 3x^2$, $0 \leq x \leq 1$. Find a sequence $a_n$ such that the random variables $Y_n = a_n X_{(1)}$ converges in distribution to a random variable $Y$, and identify the distribution of $Y$.

# 5

# SIMULATION

## 5.1 INTRODUCTION

*Simulation* is one of the most commonly used techniques to gain information about complicated systems, but the term simulation is used to convey many different meanings. According to the *Merriam-Webster Online Dictionary*, simulation is "the imitative representation of the functioning of one system or process by means of the functioning of another." We probably think of simulation as something involving computers, but it does not have to be so. For example, airplanes flying in specific parabolic patterns are used in astronaut training to simulate weightless conditions in space. Even when we restrict our attention to computer simulation, there are many different meanings. For example, an airline pilot in training sits in a flight simulator, and a mathematician may simulate a numerical solution to a differential equation by plugging in different starting values in an algorithm. For us, however, simulation will always mean "imitating randomness," and for this reason the term *stochastic simulation* is often used. The term *Monte Carlo simulation* is also common.

The main use of stochastic simulation is to approximate quantities that are difficult to obtain analytically. To take a frivolous example from the world of gambling, in roulette it is easy to specify a probability model and compute probabilities of winning, expected gains, and so on. But what about the game of blackjack? This is a card game played against a dealer where you are dealt cards one by one and after each card decide whether to stop or take another card. The dealer has one card face up that you can see. If you go above 21, you lose. If you stop below 21, the dealer draws cards

*Probability, Statistics, and Stochastic Processes*, Second Edition. Peter Olofsson and Mikael Andersson.

and must stop at or above 17. If the dealer goes over 21, you win, otherwise whoever has the higher total wins. It is difficult to find exact answers to questions such as "If I adopt a particular strategy, what is the probability that I win?" but it is easy to run a computer simulation of the game.

When we use simulations, we rely on the law of large numbers. Suppose, for instance, that we wish to find an approximate value of $p$, the probability of winning in blackjack using some particular strategy. We run a number $n$ of simulated rounds of the game and observe wins in $X$ of them. The relative frequency $X/n$ is our approximation of $p$, and we know by the law of large numbers that this gets close to $p$ for large $n$ (and "large $n$" is something we can often guarantee in simulation studies). If we are interested in the expected gain, $\mu$, save the gain in each round to obtain a sequence $Y_1, Y_2, \ldots, Y_n$ of gains and use the sample mean $\bar{Y}$ to approximate $\mu$.

We have already seen how simulated data can be used to illustrate concepts such as pmf, pdf, and expected value. Such data can be generated for a wide variety of distributions by using ready-made routines in any of the major mathematical or statistical software packages (Matlab has been used in our examples). In this chapter, we take a closer look at how such simulated data are generated.

## 5.2 RANDOM NUMBER GENERATION

The most fundamental object in simulation is the standard uniform distribution. Even simple calculators can often generate what are usually called *random numbers*, which are precisely simulated observations from the standard uniform distribution. So how do we simulate the standard uniform distribution?

There are several issues, and we will not address them all. An immediate problem that comes to mind is that $[0, 1]$ is an entire interval, but a computer has only finite precision. Hence, we must be satisfied with values on some lattice $\{0, \frac{1}{m}, \ldots, \frac{m-1}{m}, 1\}$, which is not too serious a restriction if $m$ is large. One way to do this is to generate random integers between 0 and $m$ and divide them by $m$. In other words, if $Y$ has the discrete uniform distribution on $\{0, 1, \ldots, m\}$, then $U = Y/m$ has the discrete uniform distribution on $\{0, \frac{1}{m}, \ldots, \frac{m-1}{m}, 1\}$, which for large $m$ is an acceptable approximation. But this only shifted the problem to how to generate random integers. How do we do that?

Here is where we must admit that random number generators seldom produce "truly" random numbers. Instead, deterministic algorithms are used that produce sequences that "look random." For this reason, the term *pseudorandom* numbers is often used.[1] We want to achieve two main goals: that the numbers seem to have a uniform distribution and that they seem independent. One common way to generate random integers is by *congruential* random number generators (or *power residue* generators). These start with a value $Y_0$, the *seed*, and generate a sequence of integers by computing the next from the previous according to the formula

$$Y_{n+1} = aY_n + b \pmod{(m+1)}$$

[1] Incidentally, *Merriam-Webster* also gives the alternative meaning of simulation as "a sham object."

where $a$, $b$, and $m$ are fixed integers. Note that the sequence is periodic since once a value is repeated, the entire sequence will be repeated. Let us now take $m = 19$, $a = b = 1$, and $Y_0 = 0$. We get

$$\begin{aligned} Y_1 &= 1 \times Y_0 + 1 \text{ (mod 20)} = 1 \\ Y_2 &= 1 \times Y_1 + 1 \text{ (mod 20)} = 2 \\ &\vdots \\ Y_{19} &= 1 \times Y_{18} + 1 \text{ (mod 20)} = 19 \\ Y_{20} &= 1 \times Y_{19} + 1 \text{ (mod 20)} = 20 \text{ (mod 20)} = 0 \end{aligned}$$

which gives the sequence $0, 1, 2, \ldots, 19, 0, 1, 2, \ldots, 19, 0, 1, 2, \ldots,$ where each number in the long run shows up with relative frequency $\frac{1}{20}$, so the distribution would look uniform. However, the observations do not look independent. Indeed, if we are presented this sequence, we would quickly figure out the algorithm that produced it. Not good. Let us instead try $m = 19$, $a = 5$, $b = 3$, and $Y_0 = 0$. We now get

$$\begin{aligned} Y_1 &= 5 \times 0 + 3 \text{ (mod 20)} = 3 \\ Y_2 &= 5 \times 3 + 3 \text{ (mod 20)} = 18 \\ Y_3 &= 5 \times 18 + 3 \text{ (mod 20)} = 93 \text{ (mod 20)} = 13 \\ Y_4 &= 5 \times 13 + 3 \text{ (mod 20)} = 8 \\ Y_5 &= 5 \times 8 + 3 \text{ (mod 20)} = 3 \end{aligned}$$

which gives the sequence $0, 3, 18, 13, 8, 3, 18, 13, 8, 3, \ldots,$ where the pattern $3, 18, 13, 8$ is repeated indefinitely. This is an improvement since there is no immediate clear structure for short pieces of the sequence. On the other hand, when a period is completed, most numbers between 0 and 19 have not shown up at all, so we do not get the uniform distribution. A problem here is that the period is too short.

These calculations illustrate some potential problems with random number generation. We will not address this further but mention that nice results can be obtained from number theory that give criteria for how to choose $m$, $a$, and $b$ to avoid these problems and get a good random-looking sequence. For practical purposes, $m$ must, of course, be much larger than in the two examples above.

Assuming thus that we can generate observations from the standard uniform distribution, how can we transform these to observations from other distributions? In the following sections, we will investigate this.

## 5.3 SIMULATION OF DISCRETE DISTRIBUTIONS

Let us start with the simplest of discrete random variables, indicators. Recall that an indicator function $I_A$ assumes the value 1 if the event $A$ occurs and 0 otherwise. Suppose that the probability of $A$ is $p$ so that $I_A$ assumes the values 0 or 1 with probabilities $1 - p$ or $p$, respectively. How can we generate an observation on $I_A$ if we are given $U \sim \text{unif}[0, 1]$? The idea is simple. Let

$$I_A = \begin{cases} 1 & \text{if} \quad U \leq p \\ 0 & \text{if} \quad U > p \end{cases}$$

which gives the correct distribution because

$$P(I_A = 1) = P(U \leq p) = p$$

which also gives $P(I_A = 0) = 1 - p$. Note that it does not matter what the event $A$ is since we are not simulating the underlying experiment directly. We are interested only in getting the right distribution for $I_A$.

Any discrete distribution can be simulated by a similar idea. If $X$ assumes the values $x_1, x_2, \ldots$ with probabilities $p_1, p_2, \ldots,$ we divide the interval $[0, 1]$ into subintervals where the $k$th subinterval has length $p_k$, and if $U$ falls there, we set $X = x_k$. It does not matter whether the range of $X$ is finite or countably infinite. To express $X$ explicitly as a function of $U$, we state the following proposition.

**Proposition 5.1.** Consider the pmf $p$ on the range $\{x_1, x_2, \ldots\}$ and let

$$F_0 = 0, \quad F_k = \sum_{j=1}^{k} p(x_j), \quad k = 1, 2, \ldots$$

Let $U \sim \text{unif}[0, 1]$ and let $X = x_k$ if $F_{k-1} < U \leq F_k$. Then $X$ has pmf $p$.

*Proof.* Note that $X = x_k$ if and only if $U \in (F_{k-1}, F_k]$, which has probability

$$P(X = x_k) = P(F_{k-1} < U \leq F_k) = F_k - F_{k-1} = p(x_k), \quad k = 1, 2, \ldots$$

as desired. If the range is finite, $\{x_1, \ldots, x_n\}$, we get $F_n = 1$. ■

The number $F_k$ is in fact $F_X(x_k)$, the distribution function of $X$ in the point $x_k$ (and we can let $x_0 = -\infty$). In the next section, we will see how this is a special case of a more general method.

For certain special discrete distributions, there are alternative and more attractive ways to simulate. We will look at one example and leave others for the Problems section.

***Example 5.1.*** Let $X \sim \text{bin}(n, p)$. We can use Proposition 5.1 to generate an observation on $X$, but we can also use the fact that the binomial distribution can be represented as a sum of indicators

$$X = \sum_{k=1}^{n} I_k$$

where the $I_k$ are i.i.d. with $P(I_k = 1) = p$. Thus, we simulate standard uniforms $U_1, \ldots, U_n$, and for each of these let $I_k = 1$ if $U_k \le p$ and add $I_k$. One advantage of this method compared to using Proposition 5.1 is that if we need to change $n$ from 100 to 200, for example, we simply simulate 100 more indicators, which is easier than having to start over and recalculate all the $F_k$. □

## 5.4 SIMULATION OF CONTINUOUS DISTRIBUTIONS

From previous results we know that if $U \sim \text{unif}[0, 1]$ and we let $X = (b - a)U + a$, then $X \sim \text{unif}[a, b]$. Hence, it is clear how to generate observations from any uniform distribution, starting from a standard uniform distribution: simply multiply each observation by $b - a$ and add $a$. As it turns out, this is a special case of a more general result that we state next.

**Proposition 5.2 (The Inverse Transformation Method).** Let $F$ be a distribution function that is continuous and strictly increasing. Further, let $U \sim \text{unif}[0, 1]$ and define the random variable $Y = F^{-1}(U)$. Then $Y$ has distribution function $F$.

*Proof.* Start with $F_Y$, the distribution function of $Y$. Take $x$ in the range of $Y$ to obtain

$$\begin{aligned} F_Y(x) &= P(F^{-1}(U) \le x) \\ &= P(U \le F(x)) = F_U(F(x)) = F(x) \end{aligned}$$

where the last equality follows since $F_U(u) = u$ if $0 \le u \le 1$. The argument here is $u = F(x)$, which is between 0 and 1 since $F$ is a cdf. ■

To generate observations from a distribution with cdf $F$, we thus find $F^{-1}$, generate i.i.d. uniform [0,1] variables $U_1, \ldots, U_n$, and let $X_k = F^{-1}(U_k)$ for $k = 1, \ldots, n$. Note that the assumptions on $F$, continuous and strictly increasing, together guarantee that $F^{-1}$ exists as a function on the entire interval [0, 1].

***Example 5.2.*** Generate an observation from an exponential distribution with parameter $\lambda$.

Here,

$$F(x) = 1 - e^{-\lambda x}, \quad x \ge 0$$

To find the inverse, as usual solve $F(x) = u$, to obtain

$$x = F^{-1}(u) = -\frac{1}{\lambda}\log(1 - u), \quad 0 \le u < 1$$

Hence, if $U \sim \text{unif}[0, 1]$, the random variable

$$X = -\frac{1}{\lambda}\log(1 - U)$$

is $\exp(\lambda)$. We can note here that since $1 - U$ is also uniform on $[0, 1]$ (see Problem 19 in Chapter 2), we might as well take $X = -\log U/\lambda$. □

***Example 5.3.*** Generate an observation $(X, Y)$ from a uniform distribution on the triangle with corners in $(0, 0)$, $(0, 1)$, and $(1, 0)$ (see Problem 92 in Chapter 3).

Our approach will be to first generate $X$, and then, given the value $X = x$, we will generate $Y$ from the conditional distribution. Since the triangle has area $\frac{1}{2}$, the joint pdf is

$$f(x, y) = 2, \quad 0 \le x \le 1, \quad 0 \le y \le 1 - x$$

which gives the marginal pdf

$$f_X(x) = 2\int_0^{1-x} dy = 2(1 - x), \quad 0 \le x \le 1$$

which in turn gives the cdf

$$F_X(x) = \int_0^x 2(1 - t)dt = 2x - x^2, \quad 0 \le x \le 1$$

Next we find the inverse, $F^{-1}$. Solve

$$u = 2x - x^2$$

to obtain

$$x = F^{-1}(u) = 1 - \sqrt{1 - u}, \quad 0 \le u \le 1$$

Note that solving the quadratic gives two solutions but only the one with "−" is correct (why?). We also need the conditional distribution for $Y$ given $X = x$, which is

$$f_Y(y|x) = \frac{f(x, y)}{f_X(x)} = \frac{1}{1 - x}, \quad 0 \le y \le 1 - x$$

that is, $Y|X = x \sim \text{unif}[0, 1 - x]$. Now let $U$ and $V$ be independent and uniform on $[0, 1]$, and let

$$X = 1 - \sqrt{1 - U}$$
$$Y = V(1 - X)$$

to get a pair $(X, Y)$ with the desired distribution. Figure 5.1 shows the outcome of 100 generated such pairs $(X, Y)$. Note that we cannot use the same random number $U$ for both $X$ and $Y$. This would give the correct *marginal* distributions of $X$ and $Y$

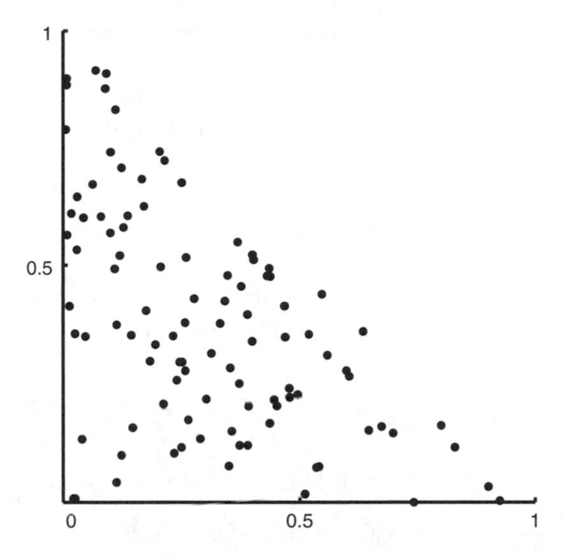


**FIGURE 5.1** One hundred simulated observations, uniform on a triangle.

but not the correct *joint* distribution since $Y$ would become a deterministic function of $X$. Also note that since $X$ and $Y$ are dependent, we need to pair each $X$ value with its corresponding $Y$ value. □

The assumptions on $F$ are that it is continuous and strictly increasing. If there are parts where $F$ is constant, these parts correspond to values that $X$ cannot assume, so this is not a problem. If there are points where $F$ jumps, these are points that $X$ can assume with strictly positive probability. In particular, if $X$ is a discrete random variable, $F$ consists entirely of jumps and constant parts, but also remember that there are random variables that are mixtures of discrete and continuous parts. For a general cdf $F$, we can introduce the generalized inverse, or *quantile function*, $F^{\leftarrow}$, defined by

$$F^{\leftarrow}(u) = \inf\{x : F(x) \geq u\}$$

This has the property that

$$F^{\leftarrow}(u) \leq x \quad \Leftrightarrow \quad u \leq F(x)$$

and in the case of strictly increasing $F$, we have $F^{-1} = F^{\leftarrow}$. It can be shown that if $U \sim \text{unif}[0, 1]$ and we let $X = F^{\leftarrow}(U)$, then $X$ has cdf $F$. In principle, this solves the problem of simulating any distribution, discrete or continuous. For example, the random variable $X$ in Proposition 5.1 is merely $F^{\leftarrow}(U)$ in disguise.

If we do not know the functional form of the cdf $F$, we cannot use the inverse transformation method. This is, for example, the case for the normal distribution and since we have pointed out the importance of this distribution, we would certainly like to have a way for generating observations from it. We next describe a clever method to generate observations from a continuous distribution that requires that we know the pdf but not the cdf.

We want to generate observations on a continuous random variable $X$ that has pdf $f$. Suppose that there is another random variable $Y$ with pdf $g$, from which we know how to generate observations and that there is a constant $c > 1$ such that

$$f(x) \leq cg(x) \quad \text{for all } x$$

We can then use consecutive generated observations on $Y$ to create observations on $X$ according to a certain algorithm, which we state as a proposition.

**Proposition 5.3 (The Rejection Method).**

1. Generate $Y$ and $U \sim \text{unif}[0, 1]$ independent of each other.
2. If $U \leq \dfrac{f(Y)}{cg(Y)}$, set $X = Y$. Otherwise return to step 1.

The random variable $X$ generated by this algorithm has pdf $f$.

When the criterion is satisfied in step 2 and we set $X = Y$, we say that we *accept* $Y$ and otherwise *reject* it. Hence the name of the method.

*Proof.* Let us first make sure that the algorithm terminates. The probability in any given step 2 to accept $Y$ is, by Corollary 3.2,

$$\begin{aligned} P\left(U \leq \frac{f(Y)}{cg(Y)}\right) &= \int_R P\left(U \leq \frac{f(y)}{cg(y)}\right) g(y)dy \\ &= \int_R \frac{f(y)}{cg(y)} g(y)dy = \frac{1}{c}\int_R f(y)dy = \frac{1}{c} \end{aligned}$$

where we used the independence of $U$ and $Y$ and the fact that $U \sim \text{unif}[0, 1]$. Hence, the number of iterations until we accept a value has a geometric distribution with success probability $1/c$. The algorithm therefore always terminates in a number of steps with mean $c$ from which it also follows that we should choose $c$ as small as possible.

Next we turn to the question of why this gives the correct distribution. To show this, we will show that the conditional distribution of $Y$, given acceptance, is the same as the distribution of $X$. Recalling the definition of conditional probability and the fact that the probability of acceptance is $1/c$, we get

$$P\left(Y \leq x \,\middle|\, U \leq \frac{f(Y)}{cg(Y)}\right) = c\,P\left(Y \leq x, U \leq \frac{f(Y)}{cg(Y)}\right)$$

By independence, the joint pdf of $(U, Y)$ is $f(u, y) = g(y)$, and the above expression becomes

$$c\,P\left(Y \le x, U \le \frac{f(Y)}{cg(Y)}\right) = c\int_{-\infty}^{x}\int_{0}^{f(y)/cg(y)} g(y)du\,dy$$

$$= c\int_{-\infty}^{x} \frac{f(y)}{cg(y)} g(y)dy \;=\; P(X \le x)$$

which is what we wanted to prove. ■

Let us now see how this applies to the normal distribution.

***Example 5.4.*** Let $X \sim N(0, 1)$ so that $X$ has pdf

$$\varphi(x) = \frac{1}{\sqrt{2\pi}} e^{-x^2/2}, \quad x \in R$$

We will use the pdf of an exponential distribution as the function $g$. However, the normal distribution can take on values on the entire real line and the exponential, only positive values, so we have to make some adjustments. Since the standard normal distribution is symmetric around 0, we can get the right distribution of $X$ by first generating a value of $|X|$ and then choosing "+" or "−" with equal probabilities. According to Problem 85 in Chapter 2, the pdf of $|X|$ is

$$f(x) = 2\varphi(x) = \frac{2}{\sqrt{2\pi}} e^{-x^2/2}, \quad x \ge 0$$

Now let $Y \sim \exp(1)$. By Example 5.2, we know how to generate observations on $Y$ that has pdf

$$g(x) = e^{-x}, \quad x \ge 0$$

and we will proceed to find the constant $c$. Note that

$$\frac{f(x)}{g(x)} = \sqrt{\frac{2}{\pi}} \exp\left(\frac{-(x^2 - 2x)}{2}\right) = \sqrt{\frac{2e}{\pi}} \exp\left(\frac{-(x-1)^2}{2}\right)$$

after completing the square (note that we use "exp" in two different meanings: the exponential *distribution* and the exponential *function*; do not confuse these). Since the second factor is at most 1, we can take $c = \sqrt{2e/\pi}$ to obtain $f(x) \le cg(x)$. To apply the method, we follow the two steps:

1. Generate $U \sim \text{unif}[0, 1]$ and $Y \sim \exp(1)$ independent of each other.
2. If $U \le \exp(-(Y-1)^2/2)$, set $|X| = Y$. Otherwise repeat step 1.

Finally, to choose the sign of $X$, generate $V \sim \text{unif}[0, 1]$. If $V \le \frac{1}{2}$, set $X = |X|$, otherwise set $X = -|X|$.

The mean number of steps until acceptance is $c \approx 1.32$, so the algorithm terminates quickly. There are ways to improve the efficiency of the method, but we will not address them here. □

We conclude this section with another way to simulate the standard normal distribution.

***Example 5.5 (Box–Muller Method).*** Recall from Example 3.26 that if $X$ and $Y$ are independent and have standard normal distributions, then their polar coordinates $R$ and $\Theta$ are independent with $R^2 \sim \exp(\frac{1}{2})$ and $\Theta \sim \text{unif}[0, 2\pi]$. Since we know how to generate the uniform and exponential distributions, we should be able to use this to generate observations on $X$ and $Y$.

Suppose that we have $U$ and $V$ independent uniform on $[0, 1]$. The inverse transformation method gives that $-2 \log U \sim \exp(\frac{1}{2})$, and hence observations on $R$ and $\Theta$ are generated by

$$R = \sqrt{-2 \log U}$$
$$\Theta = 2\pi V$$

and since $X = R \cos \Theta$ and $Y = R \sin \Theta$, we get

$$X = \sqrt{-2 \log U} \cos(2\pi V)$$
$$Y = \sqrt{-2 \log U} \sin(2\pi V)$$

which are independent, and each has the standard normal distribution. Hence, with this method we get two observations at a time, so $(X, Y)$ is bivariate normal with correlation $\rho = 0$. In Problem 17, we investigate how to generate observations on the bivariate normal distribution in general.

Note the somewhat surprising fact that $X$ and $Y$ are independent even though they are both functions of the same random variables $U$ and $V$ (but recall Problem 95(c) in Chapter 3). □

## 5.5 MISCELLANEOUS

A nice use of simulation is that it can help us suggest what an analytical formula should look like, which we can then set out to formally prove. We look at an example of this.

***Example 5.6.*** Recall the gambler's ruin problem, where two players, Ann and Bob, take turns flipping a fair coin and the winner in each round gets a dollar from the other player. Suppose that the players start with $a$ and $b$ dollars, respectively. How long can we expect the game to last?

**TABLE 5.1 Average Simulated Duration Times $\bar{T}$ Until Ruin and Their Standard Deviations $s_{\bar{T}}$ for Different Values of $a$ and $b$**

| $a$ | $b$ | $\bar{T}$ | $s_{\bar{T}}$ | $ab$ |
|---|---|---|---|---|
| 2 | 3 | 6.5 | 0.7 | 6 |
| 3 | 4 | 11.0 | 1.1 | 12 |
| 3 | 5 | 15.4 | 1.5 | 15 |
| 4 | 6 | 22.2 | 2.0 | 24 |

Let $T$ be the duration of the game. If we denote Ann's fortune after the $n$th round by $S_n$, we thus have

$$T = \min\{n : S_n = -a \text{ or } S_n = b\}$$

and we want to find $E[T]$. To use the definition, we need the distribution of $T$, and this is quite tricky to find. (Just try!) However, we can investigate the problem with a computer simulation for different values of $a$ and $b$ and see if we can guess a formula for $E[T]$. Table 5.1 shows average values of $T$ obtained from simulations for different values of $a$ and $b$. It seems that $E[T] = ab$ is a good guess. Of course, this is not a proof, but now that we have an expression, we can try to prove it.

Let us follow the idea from Example 1.58 and condition on the first flip or in the random-walk interpretation, the first step $S_1$. Let $\mu_{a,b}$ be the expected number of rounds until the game is over when the initial fortunes are $a$ and $b$. After the first round, we are at either $-1$ or 1 with equal probabilities, so the law of total expectation gives

$$\begin{aligned}\mu &= E[T|S_1 = -1]P(S_1 = -1) + E[T|S_1 = 1]P(S_1 = 1)\\ &= (1 + \mu_{a-1,b+1})\frac{1}{2} + (1 + \mu_{a+1,b-1})\frac{1}{2}\\ &= 1 + \frac{1}{2}(\mu_{a-1,b+1} + \mu_{a+1,b-1})\end{aligned}$$

and since

$$1 + \frac{1}{2}\Big((a-1)(b+1) + (a+1)(b-1)\Big) = ab$$

our suggested expression satisfies the equation. Thus, the expected duration of the game is the same in the case $a = 1, b = 100$ as in the case $a = 10, b = 10$. Since the game can be over already after one round in the first case (and will be so half of the time) but there must be at least 10 rounds (and most likely many more) in the second, we realize that although the means are the same, the distributions in the two cases must be different. □

Finally, an example that shows how simulation can help us guess better than random.

***Example 5.7.*** In an urn there are two slips of paper, each with a real number written on it. You do not know how the numbers have been chosen, only that they are different. You pick a slip at random, look at the number, and are asked to guess whether it is the larger or smaller of the two. Can you do this and be correct with a probability strictly greater than $\frac{1}{2}$?

Surprisingly, you can! Call the two numbers $a$ and $b$ and suppose that $a < b$. Call the number you picked $X$ so that $X$ equals $a$ or $b$ with probability $\frac{1}{2}$ each. Now generate an observation $Y$ from a continuous distribution with range $R$, for example, a normal distribution, independent of $X$. Your rule to decide is to pretend that the number on the other slip in the urn is $Y$. Hence, if $X < Y$, you guess that you have the smaller number; if $X > Y$, you guess that you have the larger. It is easy to realize that you will guess correctly if either $X = a$ and $Y > a$, or if $X = b$ and $Y < b$. (Since $Y$ is continuous, we rule out the possibility that $Y$ is *exactly* equal to $a$ or $b$. In the actual simulation, we are limited by finite precision, but with sufficiently many decimals in our observation on $Y$, it is very unlikely that $Y$ equals $a$ or $b$.) Let $F$ be the cdf of $Y$. Since $X$ and $Y$ are independent, the probability to guess correctly is

$$
\begin{aligned}
p &= P(X = a, Y > a) + P(X = b, Y < b) \\
&= P(X = a)P(Y > a) + P(X = b)P(Y < b) \\
&= \frac{1}{2}(1 - F(a) + F(b)) > \frac{1}{2}
\end{aligned}
$$

since $F(b) > F(a)$. The value of $p$ of course depends on $a$, $b$, and $F$, but in any case, the probability of guessing correctly is always strictly greater than $\frac{1}{2}$. Since the simulation of $Y$ has nothing to do with the problem of choosing and guessing, this is really mysterious, isn't it? □

## PROBLEMS

### Section 5.3. Simulation of Discrete Distributions

1. A family with three children is chosen at random and the number $X$ of daughters is counted. Show how to simulate an observation on $X$ based on $U \sim \text{unif}[0, 1]$.
2. A girl who has two siblings is chosen at random and the number $X$ of her sisters is counted. Show how to simulate an observation on $X$ based on $U \sim \text{unif}[0, 1]$.
3. Let $Y \sim \text{Poi}(\lambda)$ and suppose that you can simulate $X_1, X_2, \ldots$ that are independent exp(1). Suggest how the $X_k$ can be used to simulate observations on $Y$.
4. Suppose that you know how to generate the uniform and the Poisson distribution. Describe how you can use this to simulate a Poisson process on a given interval $[0, t]$. *Hint:* The order statistic property.

5. Let $X$ have a negative binomial distribution with parameters $r$ and $p$. Describe how to simulate observations on $X$. *Hint:* Problem 52, Chapter 3.

**Section 5.4. Simulation of Continuous Distributions**

6. Let $U \sim \text{unif}[0, 1]$ and let $X = (b - a)U + a$. From previous discussions, we know that $X \sim \text{unif}[a, b]$, which can be used to simulate values of $X$. Show that this is a special case of the inverse transformation method.
7. The random variable $X$ has pdf $f(x) = 3x^2, \quad 0 \leq x \leq 1$. Describe how to generate an observation on $X$ based on $U \sim \text{unif}[0, 1]$.
8. Let $X$ have a Cauchy distribution (see Section 2.8.3). **(a)** Describe how to generate an observation on $X$ based on $U \sim \text{unif}[0, 1]$. **(b)** Use the result from (a) to create repeated samples of some large size, for example, $n = 100,000$, and each time compute the sample mean. What do you observe and why?
9. Let $X$ have a Weibull distribution (see Section 2.10). Describe how to generate an observation on $X$ based on $U \sim \text{unif}[0, 1]$.
10. Let $X$ have a geometric distribution with success probability $p$. Demonstrate how you can simulate an observation on $X$ based on $U \sim \text{unif}[0, 1]$ using Problem 74 in Chapter 2.
11. The mixed random variable $X$ in Example 2.44 has cdf $F(x) = 1 - 0.8e^{-x}, x \geq 0$. Describe how to generate an observation on $X$ based on $U \sim \text{unif}[0, 1]$.
12. Suppose that you have a random number generator, a fair coin, and Table A.1. Describe how you can use these to generate observations on the standard normal distribution.
13. Let $X \sim \exp(1)$. Use simulation to determine an approximate value of $E[\log X]$.
14. Let $X \sim N(0, 1)$. Use simulation to determine an approximate value of $E[\sin X]$.
15. Let $X$, $Y$, and $Z$ be independent and uniform $[0, 1]$. Use simulation to find an approximate value of $P(X + Y + Z \leq 2.5)$.
16. Let $g: [0, 1] \rightarrow R$ be a function whose integral $I = \int_0^1 g(x)dx$ is impossible to compute explicitly. How can you approximate $I$ by simulation of standard uniforms $U_1, U_2, \ldots$?
17. The Box–Muller method gives you a pair $(X, Y)$ of independent standard normals. Describe how you can use these to generate observations on a bivariate normal distribution with parameters $\mu_1, \mu_2, \sigma_1^2, \sigma_2^2$, and $\rho$.
18. Describe how to generate observations on a pair $(X, Y)$ that is uniform on the unit disk $\{(x, y) : x^2 + y^2 \leq 1\}$. *Hint:* Use polar coordinates.
19. Let $X$ have pdf $f(x) = 4x^2e^{-x^2}/\sqrt{\pi}, \quad x \geq 0$. Apply the rejection method to generate observations on $X$.

# 6

# STATISTICAL INFERENCE

## 6.1 INTRODUCTION

In the previous chapters, we developed a theory of probability that allows us to model and analyze random phenomena in terms of random variables and their distributions. While developing this theory, we often referred to real-world observations and data sets, for example, in the assumption that the tropical cyclones in Example 2.36 follow a Poisson distribution with mean 15. Although we might be able to argue that the distribution should be Poisson from purely physical and meteorological principles point of view, where did the number 15 come from? It is simply the average number of cyclones per year that has been observed during the years 1988–2003, so we used this measured value as our parameter. This is a typical situation in any application of probability theory. We formulate a model by making assumptions about distributions and their parameters, but in order to be able to draw any useful conclusion, we need data. In this chapter, we outline the field of *statistical inference*, or *statistics* for short, which ties together probability models and data collection.

## 6.2 POINT ESTIMATORS

Suppose that we manufacture lightbulbs and want to state the average lifetime on the box. Let us say that we have the following five observed lifetimes (in hours):

$$983, \; 1063, \; 1241, \; 1040, \; 1103$$

*Probability, Statistics, and Stochastic Processes*, Second Edition. Peter Olofsson and Mikael Andersson.

which give us the average 1086. If this is all the information we have, it seems reasonable to state 1086 as the average lifetime (although "at least 1000 h" might sound better from a commercial point of view). Let us now describe this in terms of random variables.

Let the random variable $X$ be the lifetime of a lightbulb and let $\mu = E[X]$. Here $\mu$ is an *unknown parameter*. We decide to repeat the experiment to measure a lifetime five times and will then get an outcome on the five random variables $X_1, \ldots, X_5$ that are i.i.d. We now *estimate* $\mu$ by

$$\bar{X} = \frac{1}{5}\sum_{k=1}^{5} X_k$$

which we recall from Section 4.1 as the sample mean. We can think of this as the way we would describe it before actually performing the experiment. The $X_k$ are random and so is $\bar{X}$. After the experiment, we get observed values, and with the values given above we get the outcome $\bar{X} = 1086$. We now generalize the idea of this example, for which we need the following definition.

***Definition 6.1.*** If the random variables $X_1, \ldots, X_n$ are i.i.d., we refer to them collectively as a (*random*) *sample*.

Note that it is the entire collection $X_1, \ldots, X_n$ that is called "a sample," which may be a bit different from the way the word "sample" is used in everyday language. In practice, it is not necessarily the case that observations are independent or have the same distribution, and we will later use the term "sample" in such more general cases as well. For now, however, we will stick to the i.i.d. observations.

Suppose now that we want to use a random sample to gain information about an unknown parameter $\theta$. The following definition is central.

***Definition 6.2.*** A random variable $\widehat{\theta}$, which is a function of a random sample and is used to estimate an unknown parameter $\theta$, is called an *estimator* of $\theta$. The observed value of $\widehat{\theta}$ is called an *estimate* of $\theta$.

Thus, if we have a sample $X_1, \ldots, X_n$, an estimator is a random variable of the type $g(X_1, \ldots, X_n)$ for some function $g : R^n \to R$. It is important to understand that a parameter $\theta$ is a fixed but unknown number and an estimator $\widehat{\theta}$ is a random variable to be computed from a sample. We will stick to the distinction between estimator and estimate, although these terms are sometimes used interchangeably in the statistics literature.

In the lightbulb example, we have the parameter $\mu$ and the function $g(X_1, \ldots, X_n) = (X_1 + \cdots + X_n)/n$, which gives the estimator $\widehat{\mu} = \bar{X}$ and the estimate 1086. If we repeat the experiment with five new lightbulbs, the estimator is the same but the estimate will change.

It is natural to estimate the mean $\mu$ by the sample mean $\bar{X}$, but how do we know that there are no other estimators that are better? What does "better estimator" mean, anyway? The intuitive general criterion for a good estimator $\widehat{\theta}$ is simply that it is "as close to $\theta$ as possible." How can we formalize this intuition?

Since an estimator $\widehat{\theta}$ is a random variable, it has an expected value $E[\widehat{\theta}]$. Since this is a number and we use the estimator to estimate the unknown number $\theta$, it seems reasonable to require that the two are equal.

***Definition 6.3.*** The estimator $\widehat{\theta}$ is said to be *unbiased* if

$$E[\widehat{\theta}] = \theta$$

If $E[\widehat{\theta}] \neq \theta$, $\widehat{\theta}$ is said to be *biased*. Thus, an unbiased estimator "aims at the true value of $\theta$" or is "correct on average." For an illustration of the idea, consider Figure 6.1, where histograms are given for unbiased and biased estimators, respectively. Each histogram is computed from 1000 values of $\widehat{\theta}$, and each such value is computed from a new sample of a fixed size $n$.

From Section 4.1, we know that $E[\bar{X}] = \mu$, and $\bar{X}$ is thus an unbiased estimator of $\mu$. It is not the only unbiased estimator of the mean $\mu$, though. We could also use $X_1$, the first observation, and disregard all the others. Since $E[X_1] = \mu$, this is also an unbiased estimator, and in the lightbulb example we get the estimate $X_1 = 983$. Intuitively $\bar{X}$ ought to be a better estimator than $X_1$ since it uses all the information in the sample. To extend this argument, if we would increase the number of

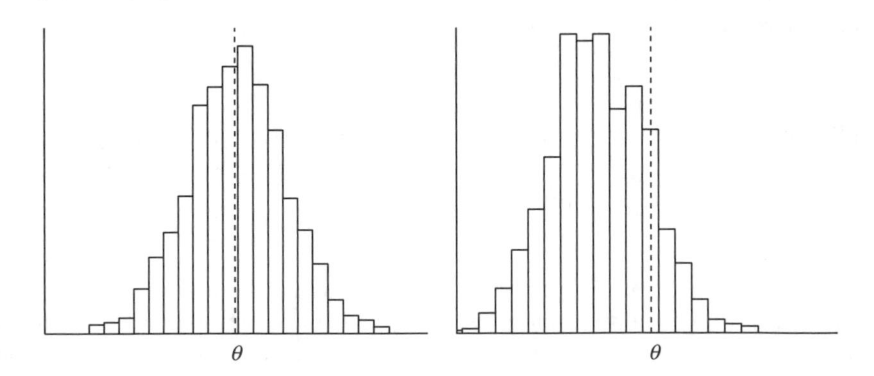


**FIGURE 6.1** A parameter $\theta$ and histograms of repeated observations on its estimator $\widehat{\theta}$. On the left, the estimator is unbiased and on the right, it is biased with $E[\widehat{\theta}] < \theta$.

observations in the sample we would increase the amount of available information and, consequently, should obtain an even better estimator. This brings us to another important property of estimators.

***Definition 6.4.*** The estimator $\widehat{\theta}_n$ based on the sample $X_1, \ldots, X_n$ is said to be *consistent* if

$$\widehat{\theta}_n \xrightarrow{P} \theta$$

as $n \to \infty$.

This property can often be difficult to verify for a given estimator, so the following result is very useful.

**Proposition 6.1.** Let $\widehat{\theta}_n$ be an unbiased estimator based on the sample $X_1, \ldots, X_n$. If

$$\text{Var}[\,\widehat{\theta}_n\,] \to 0$$

as $n \to \infty$, then $\widehat{\theta}_n$ is consistent.

*Proof.* Apply Chebyshev's inequality to $\widehat{\theta}_n$ and let $c = \epsilon / \sqrt{\text{Var}[\,\widehat{\theta}_n\,]}$. This yields

$$P(|\widehat{\theta}_n - \theta| > \epsilon) \leq \frac{\text{Var}[\,\widehat{\theta}_n\,]}{\epsilon^2}$$

Now, if $\text{Var}[\,\widehat{\theta}_n\,] \to 0$ as $n \to \infty$, we see that

$$P(|\widehat{\theta}_n - \theta| > \epsilon) \to 0$$

as $n \to \infty$ for all $\epsilon > 0$, which completes the proof. ∎

Hence, unbiasedness and consistency are two desirable properties that we should look for when considering estimators. This invites the following question: If we have two unbiased and consistent estimators $\widehat{\theta}$ and $\widetilde{\theta}$, which should we choose? Intuitively, we should choose the one that tends to be closer to $\theta$, and since $E[\,\widehat{\theta}\,] = E[\,\widetilde{\theta}\,] = \theta$, it makes sense to choose the estimator with the smaller variance.

***Definition 6.5.*** Suppose that $\widehat{\theta}$ and $\widetilde{\theta}$ are two unbiased estimators of $\theta$. If

$$\mathrm{Var}[\widehat{\theta}\,] < \mathrm{Var}[\widetilde{\theta}\,]$$

then $\widehat{\theta}$ is said to be *more efficient* than $\widetilde{\theta}$.

Let us again return to the lightbulb example and the two estimators $\bar{X}$ and $X_1$. If $\sigma^2$ is the variance of an individual lifetime, we thus have

$$\mathrm{Var}[\bar{X}] = \frac{\sigma^2}{5} \quad \text{and} \quad \mathrm{Var}[X_1] = \sigma^2$$

so $\bar{X}$ is more efficient than $X_1$. Since the sample mean is a natural estimator of the mean, regardless of distribution, let us restate its properties from Section 4.1.

**Proposition 6.2.** Let $X_1, \ldots, X_n$ be a sample with mean $\mu$ and variance $\sigma^2$. Then the sample mean $\bar{X}$ has mean and variance

$$E[\bar{X}] = \mu \quad \text{and} \quad \mathrm{Var}[\bar{X}] = \frac{\sigma^2}{n}$$

There are numerous other ways to construct unbiased estimators, and it can be shown that the sample mean is the most efficient among a large class of unbiased estimators (see Problem 1).

When two estimators, $\widehat{\theta}$ and $\widetilde{\theta}$, are given, one is not necessarily more efficient than the other. The reason for this is that the variances of the estimators typically depend on unknown parameters (most notably $\theta$ itself), and it may well happen that $\mathrm{Var}[\widehat{\theta}\,] < \mathrm{Var}[\widetilde{\theta}\,]$ for some parameter values and $\mathrm{Var}[\widehat{\theta}\,] > \mathrm{Var}[\widetilde{\theta}\,]$ for others.

***Example 6.1.*** Let $X_1, \ldots, X_n$ be a random sample from a uniform distribution on $[0, \theta]$ where $\theta$ is unknown. Since the mean is $\theta/2$, one reasonable estimator is $\widehat{\theta} = 2\bar{X}$. Also, since the maximum value $X_{(n)}$ ought to be close to $\theta$ but is always smaller, another reasonable estimator is $\widetilde{\theta} = c_n X_{(n)}$, where $c_n > 1$. Determine $c_n$ so that $\widetilde{\theta}$ becomes unbiased and compare the two estimators.

We need to find $c_n$ such that

$$E[\widetilde{\theta}\,] = c_n E[X_{(n)}] = \theta$$

To compute $E[X_{(n)}]$, we recall Corollary 3.11, which states that the pdf of $X_{(n)}$ is

$$f_{(n)}(x) = nf(x)F(x)^{n-1}$$

where in our case $f(x) = 1/\theta$ and $F(x) = x/\theta$, and we get

$$f_{(n)}(x) = \frac{nx^{n-1}}{\theta^n}, \quad 0 \le x \le \theta$$

This now gives

$$E[X_{(n)}] = \int_0^\theta x f_{(n)}(x)dx = \frac{n}{\theta^n}\int_0^\theta x^n dx = \frac{n}{n+1}\theta$$

which gives $c_n = (n+1)/n$ and the estimator

$$\widetilde{\theta} = \frac{n+1}{n} X_{(n)}$$

Let us compare the two estimators. We already know that $\widetilde{\theta}$ is unbiased and since

$$E[\widehat{\theta}\,] = 2E[\bar{X}] = 2\frac{\theta}{2} = \theta$$

$\widehat{\theta}$ is also unbiased. Let us next compare the variances of the two estimators. For $\widehat{\theta}$ we have

$$\text{Var}[\widehat{\theta}\,] = 4\text{Var}[\bar{X}] = 4\frac{\theta^2}{12n} = \frac{\theta^2}{3n}$$

where we got the variance for the uniform distribution from Proposition 2.13. For $\widetilde{\theta}$ we have

$$\text{Var}[\widetilde{\theta}\,] = E[\widetilde{\theta}^2\,] - (E[\widetilde{\theta}\,])^2 = \left(\frac{n+1}{n}\right)^2 E[X_{(n)}^2] - \theta^2$$

so we need to find $E[X_{(n)}^2]$. The pdf of $X_{(n)}$ is given above, and we get

$$E[X_{(n)}^2] = \frac{n}{\theta^n}\int_0^\theta x^{n+1}dx = \frac{n}{n+2}\theta^2$$

which finally gives

$$\text{Var}[\widetilde{\theta}\,] = \left(\frac{n+1}{n}\right)^2 \frac{n}{n+2}\theta^2 - \theta^2 = \frac{\theta^2}{n(n+2)}$$

Since

$$\text{Var}[\widetilde{\theta}\,] < \text{Var}[\widehat{\theta}\,]$$

for $n \ge 2$, $\widetilde{\theta}$ is the more efficient estimator. Note that the difference in variance is quite substantial since $\widetilde{\theta}$ has $n^2$ in the denominator whereas $\widehat{\theta}$ has only $n$. □

When given two (or more) alternative estimators, Definition 6.5 thus gives a criterion to use for selecting the most efficient one. However, it does not say anything about optimality, that is, is it good enough or should we keep looking for something even better? To answer that question, we need the following result.

**Proposition 6.3.** **(Cramér–Rao Lower Bound[1])** Let $\widehat{\theta}$ be an unbiased estimator of the parameter $\theta$ based on the sample $X_1, \ldots, X_n$. Then

$$\mathrm{Var}[\widehat{\theta}] \geq \frac{1}{nI(\theta)}$$

where

$$I(\theta) = -E\left[\frac{\partial^2}{\partial\theta^2} \log f_\theta(X)\right]$$

is the *Fisher information.*[2]

The proof of Proposition 6.3 is rather involved and falls beyond the scope of this book.

This result gives us the smallest possible variance of an unbiased estimator for a given sample distribution, so if we manage to attain this, or come sufficiently close, we can be confident that we have a good estimator. We can also use it to obtain an absolute measure of efficiency.

***Definition 6.6.*** The *efficiency* of an unbiased estimator $\widehat{\theta}$ is

$$e(\widehat{\theta}) = \frac{1}{nI(\theta)\mathrm{Var}[\widehat{\theta}]}$$

We can interpret the efficiency as the ratio of the variance of the best possible unbiased estimator (if it exists) and the variance of the given $\widehat{\theta}$, and from Proposition 6.3 we get that $e(\widehat{\theta}) \leq 1$.

***Example 6.2.*** Let us assume that we have a sample $X_1, \ldots, X_n$, which is normally distributed with mean $\mu$ and variance $\sigma^2$. Proposition 6.2 says that the point estimator $\widehat{\mu} = \bar{X}$ is an unbiased estimator of $\mu$. Let us now see how efficient this estimator is.

[1]Named after the Swedish mathematician Harald Cramér (1893–1985) and the Indian statistician C. R. Rao (1920-).

[2]Named after the English statistician Sir Ronald A. Fisher (1890–1962) who, among other things, introduced the method of maximum likelihood (see Section 6.4.2).

The Fisher information of the normal distribution with respect to $\mu$ is

$$I(\mu) = -E\left[\frac{\partial^2}{\partial\mu^2}\log\left(\frac{1}{\sigma\sqrt{2\pi}}e^{-(X-\mu)^2/2\sigma^2}\right)\right]$$

$$= -E\left[\frac{\partial^2}{\partial\mu^2}\left(-\log(\sigma\sqrt{2\pi}) - \frac{(X-\mu)^2}{2\sigma^2}\right)\right] = -E\left[-\frac{1}{\sigma^2}\right] = \frac{1}{\sigma^2}$$

The Cramér–Rao Lower Bound hence becomes $1/nI(\mu) = \sigma^2/n$, which, by Proposition 6.2, is equal to the variance of $\bar{X}$. This shows that there does not exist any other unbiased estimator of $\mu$ with a higher efficiency than the sample mean. □

Efficiency is thus a criterion that we can use to choose from among estimators. To get a better idea of what the actual numbers mean, rather than using the variance, we use the standard deviation that has the correct unit of measure. For an estimator, there is a special piece of terminology.

***Definition 6.7.*** The standard deviation of an estimator, $\sigma_{\widehat{\theta}} = \sqrt{\mathrm{Var}[\widehat{\theta}]}$, is called the *standard error*.

In practice, the standard error typically depends on unknown parameters, and by estimating these, we get the *estimated standard error*, denoted by $s_{\widehat{\theta}}$. This is also sometimes referred to as the standard error if there is no risk of confusion. Let us revisit our examples.

***Example 6.3.*** In Example 6.1, suppose that we have the following 10 observations, ordered by size:

0.94, 1.56, 2.52, 3.54, 3.91, 4.16, 4.49, 6.50, 7.42, 8.69

which gives the estimates

$$\widehat{\theta} = 2\bar{X} = 8.8$$

and

$$\widetilde{\theta} = \frac{n+1}{n}X_{(n)} = \frac{11}{10} \times 8.69 = 9.6$$

The standard errors are

$$\sigma_{\widehat{\theta}} = \frac{\theta}{\sqrt{3n}} \quad \text{and} \quad \sigma_{\widetilde{\theta}} = \frac{\theta}{\sqrt{n(n+2)}}$$

which, as we can see, depend on both $\theta$ and $n$. For the estimated standard errors, we insert $n = 10$ and the values of $\widehat{\theta}$ and $\widetilde{\theta}$ to obtain

$$s_{\widehat{\theta}} = \frac{\widehat{\theta}}{\sqrt{3n}} = \frac{8.8}{\sqrt{30}} = 2.6 \quad \text{and} \quad s_{\widetilde{\theta}} = \frac{\widetilde{\theta}}{\sqrt{n(n+2)}} = \frac{9.6}{\sqrt{120}} = 0.77$$

□

***Example 6.4.*** In the lightbulb example, the estimators are $\bar{X}$ and $X_1$ and the standard errors are

$$\sigma_{\bar{X}} = \frac{\sigma}{\sqrt{5}} \quad \text{and} \quad \sigma_{X_1} = \sigma$$

which depend on the unknown parameter $\sigma$. Thus, in order to estimate the standard errors, we need to know how to estimate $\sigma$ from the sample. We address this in the next section. □

### 6.2.1 Estimating the Variance

We have learned that the sample mean is a good estimator of the mean. Now suppose that we also wish to estimate the variance $\sigma^2$. Recall the definition

$$\sigma^2 = E\left[(X - \mu)^2\right]$$

where $\mu$ is the mean. Thus, the variance is the mean of the random variable $(X - \mu)^2$ and a good estimator is the sample mean of the random variables

$$(X_1 - \mu)^2, \ldots, (X_n - \mu)^2$$

that is, we could use

$$\widehat{\sigma^2} = \frac{1}{n}\sum_{k=1}^{n}(X_k - \mu)^2$$

as our estimator of $\sigma^2$. One obvious problem is that this requires that we know the mean $\mu$, and this is rarely the situation. However, we can replace $\mu$ by its estimator $\bar{X}$ and hopefully still get a good estimator of $\sigma^2$. It turns out that this results is an estimator that is slightly biased, and instead the following is most often used.

***Definition 6.8.*** Let $X_1, \ldots, X_n$ be a random sample. The *sample variance* is defined as

$$s^2 = \frac{1}{n-1}\sum_{k=1}^{n}(X_k - \bar{X})^2$$

For computations, it is easier to use the following formula, easily proved by expanding the square and doing some simple algebra.

**Corollary 6.1.**

$$s^2 = \frac{1}{n-1}\left(\sum_{k=1}^{n} X_k^2 - n\bar{X}^2\right)$$

It may seem strange to divide by $n - 1$ when there are $n$ terms in the sum. The reason for this is that it gives an unbiased estimator.

**Proposition 6.4.** The sample variance $s^2$ is an unbiased and, if $E[X_i^4]$ is finite, consistent estimator of $\sigma^2$.

*Proof.* By Corollary 6.1, we obtain

$$E[s^2] = \frac{1}{n-1}\left(\sum_{k=1}^{n} E[X_k^2] - nE[\bar{X}^2]\right)$$

$$= \frac{1}{n-1}\left(n(\sigma^2+\mu^2) - n\left(\frac{\sigma^2}{n}+\mu^2\right)\right) = \sigma^2$$

where we used the variance formula from Corollary 2.2. The proof of consistency is left as an exercise. ∎

Dividing by $n$ would thus give an estimator that is on average too small. The intuitive reason for this is that the $X_k$ tend to be closer to $\bar{X}$ than they are to $\mu$ itself. The square root of $s^2$ is denoted by $s$ and called the *sample standard deviation*. While $s^2$ is an unbiased estimator of $\sigma^2$, $s$ is not an unbiased estimator of $\sigma$ but is nevertheless commonly used (see Problem 11).

***Example 6.5.*** Find the sample standard deviation $s$ in the lightbulb example and use it to estimate the standard errors of the estimators $X_1$ and $\bar{X}$.

Our observed sample is 983, 1040, 1063, 1103, 1241, which has an observed sample mean $\bar{X} = 1086$ and

$$\sum_{k=1}^{5} X_k^2 = 5,934,548$$

which gives

$$s^2 = \frac{1}{4}(5,934,548 - 5 \times 1086^2) = 9392$$

which finally gives $s = 96.9$. This is also the estimated standard error of $X_1$. The estimated standard error of $\bar{X}$ is

$$s_{\bar{X}} = \frac{96.9}{\sqrt{5}} = 43.3$$

which is less than half that of $X_1$. Note how the estimation of $\sigma$ achieves two things: (1) we get an estimate of the variation in individual lifetimes and (2) we get an estimate of the standard error of $\bar{X}$. □

By comparing standard errors, we can compare estimators. It is still not clear, however, what the actual value of the standard error means. It somehow measures the accuracy of our estimator and we would like to use it to get *error bounds*. In the lightbulb example, we have the estimate $\bar{X} = 1086$, which has an estimated standard error 43. Perhaps, we could summarize this as $1086 \pm 43$, but exactly what does this mean? The mean $\mu$ is unknown, so either it is in the interval [1043, 1129] or it is not. Note, however, that the interval is an outcome of the *random interval* $[\bar{X} - s_{\bar{X}}, \bar{X} + s_{\bar{X}}]$, so what we can do is to find the probability that this random interval contains $\mu$. We could then further supplement our error bounds with a probability that tells how much we believe in the bounds. Instead of taking $\pm s_X$, we could take $\pm 2s_{\bar{X}}$, $\pm 3s_{\bar{X}}$, or $\pm c\, s_{\bar{X}}$ for some other constant $c$. The probability of catching $\mu$ changes with $c$, and in the next section we introduce a systematic way to determine error bounds.

## 6.3 CONFIDENCE INTERVALS

As discussed in the previous section, it is desirable to be able to supplement an estimator with error bounds, to get an idea of its accuracy. The following definition gives the formal description.

***Definition 6.9.*** Let $X_1, \ldots, X_n$ be a random sample and $\theta$ an unknown parameter. If $T_1$ and $T_2$ are two functions of the sample such that

$$P(T_1 \leq \theta \leq T_2) = q$$

we call the interval $[T_1, T_2]$ a *confidence interval* for $\theta$ with confidence level $q$. We write this as follows:

$$T_1 \leq \theta \leq T_2 \quad (q)$$

A confidence interval is thus a random interval that contains the parameter $\theta$ with probability $q$. Note that the probability statement in the definition seems to be a bit "backward" from what we are used to since there is a constant $\theta$ in the middle and random variables at the ends. Once we have numerical observations on our sample, we get numerical values of $T_1$ and $T_2$ and refer to this as an *observed confidence interval*. We often refer to the confidence level as a percentage, and instead of saying "a confidence interval with confidence level 0.95," we may say "a 95% confidence interval." We can regard a confidence interval as an estimator that is an entire interval instead of a single point, and for this reason, the distinction between *point estimation* and *interval estimation* is often made.

***Example 6.6.*** Let us return to Example 6.1 where we used the estimator

$$\widetilde{\theta} = \frac{n+1}{n} X_{(n)}$$

to estimate $\theta$ from a uniformly distributed sample on the interval $[0, \theta]$. To find a confidence interval of $\theta$, we first need to determine the distribution of the order statistic $X_{(n)}$. Since $X_{(n)}$ is the maximum of the $n$ independent random variables $X_1, \ldots, X_n$, we can write the distribution function

$$F_{X_{(n)}}(x) = P(X_{(n)} \leq x) = P(X_1 \leq x, \ldots, X_n \leq x)$$

$$= P(X_i \leq x)^n = \left(\frac{x}{\theta}\right)^n \tag{6.1}$$

The next step is to find values $x_1$ and $x_2$ such that

$$P(x_1 \leq X_{(n)} \leq x_2) = q$$

This can be done in several ways (infinitely many, actually), so let us also require the interval to be *symmetric*, that is,

$$P(X_{(n)} < x_1) = P(X_{(n)} > x_2) = \frac{1-q}{2}$$

Equation (6.1) implies that

$$\left(\frac{x_1}{\theta}\right)^n = \frac{1-q}{2}$$

whose solution is

$$x_1 = \theta \left(\frac{1-q}{2}\right)^{1/n}$$

Correspondingly, we get the upper limit as the solution to

$$1 - \left(\frac{x_2}{\theta}\right)^n = \frac{1-q}{2}$$

which is

$$x_2 = \theta \left( \frac{1+q}{2} \right)^{1/n}$$

A confidence interval is now obtained from the inequality

$$\theta \left( \frac{1-q}{2} \right)^{1/n} \leq X_{(n)} \leq \theta \left( \frac{1+q}{2} \right)^{1/n}$$

by dividing it by $\theta$ and $X_{(n)}$ and finally taking the reciprocal. The final result can then be expressed as

$$X_{(n)} \left( \frac{1+q}{2} \right)^{-1/n} \leq \theta \leq X_{(n)} \left( \frac{1-q}{2} \right)^{-1/n} \quad (q)$$

(Note that the upper and lower limits switch when we take the reciprocal.)

Let us now apply this to the observed sample in Example 6.3 for $q = 0.95$. The maximum observation was $x_{(n)} = 8.69$ in a sample of size $n = 10$. The lower limit will then become

$$8.69 \times 0.975^{-1/10} = 8.71$$

and the upper limit

$$8.69 \times 0.025^{-1/10} = 12.57$$

Hence, we can claim with 95% confidence that the actual value of $\theta$ lies somewhere between 8.7 and 12.6. □

This example illustrates the most common way to calculate confidence intervals, namely, to first determine the distribution of an efficient estimator (or a related statistic), use this to obtain an interval for the estimator, and finally transform this into a confidence interval. However, it is often the case that the interval is of the form

$$[\widehat{\theta} - R, \widehat{\theta} + R]$$

where $\widehat{\theta}$ is an estimator of $\theta$ and $R$ is an error bound determined such that

$$P(\widehat{\theta} - R \leq \theta \leq \widehat{\theta} + R) = q$$

We then usually write the interval in the form

$$\theta = \widehat{\theta} \pm R \quad (q)$$

where the error bound $R$ may be deterministic or random. A general observation is that we want the confidence interval to be as short as possible since this means that our estimator has high accuracy. On the other hand, we also want the confidence level $q$ to be as high as possible since this means that we have strong belief that we have caught the true parameter $\theta$ in our interval. However, these two wishes are conflicting, and there is the trade-off that higher confidence levels correspond to longer intervals. It is more important to have high values of $q$ since it would not be of much use to have a very short interval if we did not have any confidence in it. For that reason, the value of $q$ is typically determined in advance and the interval then computed. Some standard values of $q$ are 0.90, 0.95, and 0.99.

Finding a confidence interval requires computation of a probability, and this, in turn, means that we must know something about the distribution of our sample. To be able to compute the probability $P(\widehat{\theta} - R \leq \theta \leq \widehat{\theta} + R)$, we need to know the distribution of the estimator $\widehat{\theta}$ and of $R$ if it is random. We next look at one important special case, when the observations come from a normal distribution.

### 6.3.1 Confidence Interval for the Mean in the Normal Distribution with Known Variance

Suppose that $X_1, \ldots, X_n$ is a sample from a normal distribution with unknown mean $\mu$ and known variance $\sigma^2$. As usual, we estimate $\mu$ with $\bar{X}$, but how can we find a confidence interval of the form $\mu = \bar{X} \pm R$? This means that we need to find $R$ such that

$$P(\bar{X} - R \leq \mu \leq \bar{X} + R) = q$$

By transformation we obtain the equivalent inequality

$$-\frac{R}{\sigma/\sqrt{n}} \leq \frac{\bar{X} - \mu}{\sigma/\sqrt{n}} \leq \frac{R}{\sigma/\sqrt{n}}$$

where we know that the quantity in the middle is $N(0, 1)$. With $z = R\sqrt{n}/\sigma$, we get the following equation by symmetry:

$$q = P\left(-z \leq \frac{\bar{X} - \mu}{\sigma/\sqrt{n}} \leq z\right) = \Phi(z) - \Phi(-z) = 2\Phi(z) - 1$$

This implies the following proposition.

**Proposition 6.5.** If $X_1, \ldots, X_n$ is a sample from a $N(\mu, \sigma^2)$ distribution where $\sigma^2$ is known, a $100q\%$ confidence interval for $\mu$ is

$$\mu = \bar{X} \pm z\frac{\sigma}{\sqrt{n}} \quad (q)$$

where $z$ is such that $\Phi(z) = (1 + q)/2$.

***Example 6.7.*** Consider the lightbulb example and find a 95% confidence interval for $\mu$ under the assumption that $\sigma = 100$.

Since $(1+q)/2 = 0.975$, we need to find $z$ such that $\Phi(z) = 0.975$. Table A.2, specifically constructed for confidence intervals, gives us $z = 1.96$, which yields the confidence interval

$$\mu = 1086 \pm 1.96 \times \frac{100}{\sqrt{5}} = 1086 \pm 88 \quad (0.95)$$

□

This result is mostly of theoretical interest since, in most practical situations, the variance is usually not known beforehand. It is possible, however, to generalize this for unknown variance, but it requires some additional mathematics and will be deferred to the next chapter.

### 6.3.2 Confidence Interval for an Unknown Probability

Suppose that we are interested in the probability $p$ of some event $A$, repeat the experiment $n$ times, and observe $A$ in $X$ of these. Then $X \sim \text{bin}(n, p)$, and a good estimator is the relative frequency $\widehat{p} = X/n$ (in Section 4.2 denoted $f_n$). Thus, by Equation (4.2), a consequence of the normal approximation to the binomial distribution is that

$$\widehat{p} \overset{d}{\approx} N\left(p, \frac{p(1-p)}{n}\right)$$

which means that

$$P\left(-z \leq \frac{\widehat{p} - p}{\sqrt{p(1-p)/n}} \leq z\right) \approx q$$

where $\Phi(z) = (1+q)/2$. Although we could get an interval for $p$ from the preceding expression, we make yet another simplifying approximation and replace $p$ by $\widehat{p}$ in the denominator, which gives

$$P\left(-z \leq \frac{\widehat{p} - p}{\sqrt{\widehat{p}(1-\widehat{p})/n}} \leq z\right) \approx q$$

which in turn gives the following result.

**Proposition 6.6.** An approximate $100q\%$ confidence interval for $p$ is given by

$$p = \widehat{p} \pm z\sqrt{\frac{\widehat{p}(1-\widehat{p})}{n}} \quad (\approx q)$$

where $\Phi(z) = (1+q)/2$.

Let us comment on the substitution of $\widehat{p}$ for $p$. Since

$$\frac{\widehat{p}-p}{\sqrt{\widehat{p}(1-\widehat{p})/n}} = \frac{\widehat{p}-p}{\sqrt{p(1-p)/n}}\sqrt{\frac{p(1-p)}{\widehat{p}(1-\widehat{p})}}$$

where the first factor is approximately $N(0, 1)$ and by Corollary 4.2 the second factor is approximately 1, the product should also be approximately $N(0, 1)$. We are not explaining the exact asymptotic results that are needed here, but hope this gives some intuition as to why the substitution is valid. We cannot substitute $\widehat{p}$ for $p$ just anywhere; if we do it in the numerator, for instance, we get 0.

An application where the approximation works well is political opinion polls. Let $p$ be the unknown proportion of supporters of a particular candidate, draw a sample of $n$ individuals, and ask if they favor this candidate. If this number is $X$, we can estimate $p$ by $\widehat{p} = X/n$ as usual. Now, the distribution of $X$ is not exactly binomial since we sample without replacement but by Proposition 4.3, on the binomial approximation to the hypergeometric, we can assume that $X$ is binomial for all practical purposes.[3] The total population size $N$ is typically on the order of many millions; the sample size $n$, typically around 1000. The confidence interval is, as above

$$p = \widehat{p} \pm z\sqrt{\frac{\widehat{p}(1-\widehat{p})}{n}} \quad (\approx q)$$

where the quantity $\pm z\sqrt{\widehat{p}(1-\widehat{p})/n}$ is referred to as the *margin of error* or the *sampling error.*

***Example 6.8.*** As an illustrative example, let us consider a historical opinion poll from October 2000, on the upcoming presidential election. In this poll, 2207 likely voters were asked whom they would vote for. The results were George Bush 47%, Al Gore 44%, Ralph Nader 2%, and Pat Buchanan 1%; the rest were either undecided or supporting other candidates. The poll was reported to have a ±2% margin of error.

Let us take confidence level $q = 0.95$, which gives $z = 1.96$. The four confidence intervals are

$$p_{\text{Bush}} = 0.47 \pm 1.96\sqrt{\frac{0.47(1-0.47)}{2207}} = 0.47 \pm 0.02$$

$$p_{\text{Gore}} = 0.44 \pm 1.96\sqrt{\frac{0.44(1-0.44)}{2207}} = 0.44 \pm 0.02$$

[3]This "two-step" approximation suggests that we can approximate the hypergeometric distribution directly by the normal, without going via the binomial distribution. This is indeed true, and we could then use the slightly lower variance of the hypergeometric distribution instead. In the given example, there is nothing to gain from this.

$$p_{\text{Nader}} = 0.02 \pm 1.96\sqrt{\frac{0.02(1-0.02)}{2207}} = 0.02 \pm 0.006$$

$$p_{\text{Buchanan}} = 0.01 \pm 1.96\sqrt{\frac{0.01(1-0.01)}{2207}} = 0.01 \pm 0.004$$

or in terms of percentage points, $p_{\text{Bush}} = 47 \pm 2$ and so on. Note how the margins of error differ because the length of the confidence interval depends on $\widehat{p}$. When the margin of error is stated as $\pm 2\%$, this is correct for Bush's and Gore's numbers, but for the smaller numbers of Nader and Buchanan, the margin of error is significantly smaller. See Problem 21 for more on this. □

Note that the margin of error does not depend on the population size $N$, but on the sample size $n$. Obviously this is an effect of the approximation and not true in general; if we ask the 25 inhabitants of Luckenbach, Texas, whether they are feeling any pain, our margin of error will be 0. However, as long as $n$ is small relative to $N$, the size of the population does not matter. Thus, if we ask 2000 people, the poll will be as much valid in the United States as in Canada, Chile, or China.

Another point is that the margin of error also depends on the confidence level, and this is rarely reported in the media. A confidence level of 95% is supposedly the standard for the polling companies and a reasonable choice. Too much lower, and the results are not trustworthy, too much higher and the margin of error may become too large to be useful.

The theory of opinion polls is part of the area of *survey sampling*, an important subdiscipline of statistics. There are many ways to improve the estimates and confidence intervals, such as by ensuring that different population subgroups are proportionally represented in the sample, but we will not discuss this further. Instead, we will focus on a common question regarding opinion polls: If there is a change in support for a candidate between two consecutive polls, does this indicate a real change in the population? The question can be answered with a confidence interval for the difference between two probabilities, $p_1$ and $p_2$. In short, suppose that we have observed that $\widehat{p}_2 > \widehat{p}_1$. We then conclude that $p_2 > p_1$ if the confidence interval for $\widehat{p}_2 - \widehat{p}_1$ does not include 0. We call the difference *statistically significant*, a concept that will be further clarified in Section 6.5. If the interval includes 0, we cannot rule out that $p_2 \leq p_1$ and that the observed difference was due only to chance. Let us revisit the election year 2000.

***Example 6.9.*** Shortly after the October 2000 poll described in Example 6.8, another poll was taken where 2167 individuals were surveyed and the percentages were 48% for Bush, 43% for Gore, 3% for Nader, and 1% for Buchanan. Thus, two of the candidates had gained support in the second poll. Does this indicate a real change in the population?

Let us start with Bush's numbers, $p_1$ denoting the support in the population at the time of the first poll and $p_2$ the support at the time of the second poll. The estimators are $\widehat{p}_1$ and $\widehat{p}_2$, and we have

$$\widehat{p}_1 \stackrel{d}{\approx} N\left(p_1, \frac{p_1(1-p_1)}{n}\right), \quad \widehat{p}_2 \stackrel{d}{\approx} N\left(p_2, \frac{p_2(1-p_2)}{m}\right)$$

where $n$ and $m$ are the sizes of the two polls. If we assume that the second poll is done independent of the first, this means that $\widehat{p}_1$ and $\widehat{p}_2$ are independent, and thus

$$\widehat{p}_2 - \widehat{p}_1 \stackrel{d}{\approx} N\left(p_2 - p_1, \frac{p_2(1-p_2)}{m} + \frac{p_1(1-p_1)}{n}\right)$$

and we can construct a confidence interval for $p_2 - p_1$ with methods similar to those used before. If we thus substitute $p_1$ and $p_2$ with their estimators in the expression for the variance, we get the confidence interval

$$p_2 - p_1 = \widehat{p}_2 - \widehat{p}_1 \pm z\sqrt{\frac{\widehat{p}_2(1-\widehat{p}_2)}{m} + \frac{\widehat{p}_1(1-\widehat{p}_1)}{n}} \quad (\approx q)$$

where as usual $\Phi(z) = (1+q)/2$. Again taking $q = 0.95$ gives

$$p_2 - p_1 = 0.48 - 0.47 \pm 1.96\sqrt{\frac{0.48(1-0.48)}{2167} + \frac{0.47(1-0.47)}{2207}}$$

$$= 0.01 \pm 0.03$$

which is the interval $(-0.02, 0.04)$. Since this includes 0, the change is not large enough to rule out randomness and the change is thus not statistically significant. For Nader's numbers, we get

$$p_2 - p_1 = 0.03 - 0.02 \pm 1.96\sqrt{\frac{0.03(1-0.03)}{2167} + \frac{0.02(1-0.02)}{2207}}$$
$$= 0.01 \pm 0.009$$

which is an interval entirely above 0, so this change is statistically significant, albeit just barely, and roundoff errors may also play a role here. The important point, though, is that an increase by one percentage point in the polls is much more significant for Nader's numbers in the single digits than for Bush's, which are in the 40–50% range. □

Another obvious question is whether the difference between Bush's and Gore's numbers of 47% and 44% in the first poll is statistically significant. Again, this can be answered with a confidence interval for the difference between two probabilities, $p_B$ and $p_G$, but note that the estimators $\widehat{p}_B$ and $\widehat{p}_G$ are not independent since they come from the same sample. You are asked to investigate this in Problem 24.

In the media, we often hear statements like "candidate A leads candidate B in the poll," and later it is mentioned that the difference is within the margin of error. Such statements are meaningless. To "lead in the poll" in the sense that the estimated proportion is higher has no value unless the confidence interval for the difference is entirely above 0.

### 6.3.3 One-Sided Confidence Intervals

Our confidence intervals thus far have been of the type $T_1 \leq \theta \leq T_2$ or $\theta = \widehat{\theta} \pm R$, which we call *two-sided*. Sometimes, it is more desirable to have *one-sided* intervals, of the type $\theta \leq \widehat{\theta} + R$ or $\theta \geq \widehat{\theta} - R$. For instance, consider Example 6.7 where the confidence interval for the mean lifetime of a lightbulb was $1086 \pm 88$ h. Instead of claiming that the mean lifetime is between 998 and 1174, we may be interested in claiming only that it is at least 998. Thus, we are interested only in one of the confidence limits and in such a case, it is a "waste of confidence level" to construct a two-sided interval by splitting $1 - q$ in two (equal) parts. Let us say that we want an interval with only a lower bound. In the lightbulb example, this can be expressed as

$$\mu \geq \bar{X} - z\frac{\sigma}{\sqrt{n}} \quad (q)$$

where $\Phi(z) = q$. For $q = 0.95$, Table A.2 gives us $z = 1.64$, which yields the interval

$$\mu \geq 1086 - 1.64\frac{100}{\sqrt{5}} = 1086 - 73 = 1013 \quad (0.95)$$

We see that if we are interested only in a lower bound of life expectancy we can make a slightly stronger statement than before.

## 6.4 ESTIMATION METHODS

The estimators that we have come up with so far have been based on common sense, such as estimating the mean by the sample mean. In more complicated situations, there might not be an obvious estimator, and the question is how to find one. In this section, we will examine two general methods to find estimators, based on two different principles. Both methods work in great generality, and in simple cases they give estimators that are intuitively reasonable.

### 6.4.1 The Method of Moments

We have seen that the sample mean $\bar{X}$ is a natural estimator of the mean $\mu = E[X]$ and that it has good properties. Suppose instead that the parameter we wish to estimate is $\theta = E[X^2]$. Since this is the expected value of the random variable $X^2$, it is logical to estimate it by the sample mean of the squared observations:

$$\widehat{\theta} = \frac{1}{n}\sum_{k=1}^{n} X_k^2$$

In the same way, we can estimate any expected value of the type $E[X^r]$ by the corresponding sample mean of the observations raised to the $r$th power. Let us state some definitions.

***Definition 6.10.*** Let $X$ be a random variable. The $r$th *moment* of $X$ is defined as

$$\mu_r = E[X^r]$$

Hence $\mu_1 = E[X]$, $\mu_2 = E[X^2]$, and so on. Next we define the corresponding sample means to be used as estimators.

***Definition 6.11.*** $X_1, \ldots, X_n$ be a random sample. The $r$th *sample moment* is defined as

$$\widehat{\mu}_r = \frac{1}{n}\sum_{k=1}^{n} X_k^r$$

In particular, $\widehat{\mu}_1 = \bar{X}$, the sample mean. Note that the moments are parameters computed from the distribution and that the sample moments are estimators computed from the sample. Also note that for each $r$, $\widehat{\mu}_r$ has mean and variance

$$E[\,\widehat{\mu}_r] = \mu_r, \quad \mathrm{Var}[\,\widehat{\mu}_r] = \frac{1}{n}\mathrm{Var}[X^r]$$

so that each sample moment is an unbiased estimator of the corresponding moment and has a variance that is small for large $n$ (unless $\mathrm{Var}[X^r] = \infty$). Now, our parameter of interest may not be one of the moments, but if it can be expressed as a function of moments, it can be estimated by replacing moments by the corresponding sample moments. This is the main idea behind the following definition.

***Definition 6.12.*** Suppose that we can express the unknown parameter $\theta$ as a function of the first $j$ moments, $\theta = g(\mu_1, \ldots, \mu_j)$. The estimator $\widehat{\theta} = g(\widehat{\mu}_1, \ldots, \widehat{\mu}_j)$ is then called the *moment estimator* of $\theta$.

The representation $\theta = g(\mu_1, \ldots, \mu_j)$ is not necessarily unique since parameters may often be written as different functions of different moments. The convention is to start by computing the first moment $\mu_1$. If $\theta$ can be expressed as a function of $\mu_1$, we are done. If not, we go on to compute the second moment and so on, until we get the desired expression. Let us look at a few examples to illustrate the method, appropriately named the *method of moments*.

***Example 6.10.*** Let $X_1, \ldots, X_n$ be a sample from an exponential distribution with unknown parameter $\lambda$. Find the moment estimator of $\lambda$.

We start by computing the first moment

$$\mu_1 = E[X] = \frac{1}{\lambda}$$

and hence $\lambda = 1/\mu_1$. In the terminology of Definition 6.12, we have $j = 1$ and $g(x) = 1/x$. The moment estimator is therefore

$$\widehat{\lambda} = \frac{1}{\widehat{\mu}_1} = \frac{1}{\bar{X}}$$

□

In the last example, it was sufficient to find the first moment. Let us next look at a few examples where this is not the case. The first of these also illustrates the fact that we do not always need to know the distribution of the $X_k$ to find moment estimators.

***Example 6.11.*** Let $X_1, \ldots, X_n$ be any random sample with mean $\mu$ and variance $\sigma^2$. Find the moment estimators of $\mu$ and $\sigma^2$.

Since $\mu_1 = \mu$, the moment estimator of $\mu$ is $\widehat{\mu}_1 = \bar{X}$. For the variance, note that

$$\sigma^2 = E[X^2] - (E[X])^2 = \mu_2 - \mu_1^2$$

so the moment estimator is

$$\widehat{\sigma}^2 = \widehat{\mu}_2 - \widehat{\mu}_1^2 = \frac{1}{n}\sum_{k=1}^{n} X_k^2 - \bar{X}^2 = \frac{1}{n}\sum_{k=1}^{n}(X_k - \bar{X})^2$$

after some algebra. Note that this is not equal to the unbiased estimator $s^2$. Moment estimators are not necessarily unbiased but can sometimes be adjusted to be so. In this case, we can multiply $\widehat{\sigma}^2$ by $n/(n-1)$ to obtain an unbiased estimator. □

***Example 6.12.*** The following is an observed sample from a uniform distribution on $[-\theta, \theta]$, where $\theta$ is unknown. Find the moment estimate of $\theta$:

$$-6.9,\ 2.8,\ 3.4,\ 6.4,\ 6.7,\ 8.0$$

Let us first find the moment estimator for a sample $X_1, \ldots, X_n$. The first moment is $\mu_1 = E[X] = 0$, which does not help us, so we need to proceed to the second moment. We get

$$\mu_2 = \int_{-\theta}^{\theta} x^2 f(x)dx = \frac{1}{2\theta}\left[\frac{x^3}{3}\right]_{-\theta}^{\theta} = \frac{\theta^2}{3}$$

which gives

$$\theta = \sqrt{3\mu_2}$$

The moment estimator is therefore

$$\widehat{\theta} = \sqrt{3\widehat{\mu}_2} = \sqrt{\frac{3}{n}\sum_{k=1}^{n} X_k^2}$$

and in our case this becomes

$$\widehat{\theta} = \sqrt{\frac{3}{6} \times 217} = 10.4$$

which is an estimate that we would certainly have been unable to figure out by intuition alone. □

### 6.4.2 Maximum Likelihood

The method of moments from the previous section was based on the simple idea of estimating each moment by its corresponding sample moment. In this section, we consider another estimation principle, which is based on the idea of choosing the most likely parameter value for a given sample. We illustrate this in an example.

***Example 6.13.*** A digital communication system transmits 0s and 1s. We know that on average, one of the bits is sent twice as often as the other, but we do not know which one. In an attempt to decide which, we have the following four observations: 1, 1, 0, 1 (independent transmissions and no transmission error). Of course, our immediate observation is that there are more 1s than 0s and hence we ought to believe that 1 is the more common. We now formalize this intuitive reasoning as an estimation problem.

Let $p$ be the probability that 1 is sent. Then we know that $p$ is either $\frac{1}{3}$ or $\frac{2}{3}$. Hence we can view $p$ as an unknown parameter with possible values in $\{\frac{1}{3}, \frac{2}{3}\}$. Let us now compute the probability to get the outcome that we actually got. By independence

$$P(1, 1, 0, 1) = p \times p \times p \times (1 - p) \times p = p^3(1 - p)$$

If $p = \frac{1}{3}$, this equals 0.025, and if $p = \frac{2}{3}$, it equals 0.099. Since $p$ must be either of the two and $P(1, 1, 0, 1)$ is higher if $p$ is $\frac{2}{3}$, we may say that $\frac{2}{3}$ is a *more likely* value of $p$ than $\frac{1}{3}$ and choose $\frac{2}{3}$ as our estimate of $p$. □

Note the idea in the example. We look at the outcome we got and then ask which value of the parameter we think it came from. The parameter that maximizes the probability of the outcome is chosen as the estimate. Let us return to the example.

***Example 6.14.*** Now assume that we do not know anything about $p$ and wish to estimate it on the basis of the same observations and the same principle. The probability of our outcome is a function of $p$, say, $L(p)$, where

$$L(p) = p^3(1-p), \quad 0 \leq p \leq 1$$

just as above but $p$ can now be any number in $[0, 1]$. On the basis of the same principle as above, we wish to find the most likely value of $p$, and to that extent we find the maximum of $L(p)$ by the usual method of differentiation. We get

$$L'(p) = 3p^2 - 4p^3 = p^2(3-4p)$$

Setting this to 0 yields $p = 0$ or $p = \frac{3}{4}$, and since the second derivative is

$$L''(p) = p(6-12p) < 0 \quad \text{for } p = \frac{3}{4}$$

we see that $p = \frac{3}{4}$ gives the maximum (the value $p = 0$ is also unreasonable since we have 1s in our sample). Our estimate is $\widehat{p} = \frac{3}{4}$, the relative frequency of 1s. □

Let us now generalize the method from the example. First, let us restate the example in terms of random variables. Thus, let $X$ be a random variable that describes a transmitted bit. Then $X = 1$ with probability $p$ and $X = 0$ with probability $1 - p$, and the probability mass function of $X$ is therefore

$$f_p(0) = 1 - p, \quad f_p(1) = p$$

The index $p$ emphasizes that the probability mass function depends on the parameter $p$ (and it looks better to use $f$ rather than the previous $p$ for the pmf). We now realize that the function $L$ in Example 6.14 can be written as

$$L(p) = f_p(1)f_p(1)f_p(0)f_p(1)$$

the product of the probability mass function evaluated at the observed values. We can now easily generalize to the case where we have a sample $X_1, \ldots, X_n$ of 0s and 1s with probability mass function as above. The function $L$ becomes

$$L(p) = \prod_{k=1}^{n} f_p(X_k)$$

where we note that there are $X_k$ in the arguments so $L(p)$ is actually random. However, we view it as a function of $p$, and for that purpose we can view the $X_k$ as fixed. When we maximize, the maximum is attained at some point $\widehat{p}$ that must then be a function of the $X_k$ and $\widehat{p}$ is a random variable exactly as we want an estimator to be. To find out what $\widehat{p}$ is in this case, see Problem 33.

The method described above can be directly generalized to any discrete distribution. If $X_1, \ldots, X_n$ is a sample from a discrete distribution with pmf $f_\theta$, we can define $L(\theta)$ as the product $f_\theta(X_1) \cdots f_\theta(X_n)$, which again describes how likely different parameter values are to produce the sample $X_1, \ldots, X_n$. For a continuous distribution, the probability mass function is replaced by the pdf, which, as we already know, is not directly interpretable as a probability. It is still a measure of how $X$ is distributed over its range, in the sense that large values of $f$ correspond to regions where $X$ is more likely to be. Hence, we can still define the function $L(\theta)$ and interpret it as a measure of how likely a parameter value $\theta$ is to produce the sample $X_1, \ldots, X_n$. We now formalize this.

***Definition 6.13.*** Let $X_1, \ldots, X_n$ be a random sample from a distribution that has pmf or pdf $f_\theta$. The function

$$L(\theta) = \prod_{k=1}^{n} f_\theta(X_k)$$

is called the *likelihood function*. The value $\widehat{\theta}$ where $L$ attains its maximum is called the *maximum-likelihood estimator* (MLE) of $\theta$.

To find the MLE, we thus view $X_1, \ldots, X_n$ as fixed and find the maximum of the function $L(\theta)$ by using common techniques from calculus. As it is most often easier to maximize a sum than a product, we define the *log-likelihood function* as

$$l(\theta) = \log L(\theta)$$

and maximize this instead of $L$. Since the logarithm is a strictly increasing function, $l$ and $L$ attain maximum for the same argument. To find the MLE, the following algorithm often works:

1. Find $L(\theta)$ and $l(\theta) = \log L(\theta)$.
2. Differentiate $l(\theta)$ with respect to $\theta$ and set equal to 0.
3. Solve for $\theta$ and denote the solution $\widehat{\theta}$.
4. Check that $l''(\widehat{\theta}) < 0$ to ensure maximum. If this holds, then $\widehat{\theta}$ is the MLE.

Although $L(\theta)$ could have several local maxima, it can be shown that the MLE exists and is unique under some fairly general assumptions. The second derivative check in step 4 is also often superfluous, as the function $l(\theta)$ is often strictly concave. We will not address these issues here and instead turn to some examples.

***Example 6.15.*** Let $X_1, \ldots, X_n$ be a sample from an exponential distribution with unknown parameter $\lambda$. Find the MLE of $\lambda$.

The pdf is

$$f_\lambda(x) = \lambda\, e^{-\lambda x}, \quad x \geq 0$$

which gives likelihood function

$$L(\lambda) = \prod_{k=1}^{n} \lambda\, e^{-\lambda X_k} = \lambda^n \exp\left(-\lambda \sum_{k=1}^{n} X_k\right)$$

The log-likelihood function is

$$l(\lambda) = \log L(\lambda) = n \log \lambda - \lambda \sum_{k=1}^{n} X_k$$

which has derivative

$$\frac{d}{d\lambda} l(\lambda) = \frac{n}{\lambda} - \sum_{k=1}^{n} X_k = n\left(\frac{1}{\lambda} - \bar{X}\right)$$

and the equation

$$\frac{d}{d\lambda} l(\lambda) = 0$$

has solution

$$\widehat{\lambda} = \frac{1}{\bar{X}}$$

which is the MLE (the second derivative is always negative), and we note that it is the same as the moment estimator from the previous section. □

***Example 6.16.*** Let $X_1, \ldots, X_n$ be a sample from a Poisson distribution with mean $\lambda$. Find the MLE of $\lambda$.

This time the random variable is discrete with pmf

$$f_\lambda(k) = e^{-\lambda} \frac{\lambda^k}{k!}, \quad k = 0, 1, \ldots$$

which gives

$$L(\lambda) = \prod_{k=1}^{n} e^{-\lambda} \frac{\lambda^{X_k}}{X_k!} = C e^{-n\lambda} \lambda^{n\bar{X}}$$

where $C = (X_1! \cdots X_n!)^{-1}$. We get

$$l(\lambda) = \log C - n\lambda + n\bar{X} \log \lambda$$

which has derivative

$$\frac{d}{d\lambda} l(\lambda) = -n + n\frac{\bar{X}}{\lambda}$$

which set equal to 0 gives the MLE

$$\widehat{\lambda} = \bar{X}$$

the sample mean. □

Note how there is no difference between the discrete and the continuous cases in the way the method is applied. The next example is a case where differentiating the log-likelihood function does not work.

***Example 6.17.*** Let $X_1, \ldots, X_n$ be a sample from a uniform distribution on $[0, \theta]$. Find the MLE of $\theta$.

The pdf is

$$f_\theta(x) = \begin{cases} 1/\theta & \text{if } 0 \leq x \leq \theta \\ 0 & \text{otherwise} \end{cases}$$

When we write down the likelihood function, we need to remember that we view the $X_k$ as fixed and $\theta$ as the parameter. We get the following expression:

$$L(\theta) = \begin{cases} 1/\theta^n & \text{if } \theta \geq \text{ all } X_k \\ 0 & \text{otherwise} \end{cases}$$

We may first try to take the logarithm and differentiate with respect to $\theta$, but this leads us to nowhere. In Figure 6.2, we can see that $L$ attains its maximum precisely where

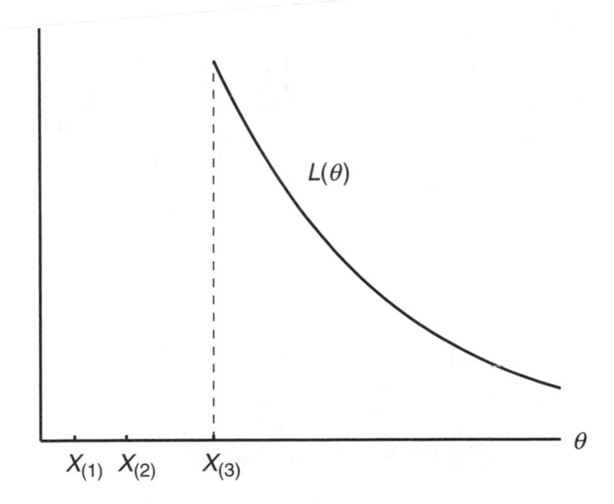


**FIGURE 6.2** The likelihood function $L(\theta)$ for a sample of size $n = 3$ from a uniform distribution on $[0, \theta]$. The maximum is attained at the largest observation $X_{(3)}$.

the largest $X$ value is. Recalling the *order statistics* from Definition 3.20, we realize that

$$\widehat{\theta} = X_{(n)}$$

the maximum value in the sample. Note that this estimator is different from the one obtained previously by the method of moments, and also recall Example 6.1. □

We will next find the MLEs of $\mu$ and $\sigma^2$ in the $N(\mu, \sigma^2)$ distribution. If one of the parameters is known, it is straightforward to estimate the other, and if both are unknown, we can view them as a two-dimensional parameter $\theta = (\mu, \sigma^2)$. In the definition of maximum likelihood, we simply maximize over two variables instead of one, everything else remaining the same. We could argue whether we should find the MLE of $\sigma$ or $\sigma^2$, but it turns out that we can find either one and then square or take the square root to find the other. This is by virtue of the following proposition, which we state without proof.

**Proposition 6.7.** If $\widehat{\theta}$ is the MLE of $\theta$ and $g$ is a one-to-one function, then $g(\widehat{\theta})$ is the MLE of $g(\theta)$.

***Example 6.18.*** Let $X_1, \ldots, X_n$ be a random sample from a $N(\mu, \sigma^2)$ distribution where both $\mu$ and $\sigma^2$ are unknown. Find the MLEs of $\mu$ and $\sigma^2$.

By virtue of Proposition 6.7, we can view the likelihood function as a function of $\mu$ and $\sigma$, rather than $\mu$ and $\sigma^2$. The likelihood function is

$$\begin{aligned} L(\mu, \sigma) = \prod_{k=1}^{n} f_{\mu,\sigma}(X_k) &= \prod_{k=1}^{n} \frac{1}{\sigma\sqrt{2\pi}} e^{-(X_k-\mu)^2/2\sigma^2} \\ &= \left(\frac{1}{\sqrt{2\pi}}\right)^n \frac{1}{\sigma^n} \exp\left(-\frac{1}{2\sigma^2}\sum_{k=1}^{n}(X_k - \mu)^2\right) \end{aligned}$$

and the log-likelihood function is

$$l(\mu, \sigma) = -n \log \sqrt{2\pi} - n \log \sigma - \frac{1}{2\sigma^2}\sum_{k=1}^{n}(X_k - \mu)^2$$

We set the partial derivative with respect to $\mu$ equal to 0 to obtain

$$\frac{\partial l}{\partial \mu} = \frac{1}{\sigma^2}\sum_{k=1}^{n}(X_k - \mu) = \frac{1}{\sigma^2}\left(\sum_{k=1}^{n} X_k - n\mu\right) = 0$$

which gives $\widehat{\mu} = \bar{X}$. The partial derivative with respect to $\sigma$, with $\mu$ replaced by $\bar{X}$, gives

$$\frac{\partial l}{\partial \sigma} = \frac{1}{\sigma}\left(-n + \frac{1}{\sigma^2}\sum_{k=1}^{n}(X_k - \bar{X})^2\right) = 0$$

which gives the MLEs

$$\widehat{\mu} = \bar{X}$$

$$\widehat{\sigma^2} = \frac{1}{n}\sum_{k=1}^{n}(X_k - \bar{X})^2$$

which are the same as the moment estimators. By virtue of Proposition 6.7, the MLE of the standard deviation $\sigma$ is

$$\widehat{\sigma} = \sqrt{\frac{1}{n}\sum_{k=1}^{n}(X_k - \bar{X})^2}$$

□

The maximum-likelihood principle is central to statistics and can be used to solve various estimation problems, even for quite complex models. Besides being general, it can also be shown to produce good estimates. In fact, MLEs are in a sense optimal, at least asymptotically, which the following result formalizes.

**Proposition 6.8.** Let $\widehat{\theta}_n$ be the MLE of $\theta$ based on the sample $X_1, \ldots, X_n$ and assume that the Fisher information $I(\theta)$ exists. Then

(i) $E[\widehat{\theta}_n] \to \theta$
(ii) $\widehat{\theta}_n$ is consistent
(iii) $e(\widehat{\theta}_n) \to 1$
(iv) $\sqrt{nI(\theta)}(\widehat{\theta}_n - \theta) \xrightarrow{d} N(0, 1)$

as $n \to \infty$.

The first and the third part says that an MLE is *asymptotically unbiased* and *asymptotically efficient*, which implies that, for large samples, there is (practically) no better estimator. The last part is very useful when deriving approximate confidence intervals based on MLEs. It can be reformulated for large $n$ as

$$\widehat{\theta}_n \overset{d}{\approx} N\left(\theta, \frac{1}{nI(\theta)}\right)$$

which can be used in the same way as in Section 6.3.1 to obtain the interval

$$\theta = \widehat{\theta}_n \pm \frac{z}{\sqrt{nI(\widehat{\theta}_n)}} \quad (\approx q)$$

where $z$ satisfies $\Phi(z) = (1+q)/2$. Note that $I(\theta)$ is replaced by $I(\widehat{\theta}_n)$ in the error bound. Since $\theta$ is unknown, we cannot compute $I(\theta)$, but if $n$ is sufficiently large we know that $\theta \approx \widehat{\theta}_n$ and, consequently, that $I(\theta) \approx I(\widehat{\theta}_n)$. One has to be aware that this interval is probably slightly narrower than an exact interval or, correspondingly, that the true confidence level is lower than $q$ since $\text{Var}(\widehat{\theta}_n) \geq 1/nI(\theta)$, but for large $n$ this will be negligible.

***Example 6.19.*** Let us apply this to Example 6.15 to derive an approximate confidence interval for $\lambda$ in the exponential distribution. The Fisher information in this case becomes

$$I(\lambda) = -E\left[\frac{\partial^2}{\partial\lambda^2}\log(\lambda e^{-\lambda X})\right] = \frac{1}{\lambda^2}$$

Replacing $\lambda$ with $\widehat{\lambda} = 1/\bar{X}$ yields

$$\lambda = \frac{1}{\bar{X}} \pm \frac{z}{\sqrt{n\bar{X}^2}} = \frac{1 \pm z/\sqrt{n}}{\bar{X}} \quad (\approx q)$$

However, the exponential distribution is notoriously difficult to approximate by the normal distribution because of its heavily skewed nature, so $n$ would have to be really large for this interval to be accurate. In this case, it is actually preferable to use the central limit theorem

$$\bar{X} \overset{d}{\approx} N\left(\mu = \frac{1}{\lambda}, \frac{\sigma^2}{n} = \frac{1}{n\lambda^2}\right)$$

to derive an approximate confidence interval for $\mu$ as

$$\mu = \bar{X} \pm \frac{z}{\sqrt{n\widehat{\lambda}^2}} = \bar{X} \pm z\frac{\bar{X}}{\sqrt{n}} \quad (\approx q)$$

and then taking the reciprocal to obtain

$$\frac{1}{\bar{X}(1+z/\sqrt{n})} \leq \lambda \leq \frac{1}{\bar{X}(1-z/\sqrt{n})} \quad (\approx q)$$

□

### 6.4.3 Evaluation of Estimators with Simulation

Estimators derived using the method of moments or maximum-likelihood can have complicated expressions as functions of the sample. If, for example, we want to check unbiasedness or compute the standard error for quality assessment, this may be difficult or even impossible. One way to evaluate such estimators is to use simulation.

***Example 6.20.*** Let $X_1, X_2, \ldots, X_n$ be a random sample from a distribution with pdf

$$f(x) = \theta x^{\theta-1}, \quad 0 \leq x \leq 1$$

where $\theta$ is an unknown parameter. Find the moment estimator of $\theta$ and examine its properties.

The first moment is

$$\mu = \int_0^1 x f(x) dx = \theta \int_0^1 x^\theta dx = \frac{\theta}{1+\theta}$$

which gives

$$\theta = \frac{\mu}{1-\mu}$$

and the moment estimator

$$\widehat{\theta} = \frac{\bar{X}}{1-\bar{X}}$$

It is difficult to compute the mean and variance of $\widehat{\theta}$, so let us instead examine its properties by simulation. To do so, we first choose a value of $\theta$ and a sample size $n$, then run repeated simulations of samples $X_1, \ldots, X_n$ and each time compute $\widehat{\theta}$. This gives us a sample $\widehat{\theta}_1, \ldots, \widehat{\theta}_N$, where $N$ is the number of simulated samples. Now we can use the sample mean (let us call it $\widetilde{\theta}$ to avoid putting a bar on top of a hat)

$$\widetilde{\theta} = \frac{1}{N} \sum_{k=1}^{N} \widehat{\theta}_k$$

and sample variance

$$s^2 = \frac{1}{N-1} \sum_{k=1}^{N} (\widehat{\theta}_k - \widetilde{\theta})^2$$

to estimate $E[\widehat{\theta}]$ and $\text{Var}[\widehat{\theta}]$ (where as usual it is more informative to take the square root to get the standard deviation $s$). Note that each observation on $\widehat{\theta}$ is computed from $n$ observations, so we simulate a total of $N \times n$ $X$ values. We also need to repeat this for various values of $\theta$ and $n$.

To simulate observations on $X$, we use the inverse transform method from Proposition 5.2. The cdf of $X$ is

$$F(x) = \int_0^x f(t) dt = \theta \int_0^x t^{\theta-1} dt = x^\theta, \quad 0 \leq x \leq 1$$

**TABLE 6.1 Estimated Means $\widetilde{\theta}$ and Standard Deviations $s$ of the Estimator $\widehat{\theta}$, Based on Simulations of Sample Size $n$ and True Parameter Value $\theta$**

| $n$ | $\theta$ | $\widetilde{\theta}$ | $s$ |
|---|---|---|---|
| 5 | 2 | 2.32 | 1.33 |
| 10 | 2 | 2.18 | 0.85 |
| 50 | 2 | 2.01 | 0.30 |
| 100 | 2 | 2.02 | 0.21 |
| 1000 | 2 | 2.00 | 0.07 |

the inverse of which is

$$F^{-1}(x) = x^{1/\theta}, \quad x \geq 0$$

Let us take $\theta = 2$. If $U_1, \ldots, U_n$ are i.i.d. standard uniform, our sample is

$$X_1 = \sqrt{U_1}, \ldots, X_n = \sqrt{U_n}$$

and from this we compute

$$\widehat{\theta} = \frac{\bar{X}}{1 - \bar{X}}$$

This gives us one observation on $\widehat{\theta}$, and we repeat the procedure $N$ times to get our sample of observations on $\widehat{\theta}$. Let us take $N = 1000$ and $n = 10$. A Matlab simulation gave the estimated mean $\widetilde{\theta} = 2.18$, which we compare with the true mean 2, and it looks pretty good. The standard error was $s = 0.85$. Make sure not to confuse the two sample sizes; $n$ is the "real" sample size and $N$ is the number of simulations that we decide to run. To investigate the effect of the sample size $n$, simulations were run with $n = 5, 10, 50, 100$, and 1000. The results are summarized in Table 6.1.

We can see that $\widehat{\theta}$ tends to overestimate $\theta$ a little but that it is pretty accurate and becomes more so when the value of $n$ is larger. This last observation is not surprising since $\widehat{\theta}$ is based on $\bar{X}$, and we can invoke the law of large numbers and Corollary 4.2. These results are of no help when it comes to smaller values of $n$, though. Figure 6.3 gives histograms for the cases $n = 10$ and $n = 1000$. We will stop here, but to continue the investigation we would need to run simulations for many different values of $\theta$. If we have reason to believe that $\theta$ is in some particular range, this can help us reduce the number of simulations. □

### 6.4.4 Bootstrap Simulation

The simulation method presented in the previous section works only when we have full knowledge about the underlying distribution of an actual sample, something that

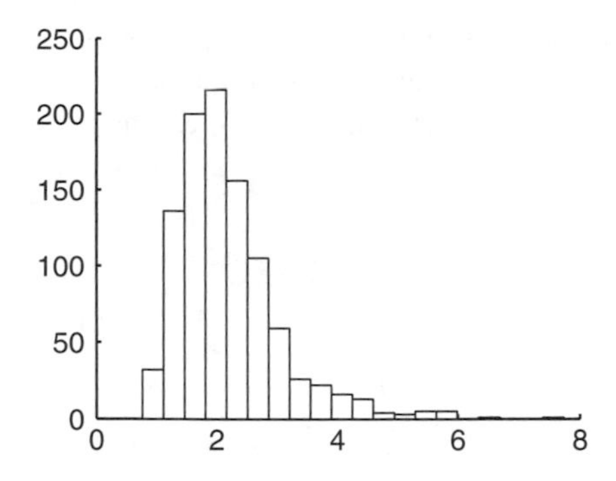


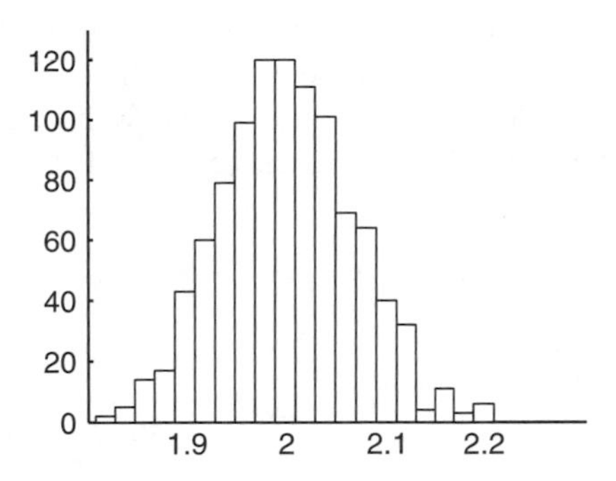


**FIGURE 6.3** Histograms from simulations of the estimator $\widehat{\theta}$ for $\theta = 2$, $n = 10$ and $\theta = 2$, $n = 1000$, respectively. Note the different scales on the $x$-axis.

is rarely true in reality. Let us assume that we have the observed sample $x_1, \ldots, x_n$ and we want to say something about the point estimator $\widehat{\theta} = g(X_1, \ldots, X_n)$. The general idea that the properties of estimators may be evaluated by generating new random samples can still be used, but instead of the true distribution we use the *empirical distribution function*

$$\widehat{F}_n(x) = \frac{1}{n} \sum_{i=1}^{n} I_{\{x_i \leq x\}}$$

where $I_A$ denotes the indicator of the event $A$. In practice, it simply means that a simulated sample $X_1, \ldots, X_n$ is obtained by picking $n$ values at random from the observed sample with replacement. Finally, we obtain a simulated sample of values of $\widehat{\theta}$ by repeating this procedure $N$ times as in the previous section. If $n$ is large, the empirical distribution $\widehat{F}_n(x)$ approximates the true distribution $F(x)$, which means the obtained sample of $\widehat{\theta}$ values will reflect the true distribution reasonably well. This method of simulating random samples from an unknown distribution is called *bootstrap simulation*[4].

***Example 6.21.*** The concentrations of antibodies against pertussis (whooping cough) in the blood of umbilical cords of 30 randomly chosen mothers in Sweden from 2007 are presented below.

| | | | | | | | | | |
|---|---|---|---|---|---|---|---|---|---|
| 0.0 | 0.0 | 0.0 | 0.0 | 1.7 | 2.3 | 2.9 | 3.3 | 4.0 | 4.5 |
| 5.3 | 6.1 | 6.7 | 7.9 | 8.9 | 10.3 | 11.1 | 12.2 | 13.1 | 14.4 |
| 16.0 | 17.4 | 21.0 | 24.4 | 29.4 | 33.5 | 42.1 | 52.2 | 93.9 | 353.7 |

[4]Using a distribution we know nothing about is like lifting ourselves by the bootstraps much like the eighteenth century nobleman and adventurer Baron Münchhausen.

Let us say that we want to estimate the mean concentration and give a 95% confidence interval based on the sample. The observed sample mean and sample variance are $\bar{x} = 26.6$ and $s^2 = 4201$. Since we have a rather large sample size, we can hope that the approximate confidence interval

$$\mu = \bar{x} \pm z\frac{s}{\sqrt{n}} = 26.6 \pm 1.96\sqrt{\frac{4201}{30}} = 26.6 \pm 23.2$$

or

$$3 \leq \mu \leq 50 \qquad (95\%)$$

is somewhat reliable. Now, let us instead generate $N = 1000$ independent bootstrap samples of size 30 by resampling the original data set and calculate the sample mean of each one. This produces a sample $\bar{x}_1, \ldots, \bar{x}_N$ that hopefully represents the true sampling distribution of $\bar{X}$. We then get the bootstrap mean as

$$\bar{\bar{x}} = \frac{1}{N}\sum_{i=1}^{N} \bar{x}_i = 26.8$$

and bootstrap variance as

$$s_{\bar{x}}^2 = \frac{1}{N-1}\sum_{i=1}^{N}(\bar{x}_i - \bar{\bar{x}})^2 = 141$$

The latter quantity corresponds to the standard error, which also can be estimated as

$$\frac{s^2}{n} = \frac{4201}{30} = 140$$

which is reassuring. However, since the simulated means can be regarded as a random sample from the underlying sampling distribution of $\bar{X}$, we can calculate a confidence interval by taking the 2.5 and 97.5 percentiles, that is, the 25th smallest value $\bar{x}_{(25)}$ and the 25th largest value $\bar{x}_{(975)}$. In this case, we get the bootstrap confidence interval

$$10 \leq \mu \leq 55 \qquad (95\%)$$

Note that this yields an asymmetric interval, which reflects the asymmetrical nature of the observed sample.

Since this procedure can be applied to any estimator, we can also get a confidence interval for the variance. Again, generate $N = 1000$ bootstrap samples as before (or recycle the old ones) and calculate the sample variances $s_1^2, \ldots, s_N^2$ for all of them. It turns out that we get the bootstrap mean

$$\bar{s^2} = \frac{1}{N}\sum_{i=1}^{N} s_i^2 = 4074$$

and bootstrap variance

$$s_{s^2}^2 = \frac{1}{N-1} \sum_{i=1}^{N} (s_i^2 - \bar{s^2})^2 = 1.216 \times 10^7$$

Again, the second value is an estimate of the sampling variance of $s^2$, which usually is difficult to obtain analytically. Finally, we find the 25th smallest $s_{(25)}^2$ and 25th largest $s_{(975)}^2$ to get a 95% confidence interval as

$$108 \leq \sigma^2 \leq 11100 \qquad (95\%)$$

□

In Section 6.4.3, we could increase the number of simulations to achieve any desired precision in our estimates since we used the exact sample distribution. Here, we are limited by the size of our original observed sample, and increasing the number of bootstrap samples beyond some level does not improve the results very much. In this example, we used $N = 1000$, which is quite sufficient for a sample of size $n = 30$. It makes no sense to put a lot of simulation power into a bootstrap analysis if the results refer only to the empirical distribution, which is just an approximation of the true underlying distribution.

## 6.5 HYPOTHESIS TESTING

We have learned to find estimators and confidence intervals for unknown parameters, based on observed data. Once this has been done, we are also often interested in drawing some particular conclusion about the underlying distribution. Let us look at a simple example to illustrate the main problem and idea.

***Example 6.22.*** Suppose that we have a coin and want to test whether it is fair. To do so, we flip it 100 times and count the number of heads. How can we decide if the coin is fair?

First, if the coin is fair, the expected number of heads is 50. Now, we do not require to get exactly 50 heads since there is variability due to randomness, so the question is whether our outcome is easily explained by such natural variability or indicates that the coin is unfair. What if we get 55 heads? What if we get 70, 80, or 100 heads? Certainly all of these outcomes are *possible*, so the only criterion we can use is how *probable* they are, and to assess this, we need to invoke the distribution of the number of heads.

Let us first restate the problem in terms of the unknown parameter $p$, the probability of getting heads in a single flip. Our estimate of $p$ is the observed relative frequency $\widehat{p}$, and if the coin is fair, then $p = 0.5$. The question is now whether the value of $\widehat{p}$ deviates too much from 0.5 for us to believe in fairness. We need to quantify "too much" and decide how much deviation we can accept. Let us say that an outcome that has probability as small as 0.05 is acceptable but not any smaller. Thus, we first assume that $p = 0.5$ and will change our mind and decide that the coin is unfair if $\widehat{p}$ deviates from 0.5 by more than $d$, where

$$P(|\,\widehat{p} - 0.5\,| \geq d) = 0.05$$

By the central limit theorem, we know that $\widehat{p}$ has an approximate normal distribution with mean 0.5 and standard deviation $\sqrt{0.5(1-0.5)/100} = 0.05$. Thus, we choose $d$ such that

$$\begin{aligned} 0.05 = P(|\,\widehat{p} - 0.5\,| \geq d) &= 1 - P\left(-\frac{d}{0.05} \leq \frac{\widehat{p} - 0.5}{0.05} \leq \frac{d}{0.05}\right) \\ &= 2\left(1 - \Phi\left(\frac{d}{0.05}\right)\right) \end{aligned}$$

which gives $\Phi(d/0.05) = 0.975$, which in turn gives $d = 1.96 \times 0.05 = 0.098$. Thus, if $\widehat{p}$ is less than 0.402 or greater than 0.598, we decide that this is not because of random variation but rather that the coin is not fair. In this case, we say that we *reject the hypothesis of fairness*. In terms of the number of heads, this means that we reject the hypothesis of fairness if there are $\leq 40$ or $\geq 60$ heads. There is a possibility that such a conclusion is wrong and we know that the probability of this is 0.05. We are thus taking a 5% risk of classifying a fair coin as unfair. □

This example illustrates the main idea in *hypothesis testing*. We set up a hypothesis about an unknown parameter and test it by estimating the parameter. According to how extreme the estimate comes out, we either *reject* or *accept* the hypothesis.

Note how this is reminiscent of mathematical proof by contradiction. In such a proof, we start by making an assumption, and if we arrive at a contradiction, we conclude that the assumption was false. In hypothesis testing we make an assumption, and if we observe a very unlikely outcome, we decide that the assumption was false, realizing that there is a small risk that we are wrong.

Let us introduce some concepts and notation. Suppose that $\theta$ is an unknown parameter and we want to decide whether $\theta$ equals some specific value $\theta_0$. We then formulate the *null hypothesis* that $\theta = \theta_0$, written

$$H_0 : \theta = \theta_0$$

In conjunction with the null hypothesis, we also have an *alternative hypothesis*, denoted $H_A$. This could, for example, be

$$H_0 : \theta = \theta_0 \quad \text{versus} \quad H_A : \theta > \theta_0$$

Such an alternative hypothesis is said to be *one-sided* (as is $H_A : \theta < \theta_0$). A test against a *two-sided* alternative has the form

$$H_0 : \theta = \theta_0 \quad \text{versus} \quad H_A : \theta \neq \theta_0$$

It has to be decided in each case whether a one-sided or two-sided alternative hypothesis is reasonable. To test $H_0$ versus $H_A$, we need the following concepts.

***Definition 6.14.*** A *test statistic* $T$ is a function of the sample, used to test $H_0$. The *significance level* $\alpha$ and the *critical region* $C$ are determined such that

$$P(T \in C) = \alpha$$

under the assumption that $H_0$ is true. If $T \in C$, we *reject* $H_0$ in favor of $H_A$.

The critical region is often an interval of the form $C = (-\infty, c\,]$ or something similar. The number $c$ is then called the *critical value*. The test statistic[5] is typically based on an estimator $\widehat{\theta}$. It could be $\widehat{\theta}$ itself or some function of $\widehat{\theta}$. Two things are important for the test statistic: it indicates whether we should believe more in $H_A$ than in $H_0$ and its distribution is completely known. The significance level $\alpha$ is the risk we are willing to take to reject a hypothesis that is in fact true, and some standard values of $\alpha$ are 0.05, 0.01, and 0.001 or as percentage points, 5%, 1%, and 0.1%. If we cannot reject $H_0$, we say that we *accept* it. This does not mean that we *prove* it, it is only that the data do not support a rejection. In the coin flip example, even if we cannot reject fairness, the coin may still be unfair but to such a small extent that it is hard to detect. This is a typical problem in hypothesis testing, and we will return to it later.

In the coin flip example above, we have the null and alternative hypotheses

$$H_0 : p = 0.5 \quad \text{versus} \quad H_A : p \neq 0.5$$

Here we want a two-sided alternative since we are not interested primarily in any particular "direction of unfairness." The estimator of $p$ is $\widehat{p}$, the relative frequency. Large deviations of $\widehat{p}$ from 0.5 indicate that $H_0$ is not true. Rather than using $\widehat{p}$ directly, let us use the test statistic $T$ defined as

$$T = \frac{\widehat{p} - 0.5}{0.05}$$

which is approximately $N(0, 1)$ if $H_0$ is true and we proceed to find the critical region $C$. We should reject $H_0$ if the observed value of $T$ deviates too much from 0. The critical region should be of the form $C = (-\infty, -c\,] \cup [\,c, \infty)$, and if we choose significance level 0.05, we get

$$0.05 = P(T \in C) = 2(1 - \Phi(c))$$

which gives $\Phi(c) = 0.975$, which finally gives $c = 1.96$. Thus, we reject $H_0$ in favor of $H_A$ on the 5% level if $|T| \geq 1.96$.

[5]Generally, the term *statistic* is used for any function of a random sample. Thus, an estimator is a statistic used to estimate an unknown parameter, and we have previously introduced order statistics.

***Example 6.23.*** The IQ of a randomly chosen individual in a population is, as pointed out in Example 2.41, constructed to be normally distributed with mean 100 and standard deviation 15. Assume that a group of eight pupils have tested a new pedagogical method, designed to increase IQ, for one semester. To evaluate the method, they take an IQ test after the study period, which yields the results

$$87,\ 92,\ 97,\ 110,\ 115,\ 120,\ 121,\ 122$$

The main question is now: Does this sample provide enough evidence to conclude that the method works? Clearly, five out of the eight pupils recorded IQs above average, some of them quite far above. On the other hand, can we call the test successful when three pupils ended up below average? Since we want to draw general conclusions about the method and not just how these eight pupils performed, we can consider this as an observed sample from some underlying distribution. But which distribution? If the method does not work, that is, it does not affect the IQ, we would expect the sample to be normally distributed with parameters $\mu = 100$ and $\sigma = 15$. On the other hand, if it had been effective, we would expect that the pupils had increased their IQs by some quantity. If we assume that the individual variation is the same, we can then assume that it is normally distributed with some $\mu$ larger than 100 and $\sigma = 15$. To formalize this, we want to test the hypotheses

$$H_0 : \mu = 100 \quad \text{versus} \quad H_A : \mu > 100$$

Since we estimate $\mu$ by the sample mean $\bar{X}$, we can use that as a test statistic or, equivalently,

$$Z = \frac{\bar{X} - \mu_0}{\sigma/\sqrt{n}} = \frac{\bar{X} - 100}{15/\sqrt{8}}$$

Now, we are looking for evidence to reject $H_0$ in favor of $H_A$, so we should only reject if $Z$ is large enough. Under the assumption that $H_0$ is true, $Z \stackrel{d}{=} N(0, 1)$, which means that

$$\alpha = P(Z > c) = 1 - \Phi(c)$$

For $\alpha = 0.10$, for example, we get that $c = 1.28$. In this case, the value of the test statistic becomes

$$Z = \frac{108 - 100}{15/\sqrt{8}} = 1.51$$

which means that we can reject the null hypothesis and claim that the method indeed works. □

The IQ example illustrates a typical situation in hypothesis testing—it is often the *alternative* hypothesis that we want to prove, and we therefore set up a null hypothesis that we wish to reject in favor of the alternative. From a scientific point of view, it is easier to *falsify* a hypothesis than to prove it, and the philosophy of hypothesis testing reflects this principle.[6] If we cannot reject the null hypothesis, we say that we accept it, which does not mean that it is proven, only that we cannot show it to be false. In legal terminology, the null hypothesis is "innocent until proven guilty." The two examples illustrate the general procedure to perform a hypothesis test:

1. State the null and alternative hypotheses, $H_0$ and $H_A$.
2. Find the test statistic $T$ and decide for which type of values (large, small, positive, negative, ...) it rejects $H_0$ in favor of $H_A$. Make sure that the distribution of $T$ is completely known under the assumption that $H_0$ is true (often expressed "under $H_0$").
3. Choose a significance level $\alpha$. Find the critical region $C$ by assuming that $H_0$ is true and set $P(T \in C) = \alpha$. The general form of $C$ was determined in 2, now you get the numbers.
4. Compute $T$ and compare with the critical region. If $T \in C$, reject $H_0$ in favor of $H_A$, otherwise accept $H_0$.

Although the method of hypothesis testing seems somewhat different from calculating confidence intervals, they are in fact two sides of the same coin, which can be formalized as follows.

**Proposition 6.9 (Correspondence Theorem).** Let $X_1, \ldots, X_n$ be a random sample and $\theta$ an unknown parameter.

(i) Let $I$ be a confidence interval for $\theta$ such that $P(\theta \in I) = q$. Then $\theta_0 \notin I$ is a rejection rule for the test of $H_0 : \theta = \theta_0$ on significance level $1 - q$.

(ii) Let $C$ be a critical region for the test statistic $T$ such that $P(T \in C) = \alpha$ under the assumption that $H_0 : \theta = \theta_0$ holds. Then $I = \{\theta_0 : T \notin C\}$ is a confidence interval of $\theta$ on confidence level $1 - \alpha$.

The proof is rather straightforward and is left as an exercise.

More informally, the correspondence theorem says that a confidence interval consists of exactly those parameter values that cannot be rejected in a hypothesis test and vice versa. This means that a hypothesis test can always be carried out by first calculating a confidence interval and then checking if $\theta_0$ is included, and if not, $H_0$ can be rejected. However, it is the second part that is most useful because there are situations where it is easier to come up with a good hypothesis test than a confidence interval, so then the former can be used to construct the latter. This will be used in

[6] Recall our discussion in Example 1.33 regarding the hypothesis "All swans are white." To falsify this, we only need to observe a single black swan, but to prove it we would need to observe all swans and note that they are white. As long as all hitherto observed swans are white, we may say that we accept the hypothesis but it has not been proved.

Section 6.9 where some alternative methods of hypothesis testing will be introduced and where it is not obvious how confidence intervals for the parameters involved should be constructed.

### 6.5.1 Large Sample Tests

In Section 6.4, we introduced an approximate method of calculating confidence intervals for MLEs. The same result can also be used to obtain an approximate test procedure.

**Proposition 6.10.** Let $\widehat{\theta}$ be an MLE of the unknown parameter $\theta$ based on the large sample $X_1, \ldots, X_n$. We wish to test the null hypothesis

$$H_0 : \theta = \theta_0 \quad \text{versus} \quad H_A : \theta \neq \theta_0$$

The test statistic is

$$Z = \sqrt{nI(\theta_0)}(\widehat{\theta} - \theta_0)$$

where $I(\theta_0)$ is the Fisher information at $\theta_0$, and we reject $H_0$ on level $\approx \alpha$ if

$$|Z| \geq c$$

where $\Phi(c) = 1 - \alpha/2$.

*Proof.* Proposition 6.8 implies that

$$Z = \sqrt{nI(\theta_0)}(\widehat{\theta} - \theta_0) \overset{d}{\approx} N(0, 1)$$

for large $n$ under the condition that $H_0$ holds. The approximate significance level is then

$$P(|Z| \geq c) = P(Z \leq -c) + P(Z \geq c) \approx \Phi(-c) + 1 - \Phi(c)$$

$$= 2(1 - \Phi(c)) = \alpha$$

■

***Example 6.24.*** Let $X_1, \ldots, X_n$ be a random sample from the Poisson distribution with parameter $\lambda$. We want to test the hypotheses

$$H_0 : \lambda = \lambda_0 \quad \text{versus} \quad H_A : \lambda \neq \lambda_0$$

on the 5% level under the assumption that $n$ is large. In Example 6.16, the MLE of $\lambda$ was derived as $\widehat{\lambda} = \bar{X}$. The Fisher information for the Poisson distribution is

$$I(\lambda) = -E\left[\frac{\partial^2}{\partial\lambda^2}\log\left(e^{-\lambda}\frac{\lambda^X}{X!}\right)\right] = -E\left[-\frac{X}{\lambda^2}\right] = \frac{1}{\lambda}$$

This yields the test statistic

$$Z = \frac{\sqrt{n}(\widehat{\lambda} - \lambda_0)}{\sqrt{\lambda_0}}$$

and the rejection rule $|Z| \geq 1.96$. □

### 6.5.2 Test for an Unknown Probability

The coin flip example also fits into a general test situation. Suppose that we want to test whether an unknown probability $p$ equals a specific value $p_0$. We can then test $H_0 : p = p_0$ based on the estimator $\widehat{p}$.

**Proposition 6.11.** Let $p$ be an unknown probability, estimated by the relative frequency $\widehat{p}$ based on $n$ independent trials. We wish to test the null hypothesis

$$H_0 : p = p_0 \quad \text{versus} \quad H_A : p \neq p_0$$

The test statistic is

$$T = \frac{\widehat{p} - p_0}{\sqrt{p_0(1 - p_0)/n}}$$

and we reject $H_0$ on level $\approx \alpha$ if

$$|T| \geq c$$

where $\Phi(c) = 1 - \alpha/2$.

*Proof.* Under $H_0$, we know that

$$\widehat{p} \overset{d}{\approx} N\left(p_0, \frac{p_0(1 - p_0)}{n}\right)$$

and hence $T \overset{d}{\approx} N(0, 1)$ and

$$\alpha = P(|T| \geq c) \approx 2(1 - \Phi(c))$$

■

For one-sided alternatives, we reject if $T \geq c$ (or $T \leq -c$), where $\Phi(c) = 1 - \alpha$, in an analogy with previous results. Note that we are using the normal approximation, so the significance level can be reported to be only $\approx\alpha$. If there is a reason to suspect that the normal approximation does not work well, a test could be derived from the binomial distribution instead.

***Example 6.25.*** A company makes plastic paper clips and the manufacturing process needs to be adjusted if the probability that a clip is defective exceeds 10%. To investigate whether adjustment is needed, 500 randomly selected clips are checked and 55 of these are found to be defective. Test on the 5% level whether adjustment is needed.

The hypotheses are

$$H_0 : p = 0.1 \quad \text{versus} \quad H_A : p > 0.1$$

and we use the test statistic

$$T = \frac{\widehat{p} - 0.1}{\sqrt{0.1 \times 0.9/500}}$$

and reject if $T \geq c$, where $\Phi(c) = 0.95$ (one-sided alternative). Table A.2 gives $c = 1.64$, and since we have observed $\widehat{p} = 55/500 = 0.11$, we get $T = 0.75$ and cannot reject $H_0$. There is no need for adjustment. □

## 6.6 FURTHER TOPICS IN HYPOTHESIS TESTING

In this section, we will address some topics related to testing methodology, potential problems and errors, and quality criteria for hypothesis tests. We start with a closer examination of significance levels.

### 6.6.1 *P*-Values

The choice of significance level is quite arbitrary, although certain values have emerged as being typical. Still, it is not always satisfactory to fix $\alpha$ in advance. Let us revisit Example 6.23 about the IQ test to see an example of this. We chose $\alpha = 0.10$, which gave the critical value $c = 1.28$, and since we observed $T = 1.51$, we could reject $H_0$. Now, 1.51 is not much higher than 1.28, so we might not have been able to reject also on the level $\alpha = 0.05$. For this $\alpha$, Table A.2 gives $c = 1.64$, so we cannot indeed reject on the 5% level. There is still some room for improvement, and we realize that we can reject on any level as long as the corresponding $c$ does not exceed 1.51. With $c = 1.51$ and Table A.1, we get

$$P(Z \geq 1.51) = 1 - \Phi(1.51) = 0.066$$

which is the lowest significance level on which we can reject $H_0$ from the observed data. By stating this number instead of merely testing on some fixed significance level, we have given a measure of how strong evidence the data give against $H_0$.

***Definition 6.15.*** The $P$-value of a test is the lowest significance level on which we can reject $H_0$ for a given data set.

Thus, if we compute the $P$-value $p$, we reject on level $\alpha$ if $p \leq \alpha$. It is common to state the $P$-value when a hypothesis test is performed. This is desirable since the reader of a scientific report can in this way evaluate the strength of evidence against a hypothesis, rather than just being told that it was rejected on a particular level. The reason for using significance levels is historical; before the days of computers, values needed to be computed by hand and tabulated for easy access and in the process the standard significance levels emerged. There is nothing scary about 5% as opposed to 4.7% or 5.2% but in practice, conclusions must be drawn and actions taken, and using nice round numbers such as 1% or 5% is easier on the ear and mind.

***Example 6.26.*** Consider Example 6.25, where 55 out of 500 examined clips were defective and we could not reject $H_0$ on the 5% level. Since the estimate is 0.11, above the limit of 10% defective, there might still be some suspicion that the process is not correct. To get a measure of how bad it is, compute the $P$-value.

The observed value of the test statistic was $T = 0.75$ and since $T \sim N(0, 1)$ and the test is one-sided and rejects large values of $T$, the $P$-value is

$$p = P(T \geq 0.75) = 1 - \Phi(0.75) = 0.23$$

which is not likely small enough to arouse suspicion. □

### 6.6.2 Data Snooping

For a hypothesis test to be meaningful, it is important that the hypotheses be formulated before the data are analyzed as the following example illustrates.

***Example 6.27.*** On the Texas Lottery Web site, number frequencies for the various games are published. When I looked at the number frequencies for "Pick 3" (recall Example 1.25) during the year 2003, I noticed that the number 7 seemed to be overrepresented on Wednesdays. Out of 150 numbers drawn on a Wednesday, 22 were 7s. Are 7s overrepresented? Test on the 5% level.

Let $p$ be the probability that 7 is drawn on a Wednesday. If the drawing procedure is fair, we should have $p = 0.1$, and we test

$$H_0 : p = 0.1 \quad \text{versus} \quad H_A : p > 0.1$$

We have $\widehat{p} = 0.147$, and from Section 6.5.2 we get the test statistic

$$T = \frac{0.147 - 0.1}{\sqrt{0.1(1 - 0.1)/150}} = 1.91$$

The critical value is the value $c$ for which $\Phi(c) = 0.95$, which gives $c = 1.64$. We reject $H_0$ and conclude that Wednesdays give more 7s. □

Why are 7s overrepresented? Note that Wednesday is the third day of the week and since the numbers 3 and 7 are often assumed to have mystical powers, maybe we are onto something here.

Sorry to get your hopes up, but we are not about to prove numerology to be helpful in playing the lottery. This is a typical example of *data snooping*: to first look at the data and then formulate and test the hypothesis. The $P$-value, that is, the probability of getting at least twenty-two 7s on a Wednesday, is

$$P(T \geq 1.91) = 0.028$$

which is certainly quite low. However, the probability of getting at least twenty-two 7s on *some* day of the week is

$$1 - (1 - 0.028)^{12} = 0.29$$

(numbers are drawn twice a day Monday–Saturday). Now, this is not a very small probability. The probability that *some* number is overrepresented *some* day of the week is even larger. Thus, if we look at the lottery data, it is not at all unlikely that we find some number that is overrepresented on some day of the week, and after we have found it, we can easily prove the claim in a hypothesis test. If we look specifically for 7 and Wednesday, we can go through the last 10 years and have a 25% chance of finding at least 1 year with overrepresentation. In this case, I was lucky enough to find it the first year I looked at; now what are the odds of that...?

This example may be frivolous but the problem that it illustrates is real. If a hypothesis is formulated after examination of a data set, then it cannot be tested on the same data set. Indeed, if you keep generating various hypotheses from the same data set, eventually you will find one that you can reject, or as statisticians put it: "If you torture data long enough, it will confess." This is also related to the problem of *multiple hypothesis testing*, which we address in Section 6.6.4.

### 6.6.3 The Power of a Test

When a hypothesis test is performed, the significance level is the probability that a true null hypothesis is rejected. This is not the only error that we can commit, however; it could also happen that the null hypothesis is false and should have been rejected but that our test fails to do so. These two types of error are referred to as *type I* and *type II errors*, respectively. By choosing a low significance level, we have ensured a low probability of committing a type I error but what about the other type, to accept a false null hypothesis?

Let us consider Example 6.22, to test whether a coin is fair on the basis of 100 flips and the relative frequency $\widehat{p}$. The hypotheses are

$$H_0 : p = 0.5 \quad \text{versus} \quad H_A : p \neq 0.5$$

the test statistic is $T = (\widehat{p} - 0.5)/0.05$, and we reject $H_0$ on the 5% level if $|T| \geq 1.96$, which is to say that $\widehat{p} \leq 0.402$ or $\widehat{p} \geq 0.598$. If $H_0$ is true, the probability that we reject it in error is 0.05.

So far so good, but now suppose that $H_0$ is false and $H_A$ is true. What is the probability that we fail to reject $H_0$? This cannot be answered immediately since $H_A$ specifies a whole range of parameter values, not just one value as $H_0$ does. Thus, the probability of rejecting $H_0$ depends on "how false" it is or, more precisely, which of the values in the range of $H_A$ is the true parameter value. Let us, for example, assume that the true value of $p$ is 0.6. Then $\widehat{p}$ is approximately normal with mean 0.6 and variance $0.6(1 - 0.6)/100 = 0.0024$, and the probability of rejecting $H_0$ is

$$\begin{aligned} 1 - P(0.402 \leq \widehat{p} \leq 0.598) &= 1 - \left(\Phi\left(\frac{0.598 - 0.6}{\sqrt{0.0024}}\right) - \Phi\left(\frac{0.402 - 0.6}{\sqrt{0.0024}}\right)\right) \\ &= 1 - (\Phi(-0.04) - \Phi(-4.04)) = 0.52 \end{aligned}$$

If instead the true value is $p = 0.7$, a similar computation gives the probability to reject $H_0$ as 0.99, which is higher, since it is easier to reject $H_0$ the farther the true $p$ is from 0.5. Generally, if the true value is $p$, the probability of rejecting $H_0$ is

$$1 - \left(\Phi\left(\frac{0.598 - p}{\sqrt{p(1-p)/100}}\right) - \Phi\left(\frac{0.402 - p}{\sqrt{p(1-p)/100}}\right)\right)$$

which we note is a function of $p$. This function is called the *power function*, or simply the *power*, of the test. If we denote it by $g(p)$, we thus have $g(0.5) = 0.05$ (that is how the significance level is defined), $g(0.6) = 0.52$, and $g(0.7) = 0.99$. Let us state a general definition.

***Definition 6.16.*** Suppose that we test the null hypothesis $H_0 : \theta = \theta_0$. The function

$$g(\theta) = P\,(\text{reject } H_0 \text{ if the true parameter value is } \theta)$$

is called the *power function* of the test.

Note that $g(\theta_0) = \alpha$, the significance level. If we have a one-sided alternative, for example, $H_A : \theta > \theta_0$, we want the power function to increase as quickly as possible as soon as $\theta$ increases above $\theta_0$. Below $\theta_0$ we want the power function to take on low values. If we instead have a two-sided alternative, $H_A : \theta \neq \theta_0$, we want the power function to increase on both sides of $\theta_0$. See Figure 6.4 for an illustration.

The power function can be used to choose between tests. Suppose that we have two test procedures that both have significance level $\alpha$. If the power functions $g_1$ and $g_2$ are such that $g_1(\theta) > g_2(\theta)$ for all $\theta$ such that $H_A$ is true, we say that the first test is *more powerful* than the second and is preferable since it is more likely to detect deviations from $H_0$. If we have two test procedures, it is not necessarily the case that

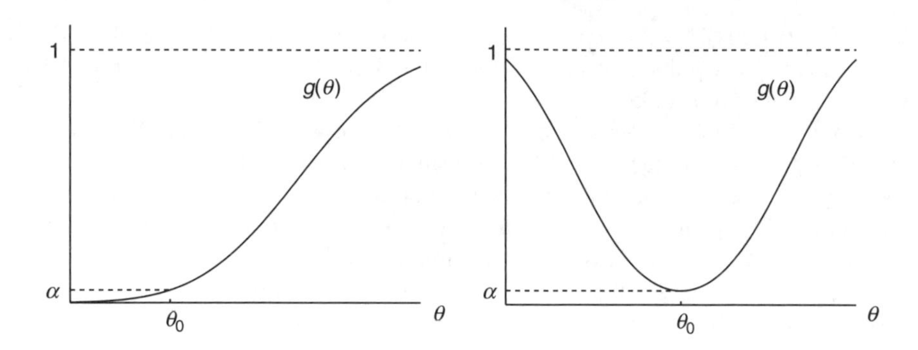


**FIGURE 6.4** Power functions for test of $H_0 : \theta = \theta_0$. To the left, the alternative is one-sided, $\theta > \theta_0$, and to the right, the alternative is two-sided.

one is more powerful than the other since we can have $g_1(\theta) > g_2(\theta)$ for some values of $\theta$ and $g_1(\theta) < g_2(\theta)$ for others (see Problem 59). In particular, if one test is more powerful than any other test, it is said to be *uniformly most powerful* and according to what we just stated, such a test does not necessarily exist. The power function can also be used to determine sample sizes (see Problem 60).

### 6.6.4 Multiple Hypothesis Testing

By carefully chosing the appropriate level of significance, we can, as pointed out in the previous section, control the risk of rejecting a true null hypothesis in a statistical hypothesis test. This works well when we have one single parameter of interest, but in many situations we typically have many different parameters and, consequently, many different hypotheses we want to test.

Let us say that we want to investigate the effect of a new drug on the human body and therefore administer the drug to a group of randomly selected individuals and monitor their well-being by measuring blood pressure, heart rate, body temperature, oxygen uptake, amount of sleep, you name it. For comparison, we measure the same variables by monitoring a control group of comparable individuals that did not get the drug or, preferably, a placebo to minimize the well-known placebo effect. The natural next step is to carry out a series of tests for each measured variable separately to investigate whether the drug had any significant effect on the body. The problem is that if we choose significance level $\alpha$ for each single test, the overall risk of rejecting at least one true null hypothesis usually becomes much larger than we want.

Let us, for example, say that we carry out 20 independent tests on level $\alpha = 0.05$, we get the multiple significance level

$$\alpha_m = 1 - P(A_1 \cap \ldots \cap A_{20}) = 1 - \prod_{i=1}^{20} P(A_i) = 1 - (1 - 0.05)^{20} = 0.64$$

where $A_i$ denotes the event of accepting the null hypothesis in test $i$. Hence, even if not one single null hypothesis is true, it is quite likely that we will still reject at least one of them and risk drawing wrong conclusions.

As long as we carry out independent tests, it is easy to choose $\alpha$ so that we get the right level of $\alpha_m$ (Problem 61), but often the tests are dependent as in the drug example above where we measure different variables but on the same individuals. However, there is a simple procedure called *Bonferroni correction* that guarantees that the multiple significance level at least does not exceed a desired value.

Assume that we want to carry out a sequence of $n$ dependent tests, each on the same significance level $\alpha$, so that the multiple significance level $\alpha_m$ does not exceed the value $\alpha'$. The inequality in Problem 10 in Chapter 1 yields that

$$\begin{aligned}\alpha_m &= 1 - P(A_1 \cap \ldots \cap A_n) = P(A_1^c \cup \ldots \cup A_n^c) \\ &\leq P(A_1^c) + \ldots + P(A_n^c) = n\alpha\end{aligned}$$

Consequently, if we choose $\alpha = \alpha'/n$, we guarantee that $\alpha_m \leq \alpha'$.

The advantage of this procedure is that it always works no matter what kind of tests we are performing, but the major disadvantage is that it is quite conservative especially for large $n$. Again, let us say that we wish to carry out 20 different test and that the multiple significance level should not exceed 5%. Then we end up with $\alpha = 0.05/20 = 0.0025$, which means that it is much more difficult to reject *any* null hypothesis, even the false ones.

Luckily, there is an extension of this method called the *Bonferroni–Holm correction* that remedies this to some extent. First, calculate the $p$-value for every single test and order them as $p_{(1)} \leq \ldots \leq p_{(n)}$. Then, reject the null hypothesis of the test with the smallest $p$-value if $p_{(1)} \leq \alpha'/n$. If this is true, move on and reject the null hypothesis of the test with the second smallest $p$-value if $p_{(2)} \leq \alpha'/(n-1)$. If this is also true, reject the null hypothesis of the test with the third smallest $p$-value if $p_{(3)} \leq \alpha'/(n-2)$, until you get to a test whose null hypothesis cannot be rejected, where you stop and accept all other null hypotheses.

## 6.7 GOODNESS OF FIT

In the previous sections, we tested hypotheses that were stated in terms of unknown parameters. To be able to do this, we had to assume that our observations came from some specific distribution, for example, normal or binomial. Sometimes, this assumption is precisely what we wish to test: whether our data actually do come from a specific distribution. In this section, we will look at one method of doing this. Let us start with an example.

***Example 6.28.*** A transmitter sends 0s and 1s and is supposed to do so such that 1s are twice as likely as 0s. It has been observed that out of 1500 independently transmitted bits, 470 were 0s and 1030 were 1s. Does this support the claim that 1s are twice as likely as 0s?

We can describe the null hypothesis as a hypothesis about an entire distribution as

$$H_0 : \text{ the distribution is } \left(\frac{1}{3}, \frac{2}{3}\right) \text{ on } \{0, 1\}$$

against the alternative that $H_0$ is not true. Let us start by arguing intuitively. Suppose that we have observed $X$ 0s and $Y$ 1s. We should then reject $H_0$ if the observed frequencies $X$ and $Y$ deviate too much from the expected frequencies of 500 and 1000, respectively. The deviation of $X$ from 500 can be measured by the squared difference

$$(X - 500)^2$$

but since we would expect this to be quite large even by pure chance because the number of trials is large, we account for this by also dividing by the expected number, 500. Treating $Y$ in a similar manner suggests the summary measure $Z^2$ of the deviation, defined as

$$Z^2 = \frac{(X - 500)^2}{500} + \frac{(Y - 1000)^2}{1000}$$

(we will soon see why we denote it as a square). We should now reject $H_0$ for large values of $Z^2$ and need to figure out its distribution. Since $Y = 1500 - X$, we get

$$(Y - 1000)^2 = (1500 - X - 1000)^2 = (X - 500)^2$$

which gives

$$Z^2 = \frac{(X - 500)^2}{500} + \frac{(X - 500)^2}{1000} = \frac{(X - 500)^2}{1000/3}$$

and since $X \sim \text{bin}(1500, 1/3)$, we know that

$$X \stackrel{d}{\approx} N\left(500, \frac{1000}{3}\right)$$

and hence the quantity $Z^2$ is the square of a random variable that is approximately $N(0, 1)$. Thus, we reject $H_0$ on level $\alpha$ if $Z^2 \geq x$ where $\Phi(\sqrt{x}) = 1 - \alpha$. With $\alpha = 0.05$, Table A.2 gives $x = 1.96^2 = 3.84$. Our observed value of $Z^2$ is

$$\frac{(470 - 500)^2}{500} + \frac{(1030 - 1000)^2}{1000} = 2.70$$

and since this is less than 3.84, we cannot reject $H_0$. The observed values of (470, 1030) are close enough to the expected (500, 1000) to fit the suggested distribution well. □

Note how the test statistic in this example is the sum of two squared random variables but how we can rewrite it as the square of one random variable, which is approximately $N(0, 1)$. This illustrates a general principle and result. Suppose that our observations are such that there are $r$ possible different outcomes (which need not necessarily be numerical) and that we wish to test for a particular distribution. We calculate the squared differences between observed and expected frequencies and divide by the expected frequencies for each of the $r$ outcomes. When we add these together, the sum can be rewritten as the sum of the squares of $r - 1$ independent $N(0, 1)$ variables. The general proof is beyond the scope of this text, but we state the result.

**Proposition 6.12.** Suppose that in an experiment, an observation can fall into any of the $r$ different categories, with probabilities $p_1, \ldots, p_r$. If the experiment is repeated $n$ times and $X_k$ denotes the number of observations that fall into category $k$, for $k = 1, \ldots, r$, then

$$\sum_{k=1}^{r} \frac{(X_k - np_k)^2}{np_k} \stackrel{d}{\approx} \chi^2_{r-1}$$

an approximate chi-square distribution with $r - 1$ degrees of freedom.[7]

Note that the random vector $(X_1, \ldots, X_r)$ has a multinomial distribution with parameters $(n, p_1, \ldots, p_r)$, and a more formal statement of the proposition would be an asymptotic result regarding such a distribution. The random variable in the proposition is often denoted $\chi^2$ and written in the alternative form

$$\chi^2 = \sum_{k=1}^{r} \frac{(O_k - E_k)^2}{E_k} \tag{6.2}$$

where "$O$" and "$E$" stand for "observed" and "expected," respectively. As a rule of thumb, the approximation is valid whenever $E_k \geq 5$ for all $k$. As indicated above, our interest in the proposition is to test the hypothesis of some specified distribution, and next we state the test.

**Corollary 6.2.** *To test the null hypothesis*

$$H_0: \text{ the distribution is } (p_1, \ldots, p_r)$$

*against the alternative that $H_0$ is not true, we use the test statistic $\chi^2$ above and reject $H_0$ on level $\alpha$ if*

$$\chi^2 \geq x$$

*where $F_{\chi^2_{r-1}}(x) = 1 - \alpha$.*

[7]The chi-square distribution will be defined and treated in more detail in Chapter 7. For the present exposition, it suffices to know that critical values can be obtained from Table A.4.

This is one example of the so-called *goodness-of-fit test*, where we test whether a proposed distribution fits the observed data well. The intent is not necessarily to reject the null hypothesis, but we may also wish to argue that a particular distribution is reasonable. The test is one-sided since we reject the null hypothesis for large deviations from what is expected. Sometimes, however, it can be argued that we should also reject for small values, since these could indicate that the fit is "too good to be true" and that data may have been manipulated to fit the desired distribution.[8] We will not address this in the text, but see the Problems section.

***Example 6.29.*** Recall that in the lottery game "Pick 3," the winning three-digit number is chosen by picking at random three times from 0, 1, ..., 9. The following tabulation shows the observed and expected number frequency for 2070 drawn numbers. Is the drawing procedure fair? Test on the 5% level.

| Number | 0 | 1 | 2 | 3 | 4 | 5 | 6 | 7 | 8 | 9 |
|---|---|---|---|---|---|---|---|---|---|---|
| Observed | 182 | 201 | 211 | 184 | 212 | 199 | 209 | 241 | 214 | 217 |
| Expected | 207 | 207 | 207 | 207 | 207 | 207 | 207 | 207 | 207 | 207 |

If the drawing procedure is fair, each integer should have probability $1/10$, and our null hypothesis is

$$H_0 : \text{ the distribution is } \left(\frac{1}{10}, \ldots, \frac{1}{10}\right)$$

and the chi-square statistic

$$\chi^2 = \sum_{k=1}^{10} \frac{(O_k - E_k)^2}{E_k} = \frac{(182 - 207)^2}{207} + \cdots + \frac{(217 - 207)^2}{207} = 12.6$$

which we compare with the value $x$, which is such that $F_{\chi^2_9}(x) = 0.95$, giving $x = 19.92$, and we cannot reject $H_0$. The $P$-value is $P(\chi^2 \geq 12.6) = 1 - F_{\chi^2_9}(12.6)$ that can be computed in Matlab as "1-cdf( 'chi2', 12.6, 9)" and equals 0.18; thus, the drawing procedure shows no signs of unfairness (but see Problem 63). □

If the expected frequency in a class is too small, this can be fixed by lumping classes together. For instance, if we have only 30 observations in the Texas Lottery example above, the expected frequencies are 3, which is below the number 5 in our rule of thumb. If we lump the classes together two by two, we get the five classes $\{0, 1\}, \{2, 3\}, \ldots, \{8, 9\}$, each with expected frequency 6, so we can go ahead and

[8]A famous example is Gregor Mendel's experiments with garden peas, which led him to discover the fundamental laws of genetics. His data fit his hypotheses so well that foul play seems hard to rule out. It has been suggested that an overzealous gardener who knew what results Mendel wanted, manipulated the data.

do the goodness-of-fit test on these classes and the resulting chi-square statistic has 4 degrees of freedom. This idea of lumping classes together makes it possible to do goodness-of-fit tests also for distributions with infinite range. Let us look at an example.

***Example 6.30.*** A "great earthquake" is defined as one that has magnitude $\geq 8.0$ on the Richter scale. Suppose that it has been claimed that the number of great earthquakes worldwide is on average 1.5 per year. The tabulation below gives the numbers of great earthquakes between 1969 and 2001. Do the data support the claim? Test on the 5% significance level.

| Number of earthquakes | 0 | 1 | 2 | 3 | 4 | 5 | ... |
|---|---|---|---|---|---|---|---|
| Number of years | 15 | 13 | 4 | 1 | 0 | 0 | ... |

Thus, there were 15 years with no earthquakes, 13 years with one earthquake, and so on. No year had more than three earthquakes. If we assume that earthquakes occur according to a Poisson process, the claim is that the number of earthquakes in a given year is Poi(1.5). Let us thus do a goodness-of-fit test of the null hypothesis

$$H_0 : \text{ data come from a Poi(1.5) distribution}$$

against the alternative that $H_0$ is false. Since the range is infinite, we have to lump classes together, so let us look at the expected frequencies. The probabilities are

$$p_k = e^{-1.5}\frac{1.5^k}{k!}, \quad k = 0, 1, \ldots$$

and with $n = 33$, we get the following expected frequencies

$$E_0 = 7.4, \quad E_1 = 11.0, \quad E_2 = 8.3, \quad E_3 = 4.1, \quad E_4 = 1.6$$

and there is a rapid decline as we continue. To get expected frequencies that are at least 5, we create a new class, "$\geq 3$." If $X \sim$ Poi(1.5), this new class has an expected frequency

$$nP(X \geq 3) = n(1 - P(X \leq 2)) = 33(1 - (p_0 + p_1 + p_2)) = 6.3$$

The four classes with their observed and expected frequencies are

| Class | 0 | 1 | 2 | $\geq 3$ |
|---|---|---|---|---|
| Observed | 15 | 13 | 4 | 1 |
| Expected | 7.4 | 11.0 | 8.3 | 6.3 |

The chi-square statistic is $\chi^2 = 14.9$ that we compare with the number $x$, which is such that $F_{\chi^2_3}(x) = 0.95$, and a look at Table A.4 reveals that $x = 7.82$, and hence we reject $H_0$ on the 5% level. The $P$-value is

$$P(\chi^2 \geq 14.9) = 1 - F_3(14.9) = 0.002$$

and thus the fit to a Poisson distribution with mean 1.5 is poor. □

In the last example, we rejected the hypothesis that the data came from a Poi(1.5) distribution. Since the alternative in a goodness-of-fit test is simply "$H_0$ is false," this means that it could be the claim of a Poisson distribution that was rejected or only that it did not have the right mean. Suppose that we want to test whether the data come from *some* Poisson distribution but not with a specified mean.

The idea is to simultaneously estimate the mean $\lambda$ from the data and test $H_0$ with the estimated values of the probabilities (which we recall are all functions of $\lambda$). We can think of this as testing the Poisson distribution that fits the data the best, and if this is rejected, so is any other Poisson distribution. Here is the general result.

**Proposition 6.13.** Under the assumptions of Proposition 6.12, suppose that the probabilities depend on an unknown parameter, say, $p_1(\theta), \ldots, p_r(\theta)$. If $\widehat{\theta}$ is the MLE of $\theta$ and the $p_k(\theta)$ satisfy certain technical conditions, then

$$\sum_{k=1}^{r} \frac{(X_k - np_k(\widehat{\theta}))^2}{np_k(\widehat{\theta})} \stackrel{d}{\approx} \chi^2_{r-2}$$

a chi-square distribution with $r - 2$ degrees of freedom.

The technical conditions are assumptions of differentiability of the $p_k(\theta)$, which are always satisfied in our examples. A more general result is that if there are $j$ unknown parameters, we replace them by their MLEs and the resulting distribution is chi-square with $r - 1 - j$ degrees of freedom. Thus, we lose one degree of freedom for each estimated parameter. In Equation (6.2), the $E_k$ are now *estimated* expected frequencies, $E_k = np_k(\widehat{\theta})$, and for the goodness-of-fit test, we should make sure that all the $np_k(\widehat{\theta}) \geq 5$. We should also point out that it is not necessary to use the MLEs; the result is true for any estimators that satisfy certain asymptotic properties. We will, however, stick to MLEs.

***Example 6.31.*** In the earthquake example above, test whether the data come from a Poisson distribution.

The pmf for a Poisson distribution is

$$p_k(\lambda) = e^{-\lambda} \frac{\lambda^k}{k!}$$

where we need to estimate $\lambda$. From Section 6.4 we know that the MLE of $\lambda$ is $\widehat{\lambda} = \bar{X}$. We have the observed value

$$\widehat{\lambda} = \frac{0 \times 15 + 1 \times 13 + 2 \times 4 + 3 \times 1}{33} = 0.73$$

which gives the estimated probabilities

$$p_k(\widehat{\lambda}) = e^{-0.73} \frac{0.73^k}{k!}, \quad k = 0, 1, 2, \ldots$$

and estimated expected frequencies

$$E_k = 33\, e^{-0.73} \frac{0.73^k}{k!}, \quad k = 0, 1, 2, \ldots$$

Again, we need to make sure to define classes such that the $E_k$ are at least 5 and computation gives the following classes and numbers.

| Class | 0 | 1 | $\geq 2$ |
|---|---|---|---|
| Observed | 15 | 13 | 5 |
| Expected | 15.9 | 11.6 | 5.5 |

This time, the chi-square statistic is

$$\chi^2 = 0.27$$

which we need to compare with $x$ which is such that $F_{\chi_1^2}(x) = 0.95$ since there are now three classes and one estimated parameter and hence $3 - 1 - 1 = 1$ degree of freedom. Table A.4 gives $x = 3.84$, and we cannot reject $H_0$. The $P$-value is $1 - F_1(0.27) = 0.60$, which means that we cannot reject on any reasonable level. The fit to a Poisson distribution is good, so the rejection in Example 6.30 was due to the fact that the suggested mean was wrong.

This is a good place to remind the reader that accepting a null hypothesis is not the same as proving it. Even if the fit to a Poisson distribution is good, the true distribution may still be something else, but the difference is too subtle to detect. All we can conclude is that it is not unreasonable to assume a Poisson distribution. It gives a good fit and unless it is rejected, we can keep using it as a working model. See Problem 66 for more. □

In the previous example, it is important to note that we do *not* test the null hypothesis that the data come from a Poisson distribution *with mean* 0.73, only that they come from *some* Poisson distribution. The estimation of $\lambda$ is *part of the test*. If we first estimate $\lambda$ by 0.73 and then test $H_0$ : Poi(0.73) with Proposition 6.2, we are guilty of data snooping since we let the data generate our hypothesis. But since we still need the estimate of $\lambda$ to do the test, what is the difference? One degree of freedom! The fact that the estimation is part of the test gives us $r - 2$ degrees of freedom instead of $r - 1$, and this makes the null hypothesis a little easier to reject. Thus, the proper test takes into account the fact that there is variability also in the estimation, whereas data snooping would make the fit look better than it really is.

The goodness-of-fit test can be used also to test for continuous distribution. All we have to do is to divide the range into classes, compute the probability of each class, and go on as before. See Problem 70 for an example.

### 6.7.1 Goodness-of-Fit Test for Independence

A special case of the goodness-of-fit test from the precious section is to test for independence between certain characteristics. If we consider two characteristics, $A$ and $B$, we can think of these as events that may occur when an object is sampled at random. There are then four possibilities: $A \cap B$, $A \cap B^c$, $A^c \cap B$, and $A^c \cap B^c$. Our null hypothesis is

$$H_0 : A \text{ and } B \text{ are independent}$$

and if we let $p = P(A)$ and $q = P(B)$, $H_0$ specifies the probabilities

| Category | $A \cap B$ | $A \cap B^c$ | $A^c \cap B$ | $A^c \cap B^c$ |
|---|---|---|---|---|
| Probability | $pq$ | $p(1-q)$ | $(1-p)q$ | $(1-p)(1-q)$ |

If $p$ and $q$ are known, we have the situation from the previous section with $r = 4$ and can do a chi-square test based on observed frequencies in the four categories. More commonly, however, the probabilities are not known and must be estimated from the data. According to the comments following Proposition 6.13, the degrees of freedom will be reduced by one for each estimated parameter. Here, we need to estimate two parameters, $p$ and $q$, and get $4 - 1 - 2 = 1$ degree of freedom. Suppose that we examine $n$ objects and classify them into each of the four categories and that we get the numbers $X_{11}, X_{12}, X_{21}$, and $X_{22}$, respectively. It is practical to display these types of data in a *contingency table*:

| | $B$ | $B^c$ |
|---|---|---|
| $A$ | $X_{11}$ | $X_{12}$ |
| $A^c$ | $X_{21}$ | $X_{22}$ |

The estimators of $p$ and $q$ are the relative frequencies

$$\widehat{p} = \frac{X_{11} + X_{12}}{n} \quad \text{and} \quad \widehat{q} = \frac{X_{11} + X_{21}}{n}$$

and the chi-square statistic

$$\chi^2 = \frac{(X_{11} - n\widehat{p}\widehat{q})^2}{n\widehat{p}\widehat{q}} + \frac{(X_{12} - n\widehat{p}(1-\widehat{q}))^2}{n\widehat{p}(1-\widehat{q})} + \frac{(X_{21} - n(1-\widehat{p})\widehat{q})^2}{n(1-\widehat{p})\widehat{q}} + \frac{(X_{22} - n(1-\widehat{p})(1-\widehat{q}))^2}{n(1-\widehat{p})(1-\widehat{q}))}$$

which has a $\chi_1^2$ distribution.

***Example 6.32.*** Recall the Berkeley admissions data from Example 1.46. Below is a contingency table of the numbers of male and female applicants in the categories "easy" and "difficult," respectively. Is choice of major independent of gender? Test on the 5% level.

| | Male | Female |
|---|---|---|
| Easy major | 1385 | 133 |
| Difficult major | 1306 | 1702 |

We have $n = 4526$, and the observed numbers are $X_{11} = 1385$, $X_{12} = 133$, $X_{21} = 1306$, and $X_{22} = 1702$. This gives the estimates

$$\widehat{p} = \frac{1385 + 133}{4526} = 0.34 \quad \text{and} \quad \widehat{q} = \frac{1385 + 1306}{4526} = 0.59$$

and the chi-square statistic

$$\chi^2 = \frac{(1385 - 4526 \times 0.34 \times 0.59)^2}{4526 \times 0.34 \times 0.59} + \frac{(133 - 4526 \times 0.34 \times 0.41)^2}{4526 \times 0.34 \times 0.41} + \frac{(1306 - 4526 \times 0.66 \times 0.59)^2}{4526 \times 0.66 \times 0.59} + \frac{(1702 - 4526 \times 0.66 \times 0.41)^2}{4526 \times 0.66 \times 0.41}$$

$$= 957.1$$

which we compare with $x$, which is such that $F_{\chi_1^2}(x) = 0.95$, giving $x = 3.84$, and we can reject independence. Since our observed value is huge, it would be interesting to compute the $P$-value. Recall the relation between the chi-square and the normal distributions to realize that our chi-square statistic is the square of a $N(0, 1)$ variable. Thus, for any $x$

$$P(\chi^2 \geq x) = 1 - P(-\sqrt{x} \leq \chi \leq \sqrt{x}) = 2(1 - \Phi(\sqrt{x}))$$

and by the approximation at the end of Section 2.7, we get

$$P(\chi^2 \geq x) \approx 2\varphi(\sqrt{x})/\sqrt{x}$$

which, with $x = 957.1$, equals $3.8 \times 10^{-210}$! Quite impressive evidence against independence.

□

The characteristics do not necessarily have to be binary (male/female, easy/difficult, etc.). Suppose that the first characteristic defines the $n_1$ categories $A_1, A_2, \ldots, A_{n_1}$ with corresponding probabilities $p_1, \ldots, p_{n_1}$ and the second characteristic defines the $n_2$ categories $B_1, B_2, \ldots, B_{n_2}$ with probabilities $q_1, \ldots, q_{n_2}$. There are then in all $n_1 n_2$ categories, and we still wish to test for independence between the two characteristics, and hence our null hypothesis is

$$H_0 : A_i \text{ and } B_j \text{ are independent for all } i \text{ and } j$$

Suppose that we have a total of $n$ classified objects; let the observed number in category $(i, j)$ be $X_{ij}$, so that the contingency table has $n_1$ rows and $n_2$ columns. Under $H_0$, the expected number of objects in category $(i, j)$ is $np_i q_j$. Assuming that the $p_i$ and $q_j$ are unknown, we estimate them by relative frequencies as

$$\widehat{p}_i = \frac{1}{n}\sum_{j=1}^{n_2} X_{ij} \quad \text{and} \quad \widehat{q}_j = \frac{1}{n}\sum_{i=1}^{n_1} X_{ij}$$

and define the chi-square statistic

$$\chi^2 = \sum_{i=1}^{n_1}\sum_{j=1}^{n_2} \frac{(X_{ij} - n\widehat{p}_i\widehat{q}_j)^2}{n\widehat{p}_i\widehat{q}_j}$$

and need to figure out how many degrees of freedom it has. In the notation of the previous section, we have $r = n_1 n_2$, and we need to estimate the $p_i$ and $q_j$. Since each set of probabilities adds to one, we have to estimate only $n_1 - 1$ of the $p_i$ and $n_2 - 1$ of the $q_j$. Thus, the resulting chi-square distribution has

$$n_1 n_2 - 1 - (n_1 - 1) - (n_2 - 1) = (n_1 - 1)(n_2 - 1)$$

degrees of freedom. We may need to lump classes together to make sure that $n\widehat{p}_i\widehat{q}_j$ are $\geq 5$ for all $i$, $j$ and also make sure that this is done in a sensible manner.

***Example 6.33.*** The following contingency table shows the classification of 1031 blood donors, according to blood type (A, B, AB, or O) and according to Rh factor (Rh+ or Rh−). Are blood type and Rh factor independent? Test on level 5%.

| | A | B | AB | O |
|---|---|---|---|---|
| Rh+ | 320 | 96 | 40 | 412 |
| Rh− | 66 | 23 | 9 | 65 |

The estimated probabilities are

$$\widehat{p}_1 = \frac{320 + 96 + 40 + 412}{1031} = 0.84, \quad \widehat{p}_2 = 1 - \widehat{p}_1 = 0.16$$

$$\widehat{q}_1 = \frac{320 + 66}{1031} = 0.37, \qquad \widehat{q}_2 = \frac{96 + 23}{1031} = 0.12$$

$$\widehat{q}_3 = \frac{40 + 9}{1031} = 0.05, \qquad \widehat{q}_4 = 1 - (\widehat{q}_1 + \widehat{q}_2 + \widehat{q}_3) = 0.46$$

and the chi-square statistic becomes

$$\chi^2 = 3.54$$

We have $n_1 = 4$ and $n_2 = 2$, which gives the number of degrees of freedom as $(n_1 - 1)(n_2 - 1) = 3$. The critical value is thus the $x$, which has $F_{\chi_3^2}(x) = 0.95$, giving $x = 7.8$, so we cannot reject independence. □

### 6.7.2 Fisher's Exact Test

As long as we have sufficient number observations, the test above is quite reliable, but what can we do if the rule of thumb $n\widehat{p}_i\widehat{q}_j \geq 5$ cannot be satisfied even after lumping classes together? For $2 \times 2$ tables, there is an alternative approach called *Fisher's exact test.*

First, let us introduce the convenient notation $X_{i\cdot} = X_{i1} + X_{i2}$, where $i = 1, 2$, for the row sums and $X_{\cdot j} = X_{1j} + X_{2j}$, where $j = 1, 2$, for the column sums in the contingency table

| | $B$ | $B^c$ |
|---|---|---|
| $A$ | $X_{11}$ | $X_{12}$ |
| $A^c$ | $X_{21}$ | $X_{22}$ |

Next, we condition on the observed row and column sums, that is, the events $\{X_{1\cdot} = x_{1\cdot}\}$ and $\{X_{\cdot 1} = x_{\cdot 1}\}$. Under the null hypothesis that the two classifications are independent, we can interpret the variable $X_{11}$ as the number of objects with property $A$ we get when randomly selecting $x_{\cdot 1}$ objects from a set of $n$ objects. This means that $X_{11}$ follows the hypergeometric distribution introduced in Section 2.5.5 with parameters $n$, $x_{1\cdot}$, and $x_{\cdot 1}$ and we get the probability mass function

$$P(X_{11} = x_{11}) = \frac{\binom{x_{1\cdot}}{x_{11}}\binom{n - x_{1\cdot}}{x_{\cdot 1} - x_{11}}}{\binom{n}{x_{\cdot 1}}} = \frac{\binom{x_{1\cdot}}{x_{11}}\binom{x_{2\cdot}}{x_{21}}}{\binom{n}{x_{\cdot 1}}} \tag{6.3}$$

Note that when $X_{11} = x_{11}$ the rest of the table is determined automatically.

If the null hypothesis is false, we get an unbalanced table with an unusually small or unusually large value of $X_{11}$. Hence, $X_{11}$ can be used as a test statistic in a test of independence and the hypergeometric distribution can be used to obtain critical values.

***Example 6.34.*** Crohn's disease is a serious chronic inflammatory disease of the intestines that may cause severe symptoms like abdominal pain, diarrhea, vomiting, and weight loss. It is believed that specialized diets may mitigate symptoms, so 20 hospitalized patients were randomly assigned to two groups of 10 patients each where one group (control) had a regular diet and the other group (treatment) were given a diet without some of the food items that were believed to aggravate the disease. After 6 months it was found that 7 of the 10 in the treatment group remained in remission compared to none in the control group, which was reduced to 8 since two patients had to undergo surgery. We can summarize the trial in the following table:

| | Remission | No Remission |
|---|---|---|
| Treatment | 7 | 3 |
| Control | 0 | 8 |

If the two diets do not affect the chance of remission, the number of patients in the treatment group that would be in remission follows the hypergeometric distribution with parameters 18, 7, and 10. The probability mass function is given in Figure 6.5.

In order to find appropriate critical values, we have to calculate the probabilities of the most extreme outcomes. Using Equation (6.3) with these parameter values

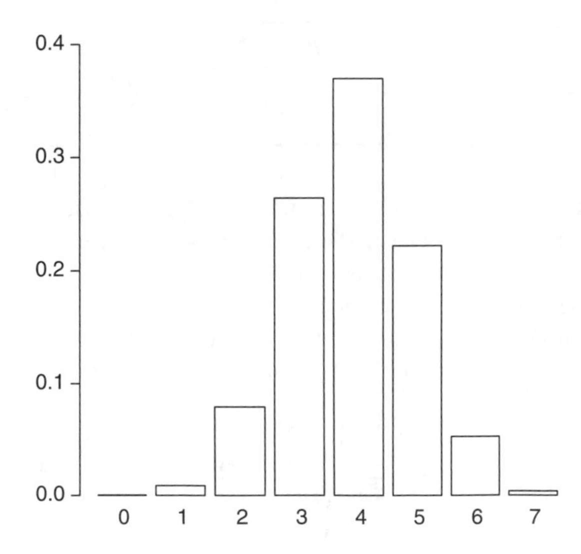


**FIGURE 6.5** The probability mass function of $X_{11}$ in Example 6.34.

yields that $P(X_{11} = 0) = 0.0003$, $P(X_{11} = 1) = 0.0088$, and $P(X_{11} = 7) = 0.0038$. A rejection region consisting of the values 0, 1, and 7 thus gives us a confidence level of $\alpha = 0.013$. Including the value with the next smallest probability (6) would add 0.053 to $\alpha$ and produce a significance level larger than 5%. In any case, since our observed value falls into the rejection region, we can reject the hypothesis that diet and remission are independent and conclude that the specialized diet indeed affects the chance of remission of patients suffering from Crohn's disease. □

## 6.8 BAYESIAN STATISTICS

The field of *Bayesian statistics* has a starting point that differs from those of the methods we have encountered so far. It is inspired by Bayes' formula, which in its simplest form (Corollary 1.7) says that

$$P(B|A) = \frac{P(A|B)P(B)}{P(A)}$$

As an example, consider Example 1.50, where there are $n$ equally suspected potential murderers. In particular, Mr Bloggs is considered to be guilty with probability $P(G) = \frac{1}{n}$. After he has been screened and found to have the murderer's genotype, the probability of his guilt is updated to

$$P(G \mid \text{same genotype}) = \frac{1}{1 + (n-1)p}$$

Now, either Mr Bloggs is guilty or he is not, and this fact does not change after his genotype has been found, so the probabilities reflect our belief in his guilt before and after screening. One way to think about this is that Mr Bloggs comes with a "guilt parameter" that is either 0 for innocent or 1 for guilty. We can describe our initial degree of belief in his guilt by assigning probabilities to the two outcomes 0 and 1. As new evidence is gathered, we update these probabilities.

The fundamental idea in Bayesian statistics is similar. By interpreting probability as a measure of the degree of belief, we can view an unknown parameter as a *random variable* instead of an unknown constant. We can then assign it a distribution that describes how likely we think different parameter values are, and after we have gathered the data, we update the distribution of the parameter to the conditional distribution given the data, using some variant of Bayes' formula. The methods we have previously developed, where parameters are viewed as fixed but unknown constants, are often referred to as *frequentist statistics*.[9]

[9] Or "classical statistics," which is somewhat ironic since these methods were developed mainly in the twentieth century, whereas early Bayesian methods, as we will soon see, were used by Laplace 100 years earlier. However, modern Bayesian methods have been proven useful relatively only recently since they often involve substantial computational problems.

***Example 6.35.*** A digital communication system transmits 0s and 1s where the probability that 1 is sent is $p$, which is either $\frac{1}{3}$ or $\frac{2}{3}$. The four observed values are $1, 1, 0, 1$. In Example 6.13, we viewed $p$ as an unknown parameter to be estimated from these observations. We will now apply the Bayesian reasoning.

We view $p$ as a random variable with range $\{\frac{1}{3}, \frac{2}{3}\}$. Thus, $p$ has a distribution and assuming that before the experiment we have no reason to believe more in any particular one of the values, it is natural to assume that $p$ is uniform on its range. Thus,

$$P\left(p = \frac{1}{3}\right) = \frac{1}{2} \quad \text{and} \quad P\left(p = \frac{2}{3}\right) = \frac{1}{2}$$

We now observe $1, 1, 0, 1$ and wish to update our distribution. This means that we compute the conditional distribution of $p$ given the observations. With $D$ denoting "data" (i.e., $1, 1, 0, 1$), Bayes' formula gives

$$\begin{aligned} P\left(p = \frac{1}{3}\,\middle|\, D\right) &= \frac{P(D|p = 1/3)P(p = 1/3)}{P(D|p = 1/3)P(p = 1/3) + P(D|p = 2/3)P(p = 2/3)} \\ &= \frac{(1/3)^3(2/3)(1/2)}{(1/3)^3(2/3)(1/2) + (2/3)^3(1/3)(1/2)} = \frac{1}{5} \end{aligned}$$

which also gives $P(p = \frac{2}{3}|D) = \frac{4}{5}$. Thus, the new distribution is $(\frac{1}{5}, \frac{4}{5})$ on the range $\{\frac{1}{3}, \frac{2}{3}\}$. The data made us believe more in $\frac{2}{3}$ than in $\frac{1}{3}$, but the latter is not ruled out. The uniform distribution we started with, before we got any data, is called the *prior distribution* (or *prior* for short) and the conditional distribution after we get the data is called the *posterior distribution.* If we were to collect more data, it would make sense to use the distribution $(\frac{1}{5}, \frac{4}{5})$ as the new prior and thus take previous measurements into account. □

The use of Bayes' formula is what gives this methodology its name. Generally, if $\theta$ is the parameter and we denote the data by $D$, we have

$$P(\theta|D) = \frac{P(D|\theta)P(\theta)}{P(D)}$$

Here $P(\theta)$ is the prior distribution, that is, the probability distribution before the experiment and $P(\theta|D)$ the posterior distribution, that is, the distribution conditioned on the data. The other probabilities are $P(D|\theta)$, the probability of the data if the parameter value is $\theta$ and $P(D)$, the unconditional probability of the data. To compute $P(D)$, we sum or integrate over $\theta$ in the numerator, depending on whether $\theta$ is discrete or continuous.

Note that we use $\theta$ to denote both the random variable and its value. Although not strictly in line with how we usually formulate probability statements, this is convenient and is also the standard notation.

In Bayesian analysis, the posterior distribution contains all the relevant information about the parameters. If we want to summarize the information with a single number, we may, for example, use the mean in the posterior distribution. Thus, computing the posterior mean can be regarded as the Bayesian equivalence of point estimation.

***Definition 6.17.*** Let $\theta$ be a parameter and $D$ a set of data. The posterior mean $E[\theta|D]$ is then called a *Bayes estimator* of $\theta$.

This is not the only way to define a Bayes estimator but the only one that we will consider. As before, we refer to this as an *estimator* if data are described in terms of random variables and an *estimate* if data are actual numerical observations on these random variables.

Note that "estimator of $\theta$" does not have the same meaning in the Bayesian as in the frequentist setting. In frequentist theory, an estimator is a random variable used to approximate the unknown constant $\theta$; in Bayesian theory it is the mean of the random variable $\theta$ conditioned on data.

***Example 6.36.*** Reconsider the previous example but suppose that we know nothing at all about $p$. Suggest a prior, then find the posterior distribution and the Bayes estimate of $p$.

If we do not know anything about $p$, we would tend to let the prior be uniform on $[0, 1]$, that is, let $p$ have pdf

$$f(p) = 1, \quad 0 \leq p \leq 1$$

and with the same data as above we get

$$f(p|D) = \frac{P(D|p)f(p)}{P(D)}$$

Here $p$ is a continuous random variable and the observations come from a discrete distribution. Thus, $P(D|p)$ and $P(D)$ are actual probabilities, and we get

$$P(D|p) = p^3(1-p)$$

and

$$P(D) = \int_0^1 P(D|p)f(p)dp = \int_0^1 p^3(1-p)dp = \frac{1}{20}$$

which gives the posterior pdf

$$f(p|D) = 20p^3(1-p), \quad 0 \leq p \leq 1$$

The mean in this distribution is

$$E[p] = \int_0^1 pf(p|D)dp = 20\int_0^1 (p^4 - p^5)dp = \frac{2}{3}$$

which is the Bayes estimate. Compare this with the frequentist estimate of the unknown parameter $p$, which is the observed relative frequency $\frac{3}{4}$. The Bayes estimate is smaller since it can be thought of as a weighted average of the prior mean, which is $\frac{1}{2}$ and the sample mean $\frac{3}{4}$. For more on this, see Problems 77 and 78. □

***Example 6.37*** **(*Laplace's Rule of Succession*).** Consider an experiment where an event occurs with unknown probability $p$. The experiment is repeated $n$ times, and the event is observed every time. Assume a uniform prior and find the posterior distribution and Bayes estimate.

With $p \sim \text{unif}[0, 1]$, we have $f(p) = 1$, $P(D|p) = p^n$, and $P(D) = \frac{1}{n+1}$, which gives posterior distribution

$$f(p|D) = \frac{P(D|p)f(p)}{P(D)} = (n+1)p^n, \quad 0 \le p \le 1$$

The mean in this distribution is

$$E[p|D] = \int_0^1 pf(p|D)dp = (n+1)\int_0^1 p^{n+1}dp = \frac{n+1}{n+2}$$

When Laplace computed this value, he interpreted it as the probability that something that has always occurred will occur once more; hence the expression "Laplace's rule of succession." The example he chose as an illustration was the event that the sun will rise again tomorrow, knowing that it has risen every day so far. Probably realizing that this was not the best chosen example, he quickly added that the probability would be much higher for "he who, seeing the principle regulating the days and seasons, realizes that nothing at the present moment can arrest the course of it." In probability terms, consecutive observations regarding sunrise are not independent. □

The uniform prior in this example is reasonable if we have no prior knowledge of $p$ whatsoever. In practice, it is likely that we know at least something. We might have previous data from some similar experiment or other information that make some values of $p$ more likely than others. A useful distribution on [0, 1] is the following.

***Definition 6.18.*** If $X$ has pdf

$$f(x) = \frac{1}{B(\alpha, \beta)}x^{\alpha-1}(1-x)^{\beta-1}, \quad 0 \le x \le 1$$

it is said to have a *beta distribution* with nonnegative parameters $\alpha$ and $\beta$.

The function $B(\alpha, \beta)$ is the *beta function*, defined as

$$B(\alpha, \beta) = \int_0^1 x^{\alpha-1}(1-x)^{\beta-1}dx$$

which for integer values of $\alpha$ and $\beta$ equals

$$B(\alpha, \beta) = \frac{(\alpha - 1)!(\beta - 1)!}{(\alpha + \beta - 1)!}$$

and can otherwise be expressed in terms of the gamma function. Note that $\alpha = \beta$ gives a beta distribution that is symmetric around 0.5, and in particular $\alpha = \beta = 1$ gives the standard uniform distribution. By computing the usual integrals and using some additional properties of the beta function, it can be shown that

$$E[X] = \frac{\alpha}{\alpha + \beta} \quad \text{and} \quad \text{Var}[X] = \frac{\alpha\beta}{(\alpha + \beta)^2(\alpha + \beta + 1)}$$

***Example 6.38.*** Reconsider Example 6.36, where we have observed $1, 1, 0, 1$ and choose as prior for $p$ a beta distribution with $\alpha = \beta = 10$. Find the posterior distribution and the Bayes estimate of $p$.

This choice of prior means that we believe that $p$ is centered at, and symmetric around, 0.5 but that we do not have strong belief in extreme deviations. The posterior distribution is

$$f(p|D) = \frac{P(D|p)f(p)}{P(D)}$$

where, as usual, $P(D|p) = p^3(1 - p)$ and

$$f(p) = \frac{p^9(1 - p)^9}{B(10, 10)}$$

which gives

$$f(p|D) = \frac{p^3(1 - p) \times p^9(1 - p)^9}{P(D)B(10, 10)} = \frac{p^{12}(1 - p)^{10}}{P(D)B(10, 10)}$$

We could compute the denominator explicitly but instead note that if it were equal to $B(13, 11)$, then this would be a beta distribution with $\alpha = 13$ and $\beta = 11$. But then this must be the case, for otherwise $f(p|D)$ would not be a pdf. Hence, we conclude that the posterior pdf is

$$f(p|D) = \frac{p^{12}(1 - p)^{10}}{B(13, 11)}$$

a beta distribution with $\alpha = 13$ and $\beta = 11$. See Figure 6.6 for a plot of the prior and posterior distributions. By the formula for the mean, the Bayes estimate is

$$E[p|D] = \frac{13}{11 + 13} \approx 0.54$$

□

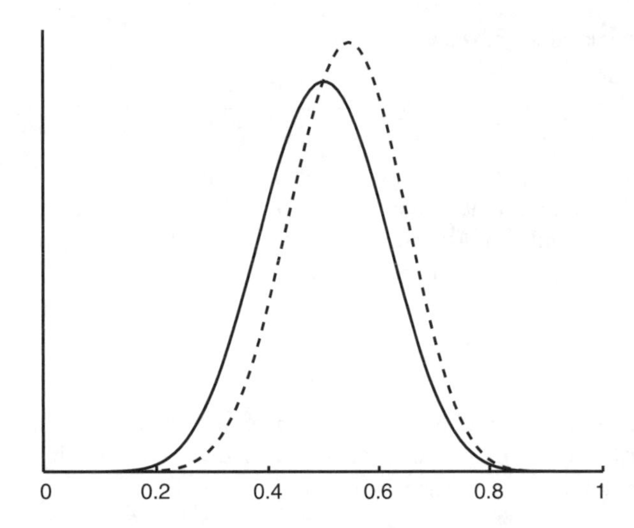


**FIGURE 6.6** The pdf's of the prior (solid) and posterior (dashed) of $p$, as in Example 6.38.

In the previous example, we recognized the principal form of the beta distribution and therefore we did not have to compute the denominator explicitly. This is typical for these types of calculations and it is common to write

$$f(p|D) \propto P(D|p)f(p)$$

that is, the posterior is proportional to the probability of the data times the prior. We can also note that $P(D|p)$ is in fact the likelihood function, so the fundamental equation in Bayesian statistics becomes

$$\text{Posterior} \propto \text{likelihood} \times \text{prior}$$

Computation of the proportionality constant $1/P(D)$ is one of the most challenging problems in Bayesian statistics. In complex models it is impossible, and sophisticated simulation methods, called *Markov chain Monte Carlo* (MCMC) *methods*, have been developed in order to get approximate solutions. For a nice introduction to such methods in a nonstatistical setting, see Häggström, *Finite Markov Chains and Algorithmic Applications* [8].

In the last example, the number of 1s in the sample has a binomial distribution. It is easy to see that if the prior of $p$ is beta, then the posterior is also beta, regardless of the sample size or the outcome. We say that the beta distribution is *conjugate* to the binomial distribution. To choose a conjugate distribution as prior has several advantages. As we saw in Example 6.38, a conjugate prior distribution means that we do not have to calculate the proportionality constant explicitly but can identify the posterior distribution by examining the functional form of the product of the prior and the likelihood with respect to the parameter. Another great advantage is that it simplifies comparisons between priors and posteriors as illustrated in Figure 6.6. Since the prior distribution is a quantification of our knowledge about a parameter before observing data, we want to see how this changes when we include the information given in a sample.

So, when we have chosen a model for the sample, how do we find the corresponding family of conjugate distributions? For most standard distributions, we can use the following result.

**Proposition 6.14.** Consider a sample $X_1, \ldots, X_n$ with pmf $p(\mathbf{x}|\theta)$ or pdf $f(\mathbf{x}|\theta)$ where $\mathbf{x} = (x_1, \ldots, x_n)$. If $p(\mathbf{x}|\theta)$ or $f(\mathbf{x}|\theta)$ can be written in the form[10]

$$c(\mathbf{x})g(\theta)^n e^{h(\theta)t(\mathbf{x})}$$

where $c$, $g$, $h$, and $t$ are arbitrary functions, the conjugate distribution can be expressed

$$f(\theta) \propto g(\theta)^a e^{h(\theta)b}$$

where $a$ and $b$ are arbitrary constants.

*Proof.* Let us consider a prior distribution of the form above. Then the posterior distribution can be written

$$f(\theta|\mathbf{x}) \propto p(\mathbf{x}|\theta)f(\theta) \propto c(\mathbf{x})g(\theta)^n e^{h(\theta)t(\mathbf{x})} \times g(\theta)^a e^{h(\theta)b}$$

$$\propto g(\theta)^{n+a} e^{h(\theta)(t(\mathbf{x})+b)} = g(\theta)^{a'} e^{h(\theta)b'}$$

where $a' = n + a$ and $b' = t(\mathbf{x}) + b$. The proof of pdf's is completely analogous. ∎

Let us assume that we do not know that the beta distribution is conjugate to the binomial distribution. The probability mass function of the binomial distribution can be written

$$p(x|p) = \binom{n}{x} p^x (1-p)^{n-x} = \binom{n}{x} (1-p)^n e^{x \log(p/(1-p))}$$

where we identify $c(x) = \binom{n}{x}$, $g(p) = 1 - p$, $t(x) = x$, and $h(p) = \log(p/(1-p))$. Hence, the prior pdf can be written as

$$f(p) \propto (1-p)^a e^{b \log(p/(1-p))} = p^b (1-p)^{a-b}$$

and by setting $a = \alpha + \beta - 2$ and $b = \alpha - 1$, we can identify this as the beta distribution. For more, see the Problems section.

[10] Any class of distributions that can be written in this form is called an *exponential family*. There are several statistical advantages in expressing distributions in this form, which we will not develop further here. It suffices to say that most standard distributions are in fact exponential families.

In the examples thus far, the data have been discrete and the parameter continuous. Let us next consider an example where both data and parameter are continuous and all probability statements must be made in terms of pdf.

***Example 6.39.*** Consider a normal distribution with unknown mean $\mu$ and known variance 1 where the prior distribution of $\mu$ is $N(0, 1)$, the standard normal. Suppose that we have observed the values $x_1, \ldots, x_n$; find the posterior distribution and the Bayes estimate.

Let $\mathbf{x} = (x_1, \ldots, x_n)$ to get the posterior pdf

$$f(\mu|\mathbf{x}) = \frac{f(\mathbf{x}|\mu)f(\mu)}{f(\mathbf{x})} \tag{6.4}$$

Let us look at the factors one by one. First note that

$$f(\mu) = \frac{1}{\sqrt{2\pi}} e^{-\mu^2/2}$$

Next, the observations are i.i.d. $N(\mu, 1)$, and we get

$$\begin{aligned} f(\mathbf{x}|\mu) = \prod_{k=1}^{n} f(x_k|\mu) &= \prod_{k=1}^{n} \frac{1}{\sqrt{2\pi}} e^{-(x_k-\mu)^2/2} \\ &= \left(\frac{1}{\sqrt{2\pi}}\right)^n \exp\left(-\frac{1}{2}\sum_{k=1}^{n}(x_k - \mu)^2\right) \end{aligned}$$

The numerator in Equation (6.4) becomes

$$f(\mathbf{x}|\mu)f(\mu) = \left(\frac{1}{\sqrt{2\pi}}\right)^{n+1} \exp\left(-\frac{1}{2}\left(\sum_{k=1}^{n}(x_k - \mu)^2 + \mu^2\right)\right)$$

where we will rewrite the exponent of the exponential. First note that

$$\begin{aligned} \sum_{k=1}^{n}(x_k - \mu)^2 + \mu^2 &= \sum_{k=1}^{n} x_k^2 - 2\mu \sum_{k=1}^{n} x_k + (n+1)\mu^2 \\ &= (n+1)\left(\frac{1}{n+1}\sum_{k=1}^{n} x_k^2 - \frac{2\mu\sum_{k=1}^{n} x_k}{n+1} + \mu^2\right) \\ &= (n+1)\left(\mu - \frac{1}{n+1}\sum_{k=1}^{n} x_k\right)^2 + R \end{aligned}$$

where $R$ is what is left over after we complete the square and depends on $n$ and $x_1, \ldots, x_n$. Since $\sum_{k=1}^{n} x_k = n\bar{x}$, we now get the posterior pdf

$$
\begin{aligned}
f(\mu|\mathbf{x}) &= \left(\frac{1}{\sqrt{2\pi}}\right)^{n+1} \frac{e^{-R/2}}{f(\mathbf{x})} \exp\left(-\frac{1}{2}(n+1)\left(\mu - \frac{n\bar{x}}{n+1}\right)^2\right) \\
&= C \exp\left(-\frac{1}{2}(n+1)\left(\mu - \frac{n\bar{x}}{n+1}\right)^2\right)
\end{aligned}
$$

where $C$ does not depend on $\mu$. Now comes the next part. We know that $f(\mu|\mathbf{x})$ is a pdf as a function of $\mu$ (remember that $\mathbf{x}$ is fixed). Also, we recognize that with $C = (n+1)/\sqrt{2\pi}$, the last expression is the pdf of a normal distribution with mean $n\bar{x}/(n+1)$ and variance $1/(n+1)$. Thus, there is no other possible value for $C$ (or $f(\mu|\mathbf{x})$ would not be a pdf), and the posterior distribution is

$$
\mu|\mathbf{x} \sim N\left(\frac{n\bar{x}}{n+1}, \frac{1}{n+1}\right)
$$

The Bayes estimate is thus

$$
E[\mu|D] = \frac{n\bar{x}}{n+1}
$$

which we can compare with the standard frequentist estimate

$$
\widehat{\mu} = \bar{x}
$$

Note how the Bayes estimate is closer to 0. This reflects the influence from the prior distribution in which the mean is 0. The posterior mean can in this case be regarded as a weighted average of the prior mean 0 and the observed sample mean $\bar{x}$, with the weights being $1/(n+1)$ and $n/(n+1)$. Note that a much larger weight is put on the sample mean and that for large $n$, $E[\mu|D] \approx \bar{x}$. This suggests that the choice of prior is less important than what the data support, and this observation reduces some of the arbitrariness in the choice of prior. See also Problem 84. □

### 6.8.1 Noninformative priors

The greatest strength of Bayesian statistics is that it enables us to incorporate prior knowledge in a statistical analysis in a mathematically stringent way. However, we also need to address the issue of what to do if we do not know anything whatsoever beforehand or if we do not want our prior belief to affect our results. In Example 6.36, we introduced a uniform prior to account for lack of information, but is this really the best choice?

If we do not know the value of a certain parameter in a given distribution it does makes sense to spread out the probability mass uniformly over the parameter space because no particular parameter value will then get a larger density than any other value. As long as the parameter space is bounded, this works well, but most parameters in the standard probability distributions are actually defined on unbounded spaces like $\lambda$ in the Poisson distribution and $\mu$ and $\sigma^2$ in the normal distribution. One way of getting around that is to choose the so-called *vague prior*, which is some distribution that gives almost equal weight to all parameter values like the uniform distribution on a really large interval and a normal distribution with a really large variance. This solves the practical problem but is still somewhat unsatisfactory from a theoretical point of view. However, it turns out that we can still use Bayes' formula if we pretend that uniform distributions on unbounded spaces do exist.

Since any uniform distribution, by definition, gives equal weight to all parameter values and, consequently, should have a constant pdf, we can express a uniform prior on $\theta$ as $f(\theta) \propto 1$. If the parameter space is unbounded, the proportionality constant cannot be positive, but let us disregard that for the moment. The posterior density can then be written

$$\text{Posterior} \propto \text{likelihood} \times \text{prior} \propto \text{likelihood}$$

and as long as the likelihood has finite and positive integral with respect to $\theta$, we have a correct posterior distribution.[11]

***Example 6.40.*** Let us return to Example 6.39 and see what happens if we choose a noninformative prior $f(\mu) \propto 1$ for $\mu$. Then, we get the posterior

$$f(\mu|\mathbf{x}) \propto f(\mathbf{x}|\mu)f(\mu) \propto f(\mathbf{x}|\mu) = \left(\frac{1}{\sqrt{2\pi}}\right)^n \exp\left(-\frac{1}{2}\sum_{k=1}^{n}(x_k - \mu)^2\right)$$
$$\propto \exp\left(-\frac{n}{2}(\mu - \bar{x})^2\right)$$

where the last step follows from the fact that

$$\sum_{k=1}^{n}(x_k - \mu)^2 = n(\mu - \bar{x})^2 + C$$

We recognize the posterior as the normal distribution $N(\bar{x}, 1/n)$, which means that the Bayes estimate becomes $E[\mu|D] = \bar{x}$. Using a noninformative prior thus produces a Bayes estimate identical to the frequentist estimate, which is logical since no prior belief should affect our results in any direction. Moreover, the posterior variance coincides with the frequentist variance of the mean $\bar{X}$, which implies that the statistical error is the same in the two approaches. □

As we have seen, the uniform distribution works in many situations as a reasonable model for the absence of prior information about parameter values, even

[11] This argument can be put in correct mathematical form by considering a sequence of correct priors $f_\nu(\theta)$ where the probability mass is gradually spread out uniformly on the parameter space as $\nu \to \infty$, for example, the normal distribution $N(0, \nu)$. The posterior can then be expressed as $f(\theta|D) = \lim_{\nu\to\infty} \frac{P(D|\theta)f_\nu(\theta)}{P(D)}$.

for unbounded parameter spaces. But that depends on what parameter we are interested in.

Let us, for example, say that we have observed a sample of lifelengths that are assumed to be exponentially distributed. One alternative is to choose a uniform prior for the parameter $\lambda$, but if we are interested in the mean lifelength it would perhaps be more natural to consider the parameter $\mu = 1/\lambda$ instead. To see that these approaches are incompatible, assume a uniform prior $f(\lambda) \propto 1$ and compute the distribution of $\mu$ as

$$f(\mu) = f(\lambda)\left|\frac{d\mu}{d\lambda}\right| \propto 1 \cdot \frac{1}{\mu^2}$$

which is definitely not uniform.

Consequently, the choice of a noninformative prior should not depend on the choice of parametrization.

**Proposition 6.15 (Jeffrey's Prior).** The prior density

$$f(\theta) \propto \sqrt{I(\theta)}$$

where

$$I(\theta) = -E\left[\frac{d^2}{d\theta^2}\log f(X|\theta)\right]$$

is the Fisher information, is invariant under parameter transformation.

What Proposition 6.15 says is essentially that if we have an alternative parameterization $\phi = g(\theta)$, it does not matter if we choose $f(\theta) \propto \sqrt{I(\theta)}$ or $f(\phi) \propto \sqrt{I(\phi)}$, we will still get the same result.

***Example 6.41.*** Let us return to the binomial distribution and see what happens if we use Jeffrey's prior for $p$. We first calculate the Fisher information

$$I(p) = -E\left[\frac{d^2}{dp^2}\log p(X|p)\right] = -E\left[-\frac{X}{p^2} - \frac{n-X}{(1-p)^2}\right] = \frac{n}{p(1-p)}$$

This yields the prior

$$f(p) \propto \sqrt{I(p)} \propto \frac{1}{\sqrt{p(1-p)}} = p^{-1/2}(1-p)^{-1/2}$$

which we identify as the beta distribution with parameters $\alpha = 1/2$ and $\beta = 1/2$. Since we know that the beta distribution is conjugate to the binomial distribution, we immediately get that the posterior is the beta distribution with parameters $\alpha = x + 1/2$

and $\beta = n - x + 1/2$. If we, for instance, apply this to Example 6.36, we would get a Bayes estimate of

$$E(p|D) = \frac{3 + 1/2}{4 + 1} = \frac{7}{10}$$

This is closer to the frequentist estimate $\frac{3}{4}$ than the Bayes estimate obtained from the uniform prior ($\frac{2}{3}$) but still not equal. □

This example clearly illustrates the problematic nature of the concept of noninformative priors. In a nutshell, introducing a prior distribution always brings some subjective information into the analysis and the two intuitive choices of uniform or Jeffrey's prior to minimize the impact of the prior sometimes coincide and sometimes differ.

### 6.8.2 Credibility Intervals

When we have obtained an estimate, the next natural step, as in the frequentist case, is to try to get an idea about the accuracy of this value. Since the posterior distribution is interpreted as our knowledge about a parameter, as a combination of our prior belief and the information contained in the observed data, the posterior standard deviation is the obvious choice of uncertainty measure analogous to the standard error. There is also a Bayesian analog to confidence interval called *credibility interval* defined as $[x_1, x_2]$, where

$$P(x_1 \leq \theta \leq x_2|D) = q$$

Note that it is the parameter and not the interval limits, conditioned on the data $D$, that is random in this framework. This makes a credibility interval more intuitive because now we can interpret $q$ as a (posterior) probability given an observed data set. In the frequentist setting, a given confidence interval is either right or wrong since $\theta$ is fixed. The confidence level only tells us what proportion of confidence intervals would be correct if we would repeat our calculations using new independent samples from the same underlying distribution, which many nonstatisticians find confusing.

***Example 6.42.*** Let us again consider the important special case where we want to estimate the probability $p$ based on a binomial response $X$ and a beta prior. If the parameters of the prior distribution are denoted $\alpha$ and $\beta$, we have seen that the posterior is also beta with parameters $\alpha + x$ and $\beta + n - x$. The posterior standard deviation is then

$$\sqrt{\text{Var}(p|X)} = \frac{1}{\alpha + \beta + n}\sqrt{\frac{(\alpha + x)(\beta + n - x)}{\alpha + \beta + n + 1}}$$

and the limits of a symmetric credibility interval can be obtained as the solution to

$$P(p < x_1|X) = P(p > x_2|X) = \frac{1 - q}{2}$$

which has to be calculated numerically. In earlier examples, we have used three different priors for the communication system example where $x = 3$ and $n = 4$. For the uniform prior, which can be seen as a beta distribution with $\alpha = \beta = 1$, we got the Bayes estimate $\frac{2}{3}$. The posterior standard deviation for this case becomes

$$\sqrt{\text{Var}(p|X)} = \frac{1}{6}\sqrt{\frac{4 \times 2}{7}} = 0.178$$

and a 95% credibility interval turns out to be [0.28, 0.95]. The informative prior with $\alpha = \beta = 10$ yields the Bayes estimate 0.54, the posterior standard deviation

$$\sqrt{\text{Var}(p|X)} = \frac{1}{24}\sqrt{\frac{13 \times 11}{25}} = 0.100$$

and the credibility interval [0.34, 0.73]. Not surprisingly, we see that an informative prior yields a smaller posterior standard deviation and, consequently, a narrower credibility interval. □

## 6.9 NONPARAMETRIC METHODS

All the methods we have developed in this chapter have been based on *parametric models*: assumptions of specific distributions and their parameters. In this section, we will investigate methods used to construct tests that make no assumptions other than that the distribution is continuous and sometimes also symmetric. Such methods are said to be *nonparametric*[12] and are widely used when it is not obvious what distribution to assume.

### 6.9.1 Nonparametric Hypothesis Testing

Suppose that we are interested in testing a hypothesis about some location parameter of a data set and do not know anything about its distribution other than that it is continuous. It is then suitable to use the median $m$ as the location parameter, recalling from Section 2.9 that this has the property

$$P(X \leq m) = P(X \geq m) = \frac{1}{2}$$

Our null hypothesis is $H_0 : m = m_0$ and if $H_0$ is true, we expect to have roughly the same number of observations above and below $m_0$, and if our sample deviates too much from this, we reject $H_0$. The test is usually described as assigning a plus sign to those observations that are above $m_0$ and a minus sign to those that are below (the

[12] A perhaps more correct term is *distribution-free*, but the term nonparametric is so well established that we will use it here.

assumption of a continuous distribution rules out observations that are exactly equal to $m_0$, at least in theory). In this formulation, let $N_+$ be the number of plus signs:

$$N_+ = \#\{k : X_k > m_0\}$$

Suppose that the alternative is two-sided, $H_A : m \neq m_0$. We then reject $H_0$ if $N_+$ is either too large or too small, and the resulting test is called the *sign test*. Let us state the test formally.

**Proposition 6.16 (The Sign Test).** Suppose that $X_1, \ldots, X_n$ is a sample from a continuous distribution with median $m$ and we wish to test

$$H_0 : m = m_0 \quad \text{versus} \quad H_A : m \neq m_0$$

We reject $H_0$ if

$$N_+ \leq k \quad \text{or} \quad N_+ \geq n - k$$

on significance level

$$\alpha = \frac{1}{2^{n-1}} \sum_{j=0}^{k} \binom{n}{j}$$

*Proof.* Under $H_0$, $N_+ \sim \text{bin}(n, \frac{1}{2})$, which means that the random variable $n - N_+$ (the number of minus signs) is also $\text{bin}(n, \frac{1}{2})$ and hence

$$P(N_+ \leq k) = P(N_+ \geq n - k) = \frac{1}{2^n} \sum_{j=0}^{k} \binom{n}{j}$$

which gives the significance level $\alpha$ in the proposition. ∎

Since the significance level can be expressed in terms of the cdf of the $\text{bin}(n, \frac{1}{2})$ distribution, it is easy to compute. For large $n$, we can use the normal approximation to the binomial, and for one-sided alternatives, the obvious adjustments are made, rejecting for only one of the inequalities $N_+ \leq k$ and $N_+ \geq n - k$ and getting significance level

$$\alpha = \frac{1}{2^n} \sum_{j=0}^{k} \binom{n}{j}$$

Previously, we set the significance level in advance, usually at 5%. Since the distribution of $N_+$ is discrete, we are seldom able to achieve those levels exactly but choose $k$ so that we come as close as possible.

***Example 6.43.*** Laboratory rats run through a maze, and the time until exit is measured. It is known that they either manage to exit relatively soon or they get lost and it takes a long time to exit, thus making intermediate times rare. The distribution of times can be assumed to be symmetric. It has been claimed that the mean exit time is more than 100 s. Use the following data to test this on level ≈5%:

26, 31, 43, 163, 171, 181, 193, 199, 206, 210

Since the distribution is symmetric, the mean $\mu$ and median $m$ are equal. The hypotheses are $H_0 : \mu = 100$ versus $H_A : \mu > 100$, and we reject $H_0$ if $N_+ \geq n - k$ where $n = 10$ and $k$ satisfies

$$\frac{1}{2^{10}} \sum_{j=0}^{k} \binom{10}{j} \approx 0.05$$

which gives $k = 2$. Thus, we reject if $N_+ \geq 8$, and since the observed value is $N_+ = 7$, we cannot reject $H_0$. □

Take a look at the data in the previous example. Since there were only seven observations above 100, the sign test could not reject $H_0$. However, the observations above 100 tend to deviate more from 100 than those below. Since the distribution is symmetric, this might indicate that the mean is more than 100 but the sign test does not take into account the values themselves, only whether they are greater than 100. We will introduce a more refined test that also takes into account the magnitudes of the deviations from the median but requires a symmetric distribution. If the distribution is symmetric, the mean and median are equal and we state hypotheses in terms of the mean $\mu$ instead. Suppose thus that we wish to test

$$H_0 : \mu = \mu_0$$

based on the observations $X_1, \ldots, X_n$, a sample from a continuous and symmetric distribution. Consider the absolute deviations from $\mu_0$

$$|X_1 - \mu_0|, \ldots, |X_n - \mu_0|,$$

and order these by size, from smallest to largest. This gives each $X_k$ a *rank* $R_k$, such that $R_k = j$ if $X_k$ has the $j$th smallest absolute deviation from $\mu_0$. Also for each $X_k$, keep track of which side of $\mu_0$ it is on by assigning to it an indicator function

$$I_k = \begin{cases} 1 & \text{if } X_k > \mu_0 \\ 0 & \text{otherwise} \end{cases}$$

Thus, for each observation $X_k$ we have a pair $(R_k, I_k)$, a rank, and an indicator for which side of $\mu_0$ it is on. The test statistic we will use is

$$W = \sum_{k=1}^{n} R_k I_k \qquad (6.5)$$

which is simply the sum of the ranks of all the observations that are above $\mu_0$. Note that $W$ ranges from 0 (all observations below $\mu_0$) to $n(n+1)/2$ (all observations above $\mu_0$). If $H_0$ is true, it is not too difficult to realize that the distribution of $W$ is symmetric with mean $n(n+1)/4$, and we reject $H_0$ if $W$ deviates too much from its mean. As usual, we need the distribution of $W$ to be able to quantify "too much." This turns out to be a nice exercise in the use of probability generating functions, and you are asked to do it in Problem 93.

**Proposition 6.17.** The probability mass function of $W$, defined above, is

$$P(W=r) = \frac{a(r)}{2^n}, \quad r = 0, 1, \ldots, \frac{n(n+1)}{2}$$

where $a(r)$ is the coefficient of $s^r$ in the expansion of $\prod_{k=1}^{n}(1+s^k)$.

By summing probabilities $P(W=r)$, we can find critical values and significance levels. For example, in a one-sided test of $H_0 : \mu = \mu_0$ against $H_A : \mu > \mu_0$, we reject if $W \geq C$ and get significance level

$$\sum_{r=C}^{n(n+1)/2} P(W=r)$$

which we can choose to be close to the desired $\alpha$ or put in the observed value of $W$ to get the $P$-value. Since the distribution of $W$ is symmetric around its mean, the identity

$$P(W=r) = P\left(W = \frac{n(n+1)}{2} - r\right)$$

can be used for computations and the significance level can instead be computed as

$$\sum_{r=0}^{n(n+1)/2-C} P(W=r)$$

**Corollary 6.3 (Wilcoxon Signed Rank Test).** *Suppose that $X_1, \ldots, X_n$ is a sample from a continuous and symmetric distribution with mean $\mu$ and we wish to test*

$$H_0 : \mu = \mu_0 \quad \textit{versus} \quad H_A : \mu \neq \mu_0$$

*With the test statistic $W$ in Equation (6.5), $H_0$ is rejected if*

$$W \leq c \quad \textit{or} \quad W \geq C$$

*where $C = n(n+1)/2 - c$. The significance level is*

$$2P(W \leq c) = 2\sum_{r=0}^{c} P(W=r)$$

If the alternative is one-sided, we make the obvious adjustments. Thus, if the alternative is $H_A : \mu < \mu_0$, we reject if $W \leq c$ and for $H_A : \mu > \mu_0$, we reject if $W \geq C$. In both cases, the significance level is $P(W \leq c)$. The test is called the *Wilcoxon signed rank test*. Table A.7 lists critical values that give significance levels approximately equal to 0.05 for one-sided and two-sided tests. Note that if we remove the ranks and only sum the indicators, we get the sign test. The assumption of a symmetric distribution makes it possible to replace the median by the mean, but even if we test for the median, it is still important to have a symmetric distribution. If this is not the case, it is natural to have larger deviations on one side of the median than the other so that such behavior does not necessarily indicate anything suspicious.

***Example 6.44.*** For the lab rats in Example 6.43, we wish to test

$$H_0 : \mu = 100 \quad \text{versus} \quad H_A : \mu > 100$$

on level $\approx 5\%$, and from Table A.7 we get $c = 11$, which gives $C = 10 \times 11/2 - 11 = 44$, so we reject $H_0$ in favor of $H_A$ if $W \geq 44$. The observed values are

$$26, \ 31, \ 43, \ 163, \ 171, \ 181, \ 193, \ 199, \ 206, \ 210$$

The absolute differences $|X_k - 100|$ are

$$74, \ 69, \ 57, \ 63, \ 71, \ 81, \ 93, \ 99, \ 106, \ 110$$

and if we order these by size we get

$$57, \ \underline{63}, \ 69, \ \underline{71}, \ 74, \ \underline{81}, \ \underline{93}, \ \underline{99}, \ \underline{106}, \ \underline{110}$$

where the underlined numbers belong to positive deviations. The ranks of the underlined values are

$$2, \ 4, \ 6, \ 7, \ 8, \ 9, \ 10$$

and if we sum these, the value of our test statistic becomes $W = 46$ and we can reject the null hypothesis, something the sign test failed to do. □

Note how the test statistic $W$ is based on the ranks of the deviations from $\mu$, not the values of the deviations themselves. Thus, in the last example, if the largest three values were instead 299, 306, and 310, the value of $W$ remains the same, so although this would be even stronger reason to reject, $W$ does not take this into account. For large sample sizes, we can use the following normal approximation for $W$. The proof is left for Problem 94.

**Proposition 6.18.** If $H_0$ is true, the mean and variance of $W$ are

$$E[W] = \frac{n(n+1)}{4}$$

$$\text{Var}[W] = \frac{n(n+1)(2n+1)}{24}$$

and for large values of $n$

$$T = \frac{W - n(n+1)/4}{\sqrt{n(n+1)(2n+1)/24}} \stackrel{d}{\approx} N(0, 1)$$

***Example 6.45.*** The density of Earth is usually reported to be 5.52 g/cm$^3$. In a famous experiment in 1798, Henry Cavendish used a clever apparatus to measure the density. Cavendish's 29 observations were

$$\begin{array}{l} 4.07,\ 4.88,\ 5.10,\ 5.26,\ 5.27,\ 5.29,\ 5.29,\ 5.30,\ 5.34,\ 5.34,\ 5.36,\ 5.39 \\ 5.42,\ 5.44,\ 5.46,\ 5.47,\ 5.50,\ 5.53,\ 5.55,\ 5.57,\ 5.58,\ 5.61,\ 5.62,\ 5.63 \\ 5.65,\ 5.75,\ 5.79,\ 5.85,\ 5.86 \end{array}$$

which have sample mean 5.42. Let $\mu$ be the true mean of Cavendish's data and test

$$H_0 : \mu = 5.52 \quad \text{versus} \quad H_A : \mu \neq 5.52$$

on the 5% level.

We will use the statistic $T$ and the normal approximation from above. With $\alpha = 0.05$, we thus reject if $|T| \geq 1.96$, and we have $n = 29$. The absolute deviations $|X_k - 5.52|$, $k = 1, \ldots, 29$, ordered by size, with positive deviations underlined are

$$\begin{array}{l} \underline{0.01},\ 0.02,\ \underline{0.03},\ \underline{0.05},\ 0.05,\ \underline{0.06},\ 0.06,\ 0.08,\ \underline{0.09},\ 0.10,\ 0.10,\ \underline{0.11} \\ 0.13,\ \underline{0.13},\ 0.16,\ 0.18,\ 0.18,\ 0.22,\ 0.23,\ \underline{0.23},\ 0.23,\ 0.25,\ 0.26,\ \underline{0.27} \\ \underline{0.33},\ \underline{0.34},\ 0.42,\ 0.64,\ 1.45 \end{array}$$

This is a good place to comment on the problem of *ties*. Note, for example, how there are two occurrences of 0.05, coming from the two measurements 5.47 and 5.57. Under our assumption of a continuous distribution, this is impossible but in real life we are limited by the accuracy of our measurements. When we put the values together, the value of the test statistic will depend on which we decide to underline, which is an arbitrary decision. Since the two 0.05s should really have the same rank, we assign them both the *mean rank* $(4 + 5)/2 = 4.5$ and proceed as before. If there are not too many ties, this is no big deal, but in general this is a problem that cannot be ignored. The test statistic and its distribution must then be adjusted to account for ties, but we will not address this issue further. Using mean ranks we get $W = 1 + 3 + 4.5 + 6.5 + 9 + 10.5 + 12 + 13.5 + 20 + 24 + 25 + 26 = 155$, which gives the test statistic

$$T = \frac{155 - 29 \times 30/4}{\sqrt{29 \times 30 \times (2 \times 29 + 1)/24}} = -1.35$$

and with $|T| = 1.35$ we cannot reject $H_0$. □

If we compare the two nonparametric tests we have seen so far, we can note that the sign test is a very crude test and it is therefore often unable to detect deviations from the null hypothesis, especially for small samples (see Problem 90). If it is reasonable to assume a symmetric distribution, the signed rank test is preferable as it takes more information into account and is thus better able to detect deviations from the null hypothesis. In the terminology of Section 6.6.3, the signed rank test is more powerful than the sign test.

If we want to test the stronger statement that the sample comes from a particular continuous distribution, we can use the so-called *Kolmogorov–Smirnov test*. Let us assume that we have a random sample $X_1, \ldots, X_n$ with continuous distribution function $F(x)$ and we want to test the hypotheses

$$H_0 : F(x) = F_0(x) \text{ for all } x$$

$$H_A : F(x) \neq F_0(x) \text{ for some } x$$

for a given $F_0(x)$.

As an estimator of $F(x)$, we use the *empirical distribution function*

$$\widehat{F}_n(x) = \frac{1}{n} \sum_{i=1}^{n} I_{\{X_i \leq x\}}$$

where $I_A$ denotes the indicator of the event $A$, and the maximum distance

$$D_n = \max_x |\widehat{F}_n(x) - F_0(x)|$$

as test statistic.

Since $\widehat{F}_n(x)$ and $F_0(x)$ are nondecreasing and $\widehat{F}_n(x)$ is a step function that only jumps at the points $X_1, \ldots, X_n$, the test statistic can be somewhat simplified as

$$D_n = \max_{1 \leq i \leq n} \left\{ \max \left( \left| \frac{i-1}{n} - F_0(X_{(i)}) \right|, \left| \frac{i}{n} - F_0(X_{(i)}) \right| \right) \right\} \tag{6.6}$$

where $X_{(1)} \leq \ldots \leq X_{(n)}$ are the order statistics. Since $F_0(X_i) \sim \text{unif}[0, 1]$ under $H_0$, we can actually rewrite (6.6) as

$$D_n = \max_{1 \leq i \leq n} \left\{ \max \left( \left| \frac{i-1}{n} - U_{(i)} \right|, \left| \frac{i}{n} - U_{(i)} \right| \right) \right\}$$

where $U_{(1)} \leq \ldots \leq U_{(n)}$ are the order statistics from a sample from the uniform distribution on $[0, 1]$. This implies that the null distribution of $D_n$ is actually independent of $F_0(x)$ and we can use the properties of the uniform distribution to derive this distribution. However, although it is possible to obtain analytical expressions for the density and distribution function for $D_n$ for different values of $n$, they are much too complex to be of any practical use, especially for large $n$. Instead, critical values for significance levels 0.01 and 0.05 can be found in Table A.9 for $n \leq 30$. For large $n$, it is possible to use the limiting distribution

$$P(\sqrt{n}D_n \leq x) \to 1 - 2\sum_{k=1}^{\infty}(-1)^{k-1}e^{-2(kx)^2} \tag{6.7}$$

as $n \to \infty$, to calculate approximate critical values or, somewhat easier, $p$-values. For instance, the condition for rejection on the 5% level becomes $D_n > 1.358/\sqrt{n}$ and on the 1% level it becomes $D_n > 1.628/\sqrt{n}$.

***Example 6.46.*** In 2009, a comparative study used different scales to test IQ scores of a group of 12 randomly selected pupils. The results for one scale were

$$78, 93, 95, 96, 99, 100, 104, 105, 110, 113, 124, 127$$

A correct IQ scale should, by definition, give a normally distributed score with mean 100 and standard deviation 15. Does the data above support this claim?

The empirical distribution function $\widehat{F}_{12}(x)$ for the data and distribution function $F_0(x)$ for the normal distribution with mean 100 and standard deviation 15 are shown in Figure 6.7 together with the test statistic $D_{12}$.

It turns out that the maximum difference is attained at $x = 93$, which yields

$$D_{12} = |\widehat{F}_{12}(93) - F_0(93)| = \Phi\left(\frac{93 - 100}{15}\right) - \frac{1}{12} = 0.237$$

From Table A.9 we get the critical value 0.375 on the 5% level, which means that we cannot reject the null hypothesis. If we use the limiting distribution above, we get the approximate critical value $1.358/\sqrt{12} = 0.392$, which is quite close to the exact value. □

### 6.9.2 Comparing Two Samples

There are two different situations for comparisons of two samples: pairwise observations and independent samples. If we have paired observations $(X_1, Y_1), \ldots, (X_n, Y_n)$ and wish to test $H_0 : \mu_1 = \mu_2$, we can base this on the differences $D_k = Y_k - X_k$ and the nice thing is that as long as the $X_k$ and $Y_k$ have the same distribution, $D_k$ has a symmetric distribution even if $X_k$ and $Y_k$ do not (see Problem 42 in Chapter 3). Thus, in this case, we can use the signed rank test to test whether the differences have mean 0.

The other two-sample situation occurs when we compare the means of two independent samples. We assume that we have two continuous distributions that have

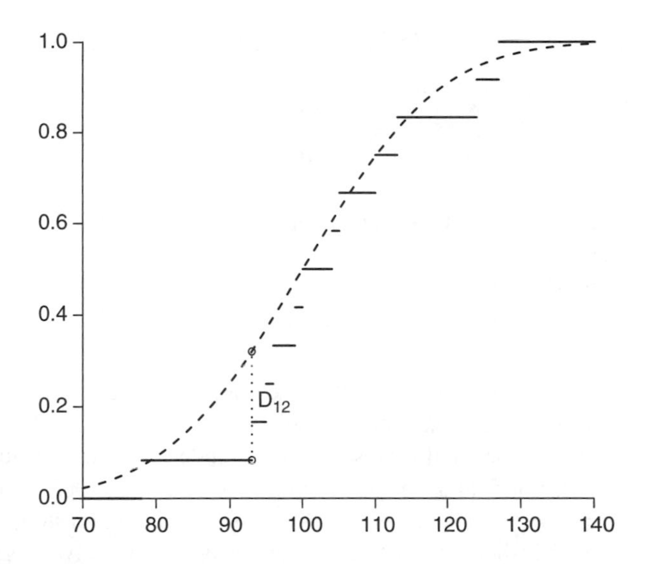


**FIGURE 6.7** The empirical distribution function (solid line) for the data in Example 6.46 and the distribution function (broken line) of the normal distribution with mean 100 and standard deviation 15. The Kolmogorov–Smirnov test statistic $D_{12}$ is also shown.

the same spread and shape and that the only possible difference is the mean (this is called a *translation model*). The distributions do not have to be symmetric. Now let $X_1, \ldots, X_m$ and $Y_1, \ldots, Y_n$ be the samples, where $m \leq n$. The fundamental idea is to put $X$ and $Y$ values together in a big sample, sort this combined sample by size, and consider the positions of the $X$ values (the smaller sample). If $H_0 : \mu_1 = \mu_2$ is true, the $X$ values should be uniformly spread within the combined sample, so if they tend to concentrate too much on either left or right, this indicates that $H_0$ is false.

Let $W$ be the sum of the ranks of $X_1, \ldots, X_m$ in the combined sample. Then $W$ ranges from $1 + 2 + \cdots + m = m(m+1)/2$ (all $X$ values smaller than all $Y$ values) to $(n+1) + (n+2) + \cdots + (n+m) = m(m+2n+1)/2$ (all $X$ values larger than all $Y$ values) and if $H_0$ is true, the distribution of $W$ is symmetric around its mean $m(m+n+1)/2$. Let us state the test based on $W$.

**Proposition 6.19 (Wilcoxon Rank Sum Test).** In the translation model above, we wish to test

$$H_0 : \mu_1 = \mu_2 \quad \text{versus} \quad H_A : \mu_1 \neq \mu_2$$

and reject $H_0$ if

$$W \leq c \quad \text{or} \quad W \geq C$$

where $C = m(m+n+1) - c$. The significance level is

$$2P(W \leq c) = 2 \sum_{r=m(m+1)/2}^{c} P(W = r)$$

The usual adjustments are done for one-sided tests. The probabilities $P(W = r)$ are more complicated to compute than in the signed rank test, and we will consider an example in Problem 98. Table A.8 can be used to find critical values for significance levels approximately equal to 0.05 for one-sided and two-sided tests.

***Example 6.47.*** When oil companies drill in the North Sea, they charter drilling rigs. After bidding and negotiating, a deal is struck with a contractor and a daily rate is set. The following data are daily rates (\$1000, rounded to integer values) for two different regions, Denmark and the Netherlands, from the year 2002. (The data were kindly provided by Dr. Patrick King of ODS-Petrodata, Inc., Houston, Texas.) Test on level $\approx 5\%$ whether there is a difference in rates between the two regions.

Netherlands: 58, 62, 63, 68, 69, 70, 77
Denmark: 50, 52, 52, 60, 60, 63, 64, 70, 82

With $\mu_D$ and $\mu_N$ as the two means, our hypotheses are

$$H_0 : \mu_D = \mu_N \quad \text{versus} \quad H_A : \mu_D \neq \mu_N$$

We have $m = 7$ and $n = 9$, and Table A.8 gives $c = 41$, which gives $C = 7(7 + 9 + 1) - 41 = 78$ (remember that the test is two-sided so the significance level equals $2P(W \leq c)$). We thus reject $H_0$ in favor of $H_A$ if $W \leq 41$ or $W \geq 78$. The data put together and ordered by size, with the Netherlands values underlined, are

$$50,\ 52,\ 52,\ \underline{58},\ 60,\ 60,\ \underline{62},\ \underline{63},\ 63,\ 64,\ \underline{68},\ \underline{69},\ \underline{70},\ 70,\ \underline{77},\ 82$$

Since there are some ties, we use mean ranks and get the sum of the Netherlands ranks

$$W = 4 + 7 + 8.5 + 11 + 12 + 13.5 + 15 = 71$$

and we cannot reject $H_0$. □

***Example 6.48.*** The drilling rigs in the previous example are of two different types: "jackup" and "semisubmersible." It is assumed that the semisubmersible rigs are more expensive to charter. Do the following data support this hypothesis?

Semisubmersible: 72, 89, 90, 94, 100, 104, 127, 155
Jackup: 50, 58, 60, 64, 68, 70, 77, 83, 103, 125

The hypotheses are, in obvious notation,

$$H_0 : \mu_S = \mu_J \quad \text{versus} \quad H_A : \mu_S > \mu_J$$

We have $m = 8$ and $n = 10$, and Table A.8 gives $c = 57$, which gives $C = 8(8 + 10 + 1) - 57 = 95$, so we reject $H_0$ in favor of $H_A$ if $W \geq 95$. The data put together with values for semisubmersible rigs underlined are

$$50,\ 58,\ 60,\ 64,\ 68,\ 70,\ \underline{72},\ 77,\ 83,\ \underline{89},\ \underline{90},\ \underline{94},\ \underline{100},\ 103,\ \underline{104},\ 125,\ \underline{127},\ \underline{155}$$

which gives the rank sum

$$W = 7 + 10 + 11 + 12 + 13 + 15 + 17 + 18 = 103$$

and we reject $H_0$ on the 5% level and conclude that the semisubmersible rigs are indeed more expensive. □

For large values of $m$ and $n$, there is a normal approximation, which is slightly more complicated than the one for the signed rank test. Since $W$ is a sum of ranks, the mean $E[W]$ is straightforward, but since the ranks are not independent, the variance requires a bit more work, using the general formula for the variance of a sum. As for the normal approximation, our central limit theorem does not apply since the summands are not independent, but more general versions of the theorem can be applied. We will not address this further but just state the result. For an application, see Problem 102.

**Proposition 6.20.** If $H_0$ is true, then

$$T = \frac{W - m(m + n + 1)/2}{\sqrt{mn(m + n + 1)/12}} \overset{d}{\approx} N(0, 1)$$

When comparing samples from two continuous distributions, we can also apply the ideas behind the Kolmogorov–Smirnov test introduced in Section 6.9.1. Let us denote the distribution function for sample $X_1, \ldots, X_m$ by $F^X(x)$ and the distribution function for sample $Y_1, \ldots, Y_n$ by $F^Y(x)$. The relevant hypotheses can now be expressed

$$H_0 : F^X(x) = F^Y(x) \quad \text{for all } x$$

$$H_A : F^X(x) \neq F^Y(x) \quad \text{for some } x$$

and we use the test statistic

$$D_{m,n} = \max_x |\widehat{F}_m^X(x) - \widehat{F}_n^Y(x)|$$

where $\widehat{F}_m^X(x)$ and $\widehat{F}_n^Y(x)$ are the empirical distribution functions for the samples. Since all empirical distribution functions are nondecreasing step functions, we realize that the maximum distance is attained at one of the points in the combined sample.

$$D_{m,n} = \max\left(\max_{1\le i\le m} |\widehat{F}_m^X(X_i) - \widehat{F}_n^Y(X_i)|, \max_{1\le j\le n} |\widehat{F}_m^X(Y_j) - \widehat{F}_n^Y(Y_j)|\right)$$

Furthermore, it turns out that it suffices to know the relative order, that is, the ranks, of the variables in the combined sample to be able to determine the value of $D_{m,n}$. Unfortunately, the actual calculations are somewhat more involved, especially for large samples, compared to the Wilcoxon rank sum test and are best left to a computer. Under the null hypothesis, we can again use the conclusion from above that the ranks are uniformly spread out on the combined sample to calculate the distribution of $D_{m,n}$ and, consequently, critical values, which are given in Table A.10. For large samples, the limiting distribution

$$P\left(\sqrt{\frac{mn}{m+n}} D_{m,n} \le x\right) \to \sum_{i=-\infty}^{\infty} (-1)^i e^{-2(ix)^2}$$

can be used to obtain critical values or $p$-values. The condition for rejecting on the 5% level can be expressed

$$D_{m,n} > 1.358\sqrt{\frac{m+n}{mn}}$$

***Example 6.49.*** Let us apply this test to Examples 6.47 and 6.48 and see if we get similar results. We first consider the daily rates of the Netherlands versus Denmark and calculate the test statistic $D_{7,9}$. This is easiest done in tabular form, which is displayed in Table 6.2.

**TABLE 6.2 The Maximal Absolute Distance Between Daily Rates in the Netherlands and Denmark (Denoted in Boldface)**

| | $\widehat{F}_7^X(x)$ | $\widehat{F}_9^Y(x)$ | $\lvert\widehat{F}_7^X(x) - \widehat{F}_9^Y(x)\rvert$ |
|---|---|---|---|
| $x < 50$ | 0 | 0 | 0 |
| $50 \le x < 52$ | 0 | 1/9 | 1/9 |
| $52 \le x < 58$ | 0 | 3/9 | 1/3 |
| $58 \le x < 60$ | 1/7 | 3/9 | 4/21 |
| $60 \le x < 62$ | 1/7 | 5/9 | **26/63** |
| $62 \le x < 63$ | 2/7 | 5/9 | 17/63 |
| $63 \le x < 64$ | 3/7 | 6/9 | 5/21 |
| $64 \le x < 68$ | 3/7 | 7/9 | 22/63 |
| $68 \le x < 69$ | 4/7 | 7/9 | 13/63 |
| $69 \le x < 70$ | 5/7 | 7/9 | 4/63 |
| $70 \le x < 77$ | 6/7 | 8/9 | 2/63 |
| $77 \le x < 82$ | 1 | 8/9 | 1/9 |
| $x \ge 82$ | 1 | 1 | 0 |

We see that the maximal distance is $D_{7,9} = 26/63 = 0.413$, which is smaller than the critical value $c = 0.667$ from Table A.10 so we cannot reject the null hypothesis using the Kolmogorov–Smirnov test either.

Again, caution is advised in the presence of ties, especially when the two empirical distribution functions compared jump simultaneously. Consider, for instance, what happens in the point $x = 63$. If daily rates had been measured more accurately, it is quite possible that $\hat{F}_9^Y(x)$ would assume the value 6/9 before $\hat{F}_7^X(x)$ jumps to 3/7. In that case the absolute difference would have been $|2/7 - 6/9| = 8/21$ before it would decrease to 5/21. In this particular example, it does not matter since the maximal absolute difference is 26/63 anyway, but in general there is a risk of underestimating $D_{m,n}$ in the presence of ties.

In Example 6.48, there are no ties, so we can be confident that the result is correct. Without going into details, it turns out that the test statistic becomes $D_{8,10} = 27/40 = 0.675$ and since the critical value from Table A.10 is $c = 0.6$, we can reject $H_0$ on the 5% level. □

### 6.9.3 Nonparametric Confidence Intervals

It is not obvious how to construct confidence intervals in the absence of a distribution assumption or even for what parameter. To approach this problem, Proposition 6.9 turns out to be quite useful. It claims that if we can formulate a hypothesis test for a specific problem, we can always translate that into a confidence interval for a required confidence level.

Let us first consider the sign test, introduced in the previous section, where we only make the assumption that a given sample comes from a continuous distribution. Proposition 6.9 basically says that we can obtain a confidence interval by including all parameter values that cannot be rejected in the corresponding hypothesis test. Let us therefore assume that we have chosen a particular $k$ so that we reject the null hypothesis $m = m_0$ if and only if $N_+ \leq k$ or $N_+ \geq n - k$, where $N_+$ is the number of positive differences $X_i - m_0$. This naturally implies that we accept the parameter value $m_0$ if and only if $k < N_+ < n - k$. The condition that $N_+$ has to be larger than $k$ means that at least $k + 1$ sample points has to be larger than $m_0$ or that $m_0 < X_{(n-k)}$, where $X_{(1)} \leq \ldots \leq X_{(n)}$ are the order statistics. By symmetry, we conclude that the lower limit can be written $m_0 > X_{(k+1)}$.

To summarize, we can express a two-sided confidence interval for the median $m$ as

$$X_{(k+1)} \leq m \leq X_{(n-k)} \qquad (q)$$

where

$$q = 1 - \frac{1}{2^{n-1}} \sum_{j=0}^{k} \binom{n}{j} \tag{6.8}$$

Note that we switched from strict to weak inequalities, which does not make a difference in this case since we have assumed a continuous distribution. As for the sign test, we cannot hope to get exactly the confidence level we want, say 5%, but have to find the value of $k$ that brings us as close as possible. If a one-sided confidence interval is required, the lower or upper limit can be dropped and the confidence level adjusted accordingly.

***Example 6.50.*** Consider once more the lab rat data of Example 6.43 and assume that we want a 95% confidence interval of the median time until exit. The test procedure in Example 6.43 can be translated directly into the one-sided confidence interval

$$m \geq X_{(3)} = 43 \qquad (\approx 0.95)$$

Since the aim of the test was to establish that the median time was larger than 100 s, it is natural to calculate only a lower bound. If we want a two-sided interval on the same level, we have to choose $k$ so that (6.8) becomes approximately 95%. In this example, it is difficult to achieve this, but the value $k = 1$ gives us $q \approx 0.98$, which is as close as we can get. Now, our interval becomes

$$31 \leq m \leq 206 \qquad (\approx 0.95)$$

□

If we can assume that the sample distribution is symmetric, we can use the Wilcoxon Signed Rank Test to obtain a, hopefully, more precise confidence interval. However, it is less straightforward to transform critical values of ranks, which this test is based on, into interval limits.

First, we have to calculate pairwise averages

$$\bar{X}_{ij} = \frac{X_i + X_j}{2} \qquad 1 \leq i \leq j \leq n$$

based on the sample. Note that we include averages where $i = j$ so that $\bar{X}_{ii} = X_i$. Then we order them as $\bar{X}_{(1)} \leq \bar{X}_{(2)} \leq \ldots \leq \bar{X}_{(n(n+1)/2)}$. In the previous section, the test statistic $W$ was defined as the sum of the ranks of the positive differences $X_i - \mu_0$. By symmetry, under the assumption that $\mu = \mu_0$, we can instead consider the sum of the ranks of the negative differences, which somewhat simplifies the following argument.

Let us now begin by considering a parameter value $\mu_0$ so that $\mu_0 < \bar{X}_{(1)}$. This means that all differences are positive and that $W = 0$. By increasing $\mu_0$ so that $\bar{X}_{(1)} < \mu_0 < \bar{X}_{(2)}$, we obtain one negative difference whose absolute value is smaller than any other so that $W = 1$.

The value $\bar{X}_{(2)}$ has to be the average of the two smallest values in the original sample (why?), so when $\mu_0$ is increased one more step, that is, $\bar{X}_{(2)} < \mu_0 < \bar{X}_{(3)}$, the difference $|X_{(1)} - \mu_0|$ becomes larger than $|X_{(2)} - \mu_0|$ and gets the rank 2 so that $W = 2$.

In the next step when $\bar{X}_{(3)} < \mu_0 < \bar{X}_{(4)}$, there are two possibilities. Either $\bar{X}_{(3)}$ is an average of two distinct values, which means that the difference $|X_{(1)} - \mu_0|$ gets rank 3, or it is equal to $\bar{X}_{(2)}$, which means that we get one more negative difference with rank 1. In either case, we get $W = 3$. By following the same kind of argument we can show that $W = k$ if and only if $\bar{X}_{(k)} < \mu_0 < \bar{X}_{(k+1)}$, where $\bar{X}_{(0)} = -\infty$ and $\bar{X}_{(n(n+1)/2+1)} = \infty$.

Since the Wilcoxon Signed Rank Test rejects the parameter value $\mu_0$ if $W \leq c$ or $W \geq n(n+1)/2 - c$, where $c$ is the critical value, we can express the confidence interval

$$\bar{X}_{(c+1)} \leq \mu \leq \bar{X}_{(n(n+1)/2-c)} \qquad (q)$$

where

$$q = 1 - 2\sum_{r=0}^{c} P(W = r)$$

***Example 6.51.*** Let us look at the data in Example 6.43 one last time. When we have a small sample, the pairwise averages can be listed in tabular form (Table 6.3).

If we want a two-sided interval, we get the critical value $c = 9$ from Table A.8, which yields

$$\bar{X}_{(10)} \leq \mu \leq \bar{X}_{(46)} \qquad (\approx 0.95)$$

or

$$101 \leq \mu \leq 195.5 \qquad (\approx 0.95)$$

It is clear that we get a narrower interval than in Example 6.50. The upper limit becomes slightly smaller but the lower limit is increased quite substantially. Actually, this is a bit misleading in this particular example since the major improvement is due more to the large gap in the data than to the methods used. However, it is possible to show that the interval based on the Wilcoxon Sign Rank test is always smaller than

**TABLE 6.3 All Pairwise Averages**

| | 26 | 31 | 43 | 163 | 171 | 181 | 193 | 199 | 206 | 210 |
|---|---|---|---|---|---|---|---|---|---|---|
| 26 | 26 | | | | | | | | | |
| 31 | 28.5 | 31 | | | | | | | | |
| 43 | 34.5 | 37 | 43 | | | | | | | |
| 163 | 94.5 | 97 | 103 | 163 | | | | | | |
| 171 | 98.5 | **101** | 107 | 167 | 171 | | | | | |
| 181 | 103.5 | 106 | 112 | 172 | 176 | 181 | | | | |
| 193 | 109.5 | 112 | 118 | 178 | 182 | 187 | 193 | | | |
| 199 | 112.5 | 115 | 121 | 181 | 185 | 190 | 196 | 199 | | |
| 206 | 116 | 118.5 | 124.5 | 184.5 | 188.5 | 193.5 | 199.5 | 202.5 | 206 | |
| 210 | 118 | 120.5 | 126.5 | 186.5 | 190.5 | **195.5** | 201.5 | 204.5 | 208 | 210 |

the interval based on the Sign Test, but the degree of improvement depends on the particular data set. □

## PROBLEMS

### Section 6.2. Point Estimators

1. Let $X_1, \ldots, X_n$ be a sample with mean $\mu$ and variance $\sigma^2$ and consider the *linear estimator*

$$L = \sum_{k=1}^{n} a_k X_k$$

where the $a_k$ are nonnegative and sum to 1. **(a)** Show that $L$ is unbiased. **(b)** Show that $L$ is consistent. **(c)** Show that $\bar{X}$ has the smallest variance among all linear estimators (compare with Problem 56 in Chapter 1).

2. Let $X_1, \ldots, X_n$ and $Y_1, \ldots, Y_m$ be two independent samples with means $\mu_1$ and $\mu_2$ and variances $\sigma_1^2$ and $\sigma_2^2$. Suggest an unbiased estimator of the difference $\mu_2 - \mu_1$ and find its standard error.

3. Let $p$ be the unknown probability of an event $A$. Repeat the experiment $n$ times and let $X$ be the number of times that $A$ occurred. Show that the relative frequency $\widehat{p} = X/n$ is an unbiased and consistent estimator of $p$ and find its standard error.

4. Suppose that we want to estimate the proportion $p$ of men with some characteristic that is considered embarrassing and not readily admitted. To deal with this, each questioned individual is given a fair coin and is asked if he has the characteristic. He is then instructed to flip the coin out of sight of the investigator. If it shows heads, he must answer truthfully, and if it shows tails he must answer "Yes." Suppose that we get $X$ out of $n$ "Yes" answers. Suggest an unbiased estimator of $p$ and compute its standard error.

5. *Capture/recapture.* To estimate the size $N$ of a fish population in a lake, $k$ fish are caught, tagged, and released. Suggest estimators of $N$ and investigate for unbiasedness if you at a later time **(a)** catch $n$ fish and get $X$ tagged, **(b)** catch fish repeatedly with replacement and get the first tagged fish in catch number $X$. **(c)** What assumptions are you making?

6. To estimate the size $S$ of an underground well, one gallon of a dye is poured into the well and is allowed to mix. Later, a water sample of one gallon is taken and the dye is found to have concentration $C$. Suggest an estimator of $S$ and investigate for unbiasedness. What assumptions are you making?

7. Consider a Poisson process with rate $\lambda$. Suppose that $X$ points are observed in an interval of length $t$. Suggest an unbiased estimator of $\lambda$ and compute its standard error.

8. Calculate the Cramér–Rao lower bound for $\lambda$ in the previous problem and show that $e(\widehat{\lambda}) = 1$.
9. How do $\bar{X}$ and $s^2$ change if a constant $a$ is added to all observations in the sample?
10. Show that $s^2$ is a consistent estimator of $\sigma^2$ if $E[X_i^4] < \infty$.
11. Show that the sample standard deviation $s$ is a biased estimator of $\sigma$. *Hint:* Since $s$ is random, $\text{Var}[s] > 0$. Now apply the variance formula from Corollary 2.2.
12. Consider the following sample of size 8 from an unknown distribution. Use the inequalities from Problem 47, Chapter 2, to get two estimated upper bounds on the probability $P(X \geq 5)$:

    1.2, 1.5, 2.2, 3.1, 3.4, 3.7, 4.0, 4.4

13. On the basis of the following data, estimate the coefficient of variation $c$ (see Problem 46 in Chapter 2). Is it likely that the data are from an exponential distribution?

    9.4, 9.6, 9.8, 10.0, 11.0, 11.2, 11.4, 11.6, 12.9

14. The following is a sample from a normal distribution.

    7.6, 9.6, 10.4, 10.7, 11.9, 14.1, 14.6. 18.5

    **(a)** Let $X$ have this normal distribution, and let $p = P(X \geq 20)$. If we estimate $p$ by the relative frequency, we just get 0. Suggest another estimate. **(b)** Let $x$ be the 95th percentile, that is, a value such that $P(X \leq x) = 0.95$. Suggest an estimate of $x$.
15. In a sample of size 100 from a normal distribution, it was observed that 23 values were below 80 and 19 above 120. Use this to find estimates of $\mu$ and $\sigma$.
16. Let $\widehat{\theta}$ be an estimator of $\theta$. The *bias* of $\widehat{\theta}$ is defined as $B = E[\widehat{\theta}] - \theta$. If $B = 0$, $\widehat{\theta}$ is unbiased, and if $B \to 0$ as $n \to \infty$, $\widehat{\theta}$ is said to be *asymptotically unbiased.* **(a)** Consider the estimator of $\sigma^2$ that is given in Example 6.18. Find its bias and show that it is asymptotically unbiased. **(b)** Consider the estimator $X_{(n)}$, the maximum observation, of $\theta$ in a unif$[0, \theta]$ distribution. Find its bias and show that it is asymptotically unbiased. **(c)** Consider a binomial distribution with known $n$ and unknown $p$, where $X$ is observed. Let $\widehat{p} = X/n$ and $\widetilde{p} = (X + 1)/(n + 1)$. Find their respective biases and investigate for asymptotic unbiasedness.

**Section 6.3. Confidence Intervals**

17. Below are two sets of IQ scores from two different universities, A and B. Find a 95% symmetric confidence interval for the difference between the means if $\sigma = 15$.

A: 106, 114, 116, 123, 124, 133
B: 99, 113, 114, 121, 126

18. A scale is known to give measurement errors that are normal with mean 0 and variance 1. A piece of metal is weighed five times, and the following weights in grams are observed:

    999.4, 999.8, 1000.4, 1000.8, 1001.0

    Find an observed symmetric confidence interval for the true weight $\mu$ with confidence level 0.95.
19. Headlines like this from 1999 are typical: "Majority favors U.S. troops in Kosovo." This was based on an opinion poll, where 547 out of 1014 said they favored troops in Kosovo. Find a 95% symmetric confidence interval for the true proportion that favors troops. Does this support the headline?
20. Here is another headline from Sweden in 1994: "Majority of Swedes support joining the European Union." This was based on a poll of 1000 people, where 505 were in favor. In order to consider it significant that a majority wanted to join, what confidence level would be needed for a symmetric confidence interval?
21. Consider the confidence interval for an unknown probability $p$. Suppose that we want confidence level 0.95 and that we want the length of the interval to be at most 0.1 (margin of error at most $\pm 0.05$). How should we choose $n$ if **(a)** we know that $p$ is at most 0.2, **(b)** we do not know anything about $p$? **(c)** In the light of (b), argue why it is reasonable to state the margin of error in an opinion poll as $\pm 1/\sqrt{n}$.
22. To answer the age-old question "Coke or Pepsi?" there are several opinion polls on the Web where people are encouraged to vote. In one such poll 145 out of 275 people preferred Coke. Should the Coca-Cola company use this in their commercials to claim that Coke is more popular? From a methodological point of view, what problems are there with online opinion polls of this kind?
23. A September 2004 opinion poll showed support for President Bush in Florida at 52%, compared to 47% in another poll earlier in the month. If the margin of error in the last poll is $\pm 3\%$, can it be argued that the change is significant since the confidence interval from the second poll is [49, 55], which is entirely above 47?
24. Consider the opinion poll in Example 6.8 where Bush had 47% and Gore 44%. Is the difference statistically significant? Let the size of the sample be $n$, and let $B$ and $G$ be the numbers of supporters of Bush and Gore, respectively, and let $p_B$ and $p_G$ be the true proportions. We need a confidence interval for $p_B - p_G$, based on the estimators $\widehat{p}_B$ and $\widehat{p}_G$, which are not independent. **(a)** Show that $\mathrm{Cov}[\widehat{p}_B, \widehat{p}_G] = -p_B p_G/n$ (recall Problem 124 in Chapter 3).

**(b)** State the approximate normal distribution of $\widehat{p}_B - \widehat{p}_G$ and use this to find a confidence interval for $p_B - p_G$ by replacing unknown probabilities with their estimators in suitable places. **(c)** With the numbers given in Example 6.8, what is the observed confidence interval? **(d)** A quick way to decide whether the difference is significant is to see if the individual confidence intervals overlap. Since Bush had $47 \pm 2$ and Gore $44 \pm 2$, which do overlap, we conclude that the difference is not significant. This is equivalent to adding the margins of error, which gives the difference $0.03 \pm 0.04$. What is the principal flaw in this? In practice, what is the difference from the interval in part (c)?

25. Assume that $X_1, \ldots X_n$ is a sample from an arbitrary distribution with unknown $\mu$ and known $\sigma$. Derive an approximate confidence interval with confidence level $q$ under the assumption that $n$ is large. *Hint:* Use the central limit theorem.

26. For large normal samples, it can be shown that

$$s \stackrel{d}{\approx} N\left(\sigma, \frac{\sigma^2}{2n}\right)$$

Use this to derive an approximate confidence interval with confidence level $q$ for $\sigma$.

27. Consider a Poisson process with rate $\lambda$. If $X$ points are observed in an interval of length $t$, argue that an approximate confidence interval for $\lambda$ with confidence level $q$ is given by

$$\lambda = X/t \pm z\sqrt{X}/t$$

where $\Phi(z) = (1+q)/2$. *Hint:* Problem 8, Chapter 4.

28. In a clinical trial for a new drug, patients were divided into two groups, one of size $n = 503$ receiving the drug and one of size $m = 430$ receiving placebo. One of the side effects studied was headache. In the drug group, 200 experienced headaches and in the placebo group, 156. Find a 95% one-sided confidence interval to investigate whether the drug increases the risk of developing headaches.

29. Let $\theta_1, \theta_2, \ldots, \theta_n$ be $n$ parameters, and let $I_1, I_2, \ldots, I_n$ be their corresponding confidence intervals, each with confidence level $q$. The *simultaneous confidence* level, $q_s$, is defined as the probability that all intervals contain their respective parameters. **(a)** Show that $q_s = q^n$ if the intervals are independent. **(b)** Suppose that a president has approval ratings that are statistically significant above 50% in 10 consecutive polls. If each poll has confidence level 0.95, what is the simultaneous confidence level? **(c)** In general, show that $q_s \geq 1 - n(1-q)$. *Hint:* Problem 13 in Chapter 1. **(d)** If you want a simultaneous confidence level of at least 0.95 and have 10 intervals that are not necessarily independent, what confidence level do you need for each individual interval?

30. *Prediction intervals.* Suppose that we have observed $X_1, \ldots, X_n$ in a normal distribution with known variance $\sigma^2$ and that we wish to predict the outcome

of a future observation, say, $X$ (remember Section 3.7.2) that is independent of the other $X_k$ and with the same distribution. A natural predictor is $\bar{X}$, and in order to measure its accuracy, we need error bounds such that

$$P(\bar{X} - R \leq X \leq \bar{X} + R) = q$$

for a given probability $q$. The interval $[\bar{X} - R, \bar{X} + R]$ is called a $100q\%$ *prediction interval* for $X$. **(a)** Show that the prediction interval is given by

$$X = \bar{X} \pm zs\sqrt{1 + \frac{1}{n}}$$

where $\Phi(z) = (1 + q)/2$. **(b)** Compare this with the $100q\%$ confidence interval for $\mu$. What is the difference in interpretation? Which interval is longer? **(c)** Consider the lightbulb data from Section 6.2. Find a 95% prediction interval for the next lifetime $X$. **(d)** Compare the 95% prediction interval with the 95% confidence interval for $\mu$. What happens as $n \to \infty$? Explain!

## Section 6.4. Estimation Methods

31. Let $X_1, \ldots, X_n$ be a random sample from a gamma distribution with unknown parameters $\alpha$ and $\lambda$. Show that the moment estimators are

$$\widehat{\lambda} = \bar{X}/(n-1)s^2 \quad \text{and} \quad \widehat{\alpha} = \bar{X}\widehat{\lambda}$$

32. Consider a Poisson process with unknown rate $\lambda$. Suppose that the $n$th point arrives at time $T$. Find the moment estimator of $\lambda$.
33. Find the MLE in Example 6.13 assuming that we have observed $X$ 1s in $n$ trials.
34. Let $X_1, \ldots, X_n$ be a sample from a uniform distribution on $(a, b)$ where the parameter $a$ is known. Find the MLE and moment estimator of $b$.
35. In the previous problem, suppose that both $a$ and $b$ are unknown. Find the MLE and moment estimator of $a$ and $b$.
36. Let $X_1, \ldots, X_n$ be a random sample from a uniform distribution on $[-\theta, \theta]$. Find the MLE of $\theta$.
37. Let $X_1, \ldots, X_n$ be a random sample from a distribution with pdf

$$f(x) = e^{-(x-\theta)}, \quad x \geq \theta$$

Find the MLE and moment estimator of $\theta$.

38. Let $X_1, X_2, \ldots, X_n$ be a random sample from a distribution with pdf

$$f_\theta(x) = \theta x^{\theta-1}, \quad 0 \leq x \leq 1$$

**(a)** Find the moment estimator of $\theta$. **(b)** Find the MLE of $\theta$. **(c)** Calculate an approximate confidence interval for $\theta$.

39. Let $X_1, X_2, \ldots, X_n$ be a random sample from a normal distribution with mean 0 and unknown variance $\theta$. Find the MLE and moment estimator of $\theta$.

40. Let $X_1, X_2, \ldots, X_n$ be a sample from a distribution with pdf

$$f(x) = ax\, e^{-ax^2/2}, \quad x \geq 0$$

**(a)** Find the moment estimator of $a$. **(b)** Find the MLE of $a$. **(c)** Calculate an approximate confidence interval for $a$.

41. Let $X_1, X_2, \ldots, X_n$ be a sample from a uniform distribution $[\theta - 1, \theta + 1]$, where $\theta$ is unknown. **(a)** Find the moment estimator of $\theta$. **(b)** Show that any point between $X_{(n)} - 1$ and $X_{(1)} + 1$ is an MLE of $\theta$. This shows that MLEs are not necessarily unique.

42. Let $X_1, X_2, \ldots, X_n$ be a sample from a distribution with pdf

$$f(x) = \frac{1}{2\sigma} e^{-|x-\mu|/\sigma}, \quad x \in R$$

**(a)** Show that $\bar{X}$ is the moment estimator of $\mu$. **(b)** Show that the sample median $X_{(k)}$ is the MLE of $\mu$ for odd $n = 2k - 1$.

## Section 6.5. Hypothesis Testing

43. Ann suspects that her coin may be manipulated so that it does not give heads and tails with equal probability when she flips it. She decides to test this by flipping the coin eight times and concludes that it is unfair if she gets all heads or all tails. Formulate hypotheses and calculate the level of significance. Find another critical region so that $\alpha \leq 0.10$.

44. A climate scientist believes that the mean number of hurricanes in the Mexican Gulf is 5 during a normal year. He wants to test if this is true for the period 2005–2008 or if the mean has increased. Assume that the number of hurricanes in a year is Poisson distributed with mean $\lambda$. **(a)** Formulate hypothesis in a statistical test. **(b)** Determine the largest critical region so that $\alpha \leq 0.10$. **(c)** In the period 2005–2008, there were 15, 5, 6, and 8 hurricanes recorded per year. Is this sufficient evidence to reject the null hypothesis? *Hint:* Problem 8(b) in Chapter 4.

45. Consider a sample $X_1, \ldots, X_n$ from a normal distribution with unknown mean $\mu$ and known variance $\sigma^2$ (compare with Problem 18). **(a)** Describe how to test the null hypothesis $H_0 : \mu = \mu_0$. **(b)** Use the data in Problem 18 to test on the 5% level if the true weight is 1000 g.

46. Show that Proposition 6.9 holds.

47. Let $X_1, \ldots, X_n$ be a sample from the exponential distribution with parameter $\lambda$. **(a)** Use Proposition 6.9 and Example 6.19 to derive a test of $H_0 : \lambda = \lambda_0$ against $H_A : \lambda \neq \lambda_0$. **(b)** Use Proposition 6.10 to derive a test of the same hypotheses. Which test do you prefer?

48. Here is a headline from the Libertarian Party Web site in 2003: "Thompson could have doubled vote in Wisconsin race, according to poll." Wisconsin

gubernatorial candidate Ed Thompson won 10.5% of the vote in 2002, and in a poll of 1000 voters shortly after the election, 23% said that they would have voted for him, had they thought he could win. Does this support the claim of (at least) doubling his vote? State the appropriate hypotheses, and test on the 5% level.

49. In a poll before the 2004 presidential election, the support for John Kerry was 78% in Washington, DC, and 38% in Texas. Suppose that both polls had sample size 1000. Can you conclude that the support in DC is at least twice as big as in Texas? How is this situation different from that in the previous problem.
50. In 2010, there were 266 persons killed in traffic accidents in Sweden compared to 358 in the previous year. Carry out a hypothesis test to see if this decrease can be considered statistically significant. What assumptions do you need to make?
51. Use Propositions 6.9 and 6.11 to derive an alternative confidence interval for the unknown probability $p$ than the one in Proposition 6.6.

**Section 6.6. Further Topics in Hypothesis Testing**

52. You are asked to decide whether males are less likely than females to call a particular toll-free number and are presented the following sequence of callers: $FFFFFM$. With $p$ as the probability of a male caller, you thus wish to test $H_0 : p = \frac{1}{2}$ versus $H_A : p < \frac{1}{2}$ and decide to compute the $P$-value. Consider the following two test methods: (a) the sample size is $n = 6$, and since the number of male callers $X$ is bin$(6, \frac{1}{2})$ under $H_0$, the $P$-value is $P(X \geq 1)$; (b) the first male caller was in the sixth call, and since the number of trials $N$ until the first male caller is geometric with $p = \frac{1}{2}$ under $H_0$, the $P$-value is $P(N \geq 6)$. Compute the two $P$-values and show that method (b) rejects on the 5% level, whereas method (a) accepts. What further information would you need in regard to how the data were collected?
53. Consider a study that compares test results of two groups of students from two universities, A and B, which are known to be of comparable quality. The null hypothesis of no difference is therefore tested versus the two-sided alternative that there is a difference. Suppose that the test statistic is $N(0, 1)$ under $H_0$ and that we test on the 5% level, which means that $H_0$ is rejected if $|T| \geq 1.96$. The observed value turned out to be 1.68 so $H_0$ cannot be rejected. However, a representative from university A notices that positive values of $T$ is in their favor and that the null hypothesis can be rejected in favor of the alternative "$H_A$ : A is better" since this test rejects on the 5% level if $T \geq 1.64$. How would you persuade the representative that publishing this would be dubious?
54. If you read a research report that claims that men are better drivers than women and that this has been confirmed in a study with a given $P$-value, you should probably double that $P$-value. Why? (Compare with the previous problem.)
55. A certain type of disease occurs in a small fraction $p$ of the population and is known to occur fairly uniformly across the United States. In 2004, each

state screens for the disease and tests on the 5% level whether it occurs in a fraction higher than $p$. A significant result is found in California. How would you persuade the Governor of California that there is no immediate reason to panic?

56. Suppose that we test and reject five different null hypotheses, each on the 5% level. If the tests are independent of each other and $k$ of the null hypotheses are true, what is the probability of rejecting *some* true null hypothesis? For which value of $k = 0, 1, \ldots, 5$ is this probability largest?
57. In July 2004, a special opinion poll was done regarding the upcoming presidential election. In this poll, the outcome in the Electoral College was targeted and a separate poll was done in each of the 50 U.S. states. If a multiple level of 5% was desired, what would the significance level have to be in each state?
58. Consider the two extreme test procedures to *always* reject the null hypothesis and to *never* reject the null hypothesis, regardless of data. What are the significance levels and power functions of the two tests?
59. Consider the test of $H_0 : p = \frac{1}{2}$ versus $H_A : p \neq \frac{1}{2}$ for an unknown probability $p$ based on $n$ repetitions of the trial. The relative frequency is $\widehat{p}$ and the test statistic is $T = 2\sqrt{n}(\widehat{p} - \frac{1}{2})$ which is approximately $N(0, 1)$ under $H_0$ (verify this). Now consider the two 5% level test procedures: (a) reject if $T \geq 1.64$ and (b) reject if $|T| \geq 1.96$. Compare the powers of the two tests and argue that (a) is more powerful than (b) for some values of the true probability $p$, whereas (b) is more powerful than (a) for other values. Which test makes more sense?
60. **(a)** Consider the test of $H_0 : p = \frac{1}{2}$ versus $H_A : p > \frac{1}{2}$ that rejects on the 5% level if $2\sqrt{n}(\widehat{p} - \frac{1}{2}) \geq 1.64$, and suppose that the true probability is 0.6. Suppose further that we want to be at least 90% certain to detect this with our test. How large must $n$ be? **(b)** Generally, suppose that the true probability is $p > 0.5$ and that we want to have at least probability $q$ to detect this. What equation in terms of the standard normal cdf $\Phi$ do we need to solve for $n$?
61. Assume that you are about to carry out two independent tests and you want the multiple significance level to be 0.05. What level of significance should you choose in each individual test? How does the result change for $m = 3, 5, 10, 20$ independent tests?
62. Show that the Bonferroni–Holm correction yields a correct multiple significance level.

## Section 6.7. Goodness of Fit

63. Consider the Texas Lottery data from Example 6.29. On a closer look, it seems that the number 7 is overrepresented. Let $p$ be the probability to get 7 and test the hypothesis $H_0 : p = 0.1$ versus $H_A : p > 0.1$ on the 5% level. Compare with the conclusion of the goodness-of-fit test in the example. Comment!
64. A store owner classifies each day as “good” or “bad,” depending on sales. Each week (6 workdays) the number of good days are counted. Use the data from 1

year below to test on the 5% level if good and bad days are equally likely. *Hint:* The number of good days in a week has a binomial distribution with $n = 6$.

| Number of good days | 0 | 1 | 2 | 3 | 4 | 5 | 6 |
|---|---|---|---|---|---|---|---|
| Number of weeks | 1 | 9 | 12 | 13 | 11 | 5 | 1 |

65. The weekly number of accidents on a particular highway was studied. Use the data below to test on the 5% level whether the number of accidents is better described by a Poisson distribution or a geometric distribution including 0.

| Number of accidents | 0 | 1 | 2 | 3 | 4 | 5 | 6 | $\geq 7$ |
|---|---|---|---|---|---|---|---|---|
| Number of weeks | 24 | 14 | 4 | 1 | 4 | 1 | 2 | 0 |

66. Consider the earthquake data in Example 6.30. Test whether the number of earthquakes in a given year has a binomial distribution with parameters $n = 25$ and $p = 0.03$. After you have done the test, what would you like to ask the person who suggested the distribution?
67. Sometimes, we want to reject also for very small values of the chi-square statistic in a goodness-of-fit test. The reason is to rule out data snooping and other manipulation that may render the fit "too good to be true." Show that such a test rejects on level $\alpha$ if

$$\chi^2 \leq x_1 \quad \text{or} \quad \chi^2 \geq x_2$$

where $F_{\chi^2_{r-1}}(x_1) = \alpha/2$ and $F_{\chi^2_{r-1}}(x_2) = 1 - \alpha/2$.
68. In Gregor Mendel's famous experiment with peas, the two types "smooth" and "wrinkled" were identified. Mendel argued that the smooth seed trait was dominant and the wrinkled was recessive (recall Section 1.6.2) and in a population of peas there should thus be 75% smooth and 25% wrinkled. Out of a total of 7324 peas, Mendel got 5474 smooth and 1850 wrinkled. **(a)** Test the dominant/recessive claim on the 1% level. **(b)** Do a two-sided test on the 1% level according to the previous problem. What do you conclude?
69. We might not want to put equal weight on "poor fit" and "too good to be true fit." Suppose that we want level 1% in a two-sided test and that we accept a close fit as long as it is not less probable than 0.1%. Give the critical values $x_1$ and $x_2$ for this test for Mendel's data and do the test.
70. Metal bars of length 100 cm are manufactured. They are first cut crudely, somewhere between 100 and 101, and then refined to achieve the desired length. The following is a set of 26 measured deviations from 100 cm. Does this cutoff waste follow a uniform distribution on [0, 1]? Test on the 5% level.

0.11, 0.18, 0.28, 0.33, 0.42, 0.42, 0.47, 0.48, 0.49, 0.49, 0.51, 0.52, 0.59
0.61, 0.62, 0.63, 0.66, 0.68, 0.74, 0.74, 0.75, 0.76, 0.78, 0.79, 0.81, 0.83

71. In the previous problem, test if the data come from *some* uniform distribution. *Hint:* The MLEs are the minimum and the maximum.

72. In a study of bats, it was investigated whether a tendency to bite people was associated with carrying rabies. [*Source:* Emerging Infectious Diseases 5:433–437 (1999).] Out of 233 bats who bit people, 69 were found to have rabies, and out of 4237 that did not bite people, 613 had rabies. Describe the data in a contingency table and test for independence on the 5% level.

73. In Example 6.32, after we have rejected independence, can we conclude from the chi-square test that females are less likely to be admitted?

74. In a study of voting behavior in the 2000 presidential election, it was investigated how voting was associated with education level. Use the contingency table below to test on the 5% level whether voting is independent of education level.

| | No Degree | High School | Some College | College Degree |
|---|---|---|---|---|
| Voter | 26 | 223 | 215 | 387 |
| Nonvoter | 168 | 432 | 221 | 211 |

75. Eight men and 10 women tried a new diet for a month. Five men but only two women lost more than 10 pounds. Use Fisher's exact test to test if the diet is equally effective for men and women.

## Section 6.8. Bayesian Statistics

76. Consider Example 6.35 and suppose that $p$ has prior distribution $P(p = \frac{1}{3}) = q$, $P(p = \frac{2}{3}) = 1 - q$. Find the posterior distribution and its mean.

77. Consider Example 6.35 where the data are 1, 1, 0, 1 and the prior uniform on [0, 1]. Compare the mode (recall Section 2.9) in the posterior distribution and the MLE of $p$.

78. Consider Example 6.35 where the data are 1, 1, 0, 1 and suppose that $p$ has prior $f(p) = 2p, 0 \leq p \leq 1$. Find the posterior distribution and the posterior mean and mode. Compare with the MLE.

79. Microchips are being produced, and there is a certain probability $p$ that a chip is defective and thus useless. To estimate $p$, 100 chips are checked and 4 of these are defective. In the archives you find the results of four previous studies, where the estimated values of $p$ are 0.05, 0.06, 0.08, and 0.10. Suggest a way to use this information to choose a prior beta distribution and then find the posterior distribution and the Bayes estimate of $p$. Compare with the frequentist estimate 0.04. From a frequentist point of view, how can the previous estimates be used?

80. When tossing a fair coin repeatedly, it turned up tails two times. The total number of tosses was, however, unknown and therefore considered as a positive integer-valued random parameter $\theta$. **(a)** Use a uniform prior on the integers $1, \ldots, 6$ and calculate the posterior distribution of $\theta$. **(b)** Calculate the mean

and variance of the posterior distribution. **(c)** Calculate a two-sided credibility interval for $\theta$ with approximate probability 70%.

81. Show that the gamma distribution is conjugate to the Poisson distribution.
82. Show that the gamma distribution is conjugate to the exponential distribution.
83. Assume that $X$ is exponentially distributed with parameter $\lambda$ and that the prior of $\lambda$ is the gamma distribution with parameters $\alpha = 2$ and $\beta = 10$. **(a)** If we make the observation $X = 5$, calculate the posterior mean and variance. **(b)** If we only get the information that $X > 5$, calculate the posterior mean and variance. Can you explain the difference?
84. Consider a normal distribution with unknown mean $\mu$ and known variance $\sigma^2$ where the prior distribution of $\mu$ is $N(\mu_0, \sigma_0^2)$. Suppose that we have observed the values $x_1, \ldots, x_n$. **(a)** Show that the posterior distribution is normal with mean and variance

$$E[\mu|D] = \frac{\sigma_0^2\sigma^2}{\sigma^2 + n\sigma_0^2}\left(\frac{\mu_0}{\sigma_0^2} + \frac{n\bar{x}}{\sigma^2}\right) \quad \text{Var}[\mu|D] = \frac{\sigma_0^2\sigma^2}{\sigma^2 + n\sigma_0^2}$$

**(b)** Express the posterior mean as a weighted average of the prior mean and the sample mean. What happens as $n \to \infty$? What happens to the posterior variance? **(c)** How do the weights in (b) depend on the variances $\sigma^2$ and $\sigma_0^2$? Explain this intuitively if we interpret a small prior variance as strong belief that the prior mean is correct.

85. Calculate Jeffrey's prior for the Poisson distribution.
86. Calculate Jeffrey's prior for the exponential distribution.
87. Let $X_1, \ldots, X_n$ be a sample from the geometric distribution with pmf

$$p(x) = \theta(1-\theta)^{x-1}, \quad x = 1, 2, \ldots$$

**(a)** Find the conjugate family of prior distributions. **(b)** Let us assume that we have observed the sample

2, 9, 4, 6, 7

Calculate posterior mean and standard deviation for a uniform prior. **(c)** Calculate posterior mean and standard deviation for Jeffrey's prior.

### Section 6.9. Nonparametric Methods

88. Below is a set of 15 IQ scores. Find parametric (based on the normal distribution) and nonparametric ≈95% observed confidence intervals for the median. For the parametric interval, assume that the scores are normal with $\sigma = 15$. Compare the lengths of the intervals and comment.

88, 90, 93, 96, 98, 106, 109, 111, 113, 113, 114, 116, 119, 126, 140

89. Below is a set of 10 measured service times (milliseconds) for a particular type of job arriving at a computer.

$$23.2,\ 27.3,\ 41.3,\ 56.6,\ 82.8,\ 83.7,\ 118.5,\ 210.8,\ 263.9,\ 621.8$$

Find the nonparametric 95% observed confidence intervals for the median.

90. Consider a sample of size $n = 5$. Show that the sign test cannot reject $H_0 : m = m_0$ on significance level 5% in a two-sided test and not on level 1% in a one-sided test. What are the smallest possible levels for which the sign test can do these rejections and what is then required of the test statistic $N_+$?

91. In Problem 89, use a one-sided sign test to test whether the median time is 100 versus the alternative that it is $<100$.

92. Below is a data set of the annual change in the last reading of the Dow index each year between 1972 and 2002 (1.06 means that it went up by 6%, 0.84 that it went down by 16%, and so on). Do a two-sided sign test to investigate if the "New Year's Eve Dow" tends to stay constant.

1.06, 0.84, 0.89, 1.30, 0.97, 0.80, 1.10, 1.07, 1.08, 0.89, 1.32, 1.06, 1.12, 1.29
1.34, 0.90, 1.14, 1.13, 1.13, 1.11, 1.30, 1.14, 1.05, 1.38, 1.25, 1.18, 1.14, 1.42
0.94, 0.95, 0.83

93. This problem outlines the proof of Proposition 6.17, where the pmf of the test statistic $W = \sum_{k=1}^{n} R_k I_k$ is given. First argue that if $H_0$ is true, the distribution of $W$ is the same as that of $U = \sum_{k=1}^{n} U_k$, where the $U_k$ are independent and $P(U_k = 0) = P(U_k = k) = \frac{1}{2}$. Next, use the fact that the pgf of $U$, which is the same as the pgf of $W$, is the product of the pgf's of the $U_k$. Then, argue that $P(W = r) = a(r)/2^n$, where $a(r)$ is the number of ways in which 0s and 1s can be assigned to the $I_k$ so that $W = r$ (and under $H_0$, all such ways are equally likely), and finally, identify $P(W = r)$ with the appropriate coefficient in the pgf of $W$.

94. Use $U$ in the previous problem to find the mean and variance of $W$. It can be shown that $W$ is asymptotically normal but more general versions of the central limit theorem than ours are needed. Why does our central limit theorem (Theorem 4.2) not apply?

95. For $W$ as above, find $P(W = 3)$ if $n = 5$.

96. A company manufactures metal plates. Each plate gives a certain amount of cutoff waste, and the process needs to be adjusted if this exceeds 100 mg per plate. Below are 11 measured weights of waste. Do they call for adjustment? Test on the 5% level.

$$88,\ 98,\ 99,\ 110,\ 118,\ 121,\ 123,\ 129,\ 136,\ 140,\ 149$$

97. We have probably all wondered from time to time whether it rained more in St. Louis or Minneapolis during the 1970s. Below are the annual total amounts

(inches) for the two cities, for the years 1970–1979. Decide what type of two-sample procedure to use (pairwise differences or independent samples) and test on the 5% level if there is a difference.

Minneapolis: 30.5, 29.4, 23.8, 21.1, 19.1, 35.1, 16.5, 34.9, 30.3, 31.0
St. Louis: 36.2, 33.7, 33.7, 39.8, 36.8, 40.2, 23.5, 43.4, 37.7, 29.5

98. Consider the Wilcoxon rank sum test with $m = 2$ and $n = 5$. Find the range of $W$ and $P(W = 7)$.
99. Execution times for a particular type of numerical computation were measured using two different algorithms, A and B. The times in milliseconds were

    A: 2, 4, 4, 8, 9, 14, 21, 25
    B: 7, 13, 25, 43, 47

    Use the rank sum test to test on level 5% whether there is a difference between the algorithms.
100. Use Kolmogorov–Smirnov's test to examine the difference in the previous problem.
101. The following data are the number of murders (in thousands) in the United States during the 1980s (starting 1984) and 1990s (*Source:* FBI Uniform Crime Reports, www.fbi.gov) ordered by size. Is there a difference between the two decades?

    1980s: 18.7, 19.0, 20.1, 20.6, 20.7, 21.5
    1990s: 15.5, 17.0, 18.2, 19.6, 21.6, 23.3, 23.4, 23.8, 24.5, 24.7

102. Recall Cavendish's density data from Example 6.45. It is known that 6 of his 29 measurements were taken before he changed his experimental apparatus. These measurements were

    5.42, 5.47, 5.50, 5.53, 5.57, 5.61

    Did the change make a difference? Test on the 5% level (use the normal approximation).

# 7

# LINEAR MODELS

## 7.1 INTRODUCTION

When the normal distribution was introduced in Section 2.7 and, particularly, when the central limit theorem was presented in Section 4.3, its importance in statistics was pointed out. Since the central limit theorem basically says that any quantity that can be seen as a sum of a large number of independent random contributions can be considered to be, at least approximately, normally distributed, it was argued that many quantities that we tend to study in practice satisfy this.

We have already looked at IQ as an example, but this is a somewhat artificial measure that is specifically constructed to be normally distributed. For another, more relevant example, consider body length of a randomly chosen individual. We can easily come up with dozens of factors that affect a person's length like parents' lengths, intake of various nutrients during childhood, exercise, sleep habits, whether the mother smoked or consumed alcohol during pregnancy, access to proper medical care, and so on.[1]

Therefore, it is quite common to make the assumption that random samples come from normal distributions with unknown mean and variance and in this chapter we will present a number of inference methods that are specifically developed to handle

[1] Age and sex are also important factors but they influence body length somewhat differently and cannot as easily be described as random contributions. For simplicity, let us assume that we consider body lengths of individuals in the same age group and of the same sex.

*Probability, Statistics, and Stochastic Processes*, Second Edition. Peter Olofsson and Mikael Andersson.

this case. Models that include normally distributed variation are usually called *linear models*, a term that hopefully will become clearer as we go along.

## 7.2 SAMPLING DISTRIBUTIONS

Before going into the various inference methods of normally distributed samples, we need to introduce some new distributions that will be useful in deriving confidence intervals and hypothesis tests. Since they can be used to describe the properties of sample means $\bar{X}$ and sample variances $s^2$, they are usually referred to as *sampling distributions*.

The first distribution is defined as follows.

***Definition 7.1.*** If the random variable $Y$ has pdf

$$f(x) = \frac{1}{2^{r/2}\Gamma(r/2)} x^{r/2-1} e^{-x/2}, \quad x \geq 0$$

then $Y$ is said to have a *chi-square distribution* with $r$ degrees of freedom, written $Y \sim \chi_r^2$.

The chi-square distribution is related to the normal distribution in the following way.

**Proposition 7.1.** Let $X_1, \ldots, X_r$ be i.i.d. random variables that are $N(0, 1)$ and let $Y = \sum_{k=1}^{r} X_k^2$. Then $Y \sim \chi_r^2$.

*Proof.* We begin by deriving the following useful equation

$$\int_0^\infty e^{-kx} x^{\alpha-1}\, dx = \int_0^\infty e^{-t} \left(\frac{t}{k}\right)^{\alpha-1} \frac{dt}{k} = \frac{\Gamma(\alpha)}{k^\alpha} \tag{7.1}$$

using the variable substitution $t = kx$ and the definition of the gamma function $\Gamma(\alpha)$ from Section 2.8.2. This can now be used to obtain the moment generating function, introduced in Section 3.11.2, of the chi-square distribution as

$$\begin{aligned} M_Y(t) = E[e^{tY}] &= \int_0^\infty e^{tx} \frac{1}{2^{r/2}\Gamma(r/2)} x^{r/2-1} e^{-x/2}\, dx \\ &= \frac{1}{2^{r/2}\Gamma(r/2)} \int_0^\infty x^{r/2-1} e^{-(1/2-t)x}\, dx \\ &= \frac{1}{2^{r/2}\Gamma(r/2)} \times \frac{\Gamma(r/2)}{(\frac{1}{2} - t)^{r/2}} = \frac{1}{(1-2t)^{r/2}} \end{aligned}$$

The next step is to derive the moment generating function of the $X_k^2$ variables as

$$\begin{aligned} M_{X_k^2}(t) &= E[e^{-tX_k^2}] = \int_{-\infty}^{\infty} e^{-tx^2} \frac{1}{\sqrt{2\pi}} e^{-x^2/2}\, dx \\ &= \frac{1}{\sqrt{1-2t}} \int_{-\infty}^{\infty} \frac{1}{(1-2t)^{-1/2}\sqrt{2\pi}} e^{-x^2/(2(1-2t)^{-1})}\, dx = \frac{1}{\sqrt{1-2t}} \end{aligned}$$

where the last equality follows from the observation that the integrand is the pdf of a normal distribution with mean 0 and variance $(1-2t)^{-1}$ and therefore integrates to one. The result now follows from Proposition 3.40 as

$$\begin{aligned} M_Y(t) &= M_{X_1^2}(t) \cdots M_{X_r^2}(t) = \left(M_{X_k^2}(t)\right)^r \\ &= \left(\frac{1}{\sqrt{1-2t}}\right)^r = \frac{1}{(1-2t)^{r/2}} \end{aligned} \tag{7.2}$$

■

It is clear from the definition and, particularly, from Proposition 7.1 that a chi-squared variable is always nonnegative. Since we mainly will use the chi-square distribution for confidence intervals and hypothesis tests, we will not go further into the theoretical properties of this distribution. Critical values can be calculated numerically using Definition 7.1 and some of them are presented in Table A.4.

The second sampling distribution is defined as follows.

***Definition 7.2.*** If the random variable $Z$ has pdf

$$f(x) = \frac{\Gamma((r+s)/2)}{\Gamma(r/2)\Gamma(s/2)} \left(\frac{r}{s}\right)^{r/2} x^{r/2-1} \left(1 + \frac{rx}{s}\right)^{-(r+s)/2}, \quad x \geq 0$$

then $Z$ is said to have an *F distribution* with $r$ and $s$ degrees of freedom, written $Z \sim F_{r,s}$.

The $F$ distribution can in turn be characterized in terms of the chi-square distribution.

**Proposition 7.2.** Let $X \sim \chi_r^2$ and $Y \sim \chi_s^2$ be independent and let

$$Z = \frac{X/r}{Y/s}$$

Then $Z \sim F_{r,s}$.

*Proof.* Let $f_X(x)$ and $f_Y(y)$ denote the pdfs of $X$ and $Y$, respectively. Then we can use Corollary 3.2 to express the cdf of $Z' = X/Y$ as

$$F_{Z'}(z) = P\left(\frac{X}{Y} \le z\right) = \int_0^\infty P(X \le yz) f_Y(y)\, dy = \int_0^\infty F_X(yz) f_Y(y)\, dy$$

Taking the derivative with respect to $z$ yields the pdf

$$\begin{aligned}
f_{Z'}(z) &= \int_0^\infty y f_X(yz) f_Y(y)\, dy \\
&= \int_0^\infty y \times \frac{1}{2^{r/2}\Gamma(r/2)} (yz)^{r/2-1} e^{-yz/2} \times \frac{1}{2^{s/2}\Gamma(s/2)} y^{s/2-1} e^{-y/2}\, dy \\
&= \frac{z^{r/2-1}}{2^{(r+s)/2}\Gamma(r/2)\Gamma(s/2)} \int_0^\infty y^{(r+s)/2-1} e^{-(1+z)y/2}\, dy \\
&= \frac{\Gamma((r+s)/2) z^{r/2-1}}{\Gamma(r/2)\Gamma(s/2)(1+z)^{(r+s)/2}}
\end{aligned}$$

where the last equality follows from Equation (7.1). Finally, using the methods of Section 2.3.2 yields the pdf of $Z$ as

$$f_Z(x) = \frac{r}{s} f_{Z'}\left(\frac{r}{s}x\right) = \frac{\Gamma((r+s)/2)}{\Gamma(r/2)\Gamma(s/2)} \left(\frac{r}{s}\right)^{r/2} x^{r/2-1} \left(1 + \frac{rx}{s}\right)^{-(r+s)/2}$$

■

Again, it is obvious that the ratio of two nonnegative chi-square variables is also nonnegative. Critical values of the $F$ distribution are given in Table A.5

We have now come to the last sampling distribution.

***Definition 7.3.*** If the random variable $Z$ has pdf

$$f(x) = \frac{\Gamma((r+1)/2)}{\sqrt{r\pi}\,\Gamma(r/2)} \left(1 + \frac{x^2}{r}\right)^{-(r+1)/2}, \quad x \in R$$

it is said to have a *t distribution* with $r$ degrees of freedom, written $Z \sim t_r$.

This can also be characterized in terms of the previous distributions.

**Proposition 7.3.** Let $X \sim N(0, 1)$ and $Y \sim \chi_r^2$ be independent and let

$$Z = \frac{X}{\sqrt{Y/r}}$$

Then $Z \sim t_r$.

*Proof.* Since $X^2$ is chi-square distributed with one degree of freedom, we realize that $V = Z^2 \sim F_{1,r}$ and, conversely, that

$$Z = \begin{cases} \sqrt{V} & \text{with probability } 1/2. \\ \\ -\sqrt{V} & \text{with probability } 1/2. \end{cases}$$

because of symmetry of the normal distribution. We can now express the cdf of $Z$ as

$$F_Z(x) = \frac{1}{2}P(-\sqrt{V} \le x) + \frac{1}{2}P(\sqrt{V} \le x) = \frac{1}{2}P(\sqrt{V} \ge -x) + \frac{1}{2}P(\sqrt{V} \le x)$$

When $x < 0$, we get

$$F_Z(x) = \frac{1}{2}P(V \ge x^2) + 0 = \frac{1}{2}(1 - F_V(x^2))$$

and if $x > 0$, we get

$$F_Z(x) = \frac{1}{2} + \frac{1}{2}P(V \le x^2) = \frac{1}{2}(1 + F_V(x^2))$$

Taking the derivative with respect to $x$ yields the pdf

$$f_Z(x) = |x| f_V(x^2) = |x| \frac{\Gamma((r+1)/2)}{\Gamma(1/2)\Gamma(r/2)} \left(\frac{1}{r}\right)^{1/2} (x^2)^{-1/2} \left(1 + \frac{x^2}{r}\right)^{-(r+1)/2}$$

$$= \frac{\Gamma((r+1)/2)}{\sqrt{r\pi}\,\Gamma(r/2)} \left(1 + \frac{x^2}{r}\right)^{-(r+1)/2}$$

where we used the property $\Gamma(1/2) = \sqrt{\pi}$. ■

Just like the standard normal distribution, the $t$ distribution is symmetric around 0. Indeed, it looks very similar to the standard normal distribution; the main difference is that it has *heavier tails* (see Figure 7.1). The difference gets smaller the larger the value of $r$, and as $r \to \infty$, the $t$ distribution converges to the normal distribution. Critical values of the $t$ distribution can be found in Table A.3.

## 7.3 SINGLE SAMPLE INFERENCE

In this chapter, we are going to develop inference methods for $\mu$ and $\sigma^2$ based on a sample $X_1, \ldots, X_n$ of i.i.d. normally distributed variables with unknown mean $\mu$ and variance $\sigma^2$. Many of the methods will be similar to those introduced in Chapter 6. What remains to be done is to derive new distributions of relevant statistics based on the sampling distributions of the previous section.

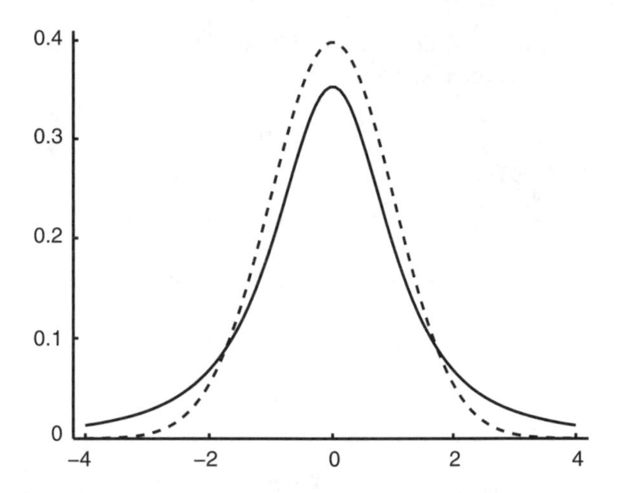


**FIGURE 7.1** The pdfs of the standard normal distribution (dashed line) and a $t$ distribution with $r = 2$ degrees of freedom.

### 7.3.1 Inference for the Variance

It turns out that it is more logical to start with the variance $\sigma^2$ in this setting. Let us go directly to the key result.

**Proposition 7.4.** Let $X_1, \ldots, X_n$ be a sample from a $N(\mu, \sigma^2)$ distribution. Then

$$\frac{(n-1)s^2}{\sigma^2} \sim \chi^2_{n-1}$$

a chi-square distribution with $n - 1$ degrees of freedom.

*Proof.* We first expand the expression

$$\begin{aligned}
&\sum_{i=1}^{n}(X_i - \mu)^2 \\
&= \sum_{i=1}^{n} \left((X_i - \bar{X}) + (\bar{X} - \mu)\right)^2 \\
&= \sum_{i=1}^{n}(X_i - \bar{X})^2 + n(\bar{X} - \mu)^2 + \sum_{i=1}^{n}(X_i\bar{X} - \bar{X}^2 - X_i\mu + \bar{X}\mu) \\
&= (n-1)s^2 + n(\bar{X} - \mu)^2 + n\bar{X}^2 - n\bar{X}^2 - n\bar{X}\mu + n\bar{X}\mu \\
&= (n-1)s^2 + n(\bar{X} - \mu)^2
\end{aligned}$$

Dividing this by $\sigma^2$ yields that

$$\frac{(n-1)s^2}{\sigma^2} + \left(\frac{\bar{X}-\mu}{\sigma/\sqrt{n}}\right)^2 = \sum_{i=1}^{n}\left(\frac{X_i-\mu}{\sigma}\right)^2$$

The second term on the left-hand side is the square of a standard normal variable and, hence, chi-square distributed with one degree of freedom. The expression on the right-hand side is the sum of the squares of $n$ independent standard normal variables and, hence, chi-square distributed with $n$ degrees of freedom. It can be shown that $\bar{X}$ and $s^2$ are independent for normal samples, so Proposition 3.40 and (7.2) together imply that

$$M(t) \times \frac{1}{\sqrt{1-2t}} = \frac{1}{(1-2t)^{n/2}}$$

where $M(t)$ is the moment generating function of $(n-1)s^2/\sigma^2$. This means that

$$M(t) = \frac{1}{(1-2t)^{(n-1)/2}}$$

which is the mgf of the chi-square distribution with $n-1$ degrees of freedom. ∎

We can use this result to find a confidence interval for $\sigma^2$. If we want confidence level $q$, we need to find $x_1$ and $x_2$ such that

$$P\left(x_1 \leq \frac{(n-1)s^2}{\sigma^2} \leq x_2\right) = q$$

that is

$$P\left(\frac{(n-1)s^2}{x_2} \leq \sigma^2 \leq \frac{(n-1)s^2}{x_1}\right) = q$$

The interval we get is not unique; in fact, there are infinitely many ways to choose $x_1$ and $x_2$ so that the last equation is satisfied for a given $q$. One additional requirement often is that the interval is *symmetric* in the sense that we are equally likely to miss the parameter to the right and to the left. Then we must choose $x_1$ and $x_2$ such that

$$P\left(\frac{(n-1)s^2}{\sigma^2} \leq x_1\right) = P\left(\frac{(n-1)s^2}{\sigma^2} \geq x_2\right) = \frac{1-q}{2}$$

(see Figure 7.2). Since $1-(1-q)/2 = (1+q)/2$, we get the following confidence interval.

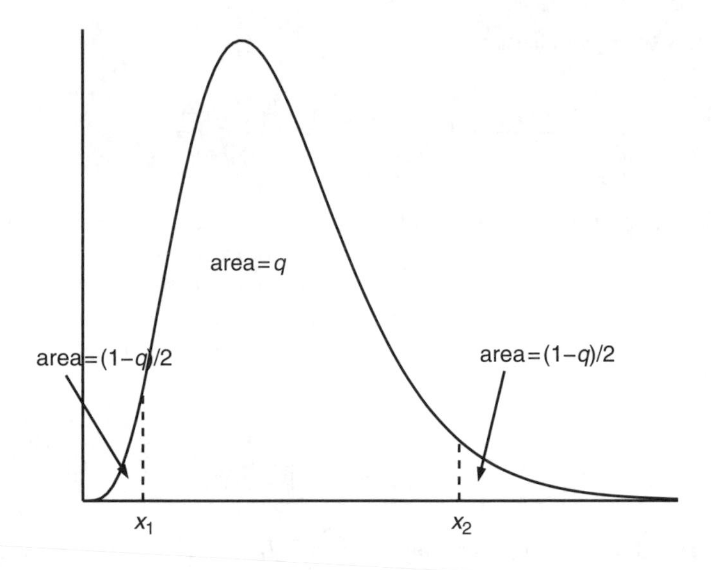


**FIGURE 7.2** The pdf of a chi-square distribution and how to choose $x_1$ and $x_2$ in order to obtain a symmetric confidence interval for $\sigma^2$ with confidence level $q$.

**Proposition 7.5.** Let $X_1, \ldots, X_n$ be a random sample from a $N(\mu, \sigma^2)$ distribution where $\mu$ is unknown. A $100q\%$ symmetric confidence interval for $\sigma^2$ is

$$\frac{(n-1)s^2}{x_2} \leq \sigma^2 \leq \frac{(n-1)s^2}{x_1} \quad (q)$$

where $F_{\chi^2_{n-1}}(x_1) = (1-q)/2$ and $F_{\chi^2_{n-1}}(x_2) = (1+q)/2$.

Note how this interval is not of the form "estimator $\pm$ something," which is because the chi-square distribution is not symmetric like the normal distribution or $t$ distribution.

***Example 7.1.*** Find the observed symmetric 95% confidence interval for the standard deviation $\sigma$ in the lightbulb example.

From Example 6.5 we have $s^2 = 9392$. With $n-1 = 4$ and $(1+q)/2 = 0.975$, Table A.4 gives $x_1 = 0.48$ and $x_2 = 11.14$, which gives $(n-1)s^2/x_2 = 4 \times 9392/11.14 = 3372$ and $(n-1)s^2/0.48 = 78,267$. These are the bounds for $\sigma^2$ and in order to get bounds for $\sigma$, take square roots to get the confidence interval

$$58 \leq \sigma \leq 280 \quad (0.95)$$

□

We can also use Proposition 7.4 to come up with a test procedure.

**Proposition 7.6.** Suppose that $X_1, \ldots, X_n$ is a sample from a normal distribution where we wish to test

$$H_0 : \sigma = \sigma_0 \quad \text{versus} \quad H_A : \sigma \neq \sigma_0$$

The test statistic is

$$X^2 = \frac{(n-1)s^2}{\sigma_0^2}$$

and $H_0$ is rejected on level $\alpha$ if

$$X^2 \leq c_1 \quad \text{or} \quad X^2 \geq c_2$$

where $F_{\chi^2_{n-1}}(c_1) = \alpha/2$ and $F_{\chi^2_{n-1}}(c_2) = 1 - \alpha/2$.

### 7.3.2 Inference for the Mean

We have already looked at confidence intervals (Example 6.7) and hypothesis tests (Example 6.23) for $\mu$ in the normal distribution when $\sigma$ is known. It turns out that we only have to modify these methods by replacing $\sigma$ with $s$ and the normal distribution with the $t$ distribution according to the following result.

**Proposition 7.7.** Let $X_1, \ldots, X_n$ be i.i.d. $N(\mu, \sigma^2)$ and let $s^2$ be the sample variance. Then

$$\frac{\bar{X} - \mu}{s/\sqrt{n}} \sim t_{n-1}$$

a $t$ distribution with $n - 1$ degrees of freedom.

*Proof.* We can rewrite the expression as follows:

$$\frac{\bar{X} - \mu}{s/\sqrt{n}} = \frac{\dfrac{\bar{X} - \mu}{\sigma/\sqrt{n}}}{\sqrt{\dfrac{(n-1)s^2}{\sigma^2} \Big/ (n-1)}}$$

where the enumerator is $N(0, 1)$ and Proposition 7.4 implies that the denominator is the square root of $\chi^2_{n-1}$ divided by the degrees of freedom. Again, the fact that $\bar{X}$ and $s^2$ are independent for normal samples and Proposition 7.3 completes the proof. ■

We can now find a confidence interval for $\mu$. The inequalities

$$\bar{X} - R \leq \mu \leq \bar{X} + R$$

are equivalent to the inequalities

$$-\frac{R}{s/\sqrt{n}} \leq \frac{\bar{X} - \mu}{s/\sqrt{n}} \leq \frac{R}{s/\sqrt{n}}$$

where we now know that the quantity in the middle has a $t$ distribution with $n - 1$ degrees of freedom. With $t = R\sqrt{n}/s$, we get the equation

$$q = P\left(-t \leq \frac{\bar{X} - \mu}{s/\sqrt{n}} \leq t\right) = F_{t_{n-1}}(t) - F_{t_{n-1}}(-t) = 2F_{t_{n-1}}(t) - 1$$

by symmetry of the $t$ distribution. This finally gives the following proposition.

**Proposition 7.8.** If $X_1, \ldots, X_n$ is a sample from a $N(\mu, \sigma^2)$ distribution where $\sigma^2$ is unknown, a $100q\%$ confidence interval for $\mu$ is

$$\mu = \bar{X} \pm t\frac{s}{\sqrt{n}} \quad (q)$$

where $t$ is such that $F_{t_{n-1}}(t) = (1 + q)/2$.

The cdf of the $t$ distribution is easily computed in many of the mathematical and statistical software packages that are available. For example, in Matlab, the command "cdf( 't', x, r )" gives the value $F_{t_r}(x)$. For your convenience, Table A.3 gives values of $t$ for various sample sizes and confidence levels. For $x$ values larger than those in the table, you may use the $N(0, 1)$ distribution as an approximation.

***Example 7.2.*** Consider the lightbulb example and find a 95% confidence interval for $\mu$.

We have $n = 5$ and, from Example 6.5, $s = 96.9$. Since $(1 + q)/2 = 0.975$ and $n - 1 = 4$, we need to find $t$ such that $F_{t_4}(t) = 0.975$, and from Table A.3 we get $t = 2.78$. This gives the confidence interval

$$\mu = 1086 \pm 2.78 \times \frac{96.9}{\sqrt{5}} = 1086 \pm 120 \quad (0.95)$$

□

By comparing this result with Example 6.7 where we assumed that $\sigma = 100$, we see that we get a considerably wider interval. This is quite typical because using an estimate for a parameter instead of the exact value increases our uncertainty, which is reflected in the result.

The symmetry of the $t$ distribution also makes it possible to design a hypothesis test.

**Proposition 7.9.** Suppose that $X_1, \ldots, X_n$ is a sample from a normal distribution where we wish to test

$$H_0 : \mu = \mu_0 \quad \text{versus} \quad H_A : \mu \neq \mu_0$$

The test statistic is

$$T = \frac{\bar{X} - \mu_0}{s/\sqrt{n}}$$

and $H_0$ is rejected on level $\alpha$ if

$$|T| \geq c$$

where $F_{t_{n-1}}(c) = 1 - \alpha/2$.

This is called the *one-sample t test*. If we test against a one-sided alternative, we should reject $H_0$ only for deviations in one direction of $T$ from 0. Thus, for the alternative $H_A : \mu > \mu_0$, we reject if $T \geq c$ where $F_{t_{n-1}}(c) = 1 - \alpha$, just as in the blood pressure example. If we instead have $H_A : \mu < \mu_0$, we should reject for large negative values of $T$. More specifically, we reject if $T \leq -c$ where $c$ satisfies

$$\alpha = P(T \leq -c) = F_{t_{n-1}}(-c) = 1 - F_{t_{n-1}}(c)$$

which again gives $F_{t_{n-1}}(c) = 1 - \alpha$. We get the following corollary.

**Corollary 7.1.** *In Proposition 7.9, if we test $H_0$ against the one-sided alternative*

$$H_A : \mu > \mu_0 \quad (\text{or } \mu < \mu_0)$$

*then $H_0$ is rejected on level $\alpha$ if*

$$T \geq c \quad (\text{or } T \leq -c)$$

*where $F_{t_{n-1}}(c) = 1 - \alpha$.*

Note how the $c$'s in the two-sided and one-sided tests are not the same. In a two-sided test, we divide the significance level equally between the two directions specified by $H_A$. In a one-sided test, we put all the significance level on one side, which makes it easier to reject in that particular direction but does not reject at all in the other direction. For example, with $n = 10$ and $\alpha = 0.05$, the two-sided test rejects if $|T| \geq 2.26$ and the one-sided test if $T \geq 1.83$ (or $T \leq -1.83$). If we want to do a one-sided test, we should not "waste significance" on the other side. Compare with the comments about one-sided versus two-sided confidence intervals in Section 6.3.3.

***Example 7.3.*** A generator is supposed to give an output voltage of 220 V. It is measured once an hour, and at the end of the day a technician decides whether adjustment is needed. Test on the 5% level if the mean voltage is 220 V, based on the following data:

$$213,\ 223,\ 225,\ 232,\ 232,\ 233,\ 237,\ 238$$

We assume that measurements are independent and follow a normal distribution. The hypotheses are

$$H_0 : \mu = 220 \quad \text{versus} \quad H_A : \mu \neq 220$$

where we choose a two-sided alternative since we are not asking for a particular direction of deviation from 220. The sample size is $n = 8$, and with $\alpha = 0.05$, Table A.3 gives $c = 2.36$ ($n - 1 = 7$, $1 - \alpha/2 = 0.975$). We have $\bar{X} = 229.1$ and $s = 8.3$, which gives the observed test statistic

$$T = \frac{229.1 - 220}{8.3/\sqrt{8}} = 3.1$$

and we reject $H_0$ and conclude that adjustment is needed. □

## 7.4 COMPARING TWO SAMPLES

A common situation is that we have two samples and are interested in making comparisons between them. One typical application is in clinical trials, where we want to determine whether a particular treatment or a new drug is better than some other standard treatment or no treatment at all. Patients are recruited and randomly assigned to two different groups, usually called *treatment* and *control*, and then treated (or not treated) accordingly. After some predetermined period of time, all patients are examined and health indicators like blood pressure or cholesterol level are measured. If it is possible to detect a statistically significant difference, we can claim that the new treatment is efficient.

### 7.4.1 Inference about Means

There are two different situations: one where the samples are independent and one where the observations come in pairs and are dependent. In the first case, we thus have two independent samples $X_1, \ldots, X_n$ and $Y_1, \ldots, Y_m$ (the sample sizes are not necessarily the same). If the $X_k$ have mean $\mu_1$ and the $Y_k$ have mean $\mu_2$, we are interested in estimating the difference $\mu_1 - \mu_2$. The estimator is $\bar{X} - \bar{Y}$ and since $\bar{X}$ and $\bar{Y}$ are independent and normally distributed, we can use this to find a confidence interval for $\mu_1 - \mu_2$. If the variances are $\sigma_1^2$ and $\sigma_2^2$, respectively, then according to what we know about linear combinations of normal distributions, we obtain

$$\bar{X} - \bar{Y} \sim N\left(\mu_1 - \mu_2, \frac{\sigma_1^2}{n} + \frac{\sigma_2^2}{m}\right)$$

where we need to estimate the variances in a way that gives us a known distribution, something that turns out to present a difficult problem. The situation simplifies if we can assume that the variances in the two samples are equal[2] so that $X_k \sim N(\mu_1, \sigma^2)$ and $Y_k \sim N(\mu_2, \sigma^2)$. If we estimate $\sigma^2$ within each sample by

$$s_1^2 = \frac{1}{n-1}\sum_{k=1}^{n}(X_k - \bar{X})^2 \quad \text{and} \quad s_2^2 = \frac{1}{m-1}\sum_{k=1}^{m}(Y_k - \bar{Y})^2$$

respectively, we can combine these to get an estimator of $\sigma^2$, based on both samples.

***Definition 7.4.*** Let $X_1, \ldots, X_n$ and $Y_1, \ldots, Y_m$ be two samples, independent of each other, with means $\mu_1$ and $\mu_2$, respectively, and the same variance $\sigma^2$. The *pooled sample variance* is then defined as

$$s_p^2 = \frac{(n-1)s_1^2 + (m-1)s_2^2}{n+m-2}$$

For a proof of the following result and further properties of $s_p^2$, see Problem 31.

**Corollary 7.2.** *The pooled sample variance $s_p^2$ is an unbiased and consistent estimator of the variance $\sigma^2$.*

Note how the pooled sample variance is a weighted average of the two sample variances $s_1^2$ and $s_2^2$, giving larger weight to the one that comes from the larger sample. Note also that neither the definition nor the result about unbiasedness assumes anything about the distributions of the random variables. If we assume normal distributions, the following result helps us with the confidence intervals that we set out to find. We use the obvious notation $s_p$ for the square root of $s_p^2$.

**Proposition 7.10.** If $X_1, \ldots, X_n$ and $Y_1, \ldots, Y_m$ are two independent samples from a $N(\mu_1, \sigma^2)$ distribution and a $N(\mu_2, \sigma^2)$ distribution, respectively, and $s_p^2$ is the pooled sample variance, then

$$\frac{\bar{X} - \bar{Y} - (\mu_1 - \mu_2)}{s_p\sqrt{\dfrac{1}{n} + \dfrac{1}{m}}} \sim t_{n+m-2}$$

a $t$ distribution with $n + m - 2$ degrees of freedom.

[2]This property is called *homoscedasticity*, a great word to throw around at cocktail parties.

Employing the usual method gives the confidence interval. We leave this as an exercise.

**Corollary 7.3.** *Under the assumptions of Proposition 7.10, the confidence interval for $\mu_1 - \mu_2$ is*

$$\mu_1 - \mu_2 = \bar{X} - \bar{Y} \pm t s_p \sqrt{\frac{1}{n} + \frac{1}{m}} \quad (q)$$

*where $F_{t_{n+m-2}}(t) = (1+q)/2$.*

***Example 7.4.*** Suppose that we have the following measured weights (in grams) from two shrimp farms, one in Louisiana and one in Arizona. (Yes, there are shrimp farms in Arizona!) Find the observed 95% confidence interval for the difference between the means:

Louisiana: 15.5, 12.7, 12.1, 14.4, 16.1, 15.0, 16.2

Arizona: 11.9, 13.3, 15.8, 11.6, 10.4, 13.6, 13.8, 12.4, 13.6, 13.0

The sample sizes are $n = 7, m = 10$; the sample means are $\bar{X} = 14.6$ and $\bar{Y} = 12.9$; and the sample variances are $s_1^2 = 2.6$ and $s_2^2 = 2.2$. This gives the pooled sample variance

$$s_p^2 = \frac{(n-1)s_1^2 + (m-1)s_2^2}{n+m-2} = \frac{6 \times 2.6 + 9 \times 2.2}{15} = 2.4$$

With $q = 0.95$ and $n + m - 2 = 15$, we get $t = 2.13$ and the observed confidence interval

$$\mu_1 - \mu_2 = 14.6 - 12.9 \pm 2.13 \times \sqrt{2.4}\sqrt{\frac{1}{7} + \frac{1}{10}}$$

$$= 1.7 \pm 1.6.$$

This interval $[0.1, 3.3]$ does not contain 0. Thus, the difference $\mu_1 - \mu_2$ is entirely above 0, which indicates that the Louisiana shrimp are on average bigger. If we had gotten an interval that included 0, this would have included both the cases $\mu_1 > \mu_2$ and $\mu_1 < \mu_2$, and no difference could have been detected. □

We can also use Proposition 7.10 to construct hypothesis tests. Since we are usually interested in determining whether there is a difference between the samples or not, the most common null hypothesis is

$$H_0 : \mu_1 = \mu_2$$

If we want to test this against a two-sided alternative hypothesis, we get the *two-sample t test*.

**Proposition 7.11.** Let $X_1, \ldots, X_n$ and $Y_1, \ldots, Y_m$ be two independent samples with means $\mu_1$ and $\mu_2$, respectively, and the same variance $\sigma^2$. We wish to test

$$H_0 : \mu_1 = \mu_2 \quad \text{versus} \quad H_A : \mu_1 \neq \mu_2$$

The test statistic is

$$T = \frac{\bar{X} - \bar{Y}}{s_p \sqrt{\dfrac{1}{n} + \dfrac{1}{m}}}$$

and we reject $H_0$ on level $\alpha$ if

$$|T| \geq c$$

where $F_{t_{n+m-2}}(c) = 1 - \alpha/2$.

The one-sided tests are obvious, as follows.

**Corollary 7.4.** *In Proposition 7.11, if we test $H_0$ against the one-sided alternative*

$$H_A : \mu_1 > \mu_2 \quad (or\ \mu_1 < \mu_2)$$

*then $H_0$ is rejected if*

$$T \geq c \quad (or\ T \leq -c)$$

*where $F_{t_{n+m-2}}(c) = 1 - \alpha$.*

***Example 7.5.*** To determine whether smoking is associated with elevated blood pressure, a group of people was divided into two categories, smokers and nonsmokers, and their blood pressures were measured. The systolic pressures for smokers were

$$128,\ 131,\ 137,\ 138,\ 139,\ 141,\ 150,\ 156$$

and for nonsmokers

$$101,\ 125,\ 129,\ 130,\ 130,\ 136,\ 138,\ 140,\ 143,\ 146$$

Test on the 5% level whether smokers had higher blood pressure.

We assume that measurements are normal. If the mean for smokers is $\mu_1$ and for nonsmokers $\mu_2$, we test

$$H_0 : \mu_1 = \mu_2 \quad \text{versus} \quad H_A : \mu_1 > \mu_2$$

We have $n = 8, m = 10$ and with $\alpha = 0.01$, Table A.3 gives $c = 1.75$ ($n + m - 2 = 16$, $1 - \alpha = 0.95$). The sample means are $\bar{X} = 140$ and $\bar{Y} = 132$ and the pooled sample variance

$$s_p^2 = \frac{7s_1^2 + 9s_2^2}{16} = 128.5$$

which gives the observed test statistic

$$T = \frac{140 - 132}{\sqrt{128.5} \times \sqrt{\dfrac{1}{8} + \dfrac{1}{10}}} = 1.49$$

and we cannot reject $H_0$. There is no clear support for the theory that smoking raises blood pressure. A practical observation is that the observed value of $T$ is fairly close to the critical value and it might be a good idea to get larger samples to be able to perhaps draw a more definite conclusion. □

If the variances are not equal, the confidence interval and the hypothesis test still work as approximations if $n$ and $m$ are roughly the same. Otherwise, certain adjustments can be made to improve the approximation, as described in the following result.

**Proposition 7.12.** If $X_1, \ldots, X_n$ and $Y_1, \ldots, Y_m$ are two independent samples from a $N(\mu_1, \sigma_1^2)$ distribution and a $N(\mu_2, \sigma_2^2)$ distribution, respectively, then

$$\frac{\bar{X} - \bar{Y} - (\mu_1 - \mu_2)}{\sqrt{\dfrac{s_1^2}{n} + \dfrac{s_2^2}{m}}} \stackrel{d}{\approx} t_\nu$$

where

$$\nu = \frac{\left(\dfrac{s_1^2}{n} + \dfrac{s_2^2}{m}\right)^2}{\dfrac{s_1^4}{n^2(n-1)} + \dfrac{s_2^4}{m^2(m-1)}}.$$

It should be quite clear how this can be used to modify the previous methods of inference.[3] Note that $\nu$ is not necessarily an integer so to be able to use Table A.3, we have to round it off to the nearest smaller integer (to be on the safe side).

To illustrate the other case, paired observations, suppose instead that we are interested in the monthly rate of growth of shrimp at one of the farms. To measure this, we take a sample of shrimp, label and weigh them, and set them aside. After

[3]The test based on Proposition 7.12 is usually referred to as *Welch's t test*.

a month, we weigh the same shrimp again and thus have a pair of weights $(X, Y)$ for each individual shrimp, where $X$ and $Y$ are obviously dependent. We get a sample $(X_1, Y_1), \ldots, (X_n, Y_n)$ of weight pairs and are interested in the difference $\mu_2 - \mu_1$. By letting $D_k = Y_k - X_k$, we can view this as a sample $D_1, \ldots, D_n$, and since $X_k$ and $Y_k$ are normal, so is $D_k$, and we are back at a one-sample problem. The mean of $D_k$ is $E[Y_k] - E[X_k] = \mu_2 - \mu_1$, and the variance is unknown and estimated by the sample variance in the $D$ sample. Note that this means that we do not need to assume that the variances of $X$ and $Y$ are equal.

***Example 7.6.*** Suppose that we have the following weights, before and after a month:

Before: 11.9, 13.3, 15.8, 11.6, 10.4, 13.6, 13.8, 12.4, 13.6, 13.0

After: 20.9, 18.1, 20.9, 13.6, 11.3, 17.2, 20.4, 16.4, 15.5, 21.5

The sample of the differences, $D_k = Y_k - X_k$, "after minus before" is

9.0, 4.8, 5.1, 2.0, 0.9, 3.6, 6.6, 4.0, 1.9, 8.5

which has sample mean $\bar{D} = 4.6$ and sample variance $s^2 = 7.5$. With $n - 1 = 9$ and $q = 0.95$, we get $t = 2.26$ and confidence interval

$$\mu_2 - \mu_1 = \bar{D} \pm t\frac{s}{\sqrt{n}} = 4.6 \pm 2.0 \quad (0.95)$$

□

### 7.4.2 Inference about Variances

Although we are mostly interested in studying changes in the mean values when comparing different groups or treatments, there are situations where changes in the variances may be relevant. Let us say that we want to compare two different measuring devices and want to determine if the measurement error, described by the variance, is smaller for one of them. Another common problem is whether we can assume equal variances so that we can use the methods described in Corollary 7.3 and Proposition 7.11.

The key result, which follows directly from Propositions 7.4 and 7.2, for the methods in this section is the following.

**Proposition 7.13.** If $X_1, \ldots, X_n$ and $Y_1, \ldots, Y_m$ are two independent samples from a $N(\mu_1, \sigma_1^2)$ distribution and a $N(\mu_2, \sigma_2^2)$ distribution, respectively, then

$$\frac{s_1^2/\sigma_1^2}{s_2^2/\sigma_2^2} \sim F_{n-1,m-1}$$

an $F$ distribution with $n - 1$ och $m - 1$ degrees of freedom.

Since the $F$ distribution, just like the chi-square distribution, is asymmetric, we have to formulate a confidence interval for $\sigma_2^2/\sigma_1^2$ as follows.

**Corollary 7.5.** *Under the assumptions of Proposition 7.13, the confidence interval for $\sigma_2^2/\sigma_1^2$ is*

$$x_1 \frac{s_2^2}{s_1^2} \leq \frac{\sigma_2^2}{\sigma_1^2} \leq x_2 \frac{s_2^2}{s_1^2} \quad (q)$$

*where $F_{F_{n-1,m-1}}(x_1) = (1-q)/2$ and $F_{F_{n-1,m-1}}(x_2) = (1+q)/2$.*

***Example 7.7.*** In Example 7.4, the sample variances were found to be $s_1^2 = 2.6$ and $s_2^2 = 2.2$. This difference was not considered to be too large so that the pooled variance, under the assumption of equal variances, was used in the calculations. Let us take a closer look at this assumption.

The sample sizes were $n = 7$ and $m = 10$, which means that we need to use the $F_{6,9}$ distribution. If we want a 95% confidence interval for the ratio of variances, Table A.5 yields the percentiles $F_{F_{6,9}}(0.181) = 0.025$ and $F_{F_{6,9}}(4.32) = 0.975$. Hence, we get the interval

$$0.181 \times \frac{2.2}{2.6} \leq \frac{\sigma_2^2}{\sigma_1^2} \leq 4.32 \times \frac{2.2}{2.6} \quad (0.95)$$

$$0.15 \leq \frac{\sigma_2^2}{\sigma_1^2} \leq 3.6 \quad (0.95)$$

Since the value one is included in the interval, we cannot exclude the possibility that the two variances are equal. □

We can also use Proposition 7.13 to construct a hypothesis test.

**Corollary 7.6.** *Under the assumptions of Proposition 7.13, we wish to test*

$$H_0 : \sigma_1^2 = \sigma_2^2 \quad versus \quad H_A : \sigma_1^2 \neq \sigma_2^2$$

*The test statistic is*

$$F = \frac{s_1^2}{s_2^2}$$

*and we reject $H_0$ on level $\alpha$ if*

$$F \leq x_1 \quad or \quad F \geq x_2$$

*where $F_{F_{n-1,m-1}}(x_1) = (1-q)/2$ and $F_{F_{n-1,m-1}}(x_2) = (1+q)/2$.*

## 7.5 ANALYSIS OF VARIANCE

In this section, we are going to take things one step further and look at statistical methods for comparing more than two independent normal samples. Of course, we can always use the methods of the previous section to carry out pairwise comparisons, but then we will end up in multiple testing problems as described in Section 6.6.4, especially if we have several samples to compare. Instead, we are going to use a more comprehensive approach where we compare the variation within and the variation between the samples to detect any differences in the means. Since we focus on the variation through variances, these methods are referred to as *Analysis of Variance* or *ANOVA*.

### 7.5.1 One-Way Analysis of Variance

Here, we assume that we have $k$ independent samples $X_{i1}, \ldots, X_{in}$ such that $X_{ij} \sim N(\mu_i, \sigma^2)$, where $i = 1, \ldots, k$ and $j = 1, \ldots, n$. Note that we make two simplifications in this model, namely, all samples are of equal size $n$ and they have the same variance $\sigma^2$. ANOVA is mostly used in *experimental design*, where individuals or other experimental units are randomly assigned to different groups and subjected to a number of different treatments and it is quite common to choose groups of equal sizes. Therefore, the variance $\sigma^2$ is interpreted as the natural individual variation, which should have nothing to do with the treatments and should therefore be the same in all groups. Actually, the assumption of equal sizes can be relaxed without much trouble, but we will not consider that case here.

The hypotheses of interest are

$$H_0 : \mu_1 = \ldots = \mu_k \quad \text{versus} \quad H_A : \mu_{i_1} \neq \mu_{i_2} \text{ for some } i_1 \text{ and } i_2$$

Now, we can obtain $k$ independent estimators of the unknown variance as

$$s_i^2 = \frac{1}{n-1} \sum_{j=1}^{n} (X_{ij} - \bar{X}_{i\cdot})^2 \quad i = 1, \ldots, k$$

where

$$\bar{X}_{i\cdot} = \frac{1}{n} \sum_{j=1}^{n} X_{ij}$$

Since they are all based on samples of size $n$, they can be pooled together as

$$s_W^2 = \frac{1}{k} \sum_{i=1}^{k} s_i^2 = \frac{1}{k(n-1)} \sum_{i=1}^{k} \sum_{j=1}^{n} (X_{ij} - \bar{X}_{i\cdot})^2 \tag{7.3}$$

which is the *within-group variance*. If the null hypothesis is true, we can regard the sample means $\bar{X}_{1\cdot}, \ldots, \bar{X}_{k\cdot}$ as a normally distributed sample with mean $\mu_i$ and variance $\sigma^2/n$. This means that we can get a second estimator of $\sigma^2$ as

$$s_B^2 = \frac{n}{k-1}\sum_{i=1}^{k}(\bar{X}_{i\cdot} - \bar{X})^2 = \frac{1}{k-1}\sum_{i=1}^{k}\sum_{j=1}^{n}(\bar{X}_{i\cdot} - \bar{X})^2 \tag{7.4}$$

where

$$\bar{X} = \frac{1}{k}\sum_{i=1}^{k}\bar{X}_{i\cdot} = \frac{1}{nk}\sum_{i=1}^{k}\sum_{j=1}^{n}X_{ij}$$

is the total sample mean. This is usually called the *between-group variance.*

The within-group variance $s_W^2$ is always an unbiased estimator of $\sigma^2$ whereas the same holds for $s_B^2$ only if the null hypothesis is true. If the alternative hypothesis is true, it is possible to show that $E[s_B^2] > \sigma^2$ (see Problem 37), so we should reject $H_0$ if $s_B^2$ is significantly larger than $s_W^2$ or, equivalently, if the test statistic

$$F = \frac{s_B^2}{s_W^2}$$

is large enough. It turns out that $s_W^2$ and $s_B^2$ are independent and chi-square distributed with $k-1$ and $k(n-1)$ degrees of freedom, respectively, under this model, which means that $F \sim F_{k-1,k(n-1)}$ and that critical values can be obtained from Table A.5.

In classical ANOVA notation, the sums in (7.3) and (7.4) and the total variation are denoted

$$\text{SSA} = \sum_{i=1}^{k}\sum_{j=1}^{n}(\bar{X}_{i\cdot} - \bar{X})^2$$

$$\text{SSE} = \sum_{i=1}^{k}\sum_{j=1}^{n}(X_{ij} - \bar{X}_{i\cdot})^2$$

$$\text{SST} = \sum_{i=1}^{k}\sum_{j=1}^{n}(X_{ij} - \bar{X})^2$$

where it can be shown that

$$\text{SST} = \text{SSA} + \text{SSE}$$

The SS in the notation stands for *Sum of Squares* with T for Total, E for Error, and A for treatment $A$.[4] The variance estimators above are also referred to as *Mean Squares* and denoted MSA and MSE, respectively. The result of an analysis of variance is

[4]In higher order ANOVA, combinations of several different treatments (usually denoted $A$, $B$, $C$, and so on) can be studied and corresponding sum of squares computed. Although only one treatment is considered in one-way ANOVA, it is often called treatment $A$.

**TABLE 7.1 An ANOVA Table for a One-Way Analysis of Variance**

| Source | DF | Sum of Squares | Mean Square | $F$-Statistic |
|---|---|---|---|---|
| Treatment $A$ | $k-1$ | SSA | $\text{MSA} = \dfrac{\text{SSA}}{k-1}$ | $F = \dfrac{\text{MSA}}{\text{MSE}}$ |
| Error | $k(n-1)$ | SSE | $\text{MSE} = \dfrac{\text{SSE}}{k(n-1)}$ | |
| Total | $kn-1$ | SST | | |

often summarized in an *ANOVA table*, whose structure is illustrated, for a one-way ANOVA, in Table 7.1. Sometimes, the $p$-value for the $F$ test is also included in the table.

***Example 7.8.*** In 2010, a large farming experiment was carried out in southern Sweden, where nine different varieties of canned peas were grown in five fields each. After harvest, the yield in metric tonnes per hectare was measured and the results were the following.

| Variety | Yields | | | | | Mean Yield | Variance |
|---|---|---|---|---|---|---|---|
| $A$ | 3.18 | 3.33 | 3.87 | 5.27 | 5.58 | 4.25 | 1.24 |
| $B$ | 2.17 | 4.55 | 4.60 | 5.15 | 5.61 | 4.42 | 1.77 |
| $C$ | 2.84 | 3.61 | 4.57 | 4.69 | 5.00 | 4.14 | 0.80 |
| $D$ | 4.52 | 4.74 | 5.74 | 5.93 | 6.18 | 5.42 | 0.55 |
| $E$ | 3.12 | 3.13 | 4.38 | 4.60 | 4.88 | 4.02 | 0.70 |
| $F$ | 1.81 | 2.83 | 3.11 | 3.80 | 4.25 | 3.16 | 0.88 |
| $G$ | 1.94 | 2.72 | 2.80 | 2.94 | 3.56 | 2.79 | 0.34 |
| $H$ | 1.75 | 2.45 | 3.10 | 3.63 | 3.92 | 2.97 | 0.78 |
| $I$ | 3.23 | 3.94 | 4.01 | 4.30 | 4.35 | 3.97 | 0.20 |

We see that the best variety $D$ yields about twice as much as the worst variety $G$ on average, but we also see that there is a large variation between the fields probably due to quality of soil, climatic conditions, drainage, and so on. Let us see if an analysis of variance can determine whether there is any difference between the varieties of peas.

Since the total mean is $\bar{X} = 3.90$, we can use Corollary 6.1 to calculate the variety sum of squares as

$$\text{SSA} = \sum_{i=1}^{9}\sum_{j=1}^{5}(\bar{X}_{i\cdot} - \bar{X})^2 = 5 \times \left(\sum_{i=1}^{9}\bar{X}_{i\cdot}^2 - 9\bar{X}^2\right) = 27.10$$

and the total sum of squares as

$$\text{SST} = \sum_{i=1}^{9}\sum_{j=1}^{5}(X_{ij} - \bar{X})^2 = \sum_{i=1}^{9}\sum_{j=1}^{5}X_{ij}^2 - 45\bar{X}^2 = 56.13$$

**TABLE 7.2 The ANOVA Table for Example 7.8**

| Source | DF | Sum of Squares | Mean Square | $F$-Statistic | $P$-Value |
|---|---|---|---|---|---|
| Variety | 8 | 27.10 | 3.39 | 4.20 | 0.001 |
| Error | 36 | 29.03 | 0.81 | | |
| Total | 44 | 56.13 | | | |

which yields the error sum of squares SSE $= 56.13 - 27.10 = 29.03$.[5] Now we get the mean squares as

$$\text{MSA} = \frac{\text{SSA}}{k-1} = \frac{27.10}{8} = 3.39$$

$$\text{MSE} = \frac{\text{SSE}}{k(n-1)} = \frac{29.03}{36} = 0.81$$

which yields the test statistic

$$F = \frac{\text{MSA}}{\text{MSE}} = \frac{3.39}{0.81} = 4.20$$

The 95% percentile of the $F$ distribution with 8 and 36 degrees of freedom is 2.21, which means that we can reject the null hypothesis on the 5% level and claim that the nine varieties vary in yield. The ANOVA is summarized in Table 7.2. □

### 7.5.2 Multiple Comparisons: Tukey's Method

Suppose that we have managed to reject the null hypothesis of equal means, as in Example 7.8, in an ANOVA. The next natural question is then: Which group or groups differ from the rest and how much? In this section, we will present the most common method of making pairwise comparisons between groups such that the joint level of significance is correct, the so-called *Tukey's method*.[6]

It is based on the following distribution.

***Definition 7.5.*** Let $X_1, \ldots, X_n$ be a normally distributed sample with mean $\mu$ and variance $\sigma^2$ and let $s^2$ be an estimator of $\sigma^2$ such that

$$\frac{ms^2}{\sigma^2} \sim \chi^2_m$$

Then

$$R_{n,m} = \frac{X_{(n)} - X_{(1)}}{s}$$

follows the *studentized range distribution* with $n$ and $m$ degrees of freedom.

[5]This is the usual order of calculations since SSA and SST are both easier to calculate than SSE.

[6]Named after the American chemist and mathematician John Tukey (1915–2000).

The cdf can be expressed as an integral in terms of the $t$ distribution and can, unfortunately, be calculated only numerically. Pairwise confidence intervals can now be obtained as follows.

**Proposition 7.14.** Let $\{X_{ij} : i = 1, \ldots, k; j = 1, \ldots, n\}$ be independent random variables such that $X_{ij} \sim N(\mu_i, \sigma^2)$. Then

$$\mu_{i_1} - \mu_{i_2} = \bar{X}_{i_1\cdot} - \bar{X}_{i_2\cdot} \pm r \frac{s_W}{\sqrt{n}}$$

are pairwise confidence intervals for $i_1 = 1, \ldots, k$ and $i_2 = 1, \ldots, k$, where $i_1 \neq i_2$, with joint confidence level $q$, where $s_W^2$ is defined in (7.3) and $F_{R_{k,k(n-1)}}(r) = q$.

Table A.6 gives 95% percentiles of the studentized range distribution.

***Example 7.9.*** In Example 7.8, we had $k = 9$ and $n = 5$, which yields the critical value $r = 4.66$. Using the within-group variation $s_W^2 = \text{MSE} = 0.81$ gives us the statistical error

$$r \frac{s_W}{\sqrt{n}} = 4.66 \times \sqrt{\frac{0.81}{5}} = 1.87$$

When we compare all group means, we find that only the three intervals

$$\mu_D - \mu_F = 5.42 - 3.16 \pm 1.87 = 2.26 \pm 1.87$$

$$\mu_D - \mu_G = 5.42 - 2.79 \pm 1.87 = 2.63 \pm 1.87$$

$$\mu_D - \mu_H = 5.42 - 2.97 \pm 1.87 = 2.45 \pm 1.87$$

do not contain the value zero. Hence, we can conclude, with 95% confidence, that variety $D$ has a mean yield larger than varieties $F$, $G$, and $H$. □

### 7.5.3 Kruskal–Wallis Test

If the assumption of normally distributed samples with equal variances is not satisfied, there is an alternative approach called *Kruskal–Wallis test*,[7] which is based on ranks like the Wilcoxon rank sum test introduced in Section 6.9.2 but using the ANOVA method of comparing sources of variation. The first step is to rank the values in all

[7]Named after the American mathematician William Kruskal (1919–2005) and economist W. Allen Wallis (1912–1998).

samples combined. If we denote the rank of observation $X_{ij}$ by $R_{ij}$, the test statistic is defined as

$$K = (kn - 1)\frac{\sum_{i=1}^{k}\sum_{j=1}^{n}(\bar{R}_{i\cdot} - \bar{R})^2}{\sum_{i=1}^{k}\sum_{j=1}^{n}(R_{ij} - \bar{R})^2} \tag{7.5}$$

where $\bar{R}$ is the mean of all ranks and $\bar{R}_{i\cdot}$ is the mean rank of group $i$. Note that the enumerator and denominator in (7.5) correspond to SSA and SST, respectively, in the one-way ANOVA. This means that we would expect a larger variation in mean ranks $\bar{R}_{i\cdot}$ if the group means are different and, consequently, should reject the null hypothesis if the test statistic $K$ is large enough. If there are no ties, it is possible to simplify (7.5) (see Problem 38) as

$$K = \frac{12}{kn^2(kn+1)}\sum_{i=1}^{k} R_{i\cdot}^2 - 3(kn+1) \tag{7.6}$$

where $R_{i\cdot} = \sum_{j=1}^{n} R_{ij}$ are the group rank sums. In most textbooks on statistics, $K \overset{d}{\approx} \chi^2_{k-1}$ is often used to obtain critical values and $p$-values for the Kruskal–Wallis test. However, this approximation is not very accurate for small $n$, especially if $k$ is large. The problem is that it is virtually impossible to calculate the exact distribution of $K$ (except for really small $n$ and $k$) due to the enormous number of combinations of ranks that has to be considered. In Table A.11, estimates of 95% percentiles are presented based on simulation methods like those presented in Section 6.4.3. Therefore, these values may not all be accurate down to the last decimal, but they at least give better critical values than the chi-square approximation.

***Example 7.10.*** There were some irregularities like different variances in the data in Example 7.8 suggesting that the conditions for carrying out an ANOVA may not have been met completely. Let us therefore apply the Kruskal–Wallis test to the data and see if we get a similar result. We first rank all 45 observations as follows.

| Variety | Ranks | | | | | Rank Sum |
|---|---|---|---|---|---|---|
| $A$ | 15 | 17 | 22 | 40 | 41 | 135 |
| $B$ | 4 | 31 | 33.5 | 39 | 42 | 149.5 |
| $C$ | 9 | 19 | 32 | 35 | 38 | 133 |
| $D$ | 30 | 36 | 43 | 44 | 45 | 198 |
| $E$ | 13 | 14 | 29 | 33.5 | 37 | 126.5 |
| $F$ | 2 | 8 | 12 | 21 | 26 | 69 |
| $G$ | 3 | 6 | 7 | 10 | 18 | 44 |
| $H$ | 1 | 5 | 11 | 20 | 23 | 60 |
| $I$ | 16 | 24 | 25 | 27 | 28 | 120 |

Note that we have one tie in the value 4.60 that occurs twice in the table and, consequently, gets the midrank 33.5. This means that (7.6) does not give the correct value of $K$ and that we should use (7.5) instead. In practice, (7.6) is actually used anyway because as long as there are not too many ties, the error is really negligible. In this case, (7.5) gives the (correct) value $K = 22.196$ while (7.6) gives $K = 22.194$. The critical value for $k = 9$ and $n = 5$ on the 5% level is 14.62 according to Table A.11, which means that we can still reject the null hypothesis. In fact, the $p$-value can be calculated to 0.0009, which is quite similar to the corresponding value in the traditional ANOVA in Example 7.8. □

## 7.6 LINEAR REGRESSION

The world is full of linear relationships. When measurements are taken, observations seldom lie on a straight line, though, as a result of measurement error and other random effects. In Section 3.8, we learned how the correlation coefficient can be used to describe the degree of linearity in a relationship between two random variables, and in Section 3.9, we saw that if $X$ and $Y$ are bivariate normal, there is a linear relationship between them in the sense that the conditional expectation of $Y$ given $X$ is a linear function of $X$. In this section, we will investigate a similar model but assume that only $Y$ is random. For fixed $x$, we assume that

$$Y = a + bx + \epsilon$$

where $a$ and $b$ are constants and

$$\epsilon \sim N(0, \sigma^2)$$

This means that deviations from the line $y = ax + b$ are normally distributed random variables, and an equivalent formulation is

$$Y \sim N(a + bx, \sigma^2)$$

We call this the *simple linear regression* model, and the line $y = a + bx$ is called the *regression line*. The $x$ values can be chosen by us or come from observations, but we think of them as fixed, not random. For each $x$ value, the corresponding $Y$ value is measured, where we assume that consecutive $Y$ values are independent random variables. Thus, we have the $x$ values $x_1, \ldots, x_n$, get a sample $Y_1, \ldots, Y_n$, where

$$Y_k = a + bx_k + \epsilon_k \sim N(a + bx_k, \sigma^2) \tag{7.7}$$

and our main objective is to estimate the intercept $a$ and slope $b$. Note that the $Y_k$ are independent but do not have the same distribution. With a slight modification of the maximum-likelihood method, we can still define the likelihood function as

$$L(a, b) = \prod_{k=1}^{n} f_{a,b}(Y_k)$$

where

$$f_{a,b}(Y_k) = \frac{1}{\sigma\sqrt{2\pi}} e^{-(Y_k - a - bx_k)^2/2\sigma^2}$$

and we can maximize over $a$ and $b$ in the usual way by taking logarithms and setting the partial derivatives to 0. The resulting estimators are stated next, leaving the proof for the Problems section.

**Proposition 7.15.** In the linear regression model, the maximum-likelihood estimators of $a$ and $b$ are

$$\widehat{b} = \frac{\sum_{k=1}^{n}(x_k - \bar{x})(Y_k - \bar{Y})}{\sum_{k=1}^{n}(x_k - \bar{x})^2}$$

$$\widehat{a} = \bar{Y} - \widehat{b}\bar{x}$$

The line $y = \widehat{a} + \widehat{b}x$ is the estimated regression line (or simply the regression line if there is no risk of confusion). There are different ways to rewrite the expression for $\widehat{b}$. We introduce one, which is suitable for computations.

**Corollary 7.7.** *Define the sums*

$$S_{xx} = \sum_{k=1}^{n}(x_k - \bar{x})^2 = \sum_{k=1}^{n} x_k^2 - \frac{1}{n}\left(\sum_{k=1}^{n} x_k\right)^2$$

$$S_{xY} = \sum_{k=1}^{n}(x_k - \bar{x})(Y_k - \bar{Y}) = \sum_{k=1}^{n} x_k Y_k - \frac{1}{n}\left(\sum_{k=1}^{n} x_k\right)\left(\sum_{k=1}^{n} Y_k\right)$$

*Then*

$$\widehat{b} = \frac{S_{xY}}{S_{xx}}$$

***Example 7.11.*** Let us look at a famous data set, Edwin Hubble's 1929 investigation of the relationship between a galaxy's distance from Earth and its recession velocity (see also Example 3.44). He got the following 24 pairs of observations, where $x$ values are distances in megaparsecs and $Y$ values are the corresponding velocities in kilometers per second. Find the estimated regression line.

Distance: 0.032, 0.034, 0.214, 0.263, 0.275, 0.275, 0.45, 0.5, 0.5, 0.63
0.8, 0.9, 0.9, 0.9, 0.9, 1.0, 1.1, 1.1, 1.4, 1.7, 2.0, 2.0, 2.0, 2.0
Velocity: 170, 290, −130, −70, −185, −220, 200, 290, 280, 200, 300 − 30
650, 150, 500, 920, 450, 500, 500, 960, 500, 850, 800, 1090

Computation of these sums gives

$$\sum_{k=1}^{24} x_k = 21.87, \quad \sum_{k=1}^{24} Y_k = 8965, \quad \sum_{k=1}^{24} x_k^2 = 29.52, \quad \sum_{k=1}^{24} x_k Y_k = 12,519$$

and

$$S_{xx} = 29.52 - \frac{21.87^2}{24} = 9.583$$

$$S_{xY} = 12,519 - \frac{21.87 \times 8965}{24} = 4348$$

which gives

$$\widehat{b} = \frac{4348}{9.583} = 454$$

$$\widehat{a} = \frac{8965}{24} - 454 \times \frac{21.87}{24} = -40.0$$

and the regression line is $y = -40.0 + 454x$, which is shown together with the data set in Figure 7.3. Note how the line intersects the $y$-axis very close to the origin. Why do you think this is the case? □

If you are familiar with the *method of least squares*, you may have noticed that the estimated regression line is precisely the line that least-squares fitting gives. (Why is this?) Our assumption of normally distributed errors enables us to further analyze the estimated regression line. Let us investigate properties of our estimators $\widehat{a}$ and $\widehat{b}$. Since the $x_k$ are fixed, the only randomness is in the $Y_k$, and since both $\widehat{a}$ and $\widehat{b}$ are linear combinations of the normally distributed $Y_k$, the estimators themselves must be normal. It is straightforward to compute their means and variances (see Problem 39).

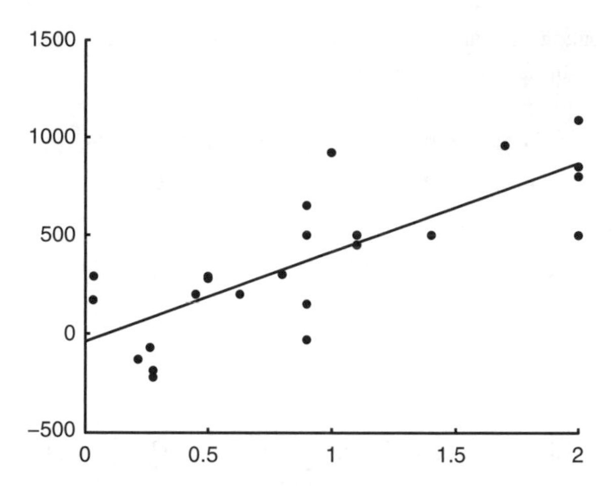


**FIGURE 7.3** Plot of Hubble's galaxy data and the estimated regression line. Distance is on the $x$-axis and recession velocity on the $y$-axis.

**Corollary 7.8.** *The estimators* $\widehat{a}$ *and* $\widehat{b}$ *have normal distributions with means and variances*

$$E[\widehat{a}] = a, \qquad \mathrm{Var}[\widehat{a}] = \frac{\sigma^2 \sum_{k=1}^{n} x_k^2}{n S_{xx}}$$

$$E[\widehat{b}] = b, \qquad \mathrm{Var}[\widehat{b}] = \frac{\sigma^2}{S_{xx}}$$

In particular, Corollary 7.8 tells us that both $\widehat{a}$ and $\widehat{b}$ are unbiased estimators and that

$$\frac{\widehat{a} - a}{\sqrt{\mathrm{Var}[\widehat{a}]}} \sim N(0, 1) \quad \text{and} \quad \frac{\widehat{b} - b}{\sqrt{\mathrm{Var}[\widehat{b}]}} \sim N(0, 1) \tag{7.8}$$

where the variances depend on $\sigma^2$, which must thus be estimated. The following estimator is used.

**Proposition 7.16.** In the linear regression model, the estimator

$$s^2 = \frac{1}{n-2} \sum_{k=1}^{n} (Y_k - \widehat{a} - \widehat{b} x_k)^2$$

is an unbiased estimator of $\sigma^2$.

Note how this resembles the sample variance from Section 6.2, in the sense that it sums the squares of the observations minus their estimated expected values. Since there are two estimated parameters, we divide by $n - 2$ instead of $n$. Another similarity is that $s^2$ is not the MLE; the MLE is obtained by dividing by $n$ rather than $n - 2$. For practical calculations, the following result is useful.

**Corollary 7.9.** *Define the sum*

$$S_{YY} = \sum_{k=1}^{n}(Y_k - \bar{Y})^2 = \sum_{k=1}^{n} Y_k^2 - \frac{1}{n}\left(\sum_{k=1}^{n} Y_k\right)^2$$

*Then*

$$s^2 = \frac{1}{n-2}\left(S_{YY} - \frac{S_{xY}^2}{S_{xx}}\right)$$

For inference about $\sigma^2$, we need the distribution of $s^2$. In the light of Proposition 7.4, the following is to be expected.

**Proposition 7.17.** In the linear regression model,

$$\frac{(n-2)s^2}{\sigma^2} \sim \chi^2_{n-2}$$

a chi-squared distribution with $n - 2$ degrees of freedom.

Note how this follows the by now familiar pattern; the number of degrees of freedom equals the number of terms in the sum, minus the number of estimated parameters. Confidence intervals for $\sigma^2$ can now be derived on the basis of the chi-square distribution (see Problem 50).

As we might expect, if the estimator $s$ replaces $\sigma$ in the expressions in Equation (7.8), we get $t$ distributions with $n - 2$ degrees of freedom instead of normal distributions. Recalling the computational formulas above, we have

$$T_a = \frac{\widehat{a} - a}{s}\sqrt{\frac{nS_{xx}}{\sum_{k=1}^{n} x_k^2}} \sim t_{n-2}$$

$$T_b = \frac{\widehat{b} - b}{s}\sqrt{S_{xx}} \sim t_{n-2}$$

which gives us the following confidence intervals.

**Corollary 7.10.** *In the linear regression model with unknown variance $\sigma^2$, confidence intervals for a and b with confidence level q are given by*

$$a = \widehat{a} \pm ts\sqrt{\frac{\sum_{k=1}^{n} x_k^2}{nS_{xx}}} \quad (q)$$

$$b = \widehat{b} \pm ts\frac{1}{\sqrt{S_{xx}}} \quad (q)$$

*where* $F_{t_{n-2}}(t) = (1+q)/2$.

***Example 7.12.*** Find 95% confidence intervals for $a$ and $b$ in Hubble's galaxy data.

From Example 7.11, we have $\widehat{a} = -40.0$, $\widehat{b} = 454$, $\sum_k x_k^2 = 29.52$, $\sum_k Y_k = 8965$, $S_{xx} = 9.583$, $S_{xY} = 4348$, and $n = 24$. We also need $\sum_k Y_k^2 = 6,516,925$ to calculate

$$S_{YY} = 6,516,925 - \frac{8965^2}{24} = 3,168,124$$

Now, we get the sample variance as

$$s^2 = \frac{1}{22}\left(3,168,124 - \frac{4348^2}{9.583}\right) = 54,328$$

which gives $s = 233$. With $q = 0.95$ and $n - 2 = 22$, we get $t = 2.07$. The confidence intervals are

$$a = -40.0 \pm 2.07 \times 233\sqrt{\frac{29.52}{24 \times 9.583}} = -40.0 \pm 173 \quad (0.95)$$

$$b = 454 \pm 2.07 \times 233\frac{1}{\sqrt{9.583}} = 454 \pm 156 \quad (0.95)$$

Note that the intervals are quite wide; in particular, the one for $a$ is all over the place. Two cautious conclusions we can draw are that $b$ is positive and that we cannot rule out that $a = 0$ (which, of course, makes sense). □

As you know by now, the step from confidence intervals to hypothesis tests is not big. Since the quantities $T_a$ and $T_b$ given above have known distributions, we can construct tests of the null hypotheses

$$H_0 : a = a_0 \quad \text{and} \quad H_0 : b = b_0$$

based on the test statistics $T_a$ and $T_b$, where we set $a = a_0$ and $b = b_0$. If the alternatives are two-sided, we reject on level $\alpha$ if $|T_a| \geq t$ and $|T_b| \geq t$, respectively, where

$F_{t_{n-2}}(t) = 1 - \alpha/2$. For one-sided tests, the usual adjustments are made, using $1 - \alpha$ instead of $1 - \alpha/2$.

***Example 7.13.*** The Old Faithful geyser (see Problem 111 in Chapter 3) is a rich source of data. Two quantities that are routinely measured are the times between eruptions (typically 30–120 min) and the length of eruptions (1–5 min). It is known that these are positively correlated, and the Yellowstone park rangers use a formula that is roughly $y = 13x + 30$ to predict the time $y$ until the next eruption, based on the length $x$ of the most recent eruption. (*Source*: The Geyser Observation and Study Association, www.geyserstudy.org.) Assume a linear regression model and estimate the regression line based on the following 20 observations, and also test whether the slope 13 is correct:

Length: 1.7, 1.7, 1.7, 1.8, 2.3, 3.1, 3.4, 3.5, 3.7, 3.9, 3.9, 4.0, 4.0, 4.0, 4.1, 4.3
4.4, 4.6, 4.7, 4.9

Time: 55, 58, 56, 42, 50, 57, 75, 80, 69, 80, 74, 68, 76, 90, 84, 80, 78, 74, 76, 76

To compute the estimators, we need the usual sums $\sum_k x_k = 69.7$, $\sum_k Y_k = 1398$, $\sum_k x_k^2 = 264.65$, and $\sum_k x_k Y_k = 5083.1$, which with $n = 20$ gives us

$$S_{xx} = 264.65 - \frac{69.7^2}{20} = 21.75$$

$$S_{xY} = 5083.1 - \frac{69.7 \times 1398}{20} = 211.1$$

Now, we get the estimators

$$\widehat{b} = \frac{211.1}{21.75} = 9.71$$

$$\widehat{a} = \frac{1398}{20} - 9.71\frac{69.7}{20} = 36.1$$

which gives the regression line $y = 9.71x + 36.1$. To test

$$H_0 : b = 13 \quad \text{versus} \quad H_A : b \neq 13$$

on level 0.05, we have $n - 2 = 18$ and $1 - 0.05/2 = 0.975$ and reject if $|T_b| \geq 2.10$, where $b = 13$. We first need $\sum_k Y_k^2 = 100,768$ and

$$S_{YY} = 100,768 - \frac{1398^2}{20} = 3048$$

to compute

$$s^2 = \frac{1}{18}\left(3048 - \frac{211.1^2}{21.75}\right) = 55.5$$

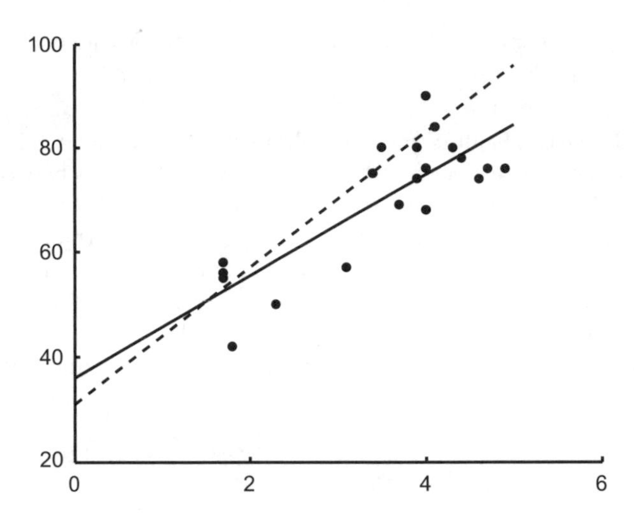


**FIGURE 7.4** The Old Faithful data with the regression line (solid) and the rangers' line (dashed).

which gives $s = 7.45$ and test statistic

$$T_b = \frac{9.71 - 13}{7.45}\sqrt{21.75} = -2.06$$

and we cannot reject $H_0$, although it is close. The rangers know what they are doing and, of course, base their estimates on much larger data sets than our 20 observations. See Figure 7.4 for the data and the two lines. □

### 7.6.1 Prediction

In the previous problem, it was mentioned that one variable (length of last eruption) is used to predict another (time until next eruption). Generally, in the linear regression model, if we observe or choose $x$, what can we say about $Y$? We know that the mean of $Y$ is $E[Y] = a + bx$, which we estimate by $\widehat{a} + \widehat{b}x$, and we can use this estimator as a predictor of $Y$ (recall Section 3.7.2). How good is it? The difference between the true value $Y$ and the predicted value $\widehat{a} + \widehat{b}x$ is

$$D = Y - \widehat{a} - \widehat{b}x = Y - \bar{Y} + \widehat{b}(x - \bar{x})$$

Note here that $\bar{x}$, $\bar{Y}$, and $\widehat{b}$ are computed from the observations $(x_1, Y_1), \ldots, (x_n, Y_n)$, $x$ is our new observed $x$ value, and $Y$ the yet unobserved value that we are trying to predict. Clearly, $E[D] = 0$, and for the variance, note that $Y$ is independent of $\bar{Y}$ and $\widehat{b}$. It is also easy to show that $\bar{Y}$ and $\widehat{b}$ are uncorrelated (see Problem 40) and hence

$$\begin{aligned}\text{Var}[D] &= \text{Var}[Y] + \text{Var}[\bar{Y}] + (x - \bar{x})^2\text{Var}[\,\widehat{b}\,] \\ &= \sigma^2 + \frac{\sigma^2}{n} + \frac{(x - \bar{x})^2\sigma^2}{S_{xx}}.\end{aligned}$$

Moreover, since $D$ is a linear combination of normal distributions, it is normal with mean 0 and the variance given above. Hence, $D/\sqrt{\text{Var}[D]} \sim N(0, 1)$, and estimating $\sigma^2$ by $s^2$ and using the computational formula for the sum of squares gives

$$T = \frac{D}{s\sqrt{1 + \dfrac{1}{n} + \dfrac{(x - \bar{x})^2}{S_{xx}}}} \sim t_{n-2}$$

a $t$ distribution with $n - 2$ degrees of freedom. Thus, we can find $t$ such that $P(-t \leq T \leq t) = q$ for our desired $q$, and since $D = Y - \widehat{a} - \widehat{b}x$ we get the following *prediction interval* (see also Problem 19).

**Corollary 7.11.** *Consider the linear regression model with estimators $\widehat{a}$ and $\widehat{b}$. If $x$ has been observed, a $100q\%$ prediction interval for the corresponding $Y$ value is given by*

$$Y = \widehat{a} + \widehat{b}x \pm ts\sqrt{1 + \frac{1}{n} + \frac{(x - \bar{x})^2}{S_{xx}}}$$

*where $F_{t_{n-2}} = (1 + q)/2$.*

Note the difference between a prediction interval and a confidence interval. A prediction interval captures the next $Y$ value for a given $x$ value; a confidence interval captures the long-term average of a large number of $Y$ values for a given $x$ value. A confidence interval for the expected value of $Y$ for a given $x$ would thus look somewhat different (see Problem 48).

Note the length of the prediction interval increases with increasing $(x - \bar{x})^2$, that is, the farther away the chosen $x$ value is from $\bar{x}$. This reflects the fact that our estimated regression line is most reliable near $\bar{x}$. Indeed, for an $x$ far outside the range of our original $x_1, \ldots, x_n$, the prediction is highly uncertain and should preferably be avoided.

***Example 7.14.*** Consider the Old Faithful example and suppose that we just observed an eruption that lasted for 1.5 min. Find the predicted time until the next eruption and a 95% prediction interval.

The estimated regression line is $y = 9.71x + 36.1$, so with $x = 1.5$ we get predicted $y$ value $9.71 \times 1.5 + 36.1 = 50.6$. For the prediction interval, we have $n = 20$, $t = 2.10$, $s = 7.5$, $\bar{x} = 69.7/20 = 3.5$, and $S_{xx} = 21.75$. We get

$$Y = 50.6 \pm 2.10 \times 7.5\sqrt{1 + \frac{1}{20} + \frac{(1.5 - 3.5)^2}{21.75}} = 50.6 \pm 17.5 \quad (0.95)$$

□

### 7.6.2 Goodness of Fit

To determine whether the linear regression model is reasonable, it is useful to examine the difference between the observed $Y$ values and those predicted by the estimated regression line. The deviations

$$E_k = Y_k - \widehat{a} - \widehat{b}x_k, \quad k = 1, \ldots, n$$

are called the *residuals*. Do not confuse the $E_k$ and the $\epsilon_k$ from above; $E_k$ is the difference between the observed $Y$ value and the *estimated* line $\widehat{a} + \widehat{b}x_k$, whereas $\epsilon_k$ is the difference between the observed $Y$ value and the *true* line $a + bx_k$, so we can think of $E_k$ as a predictor of $\epsilon_k$. If the residuals are plotted against the $x$ values, they should appear more or less randomly scattered and not display any discernible patterns, if the model is correct. In Figure 7.5a, the residuals for Hubble's galaxy data are plotted, and they look just fine. In contrast, consider plot (b) where the residuals tend to be below 0 at the ends and above in the middle, indicating a nonlinear relation. Finally, in (c), the residuals tend to increase in magnitude with increasing $x$, indicating that the variance is not constant but depends on $x$. The plots in (b) and (c) illustrate the two most common deviations from the linear regression model. Try to figure out how the data set together with the estimated regression line would look in these two cases.

Plotting the residuals provides a quick diagnostic tool to assess whether the assumed linear model is reasonable. For a more detailed analysis, it should be noted that the residual $E_k$ has mean 0 and a variance that depends on $k$. If the residuals are divided by their estimated standard deviations, these are called the *standardized residuals*. If the model is correct, the standardized residuals are approximately independent and $N(0, 1)$, which can be used for further analysis.

The situation with nonconstant variance is common and may be modeled by *weighted linear regression*, letting $Y = a + bx + \epsilon$ where $\epsilon \sim N(0, w(x)\sigma^2)$, where the $w(x)$ are weights depending on $x$.

It is also possible to apply the ANOVA method of comparing sources of variation to quantify the goodness of fit. The residuals defined above can be used to obtain the

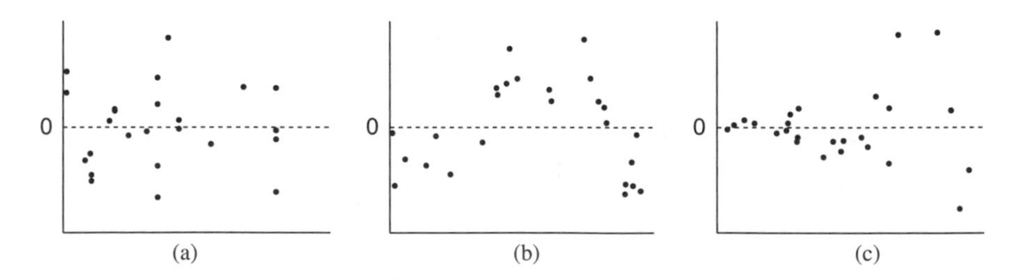


**FIGURE 7.5** Three residual plots. In (a), the residuals for Hubble's galaxy data are plotted and give no reason to suspect that the model is wrong. In (b), the "banana shape" indicates a nonlinear relationship, and in (c), the residuals indicate a nonconstant variance.

error sum of squares as

$$\text{SSE} = \sum_{k=1}^{n}(Y_k - \widehat{a} - \widehat{b}x_k)^2 = S_{YY} - \frac{S_{xY}^2}{S_{xx}}$$

and the total sum of squares is defined in a similar way as in Section 7.5 as

$$\text{SST} = \sum_{k=1}^{n}(Y_k - \bar{Y})^2 = S_{YY}$$

The difference between the total variation and the random variation

$$\text{SSR} = \text{SST} - \text{SSE} = \frac{S_{xY}^2}{S_{xx}}$$

denoted *regression sum of squares* can be interpreted as the amount of variation that can be explained by the regression model. Hence, the larger the SSR in relation to the SSE, the better because then the points will lie close to the estimated regression line and we will be able to predict $Y$ values with high accuracy. The proportion of variation explained by the model, named *coefficient of determination* and defined as

$$R^2 = 1 - \frac{\text{SSE}}{\text{SST}} = \frac{S_{xY}^2}{S_{xx}S_{YY}}$$

is the most common way of quantifying the goodness of fit of a regression model.

***Example 7.15.*** The coefficient of determination for the regression model for Hubble's data is

$$R^2 = \frac{4348^2}{9.583 \times 3,168,124} = 0.62$$

which means that most of the variation, roughly 60%, can be explained by the estimated linear relationship. The Old Faithful analysis yields a coefficient of determination of

$$R^2 = \frac{211.1^2}{21.75 \times 3048} = 0.67$$

which is even better with two-thirds of the variation explained. □

### 7.6.3 The Sample Correlation Coefficient

The simple linear regression model is useful, as we have seen, when we want to estimate the linear relationship between two measured quantities and make predictions about new observations. Sometimes, we are merely interested in estimating the strength of the connection between two variables and test whether it exists. This can be done if we assume that data can be regarded as observations from the sample of pairs $(X_1, Y_1), \ldots, (X_n, Y_n)$ and consider the correlation coefficient $\rho$, which we recall is defined as

$$\rho = \frac{E[(X - \mu_1)(Y - \mu_2)]}{\sqrt{\text{Var}[X]\text{Var}[Y]}}$$

Note that we regard both variables as random in this context while the $x$ variables previously were fixed. The difference lies mainly in the assumption of a causal relationship in the simple linear regression model. By making $Y$ random and $x$ fixed, we are implicitly assuming that the value of $x$ have a direct impact on the value of $Y$. Here, we just claim that the two variables are connected and say nothing about the direction of influence. It may even be the case that both variables are affected by a third variable not included in the model and not directly linked at all.

Anyway, the following estimator should make intuitive sense.

***Definition 7.6.*** If $(X_1, Y_2), \ldots, (X_n, Y_n)$ is a sample from a bivariate distribution, we define the *sample correlation coefficient* as

$$R = \frac{\sum_{k=1}^{n}(X_k - \bar{X})(Y_k - \bar{Y})}{\sqrt{\sum_{k=1}^{n}(X_k - \bar{X})^2 \sum_{k=1}^{n}(Y_k - \bar{Y})^2}} = \frac{S_{XY}}{\sqrt{S_{XX}S_{YY}}}$$

We use the term "sample correlation coefficient" to point out the analogy with sample mean and sample variance. If you wish to memorize a longer name, the term *product moment correlation coefficient* is also used.[8] Note that the square of the sample correlation coefficient is actually the same as the coefficient of determination introduced in Section 7.6.2, which further emphasizes the link between regression and correlation.

***Example 7.16.*** Below is a data set of 10 observations of the daily closing prices for the two U.S. stock market indices, Dow Jones Industrial Average and Nasdaq Composite Index, chosen at random from among the years of 1971 and 2003, rounded to the nearest integer (and listed in chronological order). Estimate the correlation coefficient of $X$ and $Y$.

Dow: 887, 833, 821, 961, 1259, 2176, 2820, 3442, 7289, 10715

Nasdaq: 108, 86, 74, 95, 283, 352, 430, 696, 1228, 2028

[8]Sometimes, it is also prefixed with *Pearson's* to honor its discoverer, English statistician Karl Pearson (1857–1936), one of the founders of the theory of statistics.

This is merely an exercise in computing the sums needed. Thus, we have

$$\sum_{k=1}^{10} X_k = 31203, \quad \sum_{k=1}^{10} Y_k = 5380$$

$$\sum_{k=1}^{10} X_k^2 = 1.97 \times 10^8, \quad \sum_{k=1}^{10} Y_k^2 = 6.53 \times 10^6$$

and finally

$$\sum_{k=1}^{10} X_k Y_k = 3.57 \times 10^7$$

This gives us the sum of squares

$$S_{XX} = 1.97 \times 10^8 - \frac{31203^2}{10} = 9.98 \times 10^7$$

$$S_{YY} = 6.53 \times 10^6 - \frac{5380^2}{10} = 3.63 \times 10^6$$

$$S_{XY} = 3.57 \times 10^7 - \frac{31203 \times 5380}{10} = 1.89 \times 10^7$$

Inserting these in the expression for $R$ from Definition 7.6 yields

$$R = \frac{1.89 \times 10^7}{\sqrt{(9.98 \times 10^7) \times (3.63 \times 10^6)}} = 0.995$$

which is a very high correlation. For comparison, consider the following 10 pairs of closing prices, chosen at random from the year 1971 only (the year that the Nasdaq index was started).

Dow: 941, 947, 889, 874, 850, 908, 888, 798, 829, 873

Nasdaq: 104, 104, 114, 110, 105, 110, 108, 106, 109, 105

This time the estimated correlation is $R = -0.14$, which is not only much smaller in magnitude but also negative. The reason for the difference in correlation is the relatively stable growth of the markets over a long period, such as the 32 years we first sampled from, but less stable behavior over a shorter period, such as a single year. Since both indices mirror the market as a whole but are compiled from different sets of stocks, we would expect their correlation to be positive and high for longer periods of time, but for shorter periods of time, the correlation could be anything. The two data sets are plotted in Figure 7.6. See also Problem 51. □

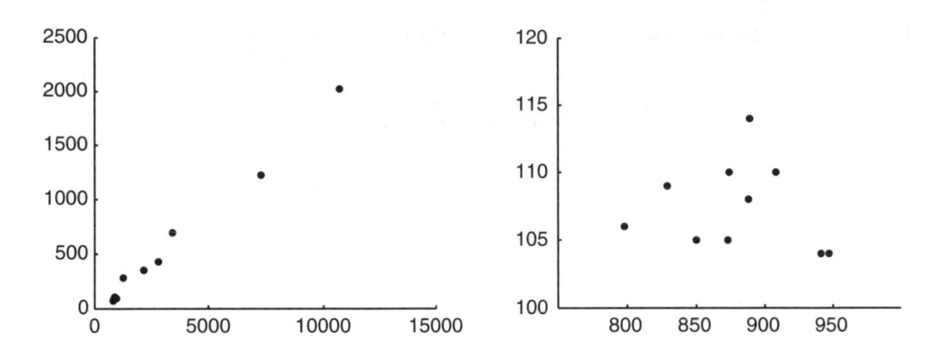


**FIGURE 7.6** Plots of Dow ($x$-axis) versus Nasdaq ($y$-axis) for 10 randomly sampled days during 1971–2003 (left) and 1971 (right).

The most common test problem regarding the correlation coefficient is to test if $\rho = 0$, if there is an association between the random variables $X$ and $Y$, and intuitively the test should be based on how far $R$ is from 0. In general, we cannot find the distribution of $R$ but if our data are from a bivariate normal distribution with $\rho = 0$, the following result gives the exact distribution for a particular function of $R$, which we can use to construct our test procedure.[9]

**Proposition 7.18.** Let $(X_1, Y_1), \ldots, (X_n, Y_n)$ be a sample from a bivariate normal distribution with $\rho = 0$, and let

$$T = R\sqrt{\frac{n-2}{1-R^2}}$$

Then

$$T \sim t_{n-2}$$

a $t$ distribution with $n - 2$ degrees of freedom.

The test procedure is now straightforward. If $R = 0$, then also $T = 0$, and as $R$ increases from $-1$ to 1, $T$ increases from $-\infty$ to $\infty$. Thus, we reject $H_0$ if $T$ is too far from 0. Let us state this formally.

**Proposition 7.19.** Let $(X_1, Y_1), \ldots, (X_n, Y_n)$ be a sample from a bivariate normal distribution where we wish to test

$$H_0 : \rho = 0 \quad \text{versus} \quad H_A : \rho \neq 0$$

[9]The result in Proposition 7.18 actually requires only that $Y$ be normal; in fact, $X$ does not even have to be random.

The test is based on the statistic $T$ above, and we reject $H_0$ on level $\alpha$ if

$$|T| \geq c$$

where $F_{t_{n-2}}(c) = 1 - \alpha/2$.

If we want to test against a one-sided alternative instead, for example, with the intention of showing a positive (or negative) correlation, we reject if $T \geq c$ (or $T \leq -c$), where $F_{t_{n-2}}(c) = 1 - \alpha$. Note that since we assume that $X$ and $Y$ are bivariate normal, a test of $\rho = 0$ is in fact a test for independence.

***Example 7.17.*** Use the two data sets from Example 7.16 to test whether the Dow and Nasdaq indices are correlated. Test on the 5% level.

We test

$$H_0 : \rho = 0 \quad \text{versus} \quad H_A : \rho \neq 0$$

where we choose a two-sided alternative since we are not testing for any particular direction of the correlation (although the one-sided alternative $\rho > 0$ would also make sense). Let us first find the critical value. We have $n = 10$, and with $\alpha = 0.05$, we get $F_{t_8}(t) = 0.95$, which gives $t = 2.31$ so that $H_0$ is rejected if $|T| \geq 2.31$. In the first data set we have $R = 0.995$, which gives

$$T = 0.995\sqrt{\frac{8}{1 - 0.995^2}} = 28.1$$

and we reject $H_0$. In the second data set, we have $R = -0.14$, which gives

$$T = \frac{8 \times (-0.14)}{\sqrt{1 - (-0.14)^2}} = -0.40$$

and we cannot reject $H_0$. The negative correlation for the short time period is not significant. □

If we wish to test $\rho = \rho_0$ for some $\rho_0 \neq 0$, we cannot use $T$. Also, if we wish to find a confidence interval for $\rho$, we cannot use $T$, either. The problem with $T$ is that it does not include the unknown parameter $\rho$, which would be necessary for both these tasks. In these cases an approximation can be used, as described in Problem 53.

### 7.6.4 Spearman's Correlation Coefficient

If data cannot be assumed to be normal, which is often the case with market data such as the above, there is an alternative approach called *Spearman's correlation coefficient*[10] based on ranks similar to the methods in Section 6.9. The idea is quite simple as in many nonparametric procedures.

***Definition 7.7.*** If $(X_1, Y_2), \ldots, (X_n, Y_n)$ is a sample from a bivariate distribution, we define *Spearman's correlation coefficient* as

$$R_S = \frac{\sum_{k=1}^{n}(r_k - \bar{r})(s_k - \bar{s})}{\sqrt{\sum_{k=1}^{n}(r_k - \bar{r})^2 \sum_{k=1}^{n}(s_k - \bar{s})^2}} = \frac{S_{rs}}{\sqrt{S_{rr}S_{ss}}}$$

where $r_k$ is the rank of $X_k$ among $X_1, \ldots, X_n$ and $s_k$ is the rank of $Y_k$ among $Y_1, \ldots, Y_n$.

If there are no ties in the data, the formula in Definition 7.7 can be simplified quite considerably.

**Proposition 7.20.** If there are no ties, Spearman's correlation coefficient can be expressed as

$$R_S = 1 - \frac{6}{n(n^2-1)} \sum_{k=1}^{n}(r_k - s_k)^2$$

Hence, what you do is rank the $X$ and $Y$ observations separately, sum the squared differences, and use the formula in Proposition 7.20. It also turns out that it is possible to test the hypothesis $H_0 : \rho = 0$ using the same test statistic and $t$ distribution as above, only that the result is approximate

$$T = R_S\sqrt{\frac{n-2}{1-R_S^2}} \overset{d}{\approx} t_{n-2}$$

for large $n$.

***Example 7.18.*** Let us return to Example 7.16 and calculate Spearman's correlation coefficient for the two data sets. When ranking the two sets of observations from the period 1971–2003, we get that

[10]Named after the English psychologist Charles Spearman (1863–1945).

| Dow: | 887 | 833 | 821 | 961 | 1259 | 2176 | 2820 | 3442 | 7289 | 10715 |
|---|---|---|---|---|---|---|---|---|---|---|
| Rank: | 3 | 2 | 1 | 4 | 5 | 6 | 7 | 8 | 9 | 10 |
| Nasdaq: | 108 | 86 | 74 | 95 | 283 | 352 | 430 | 696 | 1228 | 2028 |
| Rank: | 4 | 2 | 1 | 3 | 5 | 6 | 7 | 8 | 9 | 10 |

Since the ranks only differ in two positions and by only one unit, we get the sum of squared differences $\sum_k (r_k - s_k)^2 = 2$ and the correlation coefficient

$$R_S = 1 - \frac{6}{10 \times 99} \times 2 = 0.988$$

which is quite close to the previous result.

When looking at the data from 1971, we discover several ties for the Nasdaq data set and have to use midranks.

| Dow: | 941 | 947 | 889 | 874 | 850 | 908 | 888 | 798 | 829 | 873 |
|---|---|---|---|---|---|---|---|---|---|---|
| Rank: | 9 | 10 | 7 | 5 | 3 | 8 | 6 | 1 | 2 | 4 |
| Nasdaq: | 104 | 104 | 114 | 110 | 105 | 110 | 108 | 106 | 109 | 105 |
| Rank: | 1.5 | 1.5 | 10 | 8.5 | 3.5 | 8.5 | 6 | 5 | 7 | 3.5 |

Unfortunately, this means that the formula in Proposition 7.20 does not yield an accurate value, so here we have to use the more complicated one in Definition 7.7. After some standard calculations, we reach the result $R_S = -0.17$, which is also quite close to the previous estimate. □

## 7.7 THE GENERAL LINEAR MODEL

When comparing the one-way ANOVA model of Section 7.5 with the simple linear regression model of Section 7.6, we notice some common features. In both cases, observations are assumed to be normally distributed random variables where the means are described by deterministic parametric linear functions. Any model that satisfies these requirements is called a *General Linear Model*,[11] or GLM for short, which is really a large class of models with tremendous importance in many statistical applications. We will not expand this rich field here but rather offer a brief outline to give some feeling of the kind of problems addressed and methods used.

We begin with the formal definition, which is most conveniently expressed in matrix notation.

[11]Not to be confused with the even larger class *Generalized Linear Model*, usually abbreviated GLIM, also covering other distributions like the binomial and Poisson distributions.

***Definition 7.8.*** Let $\mathbf{Y}$ be a $n \times 1$ vector of random variables, $X$ a $n \times k$ matrix with known entries, $\boldsymbol{\beta}$ a $k \times 1$ parameter vector and $\boldsymbol{\epsilon} \sim \mathbf{N_n}(\mathbf{0}, \Sigma)$. The General Linear Model satisfies

$$\mathbf{Y} = X\boldsymbol{\beta} + \boldsymbol{\epsilon}$$

The matrix $X$ is usually called the *design matrix* because it defines the design of the model, that is, how the unknown parameters in $\boldsymbol{\beta}$ are linked to the observed values in $\mathbf{Y}$.

Using standard linear algebra, the GLM can also be expressed

$$Y_i = \sum_{j=0}^{k-1} x_{ij}\beta_j + \epsilon_i, \quad i = 1, \ldots, n \tag{7.9}$$

where $Y_i$, $\beta_j$, and $\epsilon_i$ are the components of the vectors $\mathbf{Y}$, $\boldsymbol{\beta}$, and $\boldsymbol{\epsilon}$, respectively, and $x_{ij}$ are the entries of the matrix $X$. Note that the vector $\boldsymbol{\beta}$ and the columns of $X$ are indexed from 0 to $k-1$, a common standard that will be motivated below.

In the most general form, the multivariate structure of the random error vector $\boldsymbol{\epsilon}$ makes it possible to have different variances and dependent errors, but it is quite common to assume that the covariance matrix can be written $\Sigma = \sigma^2 I_n$, where $I_n$ is the identity matrix with ones in the diagonal and zeros in all other positions. This will give us a model with independent errors and constant variance. Another common requirement is that the first column of the design matrix $X$ should consist of ones, that is, $x_{i0} = 1$ in (7.9), which is mainly a technical condition that will guarantee that parameter estimates are unique. The consequence is that all linear models get an intercept parameter $\beta_0$.

***Example 7.19 (Linear Regression).*** It is quite easy to see that the simple linear regression model (7.7) of Section 7.6 fits into this framework. Let $k = 2$ and

$$X = \begin{pmatrix} 1 & x_1 \\ 1 & x_2 \\ \vdots & \vdots \\ 1 & x_n \end{pmatrix} \qquad \boldsymbol{\beta} = \begin{pmatrix} a \\ b \end{pmatrix}$$

It is also quite straightforward to generalize this into the *Multiple Linear Regression* model

$$Y_i = \beta_0 + \beta_1 x_{i1} + \beta_2 x_{i2} + \ldots + \beta_{k-1} x_{i,k-1} + \epsilon_i \quad i = 1, \ldots, n$$

using

$$X = \begin{pmatrix} 1 & x_{11} & x_{12} & \cdots & x_{1,k-1} \\ 1 & x_{21} & x_{22} & \cdots & x_{2,k-1} \\ \vdots & \vdots & \vdots & \ddots & \vdots \\ 1 & x_{n1} & x_{n2} & \cdots & x_{n,k-1} \end{pmatrix} \qquad \boldsymbol{\beta} = \begin{pmatrix} \beta_0 \\ \beta_1 \\ \vdots \\ \beta_{k-1} \end{pmatrix}$$

□

***Example 7.20 (One-Way Analysis of Variance).*** To see that the one-way ANOVA is a linear model requires a bit more work. Recall from Section 7.5.1 that $X_{ij} \sim N(\mu_i, \sigma^2)$, which can be expressed

$$X_{ij} = \mu_i + \epsilon_{ij}, \quad i = 1, \ldots, k, \quad j = 1, \ldots, n$$

where $\epsilon_{ij} \sim N(0, \sigma^2)$ are independent. This almost fits into the form of (7.9), except that we have two indices and no intercept term.

First of all, we have to reorganize the observations $X_{ij}$ and the error terms in the vectors

$$\mathbf{Y} = (X_{11}, X_{12}, \ldots, X_{1n}, X_{21}, X_{22}, \ldots, X_{2n}, X_{31}, \ldots, X_{kn})'$$
$$\boldsymbol{\epsilon} = (\epsilon_{11}, \epsilon_{12}, \ldots, \epsilon_{1n}, \epsilon_{21}, \epsilon_{22}, \ldots, \epsilon_{2n}, \epsilon_{31}, \ldots, \epsilon_{kn})'$$

where $M'$ means the transpose of the matrix $M$.

Next, we reparameterize the group means $\mu_i$ as

$$\mu_i = \mu + \alpha_i \quad i = 1, \ldots, k$$

so that we get an intercept $\beta_0 = \mu$. Unfortunately, this adds another parameter to the model, so some kind of restriction is necessary. In standard ANOVA, it is common to require that $\sum_{i=1}^{k} \alpha_i = 0$. This means that when $\alpha_1, \ldots, \alpha_{k-1}$ have been determined, $\alpha_k = -\sum_{i=1}^{k-1} \alpha_i$ showing that the number of parameters is still $k$. Another alternative is to set one of the $\alpha_i$ to zero (most commonly $\alpha_1 = 0$ or $\alpha_k = 0$), which gives an equivalent model, but we will not consider this case here.

In order to define the design matrix, we need to introduce the so-called *dummy variables*

$$\begin{aligned} x_{i1} &= 1, \quad i = 1, \ldots, n \\ x_{i2} &= 1, \quad i = n+1, \ldots, 2n \\ x_{i,k-1} &= 1, \quad i = (k-2)n + 1, \ldots, (k-1)n \\ x_{ij} &= -1, \quad i = (k-1)n + 1, \ldots, kn \quad j = 1, \ldots, k-1 \end{aligned}$$

If we let **0** and **1** denote the $n \times 1$ vectors consisting of zeros and ones, respectively, we can express the design matrix and parameter vector as

$$X = \begin{pmatrix} \mathbf{1} & \mathbf{1} & \mathbf{0} & \cdots & \mathbf{0} & \mathbf{0} \\ \mathbf{1} & \mathbf{0} & \mathbf{1} & \cdots & \mathbf{0} & \mathbf{0} \\ \vdots & \vdots & \vdots & \ddots & \vdots & \vdots \\ \mathbf{1} & \mathbf{0} & \mathbf{0} & \cdots & \mathbf{0} & \mathbf{1} \\ \mathbf{1} & -\mathbf{1} & -\mathbf{1} & \cdots & -\mathbf{1} & -\mathbf{1} \end{pmatrix} \quad \boldsymbol{\beta} = \begin{pmatrix} \mu \\ \alpha_1 \\ \alpha_2 \\ \vdots \\ \alpha_{k-1} \end{pmatrix}$$

□

If $k = n$ we would have a square design matrix $X$ and we could get estimates of $\boldsymbol{\beta}$ as the solution to the linear matrix equation

$$\mathbf{Y} = X\boldsymbol{\beta} \tag{7.10}$$

which, if the inverse of $X$ exists, can be written $\boldsymbol{\beta} = X^{-1}\mathbf{Y}$. In practical applications, it is common to have more observations than unknown parameters (often much more), that is, $k < n$. In that case, we can reduce the dimensionality of (7.10) by multiplying with $X'$ from the left to get

$$X'\mathbf{Y} = X'X\boldsymbol{\beta}$$

If the inverse of the $k \times k$-matrix $X'X$ exists, we get the solution $\boldsymbol{\beta} = (X'X)^{-1}X'\mathbf{Y}$. In fact, we get the following important result.

**Proposition 7.21.** Assume that $(X'X)^{-1}$ exists in a GLM. Then,

$$\widehat{\boldsymbol{\beta}} = (X'X)^{-1}X'\mathbf{Y}$$

is a unique MLE of $\boldsymbol{\beta}$. It also holds that

$$\widehat{\boldsymbol{\beta}} \sim \mathbf{N_k}(\boldsymbol{\beta}, (X'X)^{-1}X'\Sigma X(X'X)^{-1}) \tag{7.11}$$

If we consider the special case where $\Sigma = \sigma^2 I_n$, we can simplify the covariance matrix in (7.11) as

$$(X'X)^{-1}X'(\sigma^2 I_n)X(X'X)^{-1} = \sigma^2(X'X)^{-1}X'X(X'X)^{-1} = \sigma^2(X'X)^{-1}$$

Then, it is possible to show that

$$s^2 = \frac{1}{n-k}||\mathbf{Y} - X\widehat{\boldsymbol{\beta}}|| = \frac{1}{n-k}\sum_{i=1}^{n}\left(Y_i - \sum_{j=0}^{k-1} x_{ij}\widehat{\beta}_j\right)^2$$

is an unbiased estimator of $\sigma^2$ such that

$$\frac{(n-k)s^2}{\sigma^2} \sim \chi^2_{n-k}$$

As a consequence of this, it follows that

$$\frac{\widehat{\beta}_j - \beta_j}{\frac{s}{\sigma}\sqrt{\text{Var}(\widehat{\beta}_j)}} = \frac{\widehat{\beta}_j - \beta_j}{s\sqrt{z_{jj}}} \sim t_{n-k}, \quad j = 1, \ldots, k$$

where $z_{jj}$ are the diagonal elements in the matrix $Z = (X'X)^{-1}$.

***Example 7.21 (Two-Way Analysis of Variance).*** Let us end this section by indicating how the theory of GLM can be used to handle higher order ANOVA models. We will consider only the two-way ANOVA here. It should be quite clear how this can be generalized to higher dimensions.

The two-way ANOVA can be written

$$X_{ijk} \sim (\mu_{ij}, \sigma^2), \quad i = 1, \ldots, a, \quad j = 1, \ldots, b, \quad k = 1, \ldots, n$$

or in alternative form

$$X_{ijk} = \mu + \alpha_i + \beta_j + \gamma_{ij} + \epsilon_{ijk}, \quad i = 1, \ldots, a, \quad j = 1, \ldots, b, \quad k = 1, \ldots, n$$

where $\epsilon_{ijk} \sim N(0, \sigma^2)$ are independent error terms. The parameters $\alpha_i$ and $\beta_j$ are called *main effects* and $\gamma_{ij}$ are called *interaction effects*. As before, we need some additional conditions to reduce the number of parameters

$$\sum_{i=1}^{a} \alpha_i = 0 \quad \sum_{j=1}^{b} \beta_j = 0 \quad \sum_{i=1}^{a} \gamma_{ij} = 0 \quad \sum_{j=1}^{b} \gamma_{ij} = 0$$

This implies that the parameter vector can be written

$$\boldsymbol{\beta} = (\mu, \alpha_1, \ldots, \alpha_{a-1}, \beta_1, \ldots, \beta_{b-1}, \gamma_{11}, \ldots, \gamma_{1,b-1}, \gamma_{21}, \ldots, \gamma_{a-1,b-1})'$$

The design matrix becomes quite messy, so we will not attempt to give a detailed description of it here. It will be a $(abn) \times (ab)$ matrix, which can be built up by introducing dummy variables much in the same way as in Example 7.20. In this particular case, it is possible to express the parameter estimators analytically as

$$\begin{aligned}
\widehat{\mu} &= \bar{X} \\
\widehat{\alpha}_i &= \bar{X}_{i\cdot\cdot} - \bar{X}, \quad i = 1, \ldots, a-1 \\
\widehat{\beta}_j &= \bar{X}_{\cdot j\cdot} - \bar{X}, \quad j = 1, \ldots, b-1 \\
\widehat{\gamma}_{ij} &= \bar{X}_{ij\cdot} - \bar{X}_{i\cdot\cdot} - \bar{X}_{\cdot j\cdot} + \bar{X}, \quad i = 1, \ldots, a-1, \quad j = 1, \ldots, b-1
\end{aligned}$$

where we use the convenient notation introduced in Section 7.5, where the dots indicate for which indices the average is taken. □

## PROBLEMS

### Section 7.2. Sampling Distributions

1. Let $X \sim \chi^2_r$ and $Y \sim \chi^2_s$ be independent. Show that $X + Y \sim \chi^2_{r+s}$.
2. Let $X \sim \chi^2_r$. Show that $X \sim \Gamma(\frac{r}{2}, \frac{1}{2})$.
3. Find the mean and variance in the $\chi^2_r$ distribution. Also find the approximate distribution for large $r$. *Hint:* Use the previous problem and Section 2.8.2.
4. Let $X \sim F_{r,s}$. Show that $X^{-1} \sim F_{s,r}$.
5. Find the mean (for $s > 2$) and variance (for $s > 4$) in the $F$ distribution. *Hint:* Use Proposition 7.2 and $\Gamma(x + 1) = x\Gamma(x)$.
6. Find the mean (for $r > 1$) and variance (for $r > 2$) in the $t$ distribution.
7. Let $X \sim t_1$. Show that $X$ is Cauchy distributed.
8. *Maxwell–Boltzmann distribution.* The velocity $V$ of a particle in an ideal gas can be represented as

$$V = \sqrt{V_x^2 + V_y^2 + V_z^2}$$

where $V_x$, $V_y$, and $V_z$ are the velocity components in a three-dimensional coordinate system. They are assumed to be independent and normally distributed with mean 0 and variance $\sigma^2 = kT/m$, where $T$ is temperature, $m$ is the mass of particles, and $k$ is the Boltzmann constant ($k = 1.381 \times 10^{-23}$). Derive the pdf of $V$. *Hint:* Use Proposition 2.8.

### Section 7.3. Single Sample Inference

9. Below are seven measurements of the ozone level (in ppm) taken at an environmental measuring station. Suppose that these have a normal distribution and find a 95% symmetric confidence interval for the mean $\mu$.
   0.06, 0.07, 0.08, 0.11, 0.12, 0.14, 0.21
10. Let $X_1, \ldots, X_n$ be a sample from a normal distribution with known mean $\mu$ and unknown variance $\sigma^2$. Use Proposition 7.1 to show that a $100q\%$ symmetric confidence interval for $\sigma^2$ can be based on $\widehat{\sigma}^2$ from Section 6.2.1 and is given by

$$n\widehat{\sigma}^2/x_2 \leq \sigma^2 \leq n\widehat{\sigma}^2/x_1 \quad (q)$$

where $F_{\chi^2_n}(x_1) = (1 - q)/2$ and $F_{\chi^2_n}(x_2) = 1 - (1 - q)/2$.
11. Below are measurement errors from an unbiased scale ($\mu = 0$). Use the previous problem to find a 95% symmetric confidence interval for its standard deviation.

   −0.08, −0.05, −0.02, 0.01, 0.02, 0.06, 0.07

12. An IQ test is being constructed, and it is desirable that the standard deviation is 15 (the mean is unknown). To assess this, the test was given to 30 randomly

selected people and the sample variance came out to be $s^2 = 340$. Find symmetric confidence intervals for $\sigma$ with confidence levels 0.90 and 0.95. Conclusions?

13. Consider the symmetric confidence interval for $\mu$ in the normal distribution, $\mu = \bar{X} \pm ts/\sqrt{n}$. How should we choose $n$ in order to limit the length of the confidence interval to $\epsilon\sigma$ with probability $q$ or greater? Express in terms of the cdf of the chi-square distribution.
14. Let $s^2$ be the sample variance and let $Y$ be a random variable that has a $\chi^2_{n-1}$ distribution. Show that **(a)** $E[s] = \sigma E[\sqrt{Y}]/\sqrt{n-1}$, and **(b)** $E[s^4] = \sigma^4 E[Y^2]/(n-1)^2$. **(c)** In (b), it can be shown that $E[Y^2] = (n-1)(n+1)$. Use this to find the variance of $s^2$ and show that $s^2$ is consistent (see Definition 6.4).
15. A company producing soda cans test whether their cans contain the intended 12 ounces. The contents of 100 cans are measured, the sample mean is 12.1, and the sample variance 0.96. State the relevant hypotheses and test on the 5% level if the mean is 12.
16. The depth of a lake at a particular point is claimed to be more than 100 m. To test this, the depth is measured repeatedly, where it is known that measurement errors have a normal distribution. On the basis of the following measurements, test on the 5% level whether the depth is more than 100 m:

    99, 101, 102, 102, 103, 103, 103

17. An electronic scale is known to give a measurement error, and to test whether it is unbiased, a 100 g weight was measured repeatedly and the following data obtained. Test on the 5% level if the scale is unbiased (i.e., if the mean is 0).

    $-1.05,\ -0.55,\ -0.01,\ 2.55,\ 3.72$

18. An environmental measuring station measures ozone levels in the air. The level is considered unhealthy if it is over 0.30 (ppm) and to determine so, the following five observations were gathered. Assume a normal distribution, state the appropriate null and alternative hypotheses (one- or two-sided?), and test on the level 5%:

    0.32, 0.35, 0.38, 0.41, 0.48

19. The vendor of a particular herbal supplement claims that it will increase your IQ. To test the claim, the supplement was given to 10 people. Before, the sample mean was 100 and after, it was 103. "I told you so!" says the vendor but test his claim on the 5% level if you are also given the sample variance of the differences, which was 318.
20. A test facility reports that they have found that a measurement device gives values that tend to be too large. The measurement errors had a sample mean of 0.5 and a sample variance of 4. How large must their sample size have been in order to draw the conclusion in a two-sided test on the 5% level?
21. Let $X_1, \ldots, X_n$ be a sample from a normal distribution with known mean $\mu$ and unknown variance $\sigma^2$. Describe how to use estimators of $\sigma^2$ and the chi-square distribution to construct tests of the null hypothesis $H_0 : \sigma^2 = \sigma_0^2$ for fixed $\sigma_0^2$.

22. The following data are from a normal distribution where it is desirable that the variance is at most 1. Do the following data cause concern that the variance is too large? Test on the 5% level.

    $-3.7, \ -1.5, \ -0.6, \ -0.4, -0.3, \ 1.0, \ 2.0$

23. An electronic scale gives measurement errors that are normal with mean 0. The scale needs to be adjusted if the standard deviation exceeds one milligram. The following data are deviations in milligrams obtained by weighing an object of known weight six times. Do the data warrant an adjustment? Test on the 5% level.

    $-0.40, \ -0.28, \ -0.27, -0.16, \ 0.05, 0.22$

24. An IQ test is constructed and the desired standard deviation is 15. The following data are 10 measured IQ scores. Test on the 5% level if the standard deviation is out of bounds.

    75, 76, 87, 90, 97, 104, 106, 107, 111, 127

**Section 7.4. Comparing Two Samples**

25. Common sea-buckthorn berries are particularly rich in vitamin E. Two genetically different types of plants were evaluated in an experiment where seven plants of type $A$ and five plants of type $B$ were cultivated. The vitamin E content (in μg/g dry weight) of the ripe berries was measured.

    Type $A$: 416, 492, 444, 404, 325, 286, 403

    Type $B$: 279, 352, 320, 385, 315

    Assume that the samples are normally distributed with equal variances and carry out a hypothesis test on the 10% level to see if the mean vitamin E content differs between the two types.

26. To investigate whether caffeine affects cholesterol levels, five patients had their cholesterol levels measured before and after taking doses of caffeine. Find a symmetric 95% confidence interval for the difference in cholesterol level based on the following data:

    Before: 162, 168, 197, 202, 225

    After: 179, 170, 196, 188, 210

27. A chemist wants to see how much an industry contributes to pollution in a nearby river. She collects 10 samples upstream and 15 samples downstream on 25 different and randomly chosen days during a 3-month period and measures the content of a certain pollutant. The sample mean and standard deviation were 13.2 and 2.8, respectively, for the upstream samples and 86.1 and 38.7, respectively, for the downstream samples. Assume that the samples are normally distributed and calculate a 95% confidence interval for the difference in mean pollution downstream and upstream.

28. A water laboratory needs new pH meters and considers two different brands. Therefore, they acquire six meters each from the manufacturers for evaluation and use them to measure the pH in a neutral solution known to have pH level 7. The sample standard deviations of each brand were 0.078 and 0.029, respectively. Is this evidence enough to claim that one brand has significantly lower measurement error than the other? What assumptions do you need to make?
29. A type of rectangular metal plate has sides $X$ and $Y$ that are normally distributed with the same variance and means $\mu_1$ and $\mu_2$, respectively. Find a 95% symmetric confidence interval for the circumference $2\mu_1 + 2\mu_2$ of a plate, based on the following two independent samples:

   $X$: 87, 90, 97, 102, 108, 110

   $Y$: 133, 147, 148, 154

30. If we have two independent samples of the same size $n$ and with the same variance, the standard confidence interval for the difference between the means is based on the estimator $\bar{Y} - \bar{X}$ of $\mu_2 - \mu_1$. However, we could also pair the observations, compute the differences $D_1, \ldots, D_n$, and base the confidence interval on $\bar{D} = \bar{Y} - \bar{X}$. Which do you think is better and why?
31. Consider two independent samples $X_1, \ldots, X_n$ and $Y_1, \ldots, Y_m$ that have the same unknown variance $\sigma^2$. Let $s_1^2$ and $s_2^2$ be the respective sample variances. **(a)** Prove that the pooled sample variance from Definition 7.4 is an unbiased estimator of $\sigma^2$. **(b)** Prove that any linear combination of the type $s_a^2 = as_1^2 + (1-a)s_2^2$ where $0 \le a \le 1$ is an unbiased estimator of $\sigma^2$. **(c)** Prove that the variance of $s_a^2$ is minimized for $a = (n-1)/(n+m-2)$ and thus that the pooled sample variance has a smaller variance than does any other linear combination of $s_1^2$ and $s_2^2$. *Hint:* By Problem 14 (c), $\text{Var}[s_1^2] = 2\sigma^4/(n-1)$ and $\text{Var}[s_2^2] = 2\sigma^4/(m-1)$.
32. In 1908, W. S. Gosset conducted a famous experiment to determine whether kiln-dried seed would give larger corn yield than would regular seed. Eleven plots were split in half and planted with regular seed on one side and kiln-dried seed on the other. (Why is this better than simply planting regular seed in one big plot and kiln-dried in another?) This gave the following 11 pairs of yields in pounds per acre:

   Regular: 1903, 1935, 1910, 2496, 2108, 1961, 2060, 1444, 1612, 1316, 1511

   Kiln-dried: 2009, 1915, 2011, 2463, 2180, 1925, 2122, 1482, 1542, 1443, 1535

   We assume these are normally distributed. State the appropriate hypotheses and test on the 5% level.
33. Let us say that you want to test if the means in two independent and normal samples are equal and you are uncertain whether the variances are equal. Then, it may seem like a good idea to first carry out a hypothesis test to see if the variances are equal. If you accept $H_0$, you carry out the test in Proposition 7.11, and if you reject $H_0$, you use Proposition 7.12 instead. What is the problem with this procedure?

## Section 7.5. Analysis of Variance

34. A biologist wanted to examine the impact of alcohol on sleep patterns. A total of 20 mice were randomly assigned to four equally large treatment groups and injected with a certain concentration of ethanol per body weight. The length of the REM sleep was then measured during a 24 h period. The result was as follows.

| Treatment | REM Sleep | | | | | Mean | Variance |
|---|---|---|---|---|---|---|---|
| 0 (control) | 89 | 73 | 91 | 68 | 75 | 79.2 | 104.2 |
| 1 g/kg | 63 | 54 | 69 | 50 | 72 | 61.6 | 89.3 |
| 2 g/kg | 45 | 60 | 40 | 56 | 39 | 48.0 | 90.5 |
| 3 g/kg | 31 | 40 | 45 | 25 | 23 | 32.8 | 90.2 |

**(a)** Carry out an ANOVA on the 5% level to see if alcohol consumption affects REM sleep. **(b)** Calculate pairwise confidence intervals for the difference in means. What groups are significantly different? **(c)** Carry out a Kruskal–Wallis test. Does the conclusion differ from (a)?

35. Consider an ANOVA with $k = 2$ and arbitrary $n$. Show that this is equivalent to the two-sample $t$ test with equal variances.
36. *Unbalanced ANOVA*. Let us assume that the $k$ treatment groups in a one-way ANOVA are of different sizes $n_1, \ldots, n_k$. Modify the balanced ANOVA and derive expressions for the sum of squares, mean squares, degrees of freedom, and $F$ test.
37. Show that $E[s_B^2] > \sigma^2$ if $\mu_{i_1} \neq \mu_{i_2}$ for some $i_1$ and $i_2$.
38. Use

$$\sum_{k=1}^{n} k = \frac{n(n+1)}{2} \quad \text{and} \quad \sum_{k=1}^{n} k^2 = \frac{n(2n+1)(n+1)}{6} \tag{7.12}$$

to derive Equation (7.6) if there are no ties in Kruskal–Wallis test.

## Section 7.6. Linear Regression

39. Verify the estimators of $a$ and $b$ in Proposition 7.15 and compute the means and variances of $\widehat{a}$ and $\widehat{b}$.
40. Show that $\bar{Y}$ and $\widehat{b}$ are uncorrelated (Problem 101 in Chapter 3). Are they independent?
41. Find 95% symmetric confidence intervals for $a$ and $b$ based on the Old Faithful data in Example 7.13. Compare with the "ranger formula" $y = 13x + 30$.
42. Test the hypothesis that $a = 0$ in Hubble's galaxy data, on significance level 0.01.

43. Consider the linear regression model without the intercept term, that is, $Y = bx + \epsilon$. **(a)** Find the MLE $\widehat{b}$ of $b$ and compute its mean and variance. **(b)** Suggest an estimator of $\sigma^2$ and a confidence interval for $b$. **(c)** Again consider Hubble's galaxy data. Argue why the model $Y = bx + \epsilon$ is reasonable there, compute $\widehat{b}$, a 95% confidence interval for $b$, and compare with the estimate obtained in Examples 7.11 and 7.12.
44. Consider the linear regression model $Y = a + bx + \epsilon$. **(a)** Suppose that $a$ is known. Find the MLE of $b$. **(b)** Suppose that $b$ is known. Find the MLE of $a$. **(c)** Compute the preceding estimates for the Old Faithful data in Example 7.13 using the rangers' values of $a$ and $b$. Compare with the estimates in the example.
45. A cell culture is growing in such a way that the total mass at time $t$ can be ideally described by the equation $y = a\, e^{bt}$, where $a$ and $b$ are constants. Because of random fluctuations, the real relationship is $Y = a\, e^{bt} L$, where $L$ is a random variable that has a lognormal distribution with $\mu = 0$ and $\sigma^2 = 1$ (see Section 2.8.1). Find estimators and 95% confidence intervals for $a$ and $b$ based on the following data, obtained from five different cultures, weighed at different times.

    Time: 1, 3, 7, 8, 10

    Weight: 1.43, 0.51, 4.57, 5.93, 1.73

46. Below are the winning times in men's 10,000 m track in the Olympic Games between the years 1952 and 2004. Times are in decimal form, so, for example, 28.4 = 28 min and 24 s. Let the years be the $x$ values and the times the $y$ values. **(a)** Find the estimated regression line and the predicted winning time in the Olympic year 2668. **(b)** Is it reasonable to assume a linear relationship?

    29.3, 28.8, 28.5, 28.4, 29.4, 27.6, 27.7, 27.7, 27.8, 27.4, 27.8, 27.1, 27.3, 27.1

47. Again consider the Old Faithful data in Example 7.13. Compute a 95% prediction interval for the time $Y$ until the next eruption if the most recent eruption was 10 min. Compare with the interval in Example 7.14. Which is longer and why?
48. Show that a confidence interval for $a + bx$, the expected value of $Y$ if $x$ has been observed, is obtained by removing the "1" under the square root in the expression for the prediction interval in Corollary 7.11.
49. Consider the following residual plots. What do they suggest about the true relation between $x$ and $Y$?

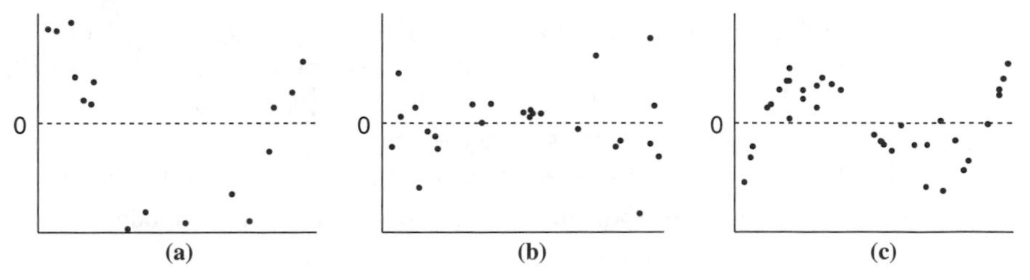

50. Consider the linear regression model. Describe how to find a symmetric $100q\%$ confidence interval for $\sigma$ based on Proposition 7.17. What does this give for Hubble's galaxy data in Example 7.11, if $q = 0.95$?
51. The following is a data set of 10 closing prices for the Dow and Nasdaq indices, chosen at random from the year 2000. Compute the sample correlation coefficient $R$ and test if $\rho = 0$. Look up the behavior of the two indices during the year 2000 and explain the value of $R$ in the light of this.

    Dow: 10,092, 11,112, 10,621, 10,714, 10,504, 10,783, 10,908 10,681, 10,707, 10,399

    Nasdaq: 4321, 4897, 3205, 3400, 3767, 3858, 3842, 3333, 3429, 3200

52. Compute Spearman's correlation coefficient $R_S$ for the data in the previous problem and test if $\rho = 0$. Do the results differ?
53. Let $\rho$ be the correlation coefficient and $R$ the sample correlation coefficient in a sample from a bivariate normal distribution, and let

$$L = \log\left(\frac{1+R}{1-R}\right)$$

    It can be shown that the following approximation holds:

$$L \overset{d}{\approx} N\left(\log\left(\frac{1+\rho}{1-\rho}\right), \frac{4}{n-3}\right)$$

    Use this to find an approximate confidence interval for $\rho$. Apply this to find the observed 95% confidence interval for the stock market data in Problem 51.
54. Consider Gosset's corn yield data from Problem 32. Use the statistic $L$ in Problem 53 to test whether the correlation between yields is greater than 0.9. Test on the 5% level.
55. To test the null hypothesis $H_0 : \rho = 0$ about the correlation coefficient, we can use either the test statistic $T$ from Proposition 7.18 or the test statistic $L$ from Problem 53. However, only one of them can be used for power calculations. Which one and why?
56. Use (7.12) to derive Proposition 7.20 if there are no ties.

### Section 7.7. The General Linear Model

57. In the multiple regression model $Y_i = \beta_0 + \beta_1 x_{i1} + \beta_2 x_{i2} + \epsilon_i$ for $i = 1, \ldots, 25$, where $\epsilon_1, \ldots, \epsilon_{25}$ are i.i.d. $N(0, \sigma^2)$, we got $\widehat{\beta}_0 = 2.71$, $\widehat{\beta}_1 = 10.20$, $\widehat{\beta}_2 = 2.07$, $s^2 = 0.180$ and the diagonal elements $z_{11} = 0.4$, $z_{22} = 0.5$, and $z_{33} = 0.02$. Calculate two-sided symmetric confidence intervals for $\beta_0$, $\beta_1$, and $\beta_2$. Use Bonferroni correction to get simultaneous confidence level at least 95%.
58. Express the two-sample model in Section 7.4 as a general linear model both for the case $\sigma_1^2 = \sigma_2^2 = \sigma^2$ and $\sigma_1^2 \neq \sigma_2^2$.

59. Use Proposition 7.21 to verify Proposition 7.15. *Hint:* The inverse of a $2 \times 2$-matrix is

$$\begin{pmatrix} a & b \\ c & d \end{pmatrix}^{-1} = \frac{1}{ad - bc} \begin{pmatrix} d & -b \\ -c & a \end{pmatrix}$$

60. Consider the two-way ANOVA in Example 7.21 and derive the sum of squares SSA and SSB for main effects, SSAB for interaction, SSE for error, and SST for total. *Hint:* Use SST = SSA + SSB + SSAB + SSE.

61. It is perfectly fine to mix categorical components, like in an ANOVA, with linear components, like in a linear regression, into a general linear model. As a simple example, consider $X_{ij} = \mu + \alpha_i + \beta x_{ij} + \epsilon_{ij}$ for $i = 1, \ldots, k$ and $j = 1, \ldots, n$, where $\sum_i \alpha_i = 0$ and $\epsilon_{11}, \ldots, \epsilon_{kn}$ are i.i.d. $N(0, \sigma^2)$. Outline the design matrix in this model.

# 8

# STOCHASTIC PROCESSES

## 8.1 INTRODUCTION

Many real-world applications of probability theory have one particular feature that data are collected sequentially in time. A few examples are weather data, stock market indices, air-pollution data, demographic data, and political tracking polls. These also have another feature in common that successive observations are typically not independent. We refer to any such collection of observations as a *stochastic process*. Formally, a stochastic process is a collection of random variables that take values in a set $S$, the *state space*. The collection is indexed by another set $T$, the *index set*. The two most common index sets are the natural numbers $T = \{0, 1, 2, \ldots\}$ and the nonnegative real numbers $T = [0, \infty)$, which usually represent discrete time and continuous time, respectively. The first index set thus gives a sequence of random variables $\{X_0, X_1, X_2, \ldots\}$ and the second, a collection of random variables $\{X(t), t \geq 0\}$, one random variable for each time $t$. In general, the index set does not have to describe time and is also commonly used to describe spatial location. The state space can be finite, countably infinite, or uncountable, depending on the application.

In order to be able to analyze a stochastic process, we need to make assumptions on the dependence between the random variables. In this chapter, we will focus on the most common dependence structure, the so-called *Markov property*, and in the next section we give a definition and several examples.

*Probability, Statistics, and Stochastic Processes*, Second Edition. Peter Olofsson and Mikael Andersson.

## 8.2 DISCRETE-TIME MARKOV CHAINS

You are playing roulette, in each round betting \$1 on odd. You start with \$10 and after each round record your new fortune. Suppose that the first five rounds gives the sequence loss, loss, win, win, win, which gives the sequence of fortunes

$$9,\ 8,\ 9,\ 10,\ 11$$

and that you wish to find the distribution of your fortune after the next round, given this information. Your fortune will be 12 if you win, which has probability $\frac{18}{38}$, and 10 if you lose, which has probability $\frac{20}{38}$. One thing we realize is that this depends only on the fact that the current fortune is \$11 and not the values prior to that. Generally, if your fortunes in the first $n$ rounds are the random variables $X_1, \ldots, X_n$, then the conditional distribution of $X_{n+1}$ given $X_1, \ldots, X_n$ depends only on $X_n$. This is a fundamental property, and we state the following general definition.

***Definition 8.1.*** Let $X_0, X_1, X_2, \ldots$ be a sequence of discrete random variables, taking values in some set $S$ and that are such that

$$P(X_{n+1} = j | X_0 = i_0, \ldots, X_{n-1} = i_{n-1}, X_n = i) = P(X_{n+1} = j | X_n = i)$$

for all $i, j, i_0, \ldots, i_{n-1}$ in $S$ and all $n$. The sequence $\{X_n\}$ is then called a *Markov chain.*

We often think of the index $n$ as discrete time and say that $X_n$ is the *state* of the chain at time $n$, where the state space $S$ may be finite or countably infinite. The defining property is called the *Markov property*, which can be stated in words as "conditioned on the present, the future is independent of the past."

In general, the probability $P(X_{n+1} = j | X_n = i)$ depends on $i$, $j$, and $n$. It is, however, often the case (as in our roulette example) that there is no dependence on $n$. We call such chains *time-homogeneous* and restrict our attention to these chains. Since the conditional probability in the definition thus depends only on $i$ and $j$, we use the notation

$$p_{ij} = P(X_{n+1} = j | X_n = i), \quad i, j \in S$$

and call these the *transition probabilities* of the Markov chain. Thus, if the chain is in state $i$, the probabilities $p_{ij}$ describe how the chain chooses which state to jump to next. Obviously, the transition probabilities have to satisfy the following two criteria:

$$\text{(a)}\ p_{ij} \geq 0 \quad \text{for all } i, j \in S, \qquad \text{(b)} \sum_{j \in S} p_{ij} = 1 \quad \text{for all } i \in S$$

***Example 8.1.*** In the roulette example above, the state space is

$$S = \{0, 1, \ldots\}$$

and if the chain is in state $i \geq 1$, it can jump to either $i-1$ or $i+1$ according to the transition probabilities

$$p_{i,i-1} = \frac{20}{38} \quad \text{and} \quad p_{i,i+1} = \frac{18}{38}$$

When $i = 0$, this means that you are ruined and cannot play anymore. Thus, you can jump to 0 but not from it. It is customary to describe this by letting $p_{00} = 1$, thus imagining that the chain performs an eternal sequence of jumps from 0 to itself. The diagram below shows a way to describe a Markov chain as a graph, which we refer to as the *transition graph*. The arrows show the possible transitions and their corresponding probabilities. Note that the sum of the numbers on the arrows going *out* from each state is 1. This is criterion (b) above.

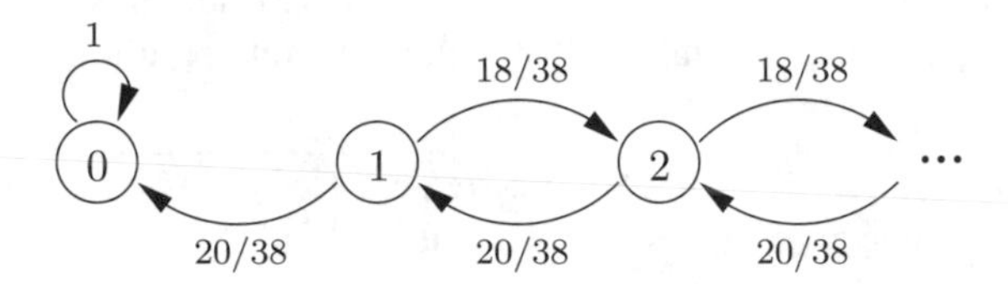


□

***Example 8.2.*** A certain gene in a plant has two alleles, $A$ and $a$ (see Section 1.6.2). Thus, its genotype with respect to this gene can be $AA$, $Aa$, or $aa$. Now suppose that a plant is crossed with itself and one offspring is selected that is crossed with itself and so on and so forth. Describe the sequence of genotypes as a Markov chain.

The state space is $S = \{AA, Aa, aa\}$, which also shows that states do not have to be numbers.[1] The Markov property is clear since the offspring's genotype depends only on the parent plant, not on the grandparent. Clearly, genotypes $AA$ and $aa$ can have only themselves as offspring and for the type $Aa$, we recall the Punnett square from Section 1.6.2 to get the following transition graph:

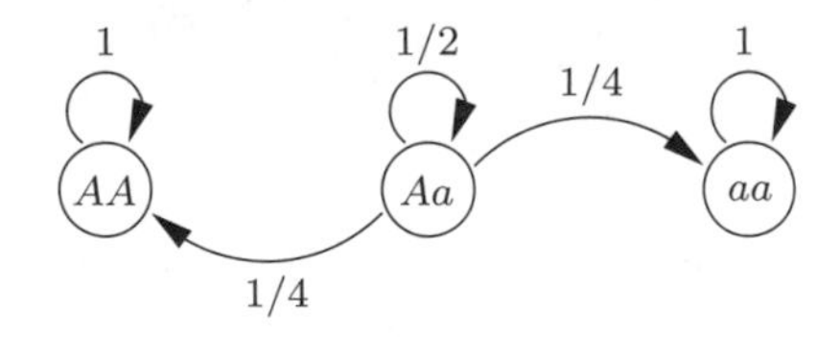


□

It is convenient to summarize the transition probabilities in the *transition matrix* $P$, which has $p_{ij}$ as its $(i, j)$th entry. Depending on the state space, the transition matrix

[1]The picky probabilist then refers to the $X_k$ as *random objects* rather than random variables. If we wish, we can rename the states $\{1, 2, 3\}$ instead, where the numbers have no role other than serving as labels.

may be finite or infinite. Thus, in the genetics example we have

$$P = \begin{pmatrix} 1 & 0 & 0 \\ 1/4 & 1/2 & 1/4 \\ 0 & 0 & 1 \end{pmatrix}$$

and in the roulette example the infinite matrix

$$P = \begin{pmatrix} 1 & 0 & 0 & 0 & 0 & \cdots \\ 20/38 & 0 & 18/38 & 0 & 0 & \cdots \\ 0 & 20/38 & 0 & 18/38 & 0 & \cdots \\ \vdots & \vdots & \vdots & \vdots & \vdots & \ddots \end{pmatrix}$$

### 8.2.1 Time Dynamics of a Markov Chain

The most fundamental aspect of a Markov chain in which we are interested is how it develops over time. The transition matrix provides us with a description of the stepwise behavior, but suppose that we want to compute the distribution of the chain two steps ahead. Let

$$p_{ij}^{(2)} = P(X_2 = j | X_0 = i)$$

and condition on the intermediate step $X_1$. The law of total probability gives

$$\begin{aligned} p_{ij}^{(2)} &= \sum_{k \in S} P(X_2 = j | X_0 = i, X_1 = k) P(X_1 = k | X_0 = i) \\ &= \sum_{k \in S} P(X_2 = j | X_1 = k) P(X_1 = k | X_0 = i) = \sum_{k \in S} p_{ik} p_{kj} \end{aligned}$$

where we used the Markov property for the second-to-last equality. We switched the order between the factors in the sum to get the intuitively appealing last expression; in order to go from $i$ to $j$ in two steps, we need to visit *some* intermediate step $k$ and jump from there to $j$. Now, recall how matrix multiplication works to help us realize from the expression above that $p_{ij}^{(2)}$ is the $(i, j)$th entry in the matrix $P^2$. Thus, in order to get the two-step transition probabilities, we square the transition matrix. Generally, define the *n-step transition probabilities* as

$$p_{ij}^{(n)} = P(X_n = j | X_0 = i)$$

and let $P^{(n)}$ be the $n$-step transition matrix. Repeating the argument above gives $P^{(n)} = P^n$, the $n$th power of the one-step transition matrix. In particular, this gives the relation

$$P^{(n+m)} = P^{(n)} P^{(m)}$$

for all $m, n$, commonly referred to as the *Chapman–Kolmogorov equations*. Spelled out coordinatewise, they become

$$p_{ij}^{(n+m)} = \sum_{k \in S} p_{ik}^{(n)} p_{kj}^{(m)}$$

for all $m, n$ and all $i, j \in S$. In words, to go from $i$ to $j$ in $n + m$ steps, we need to visit some intermediate step $k$ after $n$ steps. We let $P^{(0)} = I$, the identity matrix.

***Example 8.3.*** Find the $n$-step transition matrix in the genetics example (Example 8.2).

The state space is $S = \{AA, Aa, aa\}$, and let us start with $n = 2$. We get

$$P^{(2)} = \begin{pmatrix} 1 & 0 & 0 \\ 1/4 & 1/2 & 1/4 \\ 0 & 0 & 1 \end{pmatrix} \begin{pmatrix} 1 & 0 & 0 \\ 1/4 & 1/2 & 1/4 \\ 0 & 0 & 1 \end{pmatrix}$$

$$= \begin{pmatrix} 1 & 0 & 0 \\ 3/8 & 1/4 & 3/8 \\ 0 & 0 & 1 \end{pmatrix}$$

We now realize that 0s and 1s will remain in all powers of $P$, that the middle entry in $P^{(n)}$ is $p_{22}^{(n)} = (\frac{1}{2})^n$, and that by symmetry $p_{21}^{(n)} = p_{23}^{(n)}$. This gives the $n$-step transition matrix

$$P^{(n)} = \begin{pmatrix} 1 & 0 & 0 \\ (1-(1/2)^n)/2 & (1/2)^n & (1-(1/2)^n)/2 \\ 0 & 0 & 1 \end{pmatrix}$$

It is obvious without computations that the 0s and 1s remain unchanged; the types $AA$ and $aa$ can have offspring only of their own type. Also note how the probability to find the type $Aa$ declines rapidly with $n$, indicating that eventually this genotype will disappear. We will return to this aspect of the transition matrix. □

It should be pointed out that computation of $P^{(n)}$ is seldom this simple and may be more or less impossible if the state space is large. Even for a small state space, the computation is not trivial, as the next example shows.

***Example 8.4.*** **(ON/OFF *System*).** Consider a system that alternates between the two states 0 (OFF) and 1 (ON) and that is checked at discrete timepoints. If the system is OFF at one timepoint, the probability that it has switched to ON at the next timepoint is $p$, and if it is ON, the probability that it switches to OFF is $q$. **(a)** Describe the system as a Markov chain. **(b)** Find the $n$-step transition matrix. **(c)** Suppose that $p = \frac{3}{4}$ and $q = \frac{1}{2}$. If the system starts being OFF, what is the probability that it is ON at time $n = 3$?

For (a), the transition graph is

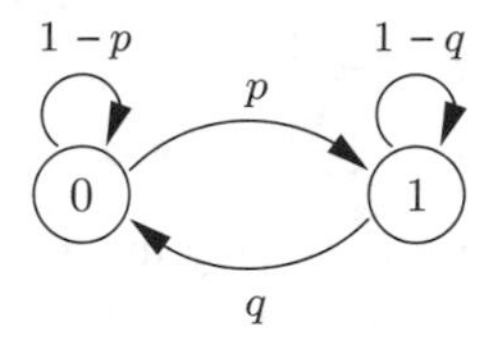


and the transition matrix

$$P = \begin{pmatrix} 1-p & p \\ q & 1-q \end{pmatrix}$$

and computation of the powers of $P$ needed for (b) is facilitated by diagonalization techniques from linear algebra. The eigenvalues of $P$ are $\lambda_1 = 1$ and $\lambda_2 = 1 - p - q$ and it can be shown that

$$P^n = \frac{1}{p+q}\begin{pmatrix} q & p \\ q & p \end{pmatrix} + \frac{\lambda_2^n}{p+q}\begin{pmatrix} p & -p \\ -q & q \end{pmatrix}$$

and you may verify that this satisfies the relation $P^{n+1} = P^n P$. To find the answer to (c), we need $p_{01}^{(3)}$, which is the $(0, 1)$ entry in $P^{(3)}$. Thus, the probability that the system is ON at time $n = 3$, given that it starts with being OFF, is

$$p_{01}^{(3)} = \frac{p}{p+q} + \frac{-p\lambda_2^3}{p+q} = \frac{3/4}{5/4} + \frac{-3/4 \times (-1/4)^3}{5/4} \approx 0.61$$

□

One interesting aspect of a Markov chain is its long-term behavior. As it turns out, there are simple and elegant asymptotic results for Markov chains that makes this easy to deal with. Before we get to those results, let us consider asymptotics in some of our examples.

***Example 8.5.*** Recall the genetics example (Example 8.2). Find the limits of the transition probabilities as $n \to \infty$.

The $n$-step transition matrix is

$$P^{(n)} = \begin{pmatrix} 1 & 0 & 0 \\ (1-(1/2)^n)/2 & (1/2)^n & (1-(1/2)^n)/2 \\ 0 & 0 & 1 \end{pmatrix}$$

and letting $n \to \infty$ gives the matrix

$$\lim_{n\to\infty} P^{(n)} = \begin{pmatrix} 1 & 0 & 0 \\ 1/2 & 0 & 1/2 \\ 0 & 0 & 1 \end{pmatrix}$$

Thus, if we start in state $AA$ or $aa$, we stay there, and if we start in state $Aa$, we eventually end up in either $AA$ or $aa$ with equal probabilities. □

***Example 8.6.*** Recall the ON/OFF system in Example 8.4. Find the limits of the transition probabilities as $n \to \infty$.

The $n$-step transition matrix is

$$P^{(n)} = \frac{1}{p+q}\begin{pmatrix} q & p \\ q & p \end{pmatrix} + \frac{\lambda_2^n}{p+q}\begin{pmatrix} p & -p \\ -q & q \end{pmatrix}$$

and since $\lambda_2 = 1 - p - q$ and thus $|\lambda_2| < 1$ (unless $p = q = 0$ or $p = q = 1$), letting $n \to \infty$ gives the matrix

$$\lim_{n\to\infty} P^{(n)} = \frac{1}{p+q}\begin{pmatrix} q & p \\ q & p \end{pmatrix}$$

Note that the rows of this matrix are identical. Thus, at a late timepoint, the probabilities that the system is OFF and ON are approximately $q/(p+q)$ and $p/(p+q)$, respectively, regardless of the initial state. Note that if $q > p$, the probability to be OFF is larger, which makes sense. □

In the previous example, the asymptotic probabilities do not depend on how the chain was started, and we call the distribution $(q/(p+q), p/(p+q))$ on the state space $\{0, 1\}$ a *limit distribution*. Compare with the genetics example where no limit distribution exists since the asymptotic probabilities depend on the initial state. A question of general interest is when a Markov chain has a limit distribution. To be able to answer this, we need to introduce some criteria that enables us to classify Markov chains.

### 8.2.2 Classification of States

The graphic representation of a Markov chain illustrates in which ways states can be reached from each other. In the roulette example, state 1 can, for example, reach state 2 in one step and state 3, in two steps. It can also reach state 3 in four steps, through the sequence 2, 1, 2, 3, and so on. One important property of state 1 is that it can reach any other state. Compare this to state 0 that cannot reach any other state. Whether or not states can reach each other in this way is of fundamental importance in the study of Markov chains, and we state the following definition.

***Definition 8.2.*** If $p_{ij}^{(n)} > 0$ for some $n$, we say that state $j$ is *accessible* from state $i$, written $i \to j$. If $i \to j$ and $j \to i$, we say that $i$ and $j$ *communicate* and write this $i \leftrightarrow j$.

If $j$ is accessible from $i$, this means that it is *possible* to reach $j$ from $i$ but not that this necessarily happens. In the roulette example, $1 \to 2$ since $p_{12} > 0$, but if the chain starts in 1, it may jump directly to 0, and thus it will never be able to visit state 2. In this example, all nonzero states communicate with each other and 0 communicates only with itself.

In general, if we fix a state $i$ in the state space of a Markov chain, we can find all states that communicate with $i$ and form the *communicating class* containing $i$. It is easy to realize that not only does $i$ communicate with all states in this class but they all communicate with each other. By convention, every state communicates with itself (it can "reach itself in 0 steps") so every state belongs to a class. If you wish to be more mathematical, the relation "$\leftrightarrow$" is an equivalence relation and thus divides the state space into equivalence classes that are precisely the communicating classes. In the roulette example, there are two classes

$$C_0 = \{0\}, \quad C_1 = \{1, 2, \ldots\}$$

and in the genetics example, each state forms its own class and we thus have

$$C_1 = \{AA\}, \quad C_2 = \{Aa\}, \quad C_3 = \{aa\}$$

In the ON/OFF system, there is only one class, the entire state space $S$. In this chain, all states communicate with each other, and it turns out that this is a desirable property.

***Definition 8.3.*** If all states in $S$ communicate with each other, the Markov chain is said to be *irreducible*.

Another important property of Markov chains has to do with returns to a state. For instance, in the roulette example, if the chain starts in state 1, it may happen that it never returns. Compare this with the ON/OFF system where the chain eventually returns to where it started (assuming that $p > 0$ and $q > 0$). We next classify states according to whether return is certain. We introduce the notation $P_i$ for the probability distribution of the chain when the initial state $X_0$ is $i$.

***Definition 8.4.*** Consider a state $i \in S$ and let $\tau_i$ be the number of steps it takes for the chain to first visit $i$. Thus

$$\tau_i = \min\{n \geq 1 : X_n = i\}$$

where $\tau_i = \infty$ if $i$ is never visited. If $P_i(\tau_i < \infty) = 1$, state $i$ is said to be *recurrent* and if $P_i(\tau_i < \infty) < 1$, it is said to be *transient*.

A recurrent state thus has the property that if the chain starts in it, the time until it returns is finite. For a transient state, there is a positive probability that the time until return is infinite, meaning that the state is never revisited. This means that a recurrent state is visited over and over but a transient state is eventually never revisited.

Now consider a transient state $i$ and another state $j$ such that $i \leftrightarrow j$. We will argue that $j$ must also be transient. By the Markov property, every visit to $j$ starts a fresh Markov chain and since $i \leftrightarrow j$, there is a positive probability to visit $i$ before coming back to $j$. We may think of this as repeated trials to reach $i$ every time the chain is in $j$, and since the success probability is positive, eventually there will be a success. If $j$ were recurrent, the chain would return to $j$ infinitely many times and the trial would also succeed infinitely many times. But this means that there would be infinitely many visits to $i$, which is impossible since $i$ is transient. Hence $j$ must also be transient.

We have argued that transience (and hence also recurrence) is a *class property*, a property that is shared by all states in a communicating class. In particular, the following holds.

**Corollary 8.1.** *In an irreducible Markov chain, either all states are transient or all states are recurrent.*

This is convenient since we can classify the entire Markov chain as transient or recurrent by checking only one state. In the case of a finite state space, there is an easy way to classify the transient and recurrent states.

**Corollary 8.2.** *Suppose that $S$ is finite. A state $i$ is transient if and only if there is another state $j$ such that $i \to j$ but $j \not\to i$.*

We omit the proof, referring instead to an intuitive argument. Every time the chain visits a transient state, there is a chance that it will never return again. In a finite state space, the only way in which this can happen is if there is some other state that can be reached but from where there is no path back. In an infinite state space, however, there is enough room for states to be transient even if they all communicate with each other. We also realize that if the state space is finite, there is not enough room for all states to be transient.

**Corollary 8.3.** *If a Markov chain has finite state space, there is at least one recurrent state.*

***Example 8.7.*** Classify the states as recurrent/transient in the ON/OFF system in Example 8.4.

To avoid trivialities, we assume that both $p$ and $q$ are strictly positive. Since the state space is finite, we can use Corollary 8.2 and note that since $i$ and $j$ communicate, they must both be recurrent. □

***Example 8.8.*** Classify the states as recurrent/transient in the roulette example (Example 8.1).

Here we must use the general definition. Let us start with state 0, which is trivially recurrent since if we start there, we are stuck there forever, that is, $\tau_0 \equiv 1$. As for state 1, if we start there and the first jump is to 0, we never return to 1, and thus $\tau_1 = \infty$ in this case. Hence, $P(\tau_1 < \infty) < 1$ and state 1 is transient. Since 1 communicates with the states 2, 3, ..., they are all transient.

The recurrent state 0 has the additional property that once the chain is there, it can never leave.[2] Such a state is called *absorbing*. □

A transient state $i$ is revisited a number of times, which has a geometric distribution with success probability $P_i(\tau_i = \infty)$ (where "success" means that the state is never revisited). This means that the expected number of returns is $1/P_i(\tau_i = \infty)$, which is finite since $P_i(\tau_i = \infty) > 0$. On the other hand, for a recurrent state the expected number of returns is infinite (since already the *actual* number is infinite). Now let

$$I_n = \begin{cases} 1 & \text{if } X_n = i \\ 0 & \text{otherwise} \end{cases}$$

and let

$$S = \sum_{n=1}^{\infty} I_n$$

the total number of returns to state $i$. We get

$$E_i[S] = \sum_{n=1}^{\infty} E_i[I_n] = \sum_{n=1}^{\infty} P_i(X_n = i) = \sum_{n=1}^{\infty} p_{ii}^{(n)}$$

which gives the following nice characterization of transience/recurrence.[3]

**Proposition 8.1.** State $i$ is

$$\text{transient if } \sum_{n=1}^{\infty} p_{ii}^{(n)} < \infty$$

$$\text{recurrent if } \sum_{n=1}^{\infty} p_{ii}^{(n)} = \infty$$

[2] A quote from *Hotel California* is sometimes given at this point in the presentation, but we resist the temptation.

[3] In the calculation, we interchanged summation and expectation, which is not always allowed when the sum is infinite. However, it can be shown that it is justified if the summands are nonnegative as in this case.

Since we have noted that it is often difficult to compute the $p_{ii}^{(n)}$, one may wonder how useful the last result is. However, we do not need to compute the exact value of the infinite sum, only determine whether it is convergent, and for this, it will suffice if we have some idea of how $p_{ii}^{(n)}$ behaves asymptotically in $n$. We will later see examples of this.

### 8.2.3 Stationary Distributions

Consider the ON/OFF system and suppose that we choose the initial state according to the probabilities $\nu_0 = P(X_0 = 0)$, $\nu_1 = P(X_0 = 1) = 1 - \nu_0$. The distribution $\boldsymbol{\nu} = (\nu_0, \nu_1)$ is called an *initial distribution* and the probability distribution of the first state $X_1$ is computed by conditioning on $X_0$, which gives

$$P(X_1 = j) = p_{0j}\nu_0 + p_{1j}\nu_1, \quad j = 0, 1$$

or in matrix notation

$$(P(X_1 = 0), P(X_1 = 1)) = \boldsymbol{\nu} P$$

Suppose in particular that we take $\nu_0 = q/(p+q)$, $\nu_1 = p/(p+q)$, the limit distribution, as initial distribution. We then get

$$P(X_1 = 0) = (1-p)\frac{q}{p+q} + q\frac{p}{p+q} = \frac{q}{p+q} = \nu_0$$

and $P(X_1 = 1) = \nu_1$. In matrix notation, $\boldsymbol{\nu} = \boldsymbol{\nu} P$, which means that the distribution does not change over time. This is an important observation, and we state the following general definition.

> ***Definition 8.5.*** Let $P$ be the transition matrix of a Markov chain with state space $S$. A probability distribution $\boldsymbol{\pi} = (\pi_1, \pi_2, \ldots)$ on $S$ satisfying
>
> $$\boldsymbol{\pi} P = \boldsymbol{\pi}$$
>
> is called a *stationary distribution* of the chain.

The entries of $\boldsymbol{\pi}$ thus satisfy

$$\pi_j = \sum_{i \in S} p_{ij}\pi_i \text{ for all } j \in S$$

and together with the condition

$$\sum_{i \in S} \pi_i = 1$$

this determines the stationary distribution. The intuition behind the probability $\pi_j$ is that it describes what proportion of time is spent in state $j$ in the long run. Other terms are *invariant distribution* and *equilibrium distribution*.

There are, however, some caveats: (1) a stationary distribution may not always exist and (2) there may be more than one. The uniqueness problem goes away if we make our usual assumption of irreducibility, an observation that we state without proof.

**Proposition 8.2.** Consider an irreducible Markov chain. If a stationary distribution exists, it is unique.

This is helpful, but we can still not guarantee that a stationary distribution exists. Things simplify if the state space is finite.

**Proposition 8.3.** If $S$ is finite and the Markov chain is irreducible, a unique stationary distribution $\boldsymbol{\pi}$ exists.

Rather than giving the proof, we examine our examples to illustrate how to compute the stationary distribution and what can go wrong if the chain is not irreducible.

***Example 8.9.*** Find the stationary distribution for the ON/OFF system in Example 8.4.

Since the chain is finite and irreducible, the stationary distribution exists and is unique. The equation $\boldsymbol{\pi} P = \boldsymbol{\pi}$ becomes

$$(\pi_0 \ \pi_1) \begin{pmatrix} 1-p & p \\ q & 1-q \end{pmatrix} = (\pi_0 \ \pi_1)$$

from which we take the first equation

$$(1-p)\pi_0 + q\pi_1 = \pi_0$$

which gives

$$\pi_1 = \frac{p}{q}\pi_0$$

The second equation is

$$p\pi_0 + (1-q)\pi_1 = \pi_1$$

which also gives $\pi_1 = (p/q)\pi_0$. To get a solution, we note that $\pi_0 + \pi_1 = 1$, which gives

$$\pi_0 \left(1 + \frac{p}{q}\right) = 1$$

which gives stationary distribution

$$\boldsymbol{\pi} = \left(\frac{q}{p+q}, \frac{p}{p+q}\right)$$

Note how $\pi_0 > \pi_1$ if $q > p$. This makes sense since if the chain is more likely to jump from 1 to 0 than the other way, in the long run it spends more time in 0. □

Note how the two equations for $\pi_0$ and $\pi_1$ in the example turned out to be the same. In general, if the state space has $r$ states, the equation $\boldsymbol{\pi} P = \boldsymbol{\pi}$ gives at most $r - 1$ linearly independent equations, and in addition to these, we also have the equation $\sum_{j \in S} \pi_j = 1$. Recalling results from linear algebra, this means that there always exists a solution to this system of equations, but unless the chain is irreducible, there may be more than one solution.

***Example 8.10.*** Find the stationary distribution in the genetics example (Example 8.2).

The chain is not irreducible, but let us still attempt to find the stationary distribution. The states are $AA$, $Aa$, and $aa$, and the equation $\boldsymbol{\pi} P = \boldsymbol{\pi}$ becomes

$$(\pi_{AA}\ \pi_{Aa}\ \pi_{aa}) \begin{pmatrix} 1 & 0 & 0 \\ 1/4 & 1/2 & 1/4 \\ 0 & 0 & 1 \end{pmatrix} = (\pi_{AA}\ \pi_{Aa}\ \pi_{aa})$$

from which we get the first equation

$$\pi_{AA} + \frac{1}{4}\pi_{Aa} = \pi_{AA}$$

which gives $\pi_{Aa} = 0$. Knowing this, the second equation gives only $0 = 0$, and the third gives $\pi_{aa} = \pi_{aa}$. Thus, any distribution of the form

$$\boldsymbol{\pi} = (\alpha,\ 0,\ 1 - \alpha)$$

where $0 \leq \alpha \leq 1$ qualifies as a stationary distribution. The evolution of the chain is simple: we choose either $AA$ or $aa$, according to $\boldsymbol{\pi}$, and whatever state we choose stays forever. Thus, we get the sequence $AA, AA, \ldots$ with probability $\alpha$ and $aa, aa, \ldots$ with probability $1 - \alpha$. Perhaps not very exciting, but a good illustration of what can happen without irreducibility. □

Things are a little more complicated if the state space is infinite. Consider, for example, the following variant of Example 1.59.

***Example 8.11.*** Recall the gambler's ruin problem in Example 1.59, where Ann starts with one dollar and Bob is infinitely wealthy. Also suppose that Ann has an infinitely wealthy and benevolent uncle who gives her a dollar to bet every time she goes broke. Describe the Markov chain and find the stationary distribution.

The state space of Ann's possible fortune is $S = \{0, 1, 2, \ldots\}$. If $i \geq 1$, the possible transitions are to states $i - 1$ and $i + 1$ and in state 0, Ann either wins and pays back her uncle's dollar or loses her uncle's dollar, stays in 0, and borrows another dollar. The transition graph is

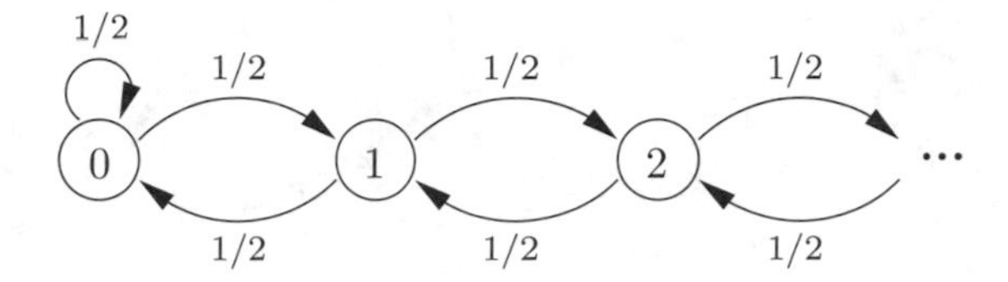


and the transition matrix is

$$P = \begin{pmatrix} 1/2 & 1/2 & 0 & 0 & 0 & \cdots \\ 1/2 & 0 & 1/2 & 0 & 0 & \cdots \\ 0 & 1/2 & 0 & 1/2 & 0 & \cdots \\ \vdots & \vdots & \vdots & \vdots & \vdots & \ddots \end{pmatrix}$$

The equation $\boldsymbol{\pi} P = \boldsymbol{\pi}$ now gives the first equation

$$\frac{1}{2}\pi_0 + \frac{1}{2}\pi_1 = \pi_0$$

which gives $\pi_1 = \pi_0$. The second equation is

$$\frac{1}{2}\pi_0 + \frac{1}{2}\pi_2 = \pi_1$$

which gives $\pi_2 = 2\pi_1 - \pi_0 = \pi_0$. The remaining equations all look the same:

$$\frac{1}{2}\pi_{n-2} + \frac{1}{2}\pi_n = \pi_{n-1}, \quad n \geq 3$$

which gives $\pi_n = 2\pi_{n-1} - \pi_{n-2} = \pi_0$. Thus, a stationary distribution must be of the form

$$\boldsymbol{\pi} = (\pi_0, \ \pi_0, \ \pi_0, \ldots)$$

which is obviously a problem since we cannot sum these probabilities to 1. If $\pi_0 = 0$, the sum is 0, and if $\pi_0 > 0$, the sum is infinite. We conclude that *no stationary distribution exists.* □

The technique used in this example to express all the $\pi_n$ in terms of $\pi_0$ is the standard way to try to find a stationary distribution. In this case, it turned out to be impossible since the $\pi_n$ did not sum to one.

***Example 8.12.*** Reconsider the previous problem under the assumption that Bob has an edge in the game, so that in each round Ann wins with probability $p < \frac{1}{2}$.

The transition graph is now

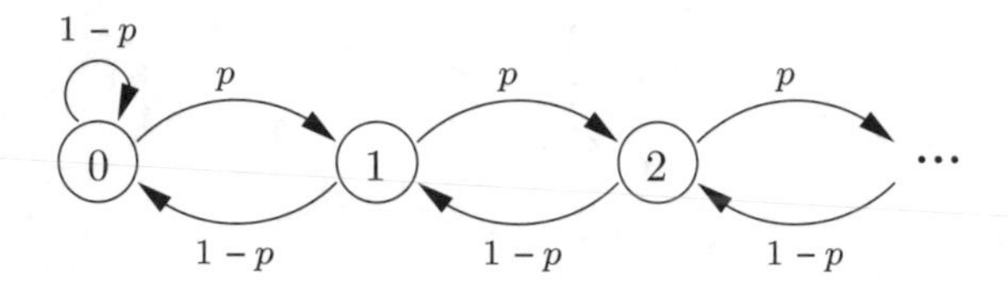


and the transition matrix

$$P = \begin{pmatrix} 1-p & p & 0 & 0 & 0 & \dots \\ 1-p & 0 & p & 0 & 0 & \dots \\ 0 & 1-p & 0 & p & 0 & \dots \\ \vdots & \vdots & \vdots & \vdots & \vdots & \ddots \end{pmatrix}$$

The equation $\boldsymbol{\pi} P = \boldsymbol{\pi}$ gives

$$(1-p)\pi_0 + (1-p)\pi_1 = \pi_0$$

which gives

$$\pi_1 = \frac{p}{1-p}\pi_0$$

The next equation is

$$p\pi_0 + (1-p)\pi_2 = \pi_1$$

which gives

$$\pi_2 = \frac{1}{1-p}(\pi_1 - p\pi_0) = \left(\frac{p}{1-p}\right)^2 \pi_0$$

The remaining equations are

$$p\pi_{n-2} + (1-p)\pi_n = \pi_{n-1}$$

and it is easily verified that

$$\pi_n = \left(\frac{p}{1-p}\right)^n \pi_0, \quad n = 0, 1, 2, \ldots$$

satisfy these equations. To find $\pi_0$, we use the condition $\sum_{j \in S} \pi_j = 1$ and get

$$1 = \pi_0 \sum_{n=0}^{\infty} \left(\frac{p}{1-p}\right)^n = \pi_0 \frac{1-p}{1-2p}$$

which gives stationary distribution

$$\pi_n = \frac{1-2p}{1-p} \left(\frac{p}{1-p}\right)^n, \quad n = 0, 1, 2, \ldots$$

□

The Markov chains are irreducible and recurrent in both examples; the only difference is that the probability $\frac{1}{2}$ was replaced by $p < \frac{1}{2}$ in the second. Why, then, does a stationary distribution exist in the second but not the first chain? We need to introduce yet another classifying property for Markov chains.

***Definition 8.6.*** Let $i$ be a recurrent state. If $E_i[\tau_i] < \infty$, then $i$ is said to be *positive recurrent*. If $E_i[\tau_i] = \infty$, $i$ is said to be *null recurrent.*

This is a more subtle distinction than between recurrence and transience. Any recurrent state is revisited infinitely many times, but only a positive recurrent state is revisited in such a way that the expected time between visits is finite (recall from Example 2.23 that a random variable can be finite and yet have an infinite expectation). Thus, recurrence/transience is distinguished by the $\tau_i$ themselves and positive/null recurrence by the expected values $E_i[\tau_i]$. It can be shown that positive recurrence is also a class property, and hence we have the following corollary.

**Corollary 8.4.** *For an irreducible Markov chain, there are three possibilities:* **(a)** *all states are positive recurrent,* **(b)** *all states are null recurrent, and* **(c)** *all states are transient.*

Now consider a finite state space. There cannot be any null recurrent states, a fact that we will not prove, but the intuition is that there simply is not enough room for very long paths of return. Also recall that there must be at least one recurrent state and hence this state must be positive recurrent. Thus, if a finite chain is irreducible, it must also be positive recurrent. If the state space is infinite, this is not true because of (b) in Corollary 8.4. The following result covers both finite and infinite state spaces.

**Proposition 8.4.** Consider an irreducible Markov chain $\{X_n\}$. Then,

$$\text{A stationary distribution } \pi \text{ exists} \Leftrightarrow \{X_n\} \text{ is positive recurrent}$$

If this is the case, $\pi$ is unique and has $\pi_j > 0$ for all $j \in S$.

The intuition behind the result is not obvious, but we can look at the last two examples for a comparison. By Proposition 8.4, all states are positive recurrent when $p < \frac{1}{2}$, something that can also be shown directly. We can think of this as there being a "pull" toward 0 and the chain settles in toward a stationary distribution. Compare this with the case $p = \frac{1}{2}$ where there is no such pull and the chain wanders around aimlessly forever. This is reflected in how the stationary distribution "tries to be uniform," but this is not possible on an infinite state space. In general, think of a positive recurrent Markov chain as one that makes a deliberate effort to revisit states, in contrast to a null recurrent chain, which just happens to revisit states without really trying.

Note that Proposition 8.4 goes in both directions. Thus, if we can find a stationary distribution of an irreducible chain, we know two things: (1) the stationary distribution is unique and (2) the chain is positive recurrent.

### 8.2.4 Convergence to the Stationary Distribution

In this section, we will state the main limit result for a Markov chain. Although the proof is not beyond the scope of the text, we will not give it; instead, we will focus on its interpretation and applications. For a proof, the interested reader may consult, for example, Grimmett and Stirzaker, *Probability and Random Processes* [7]. In Example 8.6, we found the limit distribution of the Markov chain, and let us now formally define this concept.

***Definition 8.7.*** Let $p_{ij}^{(n)}$ be the $n$-step transition probabilities of a Markov chain. If there exists a probability distribution $\mathbf{q}$ on $S$ such that

$$p_{ij}^{(n)} \to q_j \quad \text{as} \quad n \to \infty \quad \text{for all } i, j \in S$$

we call $\mathbf{q}$ the *limit distribution* of the Markov chain.

Note that the limit distribution is the same for every initial state $i \in S$. Another way to express this is that the $n$-step transition matrix $P^{(n)}$ converges to a limit matrix in which all rows are equal. The intuition behind the limit distribution is that $q_j$ describes the probability that the chain is in state $j$ at some late timepoint and that at this time, the chain has "forgotten how it started." We have seen in the examples that a limit distribution does not always exist. If it does, however, it also qualifies as a stationary distribution (see Problem 13).

The more interesting question is whether the converse is true: is the stationary distribution also the limit distribution? This would give a nice computational recipe: in order to find the limit distribution we solve $\pi P = \pi$, which is typically much easier than computing powers of the transition matrix. The following example shows that there may be a problem.

***Example 8.13.*** In the ON/OFF system in Example 8.4, suppose that $p = q = 1$, that is, that the system always changes. Find the stationary distribution and limit distribution.

The stationary distribution satisfies

$$(\pi_0\ \pi_1)\begin{pmatrix} 0 & 1 \\ 1 & 0 \end{pmatrix} = (\pi_0\ \pi_1)$$

which gives $\pi_0 = \pi_1$, so the stationary distribution is $\pi = (\frac{1}{2}, \frac{1}{2})$. In this case, it is easy to find the $n$-step transition probabilities explicitly. For example,

$$p_{00}^{(n)} = \begin{cases} 0 & \text{if } n \text{ is even} \\ 1 & \text{if } n \text{ is odd} \end{cases}$$

and similarly for the other three $n$-step transition probabilities. But this means that the $n$-step transition probabilities do not converge, and thus there *is no limit distribution*. Recall how a limit distribution forgets where the chain started; in this case, if we start in state 0, we know that the system will be in state 0 at every even timepoint and in state 1 at every odd timepoint, no matter how late. □

Thus, stationary distributions and limit distributions are not necessarily the same. What is the intuition behind these concepts? Suppose that we look at a Markov chain at some late time $n$. The stationary distribution then gives the long-term proportions of time spent in the different states *up to* time $n$. The limit distribution, on the other hand, gives the proportions of time spent in the states *at* time $n$ (so we have to think of the Markov chain being run up to time $n$ multiple times). In the previous example, let $n = 1000$. The stationary distribution $(\frac{1}{2}, \frac{1}{2})$ tells us that equal amounts of time have been spent in both states up to time $n = 1000$, regardless of the initial state. However, if we look only at precisely time $n = 1000$, the chain must be in the same state that it started in, and if we run the chain up to time $n = 1000$ from the same initial state, the proportion of time in the other state is 0. For a theoretical result that motivates the interpretation of the stationary distribution, see Problem 21.

The existence of a limit distribution is a desirable property of a Markov chain, since it means that we can get an idea of the distribution over the state space at some late, arbitrary timepoint. It turns out that the problem in the previous example is that the chain is *periodic*, in the sense that returns from a state to itself can occur only in an even number of steps.

***Definition 8.8.*** The *period* of state $i$ is defined as

$$d(i) = \gcd\{n \geq 1 : p_{ii}^{(n)} > 0\}$$

the greatest common divisor of lengths of cycles through which it is possible to return to $i$. If $d(i) = 1$, state $i$ is said to be *aperiodic*; otherwise it is called *periodic*.

The concept of a period may not be immediately clear. Let us look at two examples.

***Example 8.14.*** Find the periods of the states in the ON/OFF system with $p = q = 1$.

Since $p_{00}^{(n)} > 0$ whenever $n$ is even and 0 otherwise, the set of $n$ such that $p_{00}^{(n)} > 0$ is $\{2, 4, 6, \ldots\}$ that has greatest common divisor 2. Thus, the period of state 0 is 2, which means that the only possible return paths to state 0 have lengths that are multiples of 2. The period of state 1 is also 2. □

***Example 8.15.*** Find the period of the state 1 in the gambler's ruin example (Example 8.11).

We have $p_{11}^{(1)} = 0$, $p_{11}^{(2)} > 0$, and $p_{11}^{(3)} > 0$, and since the greatest common divisor of 2 and 3 is 1, we do not need to go any further. State 1 is aperiodic, and we note that this does *not* mean that it can reach itself in one step. See also Problem 14. □

If all states are aperiodic, we call the whole Markov chain aperiodic. It can be shown that periodicity is a class property in the sense that communicating states have the same period. Thus, if we can show that one state is aperiodic in an irreducible chain, the whole chain must be aperiodic. Aperiodicity is the last property we need to be able to say that the stationary distribution and the limit distribution coincide. The following is the main convergence theorem for Markov chains.

***Theorem 8.1.*** *Consider an irreducible, positive recurrent, and aperiodic Markov chain with stationary distribution $\pi$ and $n$-step transition probabilities $p_{ij}^{(n)}$. Then*

$$p_{ij}^{(n)} \to \pi_j \quad as \quad n \to \infty$$

*for all $i, j \in S$.*

An irreducible, positive recurrent, and aperiodic Markov chain is called *ergodic*. We have seen examples of what can go wrong when any of the three conditions are removed. Take away irreducibility, and there may be more than one stationary distribution; take away positive recurrence, and there may be none at all; take away aperiodicity, and there may be a unique stationary distribution that is not the limit distribution.

We should also point out that positive recurrence is listed as an assumption. Positive recurrence is an important characteristic of a Markov chain, describing its long-term behavior, but it is typically not checked since it is easier to find the stationary distribution. The practical way to use the theorem is thus to check irreducibility and aperiodicity and then go about solving $\pi P = \pi$. If this can be done, $\pi$ is the limit distribution, and we get positive recurrence for free.

***Example 8.16* (*Success Runs*).** A fair coin is flipped repeatedly, and at time $n$, we let $X_n$ be the length of the current run of heads. For example, if we get the sequence *HTHHHT*, we have (let $X_0 \equiv 0$)

$$X_1 = 1,\ X_2 = 0,\ X_3 = 1,\ X_4 = 2,\ X_5 = 3,\ X_6 = 0$$

Describe this sequence of *success runs* as a Markov chain and find its limit distribution.

The state space is $S = \{0, 1, 2, \ldots\}$, and from a state $i$, transitions are possible to either $i + 1$ or 0, with equal probabilities, giving the following transition graph:

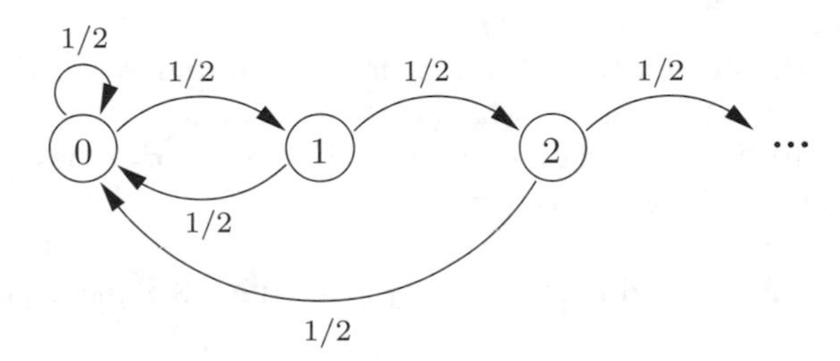


The chain is clearly irreducible, so let us look for the stationary distribution. The equation $\pi P = \pi$ becomes

$$(\pi_0, \pi_1, \ldots) \begin{pmatrix} 1/2 & 1/2 & 0 & 0 & 0 & \cdots \\ 1/2 & 0 & 1/2 & 0 & 0 & \cdots \\ 1/2 & 0 & 0 & 1/2 & 0 & \cdots \\ 1/2 & 0 & 0 & 0 & 1/2 & \cdots \\ \vdots & \vdots & \vdots & \vdots & \vdots & \ddots \end{pmatrix} = (\pi_0, \pi_1, \ldots)$$

and it is easily checked that the solution is

$$\pi_k = \frac{1}{2^{k+1}}, \quad k = 0, 1, 2, \ldots$$

which we recognize from Section 2.5.3 as a geometric distribution including 0, with success probability $p = \frac{1}{2}$. The last question is whether this also qualifies as the limit distribution and we need to check for aperiodicity. Consider state 0 and note that

$p_{00} > 0$, which means that state 0 is aperiodic. Thus, by irreducibility, the entire chain is aperiodic, Theorem 8.1 applies, and $\boldsymbol{\pi}$ is the limit distribution. □

In a Markov chain with stationary distribution $\boldsymbol{\pi}$, $\pi_i$ is the long-term frequency spent in state $i$; thus, state $i$ is revisited on average every $1/\pi_i$ steps. Now consider another state $j$. In any sequence of $N$ steps, it is visited on average $N\pi_j$ times, and we get the following nice result, which we state without formal proof. We use the notation $E_i$ for expected value when the initial state is $i$.

**Proposition 8.5.** Consider an ergodic Markov chain with stationary distribution $\boldsymbol{\pi}$ and choose two states $i$ and $j$. Let $\tau_i$ be the return time to state $i$, and let $N_j$ be the number of visits to $j$ between consecutive visits to $i$. Then

$$E_i[\tau_i] = \frac{1}{\pi_i} \quad \text{and} \quad E_i[N_j] = \frac{\pi_j}{\pi_i}$$

Note that by positive recurrence, all the $E_i[\tau_i]$ are finite and hence all the $\pi_i$ are strictly positive. The $E_i[\tau_i]$ are called the *mean recurrence times*.

***Example 8.17.*** Consider Example 8.12. Suppose that Ann wins with probability $p = \frac{1}{3}$. **(a)** If Ann just went broke, what is the expected number of rounds until she is broke again? **(b)** If Ann reaches a fortune of \$5, how many times can she expect to go broke before reaching that fortune again?

We are asking for $E_0[\tau_0]$ and $E_5[N_0]$, and by Proposition 8.5, these are

$$E_0[\tau_0] = \frac{1}{\pi_0} = \frac{1}{1/2} = 2$$

and

$$E_5[N_0] = \frac{\pi_0}{\pi_5} = \frac{1/2}{1/64} = 32$$

□

## 8.3 RANDOM WALKS AND BRANCHING PROCESSES

In this section, we will look at two special cases of Markov chains: *random walks* and *branching processes*. Although they are examples of Markov chains, their properties are such that the methods we have explored do not reach very far and we instead analyze them by methods that are suited to their particular nature.

### 8.3.1 The Simple Random Walk

Many of the examples we looked at in the previous section are similar in nature. For example, the roulette example and the various versions of gambler's ruin have in common that the states are integers and the only possible transitions are one step up or one step down. We now take a more systematic look at such Markov chains, called *simple random walks*. A simple random walk can be described as a Markov chain $\{S_n\}$ that is such that

$$S_n = \sum_{k=1}^{n} X_k$$

where the $X_k$ are i.i.d. such that $P(X_k = 1) = p$, $P(X_k = -1) = 1 - p$.

The term "simple" refers to the fact that only unit steps are possible; more generally we could let the $X_k$ have any distribution on the integers. The initial state $S_0$ is usually fixed but could also be chosen according to some probability distribution. Unless otherwise mentioned, we will always have $S_0 \equiv 0$. If $p = \frac{1}{2}$, the walk is said to be *symmetric*. It is clear from the construction that the random walk is a Markov chain with state space $S = \{\ldots, -2, -1, 0, 1, 2, \ldots\}$ and transition graph

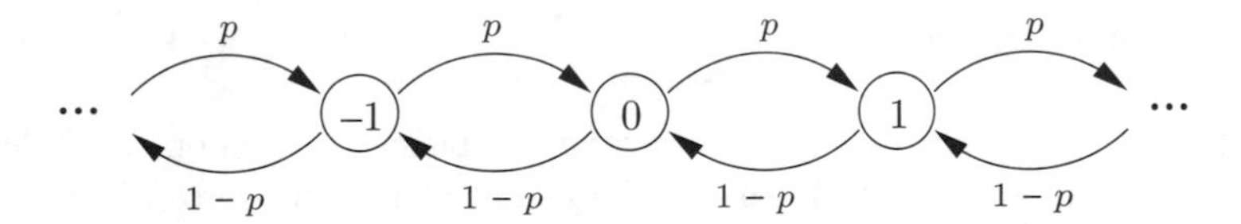


Note how the transition probabilities $p_{i,i+1}$ and $p_{i,i-1}$ do not depend on $i$, a property called *spatial homogeneity*. We can also illustrate the random walk as a function of time, as was done in Example 1.58. Note that this illustrates one particular outcome of the sequence $S_0, S_1, S_2, \ldots$, called a *sample path*, or a *realization*, of the random walk.

It is clear that the random walk is irreducible, so it has to be transient, null recurrent, or positive recurrent, and which one it is may depend on $p$. The random walk is a Markov chain where we can compute the $n$-step transition probabilities explicitly and apply Proposition 8.1. Consider any state $i$, and note first that

$$p_{ii}^{(2n-1)} = 0, \quad n = 1, 2, \ldots$$

since we cannot make it back to a state in an odd number of steps. To make it back in $2n$ steps, we must take $n$ steps up and $n$ steps down, which has probability

$$p_{ii}^{(2n)} = \binom{2n}{n} p^n (1-p)^n = \frac{(2n)!}{n!n!}(p(1-p))^n \tag{8.1}$$

and since convergence of the sum of the $p_{ii}^{(2n)}$ is not affected by the values of any finite number of terms in the beginning, we can use an asymptotic approximation of $n!$, *Stirling's formula*, which says that

$$n! \sim n^n \sqrt{n}\, e^{-n} \sqrt{2\pi}$$

where "$\sim$" means that the ratio of the two sides goes to 1 as $n \to \infty$. The practical use is that we can substitute one for the other for large $n$ in the sum in Proposition 8.1. Thus,

$$p_{ii}^{(2n)} \sim \frac{(4p(1-p))^n}{\sqrt{\pi n}}$$

and if $p = \frac{1}{2}$, this equals $1/\sqrt{\pi n}$ and the sum over $n$ is infinite. If instead $p \neq \frac{1}{2}$, $p_{ii}^{(2n)}$ is of the form $x^n/\sqrt{\pi n}$ where $|x| < 1$ and the sum over $n$ converges. This shows that the simple random walk is recurrent if $p = \frac{1}{2}$ and transient if $p \neq \frac{1}{2}$.

The next question is whether the case $p = \frac{1}{2}$ is positive recurrent or null recurrent. Repeating the argument from Example 3.35 shows that regardless of whether the first step is up or down, the expected time until return to 0 is infinite. We summarize as follows.

**Proposition 8.6.** The simple random walk is null recurrent if $p = \frac{1}{2}$ and transient if $p \neq \frac{1}{2}$.

In particular, this means that there is never a stationary distribution, so the theory for Markov chains does not give us anything further, but there are still interesting questions regarding the behavior of the random walk. Recall $\tau_1$, the time of the first visit to state 1. From Examples 1.59 and 3.35, we know that $P_0(\tau_1 < \infty) = 1$ and $E_0[\tau_1] = \infty$, if $p = \frac{1}{2}$. What about other values of $p$? If $p > \frac{1}{2}$, we must still have $P_0(\tau_1 < \infty) = 1$ but what if $p < \frac{1}{2}$? Let us use recursion and again condition on the first step. With $r = P_0(\tau_1 < \infty)$, we get

$$r = p + (1-p)r^2$$

which has solutions $r = 1$ and $r = p/(1-p)$. We can exclude $r = 1$, since if the probability to reach 1 in finite time equals 1, the same must be true for $-1$ (more likely to go down than up), but then by symmetry, the probability is also 1 that the walk gets back to 0 again in finite time, which contradicts transience. We get the following result.

**Corollary 8.5.** *The probability that the walk ever visits* 1 *is*

$$P_0(\tau_1 < \infty) = \begin{cases} 1 & \text{if } p \geq 1/2 \\ \dfrac{p}{1-p} & \text{if } p < 1/2 \end{cases}$$

Thus, if $p < \frac{1}{2}$, the probability is $p/(1-p)$ that the walk *never* visits the positive axis. If it does visit the positive axis, how far does it get? According to the following result, not very far.

**Corollary 8.6.** *The simple random walk with $p < \frac{1}{2}$ visits only finitely many states on the positive axis.*

*Proof.* Let $A$ be the event that the walk visits all positive states and $A_r$ the event that the walk eventually visits state $r$. Then

$$A = \bigcap_{r=1}^{\infty} A_r$$

where the events $A_r$ are decreasing (must have visited $r$ in order to visit $r + 1$). Moreover, in order to visit $r$, the walk must first visit 1, then 2, and so on, and by Corollary 8.2 and Proposition 1.6, we obtain

$$P(A) = \lim_{r\to\infty} P(A_r) = \lim_{r\to\infty} \left(\frac{p}{1-p}\right)^r = 0$$

since $p < \frac{1}{2}$. ■

We can continue the argument in the proof. Since the walk makes it up only to some maximum integer, it must eventually leave the positive axis for good. But then it is at $-1$, and the same argument says that it must eventually hit $-2$, never to return to $-1$ again. Thus, for any number, the random walk will eventually stay below it forever, and we have argued that the simple random walk with $p < \frac{1}{2}$ drifts toward $-\infty$. In obvious analogy, if $p > \frac{1}{2}$, the walk drifts toward $\infty$. In Problem 27, you are asked to compute the probability that the transient random walk ever returns to 0 (right now all we know is that this probability is $< 1$). In the remaining case, $p = \frac{1}{2}$, the walk must visit all states infinitely many times (why?) and must, like Ahasverus, wander aimlessly forever.

We next turn to expected values. If the walk starts in 0, what is the expected time until its first visit to 1? If $p < \frac{1}{2}$, it must be infinite, since the random variable $\tau_1$ itself may be infinite. If $p = \frac{1}{2}$, can $E_0[\tau_1]$ be finite? If it were, then the expected time to visit $-1$ would by symmetry also be finite. But then the expected time back to 0 again would also be finite, which contradicts null recurrence. Thus, if $p \leq \frac{1}{2}$, we have $E_0[\tau_1] = \infty$. It remains to investigate what happens when $p > \frac{1}{2}$. We will use a recursive approach, in the spirit of Example 3.35. Let $\mu = E_0[\tau_1]$ and condition on the first step to get the equation

$$\mu = p \times 1 + (1 - p)(1 + 2\mu)$$

which we solve for $\mu$ to get $\mu = 1/(2p - 1)$. Note, however, that also $\mu = \infty$ is a solution, so we need to argue that this can be ruled out. First note that $\tau_1 = 2n + 1$ if the walk is back at 0 at time $2n$, without having visited 1, then goes up to 1 in the following step. The probability to be back at 0 at time $2n$ without any restrictions is $p_{00}^{(2n)}$, and hence

$$P_0(\tau_1 = 2n+1) \leq p \times p_{00}^{(2n)} \sim p\frac{(4p(1-p))^n}{\sqrt{\pi n}}$$

and since $4p(1-p) < 1$, $E_0[\tau_1] = \sum_n (2n+1) P_0(\tau_1 = 2n+1)$ must be finite. We summarize as follows.

**Corollary 8.7.** *The expected time until the first visit to* 1 *is*

$$E_0[\tau_1] = \begin{cases} \dfrac{1}{2p-1} & \text{if } p > 1/2 \\ \\ \infty & \text{if } p \leq 1/2 \end{cases}$$

### 8.3.2 Multidimensional Random Walks

Let us now consider a two-dimensional simple random walk and investigate it with regard to transience/recurrence. First, how do we define it? In one dimension, if the walk is at $i$, it chooses one of the neighboring points $i-1$ and $i+1$. In two dimensions, there are different ways to define neighboring points. One way is to consider the four neighbors parallel with the axes; another, to consider the four neighboring corner points.

Regardless of definition, we will assume that the walk is symmetric so that each neighbor is chosen with probability $\frac{1}{4}$; otherwise, the walk is transient (why?). Also, since all we are interested in is transience/recurrence, it does not matter which definition we choose because in each case there are four equally likely neighbors (one case is just a 45° rotation of the other). We will go for the second version, choosing the corners with equal probabilities. The reason for this is that we can then view the two-dimensional random walk as two independent one-dimensional random walks, one on each axis. Thus, let

$$\mathbf{S}_n = (S_n^{(1)}, S_n^{(2)}), \quad n = 0, 1, 2, \ldots$$

where we let $\mathbf{S}_0 = (0, 0)$. Then $\mathbf{S}_n = (0, 0)$ if and only if $S_n^{(1)} = S_n^{(2)} = 0$ and by independence and Equation (8.1),

$$P(\mathbf{S}_n = (0,0)) = P(S_n^{(1)} = 0) P(S_n^{(2)} = 0) = \left( \binom{2n}{n} \frac{1}{4^n} \right)^2$$

and by Stirling's formula

$$P(\mathbf{S}_n = (0,0)) \sim \frac{1}{\pi n}$$

and since $\sum_n \frac{1}{n} = \infty$, we conclude that the simple random walk in two dimensions is recurrent (and must be null recurrent). But look now what happens in three dimensions. By defining

$$P(\mathbf{S}_n = (S_n^{(1)}, S_n^{(2)}, S_n^{(3)}), \quad n = 0, 1, 2, \ldots$$

where $S_n^{(1)}$, $S_n^{(2)}$, and $S_n^{(3)}$ are independent symmetric random walks and letting $\mathbf{S}_0 = (0, 0, 0)$, we get

$$P(\mathbf{S}_n = (0, 0, 0)) \sim \frac{1}{(\pi n)^{3/2}}$$

and since $\sum_n \frac{1}{n^{3/2}} < \infty$, we conclude that the walk is now transient. This might be a bit surprising, and there is no immediate intuition for why the walk always returns to the origin in one and two dimensions but not in three. We know that each of the three individual walks returns to 0 infinitely many times, but only finitely many times will they do so simultaneously. We can define the random walk in any number of dimensions and conclude the following.

**Corollary 8.8.** *The symmetric simple random walk in n dimensions is recurrent for $n = 1$ and $n = 2$, and transient for $n \geq 3$.*

One thing needs to be pointed out. In two dimensions, we argued that the definition of a neighboring point does not matter since there are four neighbors with either definition. This is not true in dimensions $n \geq 3$. For example, in three dimensions, there is a difference in choosing between the eight corner points and the six points along the axes. We chose the first definition, and the chance of return to the origin ought to be higher with the second since there are fewer choices in each step. However, it can be shown that the probability of return is still less than one and also this variant of the three-dimensional random walk is transient (see Problem 29).

### 8.3.3 Branching Processes

To steal a line of British humo(u)r from Grimmett and Stirzaker [7]: "Besides gambling, many probabilists have been interested in reproduction." In this section, we analyze a simple model for populations that are composed of individuals who reproduce independent of each other. Suppose that we start from one individual, the *ancestor*, who gets a number of children, $X$, with range $\{0, 1, 2, \ldots\}$ and pmf $p_X$. We refer to this pmf as the *offspring distribution*. Each child then reproduces independently according to the offspring distribution, their children reproduce in the same way, and so on. The resulting evolving population is an example of a *branching process* (see Figure 8.1).[4] To describe it mathematically, we let $Z_n$ be the number of

[4]In this simple form, it is usually called a *Galton–Watson* process, after the previously mentioned Sir Francis Galton, who worried about extinction of the English nobility, and Henry W. Watson, mathematician, clergyman, and mountaineer, who advocated the use of generating functions to solve the problems. In *les*

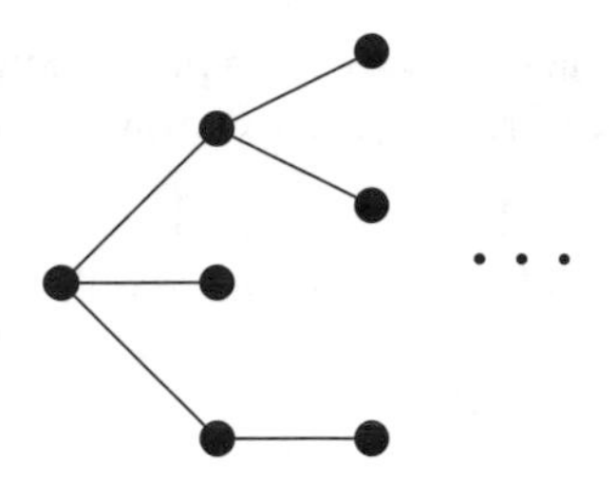

**FIGURE 8.1** A branching process with $Z_0 = 1$, $Z_1 = 3$, and $Z_2 = 3$.

individuals in the $n$th generation, and let $Z_0 \equiv 1$. The generation sizes relate to each other as

$$Z_n = \sum_{k=1}^{Z_{n-1}} X_k, \quad n = 1, 2, \ldots \tag{8.2}$$

where the $X_k$ are i.i.d. with pmf $p_X$. The formula states that in order to get the number of individuals in any generation, we go through the individuals in the preceding generation and sum their numbers of children. Note that in each generation we get a new set of $X_k$, which if needed can be indicated by a superscript, $X_k^{(n-1)}$, the number of children of the $k$th individual in the $(n-1)$st generation.

We are interested in the behavior of $Z_n$ and will focus on two issues: extinction and population growth. Note that the process $\{Z_n\}$ is a Markov chain but the transition probabilities are complicated and it is also clear that state 0 is absorbing, so the Markov chain methods that we know will not be of much help. Let us from now on exclude the two uninteresting cases $p_X(0) = 1$ and $p_X(1) = 1$. (Why are these uninteresting?) From Equation (8.2) it is clear that

$$Z_n = 0 \;\Rightarrow\; Z_{n+1} = 0 \tag{8.3}$$

in which case the population has gone extinct. Now let $E$ be the event that the population goes extinct *eventually*, that is, the event that some generation size is 0. This event can be described as

$$E = \bigcup_{n=1}^{\infty} \{Z_n = 0\}$$

and to find the probability of $E$, we will use probability generating functions. Thus, let $X$ have pgf $G$, and let $Z_n$ have pgf $G_n$. Then $G_1 = G$ (why?) and by Equation

*pays francophones*, the name of *Bienaymé* is usually also added, which is only fair since the work of I. J. Bienaymé precedes that of Galton and Watson.

(8.2) and Proposition 3.39, we get the relation

$$G_n(s) = G_{n-1}(G(s)), \quad n = 2, 3, \ldots \tag{8.4}$$

By Equation (8.3), we realize that the events $\{Z_n = 0\}$ are increasing in $n$, and Corollary 3.12 and Proposition 1.6 together give

$$P(E) = \lim_{n\to\infty} P(Z_n = 0) = \lim_{n\to\infty} G_n(0)$$

***Example 8.18.*** Consider a population of cells that may either die or reproduce, according to the following random variable:

$$X = \begin{cases} 0 & \text{with probability } 1/4 \\ 2 & \text{with probability } 3/4 \end{cases}$$

Find the probability of extinction.

We have

$$G(s) = \frac{1}{4} + \frac{3}{4}s^2$$

which gives $G(0) = \frac{1}{4}$, the probability that extinction occurs already in the first generation. For $G_2$, we use Equation (8.4) and obtain

$$G_2(s) = G(G(s)) = \frac{1}{4} + \frac{3}{4}\left(\frac{1}{4} + \frac{3}{4}s^2\right)^2$$

which gives $G_2(0) = \frac{19}{64}$. Already here we realize that it will be hard to find $P(E)$ in this way; remember that we need to find the limit of $G_n(0)$ as $n \to \infty$ and a pattern for the sequence does not readily emerge. We give up. □

The branching process in this example is about as simple as they come, and yet it is virtually impossible to find the extinction probability as the limit of $G_n(0)$. Luckily, there is a much quicker way.

**Proposition 8.7.** Consider a branching process where the offspring distribution has pgf $G$, and let $E$ be the event of extinction. Then $q = P(E)$ is the smallest solution in $[0, 1]$ to the equation $s = G(s)$.

*Proof.* Let us first show that $q$ is a solution and then that it must be smaller than any other solution. Condition on the number of children, $X$, of the ancestor and note that if

$X = k$, then there are $k$ independent branching processes that must go extinct, which has probability $q^k$ (true also for $k = 0$). This gives the following observation:

$$q = \sum_{k=0}^{\infty} P(E|X = k)P(X = k) = \sum_{k=0}^{\infty} q^k P(X = k) = G(q)$$

and we see that the extinction probability solves the equation $s = G(s)$.

Suppose that there is another solution $r \in [0, 1]$. By Problem 133 in Chapter 3, $G(s)$ is increasing, and since $r \geq 0$, we get

$$r = G(r) \geq G(0)$$

Applying $G$ again gives $r = G(r) \geq G(G(0)) = G_2(0)$ and repeating the argument gives

$$r \geq G_n(0) \quad \text{for all } n$$

But since $G_n(0) \to q$, we also get $r \geq q$, and hence $q$ is the smallest solution to the equation $s = G(s)$. ■

***Example 8.19.*** In Example 8.18, we get the equation

$$s = \frac{1}{4} + \frac{3}{4}s^2$$

which has solutions $\frac{1}{3}$ and 1. Thus, the probability of extinction is $\frac{1}{3}$. □

Note that any pgf has the property $G(1) = 1$, and hence $s = 1$ is always a solution to the equation $s = G(s)$.[5] Let us now turn to the question of population growth. More specifically, we will find the mean and variance of the $n$th-generation size $Z_n$. Equation (8.2) is central, and we can apply Corollary 3.13 to obtain the following result.

**Proposition 8.8.** Consider a branching process where the offspring distribution has mean $\mu$ and variance $\sigma^2$. Then

$$E[Z_n] = \mu^n$$

$$\text{Var}[Z_n] = \begin{cases} n\sigma^2 & \text{if } \mu = 1 \\ \dfrac{\sigma^2(\mu^n - 1)\mu^{n-1}}{\mu - 1} & \text{if } \mu \neq 1 \end{cases}$$

[5] Watson, whose work was published in 1875, found the solution $s = 1$ but overlooked the fact that there could be more solutions and erroneously concluded that extinction is always inevitable. It took yet another half century and Danish ingenuity to completely solve the problem, done by J. F. Steffensen in 1930.

*Proof.* For the mean, repeatedly apply Corollary 3.13 to Equation (8.2) and get

$$E[Z_n] = E[Z_{n-1}]\mu = E[Z_{n-2}]\mu^2 = \cdots = \mu^n$$

since $E[Z_0] = 1$. We leave it as an exercise to verify that the expression for the variance satisfies Corollary 3.13. ■

The proposition thus tells us that

$$E[Z_n] \begin{cases} \to 0 & \text{if } \mu < 1 \\ \equiv 1 & \text{if } \mu = 1 \\ \to \infty & \text{if } \mu > 1 \end{cases}$$

and that

$$\mathrm{Var}[Z_n] \begin{cases} \to 0 & \text{if } \mu < 1 \\ \to \infty & \text{if } \mu = 1 \\ \to \infty & \text{if } \mu > 1 \end{cases}$$

This suggests that the population always goes extinct if $\mu < 1$ since the mean and variance both go to 0 and $Z_n$ itself can take on only integer values. It is less clear what happens in the other two cases, but the following result gives the answer.

**Proposition 8.9.** Consider a branching process with mean number of children $\mu$. Then

$$\mu \leq 1 \Rightarrow P(E) = 1$$
$$\mu > 1 \Rightarrow P(E) < 1$$

*Proof.* If $p_X(0) = 0$, then $\mu > 1$ and $P(E) = 0$. Now suppose that $p_X(0) > 0$, and recall from Section 3.11 that $\mu = G'(1)$, the slope at the point $s = 1$. By Problem 133 in Chapter 3, $G(s)$ is convex and increasing, and since it has to increase from $G(0) = p_X(0) > 0$ to $G(1) = 1$, it must intersect the line $y = s$ if $\mu > 1$, and the intersection is the extinction probability, which is $< 1$. If instead $\mu \leq 1$, there can be no such intersection, and the extinction probability is 1. See Figure 8.2 for the two possible cases. Note that this also shows that there can never be more than two solutions in [0, 1] to the equation $s = G(s)$. ■

This result is quite remarkable. It says that whether extinction occurs for certain depends only on the mean number of children, and if this is less than or equal to 1, there will be extinction sooner or later. If $\mu > 1$, extinction may be avoided and the probability of this is found by solving the equation $s = G(s)$. The cases $\mu > 1$, $\mu = 1$, and $\mu < 1$ are called the *supercritical*, *critical*, and *subcritical*, respectively. Although extinction is certain in the last two, they exhibit some differences in behavior that motivates the distinction (see Problem 30).

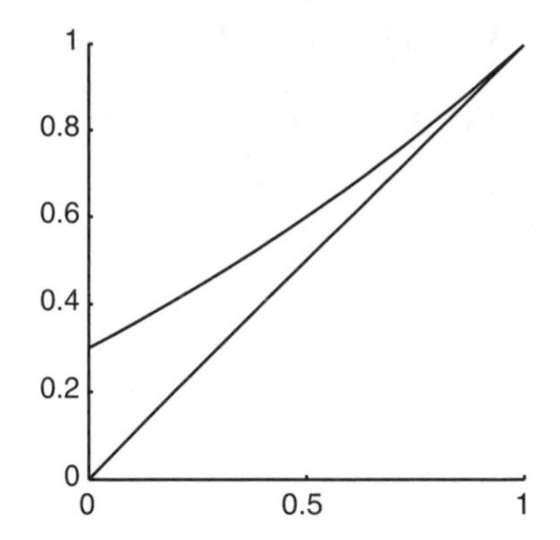


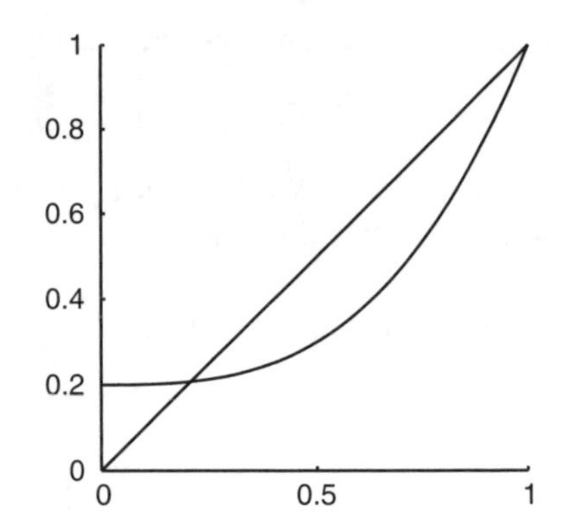


**FIGURE 8.2** Plots of the pgf $G(s)$ of the offspring distribution and the line $y = s$, in the cases $\mu = G'(1) \leq 1$ (left) and $\mu = G'(1) > 1$ (right).

***Example 8.20.*** An individual with a contagious disease enters a large city. Suppose that he passes the disease on to a number of people, who in turn pass it on to others, and so on. Suppose that each individual remains contagious for 1 day and in this day interacts with a number of people that has a Poisson distribution with mean 10 and that for each person, the probability of infection is $p$. **(a)** For which values of $p$ does the disease eventually die out? **(b)** If $p = 0.2$, what is the probability that the disease still exists in the population on day 2? **(c)** For $p = 0.2$, what is the probability that the disease eventually disappears?

The number of infected "children" of an individual has a Poisson distribution with mean $10p$, so the disease eventually dies out if $p \leq 0.1$. For part (b), we need to compute $P(Z_2 = 0) = G_2(0)$, and since

$$G(s) = e^{2(s-1)}$$

we get

$$P(Z_2 = 0) = e^{2(e^{-2}-1)} \approx 0.18$$

so the probability that the disease still exists is 0.82. For (c), we need numerically to solve the equation

$$s = e^{2(s-1)}$$

which gives $q \approx 0.2$. The spread of a disease may be adequately modeled by a branching process in its early stages but as time goes on, such a model becomes less realistic. There are several reasons for this, which we leave for the reader to ponder. □

The final question is what happens if there is no extinction. As it turns out, the only other possibility is for $Z_n$ to go to infinity. We will not give a strict proof but refer to an intuitive argument. Suppose that $Z_n$ does not go to infinity. Then there is some integer

$K$ such that $Z_n$ drops below $K$ infinitely many times. But each time it does, there is a chance that it goes extinct before the next time it drops below $K$. The probability of this is at least $p_X(0)^K$ since the "worst case" is when the population goes extinct already in the next generation. But this means that the population must become extinct sooner or later (think geometric trials here), and we have argued that if $Z_n$ does not go to infinity, the population must become extinct. In terms of probabilities

$$P(Z_n \nrightarrow \infty) \leq P(E)$$

and since the reversed inequality obviously also holds (if there is extinction, the populations size cannot go to infinity), the probabilities are equal. We have argued for the following result.

**Proposition 8.10.** Consider a branching process with extinction probability $q$. Then

$$Z_n \to \begin{cases} 0 & \text{with probability } q \\ \infty & \text{with probability } 1-q \end{cases}$$

as $n \to \infty$.

Note that "$Z_n \to 0$" is just another way of saying "extinction," since $Z_n$ takes on only integer values. We can describe this result by saying that the sequence of random variables $Z_1, Z_2, \ldots$ converges to a random variable $Z$ as $n \to \infty$, where $Z$ is either 0 or $\infty$ with probabilities $q$ or $1-q$, respectively. To be perfectly strict, we need to say that this convergence takes place *with probability* 1 since we can describe other types of sequences. For example, in a population of dividing cells let us say that $Z_1 = 2$. It could then happen that one divides and the other dies so that $Z_2 = 2$ also. If the same thing happens again, $Z_3 = 2$, and we can continue to describe a sequence $Z_1, Z_2, \ldots$ where $Z_n = 2$ for all $n$. However, it can be shown that this sequence together with all other sequences that do not converge to 0 or $\infty$ belong to an event that has probability 0. This is an example of convergence almost surely, mentioned in footnote 1 in Section 4.2.

## 8.4 CONTINUOUS -TIME MARKOV CHAINS

We have studied Markov chains in discrete time and will now turn to their counterpart in continuous time. This means that the chains stay in each state a random time that is a continuous random variable with a distribution that may depend on the state. The state of the chain at time $t$ is denoted $X(t)$, where $t$ ranges over the nonnegative real numbers. In addition to having the Markov property for the jumps, we also want the jumps to be independent of how long a time that is spent in a specific state, and in order to achieve this, we recall that there is only one continuous distribution that would ensure this property: the exponential distribution.[6] We state the following definition.

[6]We are leaving out some subtle technicalities here that the interested reader may find, for example, in Grimmett and Stirzaker, *Probability and Random Processes* [7].

***Definition 8.9.*** Let $\{X(t),\ t \geq 0\}$ be a collection of discrete random variables taking values in some set $S$ and that evolves in time as follows:

**(a)** If the current state is $i$, the time until the state is changed has an exponential distribution with parameter $\lambda(i)$.

**(b)** When state $i$ is left, a new state $j \neq i$ is chosen according to the transition probabilities of a discrete-time Markov chain.

Then $\{X(t)\}$ is called a *continuous-time Markov chain.*

Thus, a continuous-time Markov chain $\{X(t)\}$ is composed of a discrete-time Markov chain $\{X_n\}$, the *jump chain*, for the transitions and exponential random variables for the *holding times*. Recall that the holding times in a discrete Markov chain are geometric, the discrete counterpart of the exponential (see Section 2.6), so this is a natural assumption. The $\lambda(i)$ are called the *holding-time parameters*. Note that the state space is still finite or countably infinite; the discrete/continuous distinction refers to how time is measured. Let us also mention that sometimes the term *Markov process* is used in continuous time. We will not formally state the Markov property but intuitively it says that conditioned on the current state and time, where and when the chain jumps next is independent of the complete history of the chain.

Our construction also ensures that the process is *time-homogeneous*, that is, the probability $P(X(s+t) = j \,|X(s) = i)$ depends only on time through the difference $(s+t) - s = t$, and we can define the transition probabilities as

$$p_{ij}(t) = P(X(t) = j \,|X(0) = i)$$

the probability that the chain is in state $j$, $t$ time units after having been in state $i$. For each $t$, we then get a transition matrix $P(t)$ with entries $p_{ij}(t)$, $i, j \in S$, which has the following properties.

**Proposition 8.11.** Let $P(t)$ be the transition matrix for a continuous-time Markov chain with state space $S$. Then

**(a)** $P(0) = I$, the identity matrix

**(b)** $\sum_{j \in S} p_{ij}(t) = 1$ for all $i \in S$ and $t \geq 0$

**(c)** $P(s+t) = P(s)P(t)$( Chapman–Kolmogorov equations)

When you talk to mathematicians, make sure to refer to the set $\{P(t),\ t \geq 0\}$ as a *stochastic semigroup.*

*Proof.* Parts (a) and (b) are obvious. For (c), consider an element $p_{ij}(s+t)$ of $P(s+t)$ and condition on an intermediate state $k$ at time $s$ to obtain

$$\begin{aligned} p_{ij}(s+t) &= \sum_{k \in S} P_i(X(s+t) = j \mid X(s) = k) P_i(X(s) = k \mid X_0 = i) \\ &= \sum_{k \in S} p_{ik}(s) p_{kj}(t) \end{aligned}$$

which is the $(i, j)$th entry in the matrix $P(s)P(t)$. ■

One problem is that $P(t)$ is usually difficult or impossible to compute, in the same way as $P^{(n)}$ may be in the discrete case. In the discrete case, however, we know that $P^{(n)} = P^n$, so all the information we need is contained in the one-step transition matrix $P$. In the continuous case there is no analog of "one step," so we need to proceed differently in search of a more compact description.

Let the jump chain have transition probabilities $p_{ij}$ for $i \neq j$ and consider the chain in a state $i$. The holding time is $\exp(\lambda(i))$ and when it leaves, the chain jumps to state $j$ with probability $p_{ij}$. Now, if we consider the chain only when it is in state $i$ and disregard everything else, we can view the jumps from $i$ as a Poisson process with rate $\lambda(i)$. For any other state $j$, the jumps from $i$ to $j$ is then a thinned Poisson process with rate $\lambda(i)p_{ij}$. Thus, for any pair of states $i$ and $j$, we can define the *transition rate* between $i$ and $j$ as

$$\gamma_{ij} = \lambda(i) p_{ij}$$

In addition to these, we also let

$$\gamma_{ii} = -\sum_{j \neq i} \gamma_{ij}$$

and define the *generator* as the matrix $G$ whose $(i, j)$th entry is $\gamma_{ij}$. Note that once the $\gamma_{ij}$ have been inserted, the diagonal elements $\gamma_{ii}$ are chosen such that $G$ has row sums equal to 0. The generator completely describes the Markov chain, since if we are given $G$, we can retrieve the holding-time parameters as

$$\lambda(i) = -\gamma_{ii}, \quad i \in S$$

and the jump probabilities as

$$p_{ij} = -\frac{\gamma_{ij}}{\gamma_{ii}}, \quad j \neq i$$

Note that $p_{ii} = 0$ for all $i \in S$ since the $p_{ij}$ give the probability distribution when the chain leaves a state and there can be no jumps from a state to itself (see also Problem 36). Let us look at a few examples.

***Example 8.21.*** An ON/OFF system stays OFF for a time that is $\exp(\lambda)$ and ON for a time $\exp(\mu)$ ($\mu$ does not denote the mean here). Describe the system as a continuous-time Markov chain.

The holding-time parameters are $\lambda$ and $\mu$, and the only possible jumps are from 0 (OFF) to 1 (ON) and vice versa. Thus, we have

$$\gamma_{01} = \lambda, \quad \gamma_{10} = \mu$$

and after filling in the diagonal elements, we get generator

$$G = \begin{pmatrix} -\lambda & \lambda \\ \mu & -\mu \end{pmatrix}$$

We can also describe the system in a graph as follows:

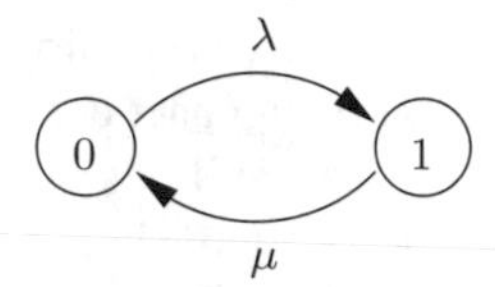


This is similar to how we described discrete-time Markov chains but the numbers on the arrows are now rates, not probabilities. The jump chain has transition matrix

$$P = \begin{pmatrix} 0 & 1 \\ 1 & 0 \end{pmatrix}$$

□

***Example 8.22.*** A continuous-time Markov chain on state space $\{1, 2, 3\}$ has generator

$$G = \begin{pmatrix} -6 & 2 & 4 \\ 1 & -2 & 1 \\ 3 & 1 & -4 \end{pmatrix}$$

Suppose that the chain is in state 1. What is the expected time until it leaves, and what is the probability that it next jumps to state 2?

The holding-time parameter in state 1 is $\lambda(1) = -\gamma_{11} = 6$, so the expected holding time is $\frac{1}{6}$. The probability to jump to state 2 is

$$p_{12} = -\frac{\gamma_{12}}{\gamma_{11}} = -\frac{2}{-6} = \frac{1}{3}$$

□

The generator now plays the role that the transition matrix did in the discrete case, and a logical question is how $G$ relates to $P(t)$. The following proposition gives the answer, where $P'(t)$ denotes the matrix of the derivatives $p'_{ij}(t)$.

**Proposition 8.12.** The transition matrix $P(t)$ and generator $G$ satisfy the *backward equations*

$$P'(t) = GP(t), \quad t \geq 0$$

and *forward equations*

$$P'(t) = P(t)G, \quad t \geq 0$$

If we spell out the backward equations elementwise, we get

$$p'_{ij}(t) = \sum_{k \in S} \gamma_{ik} p_{kj}(t), \quad i, j \in S, t \geq 0$$

and the forward equations

$$p'_{ij}(t) = \sum_{k \in S} p_{ik}(t) \gamma_{kj}, \quad i, j \in S, t \geq 0$$

*Proof.* We refer to an intuitive argument and leave out the technical details. Consider the probability $p_{ij}(t+h) = P_i(X(t+h) = j)$. The Chapman–Kolmogorov equations give

$$p_{ij}(t+h) = \sum_{k \in S} p_{ik}(t) p_{kj}(h)$$

and if $h$ is small, we have, with $T_j$ denoting the holding time in state $j$,

$$\begin{aligned} p_{jj}(h) &= P(X(h) = j | X(0) = j) \approx P(T_j > h) = e^{-\lambda(j)h} \\ &\approx 1 - \lambda(j)h = 1 + \gamma_{jj} h \end{aligned}$$

The intuition behind this is that if $h$ is small and the chain is in state $j$ at times 0 and $h$, most likely nothing happened in the interval $(0, h)$. With a similar argument, if there is a jump in $(0, h)$, we neglect the possibility of more than one jump and obtain

$$p_{kj}(h) \approx \gamma_{kj} h, \quad k \neq j$$

This gives

$$\begin{aligned} p_{ij}(t+h) &\approx p_{ij}(t)(1 + \gamma_{jj} h) + \sum_{k \neq j} p_{ik}(t) \gamma_{kj} h \\ &= p_{ij}(t) + h \sum_{k \in S} p_{ik}(t) \gamma_{kj} \end{aligned}$$

which gives

$$\frac{p_{ij}(t+h) - p_{ij}(t)}{h} = \sum_{k \in S} p_{ik}(t)\gamma_{kj}$$

and letting $h \downarrow 0$ gives the forward equations, with a similar type of argument for the backward equations. ■

It turns out that the forward equations are usually easier to solve but do not always exist (a fact that is not revealed by our intuitive argument above). In all examples and applications we consider, they do, however, exist. It is usually difficult to solve the backward and forward equations and only in simple cases can we easily find the explicit form of $P(t)$. In Problem 34, you are asked to find the solution for the simple ON/OFF system.

Since $P(0) = I$, the backward and forward equations also suggest a way to obtain the generator from $P(t)$ according to

$$G = P'(0) \tag{8.5}$$

### 8.4.1 Stationary Distributions and Limit Distributions

Just as in the discrete case, we are interested in asymptotic behavior, which is described by the limit of $P(t)$ as $t \to \infty$ and also as in the discrete case, we would like to do this via stationary distributions instead of direct calculations. A limit distribution in the continuous case is the obvious analog of the discrete case: a distribution $\mathbf{q}$ such that

$$p_{ij}(t) \to q_j, \quad \text{as } t \to \infty \quad \text{for all } i, j \in S$$

How should we define a stationary distribution? In the discrete case, it is defined through the relation $\boldsymbol{\pi} = \boldsymbol{\pi} P$, but in the continuous case there is no $P$. However, since a stationary distribution "stays forever," we also have $\boldsymbol{\pi} = \boldsymbol{\pi} P^{(n)}$ for all $n$, and we can imitate this in the continuous case.

***Definition 8.10.*** Consider a continuous-time Markov chain with transition matrix $P(t)$. A probability distribution $\boldsymbol{\pi}$ which is such that

$$\boldsymbol{\pi} P(t) = \boldsymbol{\pi} \quad \text{for all } t \geq 0$$

is called a *stationary distribution* of the chain.

The intuition is the same as in the discrete case; the probability $\pi_j$ is the proportion of time spent in state $j$ in the long run. Since we have pointed out how difficult it typically is to find $P(t)$, the definition does not give a computational recipe. Instead, we first differentiate with respect to $t$ on both sides in the definition to obtain

$$\frac{d}{dt}(\boldsymbol{\pi} P(t)) = \boldsymbol{\pi} P'(t) = \frac{d}{dt}(\boldsymbol{\pi}) = \mathbf{0}, \quad t \geq 0$$

since $\boldsymbol{\pi}$ does not depend on $t$. In particular, with $t = 0$, Equation (8.5) gives $P'(0) = G$ and we have shown the following.

**Corollary 8.9.** *The stationary distribution satisfies the equation*

$$\boldsymbol{\pi} G = \mathbf{0}$$

*where* $\mathbf{0}$ *is a vector of zeros.*

Elementwise, the equations are

$$\sum_{i \in S} \gamma_{ij} \pi_i = 0, \quad j \in S$$

and as in the discrete case, we have the additional condition that the entries in $\boldsymbol{\pi}$ sum to 1.

***Example 8.23.*** Consider the ON/OFF system in Example 8.21. Find the stationary distribution.

The equation $\boldsymbol{\pi} G = \mathbf{0}$ is

$$(\pi_0, \ \pi_1) \begin{pmatrix} -\lambda & \lambda \\ \mu & -\mu \end{pmatrix} = (0, \ 0)$$

which gives the first equation

$$-\pi_0 \lambda + \pi_1 \mu = 0$$

which in turn gives

$$\pi_1 = \frac{\lambda}{\mu} \pi_0$$

and the condition $\pi_0 + \pi_1 = 1$ gives

$$1 = \pi_0 \left(1 + \frac{\lambda}{\mu}\right)$$

and we get stationary distribution

$$(\pi_0, \ \pi_1) = \left(\frac{\mu}{\mu + \lambda}, \frac{\lambda}{\mu + \lambda}\right)$$

Note how $\pi_0 > \pi_1$ if $\mu > \lambda$, that is, when the transition rate is higher from 1 to 0 than vice versa. Also note that

$$(\pi_0, \pi_1) = \left( \frac{1/\lambda}{1/\lambda + 1/\mu}, \frac{1/\mu}{1/\lambda + 1/\mu} \right)$$

which is intuitively appealing since the expected times spent in states 0 and 1 are $1/\lambda$ and $1/\mu$, respectively. Thus, in the long run, the proportions of time spent in states 0 and 1 are $\pi_0$ and $\pi_1$.

Recall that the jump chain has stationary distribution $(\frac{1}{2}, \frac{1}{2})$, which tells us that the jump chain on average visits 0 and 1 equally many times but does not take holding times into account. □

The existence of a stationary distribution is again closely related to the concepts of irreducibility and positive recurrence. Irreducibility is only a property of how the states communicate and has nothing to do with holding times, so we call a continuous-time Markov chain irreducible if its jump chain is irreducible. As for recurrence and transience, they are defined in the analogous way, letting

$$S_i = \inf\{t : X(t) = i\}$$

where $S_i = \infty$ if $i$ is never visited. The only difference from the discrete case is that $S_i$ is now a continuous random variable and the following definition is a direct analog.

***Definition 8.11.*** If $P_i(S_i < \infty) = 1$, state $i$ is called *recurrent* and if $P_i(S_i < \infty) < 1$, state $i$ is called *transient*. If $i$ is recurrent and $E_i[S_i] < \infty$, $i$ is called *positive recurrent*; otherwise it is called *null recurrent*.

We use the notation $S$ for "sum," since $S_i$ is the sum of the holding times in all states visited before reaching $i$. We keep the notation $\tau_i$ for the return times in the jump chain. Thus, if the holding time in state $k$ is $T_k$, then $S_i$ and $\tau_i$ relate as

$$S_i = \sum_{n=0}^{\tau_i - 1} T_{X_n} \tag{8.6}$$

Now suppose that state $i$ is recurrent in the jump chain $\{X_n\}$. This means that $\tau_i$ presented above is finite, and since also the $T_k$ are finite, $S_i$ must be finite and $i$ is recurrent also in $\{X(t)\}$. Thus, if the jump chain is recurrent, so is the continuous-time chain $\{X(t)\}$. When it comes to positive recurrence, things are more complicated, as the following example shows.

***Example 8.24.*** Consider the following continuous-time version of the success run chain from Example 8.16. The holding time parameters are $\lambda(k) = 1/2^k$ for $k = 0, 1, 2 \ldots$, and the success run chain functions as the jump chain, with the exception that $p_{01} = 1$. Show that the jump chain $\{X_n\}$ is positive recurrent but $\{X(t)\}$ is not.

The state space is $S = \{0, 1, 2, \ldots\}$ and the transition matrix of the jump chain differs from that of the success run chain only in that it has $p_{01} = 1$ instead of $p_{00} = p_{01} = \frac{1}{2}$. It is easy to find the stationary distribution (see Problem 16) and this shows that $\{X_n\}$ is positive recurrent.

Now consider state 0. Equation (8.6) is extra simple since the jump chain increases by unit steps until it drops back to 0, which gives $X_n = n$ for $n < \tau_0$, and we get

$$S_0 = \sum_{n=0}^{\tau_0 - 1} T_n$$

Since the state 0 is recurrent in $\{X_n\}$, it is recurrent also in $\{X(t)\}$. However, we will show that state 0 is null recurrent in $\{X(t)\}$ and thus proceed to compute the expected value of $S_0$. To do so, we condition on $\tau_0$ and note that the range of $\tau_0$ is $\{2, 3, \ldots\}$. If $\tau_0 = k$, $S_0$ is the sum of $T_1, \ldots, T_{k-1}$ and hence

$$E_0[S_0 | \tau_0 = k] = \sum_{n=0}^{k-1} E[T_n] = \sum_{n=0}^{k-1} 2^n = 2^k - 1$$

and the distribution of $\tau_0$ is

$$P_0(\tau_0 = k) = \frac{1}{2^{k-1}}, \quad k = 2, 3, \ldots$$

since $\tau_0 = k$ precisely when there are $k - 1$ successes followed by a failure, where $k$ must be at least 2. We get

$$\begin{aligned} E_0[S_0] &= \sum_{k=2}^{\infty} E_0[S_0 | \tau_0 = k] P_0(\tau_0 = k) \\ &= \sum_{k=2}^{\infty} (2^k - 1) \frac{1}{2^{k-1}} = \sum_{k=2}^{\infty} \left(2 - \frac{1}{2^{k-1}}\right) = \infty \end{aligned}$$

which means that state 0 is null recurrent in the continuous-time chain $X(t)$. By irreducibility, the entire continuous-time chain is null recurrent. The problem is that even though the jump chain is positive recurrent, the holding times get so long that the continuous chain becomes null recurrent. □

If the holding times are instead very short, it is possible that $\{X_n\}$ is null recurrent but $\{X(t)\}$ is positive recurrent (see Problem 38). It is also possible to construct examples where the jump chain is transient but the continuous-time chain has a stationary distribution in the sense of a solution to $\boldsymbol{\pi} G = \mathbf{0}$. However, in such examples we get the unpleasant property of infinitely many jumps in finite time, and to rule out such anomalies, we always assume that the jump chain is recurrent. The following is the continuous-time analog of Proposition 8.4.

**Proposition 8.13.** Consider an irreducible continuous-time Markov chain with a recurrent jump chain. Then

$$\text{A stationary distribution } \boldsymbol{\pi} \text{ exists} \Leftrightarrow \{X(t)\} \text{ is positive recurrent}$$

The stationary distribution is unique and has $\pi_j > 0$ for all $j \in S$.

As in the discrete case, positive recurrence is an important concept for describing the behavior of the chain, but it is typically not checked directly. Instead, we look for a solution to $\boldsymbol{\pi}G = \mathbf{0}$.

We next turn to the question of convergence to the stationary distribution. In the discrete case, this was complicated by possible periodicity, but in the continuous case we have no unit step size, and thus the concept of period does not exist. We can state the convergence theorem.

***Theorem 8.2.*** *Consider an irreducible, continuous-time Markov chain with a recurrent jump chain, stationary distribution* $\boldsymbol{\pi}$*, and transition probabilities* $p_{ij}(t)$*. Then*

$$p_{ij}(t) \to \pi_j \quad as \quad t \to \infty$$

*for all* $i, j \in S$.

As in the discrete case, a continuous-time Markov chain that satisfies the assumptions in the theorem is called *ergodic*. The obvious analog of Proposition 8.5, regarding mean recurrence times and mean number of visits to intermediate states, holds in the continuous case as well. Finally, we refer to Problem 38 to see how the stationary distributions for $\{X(t)\}$ and its jump chain relate to each other.

***Example 8.25.*** Consider the ON/OFF system from Example 8.23, where the jump chain has stationary distribution $(\frac{1}{2}, \frac{1}{2})$ and the continuous-time chain has stationary distribution $(\mu/(\lambda + \mu), \lambda/(\lambda + \mu))$. These describe the proportion of jumps and the proportion of time, respectively, spent in each state in the long run. However, only the continuous-time chain also has a limit distribution. If the system starts in state 0, it forgets how it started if we consider it in real time, but not if we count the jumps, as we saw in Example 8.13. □

### 8.4.2 Birth–Death Processes

In this section, we will examine continuous-time analogs of random walks. Thus, we will consider integer-valued, continuous-time Markov chains that can only step up or down, so the only generator entries that can be positive (but do not have to be) are $\gamma_{i,i-1}$ and $\gamma_{i,i+1}$. We also restrict the state space to $S = \{0, 1, 2, \ldots\}$, the nonnegative integers. Such Markov chains are called *birth–death processes*. Let us explain why in an example.

***Example 8.26.*** Consider a population of cells. Each cell lives for a time that is $\exp(\alpha)$ and then either splits into two new cells with probability $p$ or dies with probability $1 - p$, independent of all other cells. Let $X(t)$ be the number of cells at time $t$ and describe this as a continuous-time Markov chain.

The state space is $S = \{0, 1, 2, \ldots\}$. If there are $i$ cells, the next change comes after a time that is the minimum of $i$ lifetimes that are independent and $\exp(\alpha)$. By Example 3.50, we thus have holding-time parameters

$$\lambda(i) = i\alpha, \quad i = 0, 1, 2, \ldots$$

where $\lambda(0) = 0$ means that state 0 is absorbing. This gives the transition rates

$$\gamma_{i,i+1} = i\alpha p \quad \text{and} \quad \gamma_{i,i-1} = i\alpha(1-p), \quad i = 1, 2, \ldots$$

and it is common to define the *birth rate* $\lambda_i = \gamma_{i,i+1}$ and *death rate* $\mu_i = \gamma_{i,i-1}$ [do not confuse $\lambda_i$ and $\lambda(i)$]. Since the birth and death rates are linear in $i$, this is called a *linear* birth–death process. It is also customary to denote $\lambda = \alpha p$ and $\mu = \alpha(1-p)$, the *individual* birth and death rates. The transition graph is then

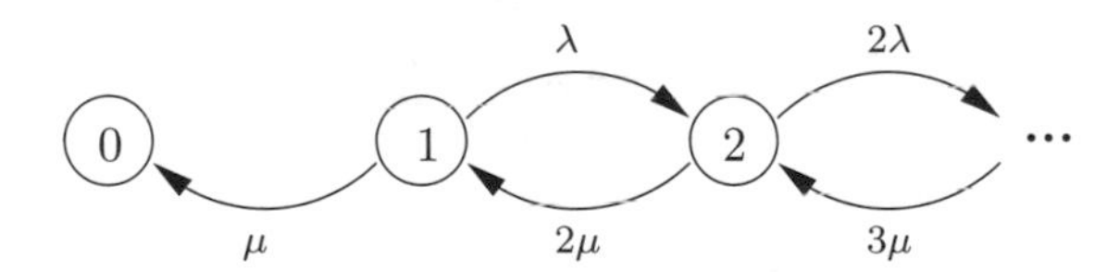


and the generator

$$G = \begin{pmatrix} 0 & 0 & 0 & 0 & 0 & \cdots \\ \mu & -(\lambda+\mu) & \lambda & 0 & 0 & \cdots \\ 0 & 2\mu & -2(\lambda+\mu) & 2\lambda & 0 & \cdots \\ 0 & 0 & 3\mu & -3(\lambda+\mu) & 3\lambda & \cdots \\ \vdots & \vdots & \vdots & \vdots & \vdots & \ddots \end{pmatrix}$$

The jump chain is the simple random walk, which we know is transient if $p > \frac{1}{2}$, the only case in which absorption in 0 can be avoided. □

***Example 8.27.*** Consider a population where the individual death rate is $\mu$ and there are no births. Instead, there is constant immigration into the population according to a Poisson process with rate $\lambda$. Describe the process, determine when a limit distribution exists, and find what it is.

The birth and death rates are

$$\lambda_i = \lambda, \quad \mu_i = i\mu, \quad i = 0, 1, \ldots$$

and we have the following transition graph:

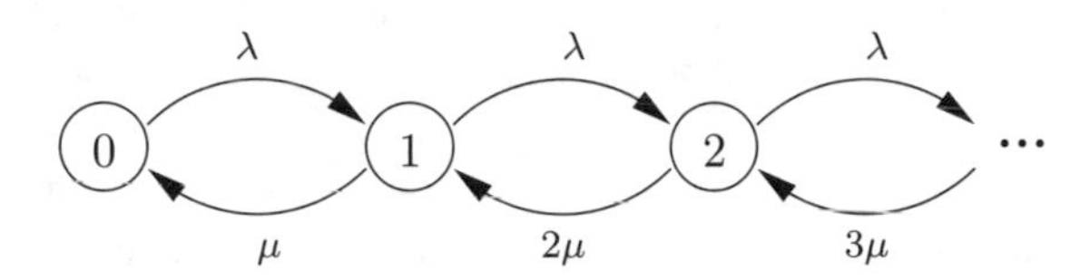


The chain is clearly irreducible, so we need only to look for a stationary distribution, which, if it exists, is also the limit distribution. The equation $\boldsymbol{\pi} G = \mathbf{0}$ becomes

$$(\pi_0, \pi_1, \ldots) \begin{pmatrix} -\lambda & \lambda & 0 & 0 & 0 & \ldots \\ \mu & -(\lambda+\mu) & \lambda & 0 & 0 & \ldots \\ 0 & 2\mu & -(\lambda+2\mu) & \lambda & 0 & \ldots \\ 0 & 0 & 3\mu & -(\lambda+3\mu) & \lambda & \ldots \\ \vdots & \vdots & \vdots & \vdots & \vdots & \ddots \end{pmatrix} = (0, 0, \ldots)$$

which gives the first equation

$$-\lambda\pi_0 + \mu\pi_1 = 0$$

which gives

$$\pi_1 = \frac{\lambda}{\mu}\pi_0$$

The next equation

$$\lambda\pi_0 - (\lambda + \mu)\pi_1 + 2\mu\pi_2 = 0$$

gives, after some algebra

$$\pi_2 = \frac{\lambda^2}{2\mu^2}\pi_0$$

The remaining equations all look the same:

$$\lambda\pi_{n-1} - (\lambda + n\mu)\pi_n + (n+1)\mu\,\pi_{n+1}, \quad n = 2, 3, \ldots$$

and it is easy to check that the general solution is

$$\pi_n = \frac{\rho^n}{n!}\pi_0, \; n = 0, 1, 2, \ldots$$

where $\rho = \lambda/\mu$. The condition $\sum_n \pi_n = 1$ gives

$$1 = \pi_0 \sum_{n=0}^{\infty} \frac{\rho^n}{n!} = \pi_0 \, e^{\rho}$$

which gives stationary distribution

$$\pi_n = e^{-\rho} \frac{\rho^n}{n!}, \quad n = 0, 1, 2, \ldots$$

which we recognize as a Poisson distribution with mean $\rho$. Note that the stationary distribution always exists. The intuitive reason for this is that the larger the population becomes, the more likely that the next event is a death rather than an immigration. Even if the immigration rate $\lambda$ is enormous compared to the individual death rate $\mu$, sooner or later there will be so many individuals that deaths start to compensate for immigration. The jump chain has transition probabilities

$$p_{i,i+1} = -\frac{\gamma_{i,i+1}}{\gamma_{ii}} = \frac{\lambda}{\lambda + i\mu}$$

and $p_{i,i-1} = i\mu/(\lambda + i\mu), i = 0, 1, \ldots$ and also note that the larger the population becomes, the more frequent the events, since the expected holding time in state $i$ is $1/(\lambda + i\mu)$. □

Since the structure of the generator is so simple in a birth–death process, it is possible to find a general formula for the stationary distribution. The general form of the generator is

$$G = \begin{pmatrix} -\lambda_0 & \lambda_0 & 0 & 0 & 0 & \cdots \\ \mu_1 & -(\lambda_1 + \mu_1) & \lambda_1 & 0 & 0 & \cdots \\ 0 & \mu_2 & -(\lambda_2 + \mu_2) & \lambda_2 & 0 & \cdots \\ 0 & 0 & \mu_3 & -(\lambda_3 + \mu_3) & \lambda_3 & \cdots \\ \vdots & \vdots & \vdots & \vdots & \vdots & \ddots \end{pmatrix}$$

from which it is easily seen that the equation $\boldsymbol{\pi} G = \mathbf{0}$ gives the first equation

$$-\lambda_0 \pi_0 + \mu_1 \pi_1 = 0$$

which gives

$$\pi_1 = \frac{\lambda_0}{\mu_1} \pi_0$$

and the remaining equations

$$\lambda_{n-2} \pi_{n-2} - (\lambda_{n-1} + \mu_{n-1}) \pi_{n-1} + \mu_n \pi_n = 0, \quad n = 2, 3, \ldots$$

and it is easy to check that these are satisfied by

$$\pi_n = \frac{\lambda_0 \lambda_1 \cdots \lambda_{n-1}}{\mu_1 \mu_2 \cdots \mu_n} \pi_0, \quad n = 1, 2, \ldots \tag{8.7}$$

Summing over $n$ now yields that a stationary distribution exists if and only if

$$1 + \sum_{n=1}^{\infty} \frac{\lambda_0 \lambda_1 \cdots \lambda_{n-1}}{\mu_1 \mu_2 \cdots \mu_n} < \infty$$

and the stationary distribution is then given by

$$\pi_0 = \left( 1 + \sum_{n=1}^{\infty} \frac{\lambda_0 \lambda_1 \cdots \lambda_{n-1}}{\mu_1 \mu_2 \cdots \mu_n} \right)^{-1}$$

and the remaining $\pi_n$ by Equation (8.7). Also see Problem 42 for a nice interpretation of the equation $\boldsymbol{\pi} G = \mathbf{0}$.

### 8.4.3 Queueing Theory

A particular class of birth–death processes arise in models for certain service systems. Let us introduce the subject of *queueing theory* with an example.

***Example 8.28*** **(*The*** $M/M/1$ ***queue*).** Customers arrive according to a Poisson process with rate $\lambda$ to a service station with one server. Service times are i.i.d. exponential with rate $\mu$ and independent of the arrivals (note that $\mu$ does not denote the mean here; the mean service time is $1/\mu$). If the server is busy, incoming customers wait in line and as soon as a service is completed, the next begins. Describe the system as a birth–death process, determining when it has a stationary distribution and what it is.

We let the state $X(t)$ be the number of customers in the system (under service and in line) at time $t$. Transition rates are given already in the problem: $\lambda_i = \lambda$ for $i \geq 0$ and $\mu_i = \mu$ for $i \geq 1$. The transition graph is

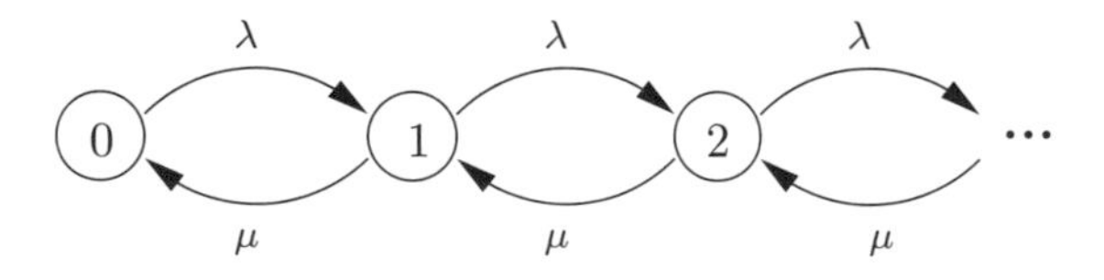


and to find the stationary distribution, note that

$$\frac{\lambda_0 \lambda_1 \cdots \lambda_{n-1}}{\mu_1 \mu_2 \cdots \mu_n} = \frac{\lambda^n}{\mu^n}$$

so with $\rho = \lambda/\mu$ we see that a stationary distribution exists if and only if

$$1 + \sum_{n=1}^{\infty} \rho^n = \sum_{n=0}^{\infty} \rho^n < \infty$$

which is to say that $\rho < 1$. Using the formula for the geometric series, it is easy to see that the stationary distribution in this case is

$$\pi_n = (1 - \rho)\rho^n, \quad n = 0, 1, 2 \ldots$$

a geometric distribution including 0. The constant $\rho$ is called the *traffic intensity* and in order for a stationary distribution to exist, this must be strictly less than one, meaning that service rates are higher than arrival rates. Only in this way can the server be efficient to regularly clear out the system.

The jump chain is the simple random walk where the probability to step up is $p = \lambda/(\lambda + \mu) = \rho/(1 + \rho)$, which means that $\rho < 1$ if and only if $p < \frac{1}{2}$. □

The system in this example is called an $M/M/1$ *queue*, where "$M$" stands for "Markov," meaning that both interarrival times (first $M$) and service times (second $M$) are exponential, which is the only way in which this system satisfies the Markov property. The "1" indicates that there is one server. Other, non-Markovian, queueing systems are, for example, $M/D/1$, where service times are deterministic and $G/G/1$, where both interarrival times and service times have some general distribution, not necessarily exponential. The analysis of such systems require other methods, and we will stick to the $M/M$ queues and analyze them as continuous time Markov chains. There are many variants of the $M/M/1$ queue. In the following three examples, we examine some of these, leaving others for the Problems section.

***Example 8.29* (*Finite Waiting Room*).** Consider the $M/M/1$ queue, but suppose that there is only room for $r$ customers in the system (the maximum queue length is $r - 1$), denoted $M/M/1/r$.

The transition graph is

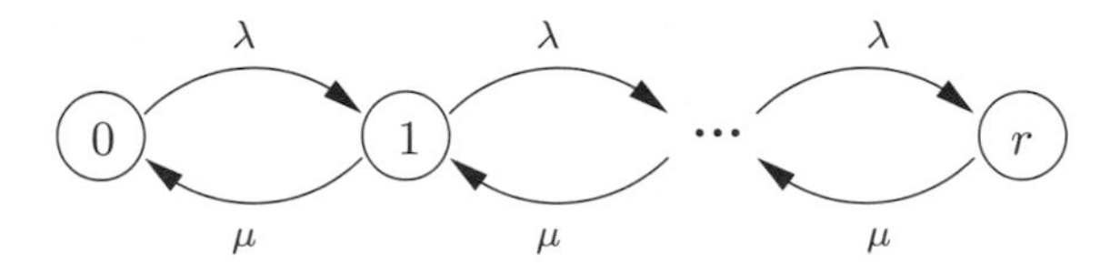


Again let $\rho = \lambda/\mu$. We have

$$\sum_{n=0}^{r} \rho^n = \begin{cases} r+1 & \text{if } \rho = 1 \\ \dfrac{1-\rho^{r+1}}{1-\rho} & \text{if } \rho \neq 1 \end{cases}$$

which gives stationary distribution

$$\pi_n = \begin{cases} \dfrac{1}{r+1}, & n = 0, 1, \ldots, r \qquad \text{if } \rho = 1 \\ \\ \dfrac{(1-\rho)\rho^n}{1-\rho^{r+1}}, & n = 0, 1, \ldots, r \qquad \text{if } \rho \neq 1 \end{cases}$$

Note how the stationary distribution is uniform if $\rho = 1$ and how the probability $\pi_r$ approaches 1 as $\rho \to \infty$. □

***Example 8.30* (*Balking*).** Consider the $M/M/1$ queue and suppose that an arriving customer who finds the system not empty joins with probability $q$ and leaves otherwise.

The only difference from the $M/M/1$ queue is that transitions from $i$ to $i+1$ where $i \geq 1$ now occur according to a thinned Poisson process with rate $\lambda q$. Thus, we have the transition graph

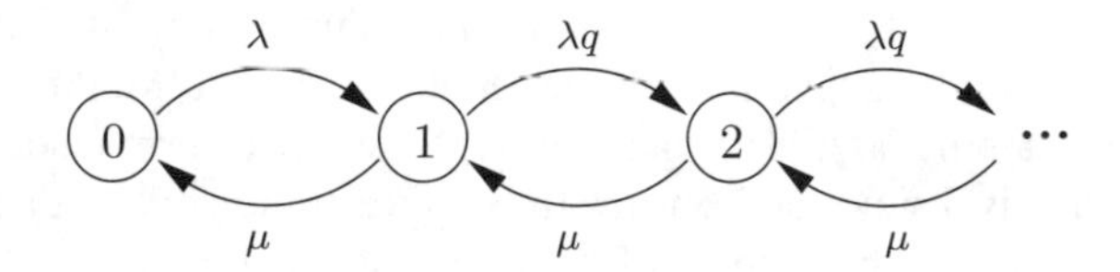


and we get

$$\frac{\lambda_0 \lambda_1 \cdots \lambda_{n-1}}{\mu_1 \mu_2 \cdots \mu_n} = \frac{\lambda^n q^{n-1}}{\mu^n}$$

where we let $\rho = \lambda q/\mu$ to obtain

$$1 + \sum_{n=1}^{\infty} \frac{\lambda_0 \lambda_1 \cdots \lambda_{n-1}}{\mu_1 \mu_2 \cdots \mu_n} = 1 + \frac{1}{q} \sum_{n=1}^{\infty} \rho^n$$

which is finite if and only if $\rho < 1$, which makes intuitive sense. Since

$$\sum_{n=1}^{\infty} \rho^n = \frac{\rho}{1-\rho}$$

the stationary distribution is given by

$$\begin{cases} \pi_0 & = \left(1 + \dfrac{\rho}{q(1-\rho)}\right)^{-1} \\ \\ \pi_n & = \dfrac{\rho^n \pi_0}{q}, \quad n = 1, 2, \ldots \end{cases}$$

Note how we redefined the traffic intensity $\rho$. We did this since we want $\rho < 1$ to be the criterion for when the system is efficient, that is, when the server manages to deal with the arrivals. □

***Example 8.31* (*More Than One Server*).** Consider the $M/M/2$ queue, which is just like the $M/M/1$ queue except that there are two servers instead of one. An arriving customer can thus get immediate service if the system is empty or if only one server is busy.

The difference this time is what happens when both servers are busy. The time for a service to be completed is now the minimum of two exponentials with rate $\mu$ and is thus $\exp(2\mu)$ (recall Example 3.50). The transition graph is

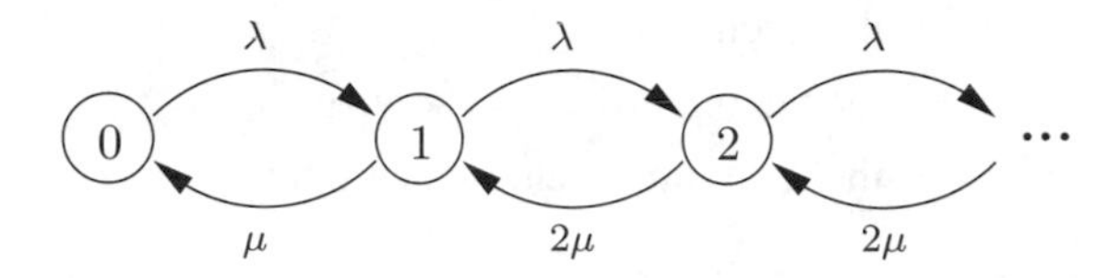


This time we get

$$\frac{\lambda_0\lambda_1\cdots\lambda_{n-1}}{\mu_1\mu_2\cdots\mu_n} = \frac{\lambda^n}{\mu(2\mu)^{n-1}} = 2\left(\frac{\lambda}{2\mu}\right)^n$$

which suggests that we define the traffic intensity as $\rho = \lambda/2\mu$. We then get

$$\pi_n = 2\rho^n\pi_0, \quad n \geq 1$$

and the stationary distribution exists if and only if $\rho < 1$. Since

$$\left(1 + 2\sum_{n=1}^{\infty}\rho^n\right) = \frac{1+\rho}{1-\rho}$$

the stationary distribution is given by

$$\begin{cases} \pi_0 = & \dfrac{1-\rho}{1+\rho} \\[2ex] \pi_n = & \dfrac{2\rho^n(1-\rho)}{1+\rho}, \quad n \geq 1 \end{cases}$$

□

### 8.4.4 Further Properties of Queueing Systems

When a queueing system has settled in to have the stationary distribution, we say that it is *in equilibrium*. Several different measures can be used to assess the efficiency of the system, often called *performance measures*. Let us examine some of them for the $M/M/1$ queue.

***Example 8.32.*** Consider the $M/M/1$ queue with $\rho < 1$ in equilibrium. **(a)** What is the expected number of customers in the system? **(b)** What is the expected queue length? **(c)** When a customer arrives, what is the probability that she does not have to wait in line? **(d)** When a customer arrives, what is her expected waiting time until service? **(e)** When a customer arrives, what is her expected total time in the system?

Let us introduce some random variables. Thus, let

$$\begin{aligned} N &= \text{the number of customers in the system} \\ Q &= \text{the queue length} \\ W &= \text{the waiting time until service} \\ T &= \text{the total time spent in the system} \end{aligned}$$

For (a), we know that $N$ has distribution

$$\pi_k = (1-\rho)\rho^k, \quad k = 0, 1, \ldots$$

the geometric distribution including 0 with success probability $1-\rho$, and from Section 2.5.3 we know that

$$E[N] = \frac{\rho}{1-\rho}$$

which answers (a). For (b), note that

$$Q = \begin{cases} 0 & \text{if } N = 0 \text{ or } N = 1 \\ N-1 & \text{if } N > 1 \end{cases}$$

and hence

$$\begin{aligned} P(Q=0) &= \pi_0 + \pi_1 \\ P(Q=k) &= \pi_{k+1}, \quad k \geq 1 \end{aligned}$$

which gives

$$\begin{aligned} E[Q] &= \sum_{k=0}^{\infty} kP(Q=k) = \sum_{k=1}^{\infty} k\pi_{k+1} \\ &= \rho \sum_{k=1}^{\infty} k(1-\rho)\rho^k = \rho E[N] = \frac{\rho^2}{1-\rho} \end{aligned}$$

The answer to (c) is simply $\pi_0 = 1-\rho$, and for (d), note that $W = 0$ if the system is empty and the sum of $N$ i.i.d. exponentials with mean $1/\mu$ if there are $N$ customers in the system (keep in mind that $\mu$ does not denote the mean but the service rate). By Corollary 3.13, we get

$$E[W] = E[N]\frac{1}{\mu} = \frac{\rho}{\mu(1-\rho)}$$

Finally, for (e), let $S$ be a service time and note that $T = W + S$ to obtain

$$E[T] = \frac{\rho}{\mu(1-\rho)} + \frac{1}{\mu} = \frac{1}{\mu(1-\rho)}$$

We summarize as follows. In the $M/M/1$ system in equilibrium, we have

$$\begin{aligned} E[N] &= \frac{\rho}{1-\rho}, & E[Q] &= \frac{\rho^2}{1-\rho} \\ E[W] &= \frac{\rho}{\mu(1-\rho)}, & E[T] &= \frac{1}{\mu(1-\rho)} \end{aligned}$$

□

There are many interesting relations between these expected values. For example,

$$\frac{E[Q]}{E[N]} = \frac{E[W]}{E[T]}$$

which is also equal to the traffic intensity $\rho$. Also, since $\mu = \rho\lambda$, we obtain

$$E[N] = \lambda E[T] \tag{8.8}$$

formulas that are intuitively reasonable. These relations hold for many queueing systems, not only the $M/M/1$. Equation (8.8) is known as *Little's formula*. Also note how all the expectations go to $\infty$ as $\rho$ approaches 1.

If there is finite waiting room, an obvious performance measure is how likely it is that the system is full, in which case arriving customers are lost. Thus, the probability that the system is full is the long-term proportion of arriving customers that are lost. Let us consider one example.

***Example 8.33.*** Consider the $M/M/1/r$ queue from Example 8.29. What proportion of customers is lost?

This is the probability that the system is full, $\pi_r$, which is

$$\pi_r = \begin{cases} \dfrac{1}{r+1} & \text{if } \rho = 1 \\[2ex] \dfrac{(1-\rho)\rho^r}{1-\rho^{r+1}} & \text{if } \rho \neq 1 \end{cases}$$

where we can note that $\pi_r \to 1$ as $\rho \to \infty$. □

## 8.5 MARTINGALES

In this section, we will introduce a class of stochastic processes called *martingales*[7] that is particularly useful in a wide variety of situations where asymptotic properties are of interest.

Let us go directly to the definition.

***Definition 8.12.*** A sequence of random variables $Y_1, Y_2, \ldots$ is called a martingale in discrete time with respect to the sequence $X_1, X_2, \ldots$ if $E[|Y_n|] < \infty$ and

$$E[Y_{n+1}|X_1, \ldots, X_n] = Y_n \tag{8.9}$$

for all $n = 1, 2, \ldots$.

The conditional expectation used in Definition 8.12 is a generalization of Definition 3.13, where we interpreted conditional expectation as a random variable. In this context, (8.9) is a random variable assuming the value $E[Y_{n+1}|X_1 = x_1, \ldots, X_n = x_n]$ whenever the event $\{X_1 = x_1, \ldots, X_n = x_n\}$ occurs.

In many applications, we can actually choose $X_k = Y_k$ for all $k = 1, 2, \ldots$ so that (8.9) takes the form

$$E[Y_{n+1}|Y_1, \ldots, Y_n] = Y_n$$

which better illustrates the fundamental property of a martingale. Basically, it says that if we have observed the process for $n$ steps, we at least know that the process in the next step on average will not deviate from the last value $Y_n$. The more general definition above is useful in situations where we can write $Y_n$ as a function of the underlying variables $X_1, \ldots, X_n$.

***Example 8.34.*** One process we considered earlier that fits nicely into this theory is the symmetric random walk from Section 8.3.1 where $X_1, X_2, \ldots$ are i.i.d. random variables where $P(X_k = 1) = P(X_k = -1) = \frac{1}{2}$ and

$$S_n = \sum_{k=1}^{n} X_k$$

The first property in Definition 8.12 is clearly satisfied since $E[|S_n|] \leq n < \infty$ and

$$E[S_{n+1}|X_1, \ldots, X_n] = E[X_{n+1} + S_n|X_1, \ldots, X_n] = E[X_{n+1}] + S_n = S_n$$

shows that $S_1, S_2, \ldots$ is a martingale with respect to $X_1, X_2, \ldots$. □

[7] The term originated in France in the seventeenth century as a class of betting strategies to increase the chances in various games of gambling.

***Example 8.35 (Branching Processes).*** Another example where martingales are useful is the branching process introduced in Section 8.3.3. Here, the variables $X_1, X_2, \dots$ represent the number of offspring of individuals in a generation and

$$Z_n = \sum_{k=1}^{Z_{n-1}} X_k, \qquad n = 1, 2, \dots$$

where $Z_0 \equiv 1$, denotes the number of individuals in generation $n$. If $\mu = E[X_k] < \infty$, Proposition 8.8 implies that $E[|Z_n|] = E[Z_n] = \mu^n < \infty$ and

$$E[Z_{n+1}|Z_1, \dots, Z_n] = E\left[\sum_{k=1}^{Z_n} X_k \middle| Z_1, \dots, Z_n\right] = Z_n \mu$$

showing that $Z_1, Z_2, \dots$ is a martingale with respect to itself, but only if $\mu = 1$. However, it is actually quite easy to construct a martingale for arbitrary $\mu$ by rescaling $Z_n$ with respect to the mean as

$$Y_n = \frac{Z_n}{\mu^n}$$

Clearly, $E[|Y_n|] = E[Y_n] = 1$ for all $n = 1, 2, \dots$ and

$$\begin{aligned} E[Y_{n+1}|Z_1, \dots, Z_n] &= E\left[\frac{1}{\mu^{n+1}} \sum_{k=1}^{Z_n} X_k \middle| Z_1, \dots, Z_n\right] \\ &= \frac{1}{\mu^{n+1}} Z_n \mu = Y_n \end{aligned}$$

showing that $Y_1, Y_2, \dots$ is a martingale with respect to $Z_1, Z_2, \dots$. □

### 8.5.1 Martingale Convergence

The main result of this section is the following.

**Proposition 8.14 (Martingale Convergence Theorem).** If $Y_1, Y_2, \dots$ is a martingale with respect to some sequence $X_1, X_2, \dots$ and $E[Y_n^2] < c < \infty$ for all $n = 1, 2, \dots$, then there exists a random variable $Y$ such that

$$Y_n \to Y \quad \text{as} \quad n \to \infty$$

with probability one.

Note that the convergence is almost surely the stronger mode of convergence mentioned in the discussion of the law of large numbers in Section 4.2. This implies that $Y_n$ also converges in probability and in distribution. Proposition 8.14 says that there exists a random variable $Y$ in the limit, but it does not say anything about its properties. If we want to find out its distribution, say, we have to use other methods, for example, limits of moment generating functions as described in the proof of Theorem 4.2.

***Example 8.36 (Strong Law of Large Numbers).*** As the first example, we are going to demonstrate how to use Proposition 8.14 to strengthen Theorem 4.1[8], which says that the sample mean $\bar{X}$ converges in probability to the mean $\mu$ as $n \to \infty$.

Let $X_1, X_2, \ldots$ be a sequence of i.i.d. random variables and let

$$Y_n = \sum_{k=1}^{n} \frac{X_k - \mu}{k}$$

Clearly, $E[Y_n] = 0$ and

$$E[Y_n^2] = \text{Var}[Y_n] = \sum_{k=1}^{n} \frac{\sigma^2}{k^2} < \sigma^2 \sum_{k=1}^{\infty} \frac{1}{k^2} = \frac{\sigma^2 \pi^2}{6}$$

so as long as $\sigma^2 < \infty$, the condition in Proposition 8.14 is satisfied. To conclude that $Y_n$ is a martingale, we have to show that $E[|Y_n|] < \infty$ for all $n$, which follows from

$$\begin{aligned} E[|Y_n|] &= E[|Y_n| \mid |Y_n| \leq 1]P(|Y_n| \leq 1) + E[|Y_n| \mid |Y_n| > 1]P(|Y_n| > 1) \\ &< 1 \times P(|Y_n| \leq 1) + E[Y_n^2] < 1 + \frac{\sigma^2 \pi^2}{6} \end{aligned}$$

Proposition 8.14 now says that there exists some $Y$ such that $Y_n \to Y$ with probability one. To obtain the strong law of large numbers, we need a mathematical result called Kronecker's lemma, which says that if we have a sequence of arbitrary real numbers $x_1, x_2, \ldots$ and positive constants $b_1, b_2, \ldots$ that increase strictly to infinity, then

$$\sum_{k=1}^{n} \frac{x_k}{b_k} \to z \quad \text{implies that} \quad \frac{1}{b_n} \sum_{k=1}^{n} x_k \to 0$$

as $n \to \infty$, where $|z| < \infty$. We have to be a bit cautious since we are dealing with random variables, but without going into details it turns out that

$$\sum_{k=1}^{n} \frac{X_k - \mu}{k} \to Y \quad \text{implies that} \quad \frac{1}{n} \sum_{k=1}^{n} (X_k - \mu) \to 0$$

as $n \to \infty$ with probability 1. Finally, we note that since $\bar{X} - \mu \to 0$ it follows that $\bar{X} \to \mu$ with probability 1. □

[8]Sometimes called the *Weak Law of Large Numbers* to distinguish it from this case.

***Example 8.37 (Branching Processes).*** Recall Example 8.35 where we showed that $Y_n = Z_n/\mu^n$ for $n = 1, 2, \ldots$ is a martingale with respect to $Z_1, Z_2, \ldots$. Proposition 8.8 implies that

$$E[Y_n^2] = (E[Y_n])^2 + \text{Var}[Y_n] = 1 + \frac{1}{\mu^{2n}} \times \frac{\sigma^2(\mu^n - 1)\mu^{n-1}}{\mu - 1}$$

$$= 1 + \frac{\sigma^2(1 - \mu^{-n})}{\mu(\mu - 1)} < 1 + \frac{\sigma^2}{\mu(\mu - 1)}$$

if $\mu > 1$. Hence, Proposition 8.14 says that there only exists a limit $Y$ for the martingale $Y_n$ in the supercritical case. Luckily, this is the most interesting case because we know from Proposition 8.9 that subcritical and critical branching processes always go extinct eventually and it is possible, using other methods, to conclude that $Y_n \to 0$ for those cases.

The properties of $Y$ are not that easy to obtain and we are not going to go much further here. It is possible, though, to show that $Y$ is continuous except for a point mass at 0 equal to the extinction probability. This means that the event $\{Y > 0\}$ is equivalent to nonextinction and if we condition on this event, we can say that $Z_n \approx Y\mu^n$ for large $n$. Essentially, the branching process grows with a more or less deterministic rate but from a random level. □

### 8.5.2 Stopping Times

Let us assume that we can interpret a martingale as the accumulated fortune of a gambler playing a fair game. The game is fair in the sense that the expected amount after each play is equal to the gambler's fortune before the play. However, if we are lucky we may win some money or if we are unlucky we may lose some money in each individual play.

Let us, for simplicity, consider the symmetric random walk as a simple model for gambling where we bet $1 and win $2 if we get heads and lose our bet if we get tails. Throughout history, innumerable attempts have been made to beat the odds, to come up with the perfect foolproof strategy to win money no matter how the dice fall or the roulette wheel spins. One of the most (in)famous is the so-called doubling strategy, where you always double your bet if you lose one play.

If it takes $n$ plays to get heads, we have lost $1 + 2 + 4 + \cdots + 2^{n-2} = 2^{n-1} - 1$ dollars along the way, but since the winning play brings in $2^{n-1}$ dollars, we will gain $1 altogether. Seems solid, doesn't it, so, what's the problem? If we let $N$ denote the number of plays it takes to get heads, we know that $p_n = P(N = n) = \left(\frac{1}{2}\right)^n$. The expected amount that we will have to bet to gain $1 is then

$$\sum_{n=1}^{\infty}(2^{n-1} - 1)p_n = \sum_{n=1}^{\infty}\left(\frac{1}{2} - \frac{1}{2^n}\right) = \infty$$

Hence, no matter how big your initial fortune is, you will most likely be ruined before you stand to win any significant amount of money.

This is an example of a strategy using a *stopping time*, which is a predetermined rule when to stop playing. The rule has to be formulated so that we know for sure, after each play, if we will stop or not and that we are not allowed to use future, unobserved observations. In mathematical terms, it can be defined as follows.

***Definition 8.13.*** A random variable $T$ that takes values in $\{1, 2, \ldots, \infty\}$ is called a stopping time for the sequence $X_1, X_2, \ldots$ if

$$P(T = n|X_1 = x_1, X_2 = x_2, \ldots) = \\ P(T = n|X_1 = x_1, X_2 = x_2, \ldots, X_n = x_n) = 0 \text{ or } 1$$

This means that if we know exactly what values the stochastic process $X_1, X_2, \ldots$ has assumed up to and including step $n$, we will know for sure if the event $\{T = n\}$ has occurred or not. Also, note that we allow $T$ to assume an infinite value, which corresponds to a situation where our rule is never met and we therefore never stop.

A poorly chosen stopping time usually means, as illustrated above, that there is a clear risk that we may have to wait for a very long time until we stop and that things may go awry before that. Exactly what conditions a useful stopping time should satisfy are covered in the following result.

**Proposition 8.15. (Optional Stopping Theorem).** Let $Y_1, Y_2, \ldots$ be a martingale and $T$ a stopping time with respect to some sequence $X_1, X_2, \ldots$. If

(i) $P(T < \infty) = 1$

(ii) $E[|Y_T|] < \infty$

(iii) $E[Y_n|T > n]P(T > n) \to 0$ as $n \to \infty$

then $E[Y_T] = E[Y_1]$.

There are several versions of the optional stopping theorem giving slightly different conditions, but those described in Proposition 8.15 are usually easy to verify. This is bad news for all gamblers since it shows that there is no (reasonable) strategy that will increase your expected fortune in a fair game.[9] The doubling strategy described above satisfies the first two conditions in Proposition 8.15 but not the third since

$$E[Y_n|T > n]P(T > n) = -(2^n - 1) \times \left(\frac{1}{2}\right)^n \to -1$$

as $n \to \infty$.

[9]It is actually even worse because casinos and other gambling venues rarely offer any fair games. The odds are usually stacked in their favor, bringing in a small but steady profit. Processes where $E[Y_{n+1}|X_1, X_2, \ldots, X_n] \leq Y_n$ are called *supermartingales* (replace $\leq$ with $\geq$ and you get *submartingales*) and have similar properties.

***Example 8.38 (Gambler's Ruin).*** As the first example of the optional stopping theorem, let us revisit Example 1.58 where Ann and Bob flipped a coin and Ann paid Bob one dollar if it turned up heads and Bob paid Ann one dollar if it turned up tails. Ann started with $a$ dollars and Bob with $b$ dollars and the question was which one would be ruined first.

If we let $X_k$ be 1 if we get heads in the $k$th flip and $-1$ otherwise, we can write Bob's total gain after $n$ coin flips as

$$S_n = \sum_{k=1}^{n} X_k$$

This is clearly a martingale with respect to $X_1, X_2, \ldots$ since

$$E[S_{n+1}|X_1, \ldots, X_n] = (S_n - 1) \times \frac{1}{2} + (S_n + 1) \times \frac{1}{2} = S_n$$

The game stops either when $S_n = a$ (Ann is ruined) or $S_n = -b$ (Bob is ruined), so

$$T = \min\{n : S_n = a \text{ or } S_n = -b\}$$

is clearly a stopping time.

We know from Section 8.3.1 that the symmetric random walk is recurrent, which means that it will hit either $a$ or $-b$ eventually. This implies both that $P(T < \infty) = 1$, verifying the first condition of Proposition 8.15, and that $P(T > n) \to 0$ as $n \to \infty$. We also know that the random variable $S_T$ only assumes the values $a$ and $-b$ so that $E[|S_T|] \le \max(a, b) < \infty$. Finally, we realize that if the event $\{T > n\}$ occurs, the martingale has not stopped at time $n$ so that $-b < S_n < a$ and the third condition follows.

Let $p$ be the probability that Bob wins all the money (and Ann gets ruined) and note that $p = P(S_T = a)$. Proposition 8.15 now implies that

$$E[S_T] = ap - b(1 - p) = E[S_1] = 0$$

which gives us that

$$p = \frac{b}{a + b}$$

We can also calculate the expected value of $T$ by considering the sequence $Y_n = S_n^2 - n$, which also is a martingale with respect to $X_1, X_2, \ldots$ since

$$E[Y_{n+1}|X_1, \ldots, X_n] = E[(S_n + X_{n+1})^2 - (n + 1)|S_n]$$

$$= S_n^2 + 2S_n E[X_{n+1}] + E[X_{n+1}^2] - (n + 1) = S_n^2 - n = Y_n$$

It is not quite as easy to verify that all conditions of Proposition 8.15 are satisfied, but we will skip those details. Hence, we know that

$$E[Y_T] = E[S_T^2 - T] = E[S_T^2] - E[T] = E[Y_1] = E[S_1^2 - 1] = 0$$

and, since we know that $S_T$ is either equal to $a$ with probability $p$ or equal to $-b$ with probability $1 - p$, we get that

$$E[T] = E[S_T^2] = a^2 p + b^2(1 - p) = \frac{a^2 b}{a + b} + \frac{b^2 a}{a + b} = ab$$

which would be very difficult to calculate in any other way. □

***Example 8.39 (Ballot Theorem).*** There has been a ballot between Ann and Bob where Ann won with $a$ votes whereas Bob got only $b$ votes. If the votes were counted one by one in random order, what is the probability that Ann was ahead of Bob the whole time?

Let $X_k$ be equal to 1 if vote number $k$ was a vote for Ann and $-1$ if the vote was for Bob. Then $S_n = \sum_{k=1}^{n} X_k$ denotes the difference between the number of votes for Ann and Bob after $n$ votes have been counted. The probability we are looking for can be expressed as $P(S_n > 0; 1 \leq n \leq N)$, where $N = a + b$ is the total number of votes.

There are several ways to solve this classic probabilistic problem,[10] but we will use the optional stopping theorem to do it here. It turns out that we can simplify the problem by going backward in time, creating the so-called *backward martingale*.

First, we note that after $n$ votes have been counted, Ann have $(n + S_n)/2$ votes and Bob have $(n - S_n)/2$ votes. This implies that

$$P(S_n = S_{n+1} + 1 | S_{n+1}) = \frac{(n + 1 - S_{n+1})/2}{n + 1}$$

$$P(S_n = S_{n+1} - 1 | S_{n+1}) = \frac{(n + 1 + S_{n+1})/2}{n + 1}$$

These conditional probabilities consider the events that we take away a vote for Bob and we take away a vote for Ann, respectively, from $n + 1$ counted votes.

Now, we define the sequence $Y_1, Y_2, \ldots, Y_N$ as

$$Y_n = \frac{S_{N-n+1}}{N - n + 1}, \qquad n = 1, 2, \ldots, N$$

[10] Introduced by the French mathematician Joseph Bertrand (1822–1900) in 1887. Martingale theory had not been developed at the time, so he solved it using combinatorical methods.

which is a martingale with respect to itself since

$$E[Y_{n+1}|Y_1, \ldots, Y_n] = E\left[\frac{S_{N-n}}{N-n}\middle| S_N, \ldots, S_{N-n+1}\right]$$

$$= \frac{1}{N-n}\{(S_{N-n+1}+1)P(S_{N-n} = S_{N-n+1}+1|S_{N-n+1})$$

$$+(S_{N-n+1}-1)P(S_{N-n} = S_{N-n+1}-1|S_{N-n+1})\}$$

$$= \frac{1}{N-n}\left(S_{N-n+1} + \frac{(N-n+1-S_{N-n+1})/2}{N-n+1}\right.$$

$$\left.-\frac{(N-n+1+S_{N-n+1})/2}{N-n+1}\right)$$

$$= \frac{S_{N-n+1}}{N-n}\left(1-\frac{1}{N-n+1}\right) = \frac{S_{N-n+1}}{N-n+1} = Y_n$$

Next, we define the stopping time $T$ as

$$T = \min\{\min\{n : Y_n = 0\}, N\}$$

Since we are going backward in time, $T$ denotes the last time the candidates were even and if this never happens, we let $T = N$, which corresponds to the first vote counted. This means that if $T < N$, the candidates were even at $T$ and $Y_T \equiv 0$. On the other hand, if $T = N$, Ann was always ahead and $Y_T \equiv 1$. Since both the martingale and the stopping time are bounded, it is obvious that all conditions in Proposition 8.15 are satisfied, so we can conclude that

$$E[Y_T] = E[Y_1] = E\left[\frac{S_N}{N}\right] = \frac{a-b}{a+b}$$

Finally, since the random variable $Y_T$ only assumes the values 0 and 1, we get

$$E[Y_T] = 0 \times P(Y_T = 0) + 1 \times P(Y_T = 1)$$

$$= 0 \times P(T < N) + 1 \times P(T = N) = P(T = N)$$

which yields the probability that Ann was always in the lead as

$$P(\text{Ann always in the lead}) = P(T = N) = \frac{a-b}{a+b}$$

□

## 8.6 RENEWAL PROCESSES

In Section 3.12, we introduced the Poisson process as a point process with i.i.d. exponentially distributed interarrival times. In this section, we will outline some theory of the most obvious generalization of this, namely, to allow the interarrival times to have any nonnegative distribution.

Let us start with the formal definition.

***Definition 8.14.*** Let $T_1, T_2, \ldots$ be i.i.d. nonnegative random variables such that $P(T_k = 0) < 1$ and $S_n = \sum_{k=1}^{n} T_k$, then

$$N(t) = \max\{n : S_n \le t\}$$

is a *renewal process* for $t \ge 0$.

The time points $S_1, S_2, \ldots$ are called the *renewals* since it is like starting all over from the beginning every time an interarrival time ends. Since $N(t) = n$ is equivalent to $S_n \le t < S_{n+1}$, we see that $N(t)$ counts the number of renewals in the interval $[0, t]$. The cdf of $T_k$ and $S_n$ will be denoted $F(t) = P(T_k \le t)$ and $F_n(t) = P(S_n \le t)$, respectively, and we let $\mu$ and $\sigma^2$ denote the mean and the variance of $T_k$.

The exact distribution of $N(t)$ is usually very difficult to derive, except for a few simple cases, but it can at least theoretically be expressed in the form

$$\begin{aligned} P(N(t) = n) &= P(S_n \le t < S_{n+1}) = P(S_n \le t) - P(S_{n+1} \le t) \\ &= F_n(t) - F_{n+1}(t) \end{aligned}$$

for $n = 0, 1, 2, \ldots$, where $F_0(t) = 1$. We can also write the mean $m(t) = E[N(t)]$, using Proposition 2.9, as

$$m(t) = \sum_{n=1}^{\infty} P(N(t) \ge n) = \sum_{n=1}^{\infty} P(S_n \le t) = \sum_{n=1}^{\infty} F_n(t)$$

The mean $m(t)$ is also called the *renewal function* and it turns out that it uniquely determines the renewal process, that is, if we are given an expression for $m(t)$, we can in principle calculate the distribution $F(t)$ of the interarrival times.

***Example 8.40.*** Let us have a look at the Poisson process expressed as a renewal process. Since the interarrival times are exponentially distributed, we know that

$$F(t) = 1 - e^{-\lambda t}$$

and, by Proposition 3.35 and Section 2.8.2, that

$$F_n(t) = 1 - e^{-\lambda t} \sum_{k=0}^{n-1} \frac{(\lambda t)^k}{k!}$$

The distribution of $N(t)$ can now be calculated as

$$P(N(t) = n) = \left(1 - e^{-\lambda t} \sum_{k=0}^{n-1} \frac{(\lambda t)^k}{k!}\right) - \left(1 - e^{-\lambda t} \sum_{k=0}^{n} \frac{(\lambda t)^k}{k!}\right) = e^{-\lambda t} \frac{(\lambda t)^n}{n!}$$

which is the Poisson distribution with mean $\lambda t$. The renewal function becomes

$$\begin{aligned} m(t) &= \sum_{n=1}^{\infty} \left(1 - e^{-\lambda t} \sum_{k=0}^{n-1} \frac{(\lambda t)^k}{k!}\right) = \sum_{n=1}^{\infty} \left(1 - e^{-\lambda t} \left(e^{\lambda t} - \sum_{k=n}^{\infty} \frac{(\lambda t)^k}{k!}\right)\right) \\ &= e^{-\lambda t} \sum_{n=1}^{\infty} \sum_{k=n}^{\infty} \frac{(\lambda t)^k}{k!} = e^{-\lambda t} \sum_{k=1}^{\infty} \sum_{n=1}^{k} \frac{(\lambda t)^k}{k!} = e^{-\lambda t} \sum_{k=1}^{\infty} \frac{(\lambda t)^k}{(k-1)!} \\ &= e^{-\lambda t} \lambda t \sum_{k=1}^{\infty} \frac{(\lambda t)^{k-1}}{(k-1)!} = \lambda t \end{aligned}$$

where we used the Taylor expansion of $e^{\lambda t}$. This is a very longwinded way to derive the mean of the Poisson process, but it at least shows that the theory works. What we have shown is that the Poisson process has a linear renewal function and, since the renewal function determines the renewal process, it is worth noting that the Poisson process is the *only* renewal process with a linear renewal function. □

For continuous interarrival times, we can use the property that the renewal process starts over at every renewal $S_n$ to obtain the following result.

**Proposition 8.16 (The Renewal Equation).** If $T_1, T_2, \ldots$ are continuous interarrival times with cdf $F(t)$ and pdf $f(t)$, the renewal function $m(t)$ satisfies

$$m(t) = F(t) + \int_0^t m(t-u) f(u)\, du$$

*Proof.* If we condition on the time of the first renewal, Proposition 3.17 says that we can write the renewal function

$$m(t) = E[N(t)] = \int_0^{\infty} E[N(t) | T_1 = u] f(u)\, du \tag{8.10}$$

Now, if the first renewal occurs after $t$, that is, $u > t$ holds, clearly $E[N(t)|T_1 = u] = 0$. On the other hand, if $u \leq t$, we count the first renewal and start the process from the beginning at time $u$. This means that

$$E[N(t)|T_1 = u] = 1 + E[N(t-u)] = 1 + m(t-u) \quad u \le t$$

which means that we can write (8.10) as

$$m(t) = \int_0^t (1 + m(t-u))f(u)\,du = F(t) + \int_0^t m(t-u)f(u)\,du$$

■

The main advantage of Proposition 8.16 is that we only need to consider the distribution of the first renewal to get the renewal function, but instead we need to solve an integral equation, which is not that easy in most cases. Still, it gives us a tool to verify whether a proposed $m(t)$ actually is a renewal function for a given renewal process.

### 8.6.1 Asymptotic Properties

Since the exact distribution of the renewal process is difficult to obtain in most cases, it would at least be interesting to see how $N(t)$ behaves as $t \to \infty$. The first result is essentially a version of the Law of Large Numbers.

**Proposition 8.17.** For $\mu > 0$, it holds that

$$\frac{N(t)}{t} \xrightarrow{P} \frac{1}{\mu} \quad \text{as } t \to \infty$$

*Proof.* Since $S_{N(t)}$ is the time of the last renewal before $t$, we note that

$$S_{N(t)} \le t < S_{N(t)+1}$$

which yields that

$$\frac{S_{N(t)}}{N(t)} \le \frac{t}{N(t)} < \frac{S_{N(t)+1}}{N(t)} \tag{8.11}$$

for $N(t) > 0$. Now, it holds that $N(t) \to \infty$ as $t \to \infty$ (why is that?), so the Law of Large Numbers implies that

$$\frac{S_{N(t)}}{N(t)} \xrightarrow{P} \mu \quad \text{as } t \to \infty$$

Furthermore, using a similar argument yields that

$$\frac{S_{N(t)+1}}{N(t)} = \frac{S_{N(t)+1}}{N(t)+1} \times \frac{N(t)+1}{N(t)} \xrightarrow{P} \mu \times 1 \quad \text{as } t \to \infty$$

We have shown that both the upper and the lower bound in (8.11) converge in probability to $\mu$ and since $t/N(t)$ is between these, it has to converge in probability to $\mu$ as well. Taking the reciprocals completes the proof. ■

The ratio $1/\mu$ is called the *renewal rate* because it denotes the average number of renewals per time unit in the long run. Note that the result holds also for infinite $\mu$, in which case we interpret the renewal rate as 0.

In light of Proposition 8.17, the following result seems quite natural.

**Proposition 8.18 (Elementary Renewal Theorem).** For $\mu > 0$, it holds that

$$\frac{m(t)}{t} \to \frac{1}{\mu} \quad \text{as } t \to \infty$$

It would seem that this is a simple consequence of Proposition 8.17, but that is actually not the case. Proving the elementary renewal theorem turns out to be rather difficult and requires some asymptotic theory that we have not introduced.

The following result is slightly more general but holds only for continuous interarrival times.

**Proposition 8.19 (Renewal Theorem).** For continuous $F(t)$ and $\mu > 0$, it holds that

$$m(t+s) - m(t) \to \frac{s}{\mu} \quad \text{as } t \to \infty$$

This result says essentially that if we slide a ruler of length $s$ on the time axis, the expected number of renewals covered by the ruler will be, at least approximately, proportional to the length of the ruler. For a fixed and large $t$, Propositions 8.18 and 8.19 can be summarized as

$$m(t) \approx \frac{t}{\mu}$$
$$m(t+s) - m(t) \approx \frac{s}{\mu}$$

which illustrates how they are connected. Proposition 8.18 can be regarded as a global average of renewals while Proposition 8.19 gives a similar local property.

There is also a central limit theorem for renewal processes.

**Proposition 8.20.** For $\mu < \infty$ and $\sigma^2 < \infty$, it holds that

$$P\left(\frac{N(t) - t/\mu}{\sigma\sqrt{t/\mu^3}} \leq x\right) \to \Phi(x)$$

as $t \to \infty$, where $\Phi(x)$ is the cdf of the standard normal distribution.

*Proof.* For a fixed $x$, let

$$t = n\mu - x\sigma\sqrt{n} \tag{8.12}$$

and consider the probability

$$P(N(t) < n) = P\left(\frac{N(t) - t/\mu}{\sigma\sqrt{t/\mu^3}} < \frac{n\mu - t}{\sigma\sqrt{t/\mu}}\right) = P\left(\frac{N(t) - t/\mu}{\sigma\sqrt{t/\mu^3}} < x\sqrt{\frac{n\mu}{t}}\right)$$

Now, if we let $n \to \infty$ we see from (8.12) that $t \to \infty$ and vice versa. Also, (8.12) implies that $\sqrt{n\mu/t} \to 1$, which means that

$$\lim_{n\to\infty} P(N(t) < n) = \lim_{t\to\infty} P\left(\frac{N(t) - t/\mu}{\sigma\sqrt{t/\mu^3}} < x\right)$$

Finally, since the events $\{N(t) < n\}$ and $\{S_n > t\}$ are equivalent, we get that

$$P(N(t) < n) = P(S_n > t) = P\left(\frac{S_n - n\mu}{\sigma\sqrt{n}} > \frac{t - n\mu}{\sigma\sqrt{n}}\right)$$

$$= P\left(\frac{S_n - n\mu}{\sigma\sqrt{n}} > -x\right) \to 1 - \Phi(-x) = \Phi(x)$$

as $n \to \infty$ by the central limit theorem. ■

***Example 8.41.*** A Geiger–Müller counter (or Geiger counter for short) is an electronic device that detects radioactive particles, usually from beta and gamma radiation. One problem is that every time a particle is registered, the counter has to be reset before it can detect new particles. These periods are called *dead periods* because any particles that arrive while the counter is reset are lost. Therefore, some appropriate adjustment needs to be done in order to avoid underestimation of the radioactive intensity.

Let us assume that the radioactive particles arrive according to a Poisson process with rate $\lambda$ and that the lengths of the dead periods $Y_1, Y_2, \ldots$ are i.i.d. random variables. Let $X_1, X_2, \ldots$ denote the times until a reset counter registers a particle and, from the properties of the Poisson process, these are independent and exponentially distributed with parameter $\lambda$. Furthermore, we assume that the lengths of the dead periods are independent of the Poisson process.

If we let the interarrival times be $T_k = X_k + Y_k$ for $k = 1, 2, \ldots$, we can define the renewals as

$$S_n = \sum_{k=1}^{n} T_k = \sum_{k=1}^{n} (X_k + Y_k)$$

In this case, we get the mean and variance of the interarrival times as

$$\mu = \frac{1}{\lambda} + \mu_Y \tag{8.13}$$
$$\sigma^2 = \frac{1}{\lambda^2} + \sigma_Y^2$$

Let us say that we have run the counter for a long time $t$ and registered $N(t)$ particle emissions.[11] Proposition 8.17 implies that

$$N(t) \approx \frac{t}{\mu} = \frac{t}{1/\lambda + \mu_Y}$$

which gives us the appropriate estimator of $\lambda$ as

$$\widehat{\lambda} = \frac{1}{\frac{t}{N(t)} - \mu_Y}$$

Having obtained an estimate for $\lambda$, we can then use Proposition 8.20 to calculate a confidence interval. First, we get an interval for $1/\mu$ with approximate confidence level $q$ as

$$\frac{1}{\mu} \approx \frac{N(t)}{t} \pm z \frac{\sigma}{\sqrt{t\mu^3}} \approx \frac{N(t)}{t} \pm z\widehat{\sigma} \frac{\sqrt{N(t)^3}}{t^2}$$

where $\Phi(z) = (1+q)/2$. In the second approximation above, we used the fact that $\mu \approx t/N(t)$ and

$$\widehat{\sigma}^2 = \frac{1}{\widehat{\lambda}^2} + \sigma_Y^2$$

Then, (8.13) can be used to transform the interval for $\lambda$.

For a practical example, let us assume that we have detected $N(t) = 5630$ particles during 1 s and the dead periods are uniformly distributed between 0 and 200 μs. Let us use 1 ms as a convenient time unit so that $\lambda$ denotes the average number of particle emissions per ms. It is clear that $\mu_Y = 0.1$, which yields the point estimate

$$\widehat{\lambda} = \frac{1}{(1000/5630) - 0.1} = 12.9$$

Now, we get

$$\widehat{\sigma}^2 = \frac{1}{12.9^2} + \frac{0.2^2}{12} = 0.0094$$

[11] If the time point $t$ ends up in a dead period, the number of detected particles is actually $N(t)+1$ since the last particle is registered before $t$ although the corresponding interarrival time ends after $t$. However, if $t$ is large, this is negligible.

which yields the 95 % interval

$$\frac{1}{\mu} \approx \frac{5630}{1000} \pm 1.96\sqrt{0.0094}\frac{\sqrt{5630^3}}{1000^2} = 5.63 \pm 0.08$$

Finally, this yields the limits for $\lambda$ as

$$\frac{1}{(1/5.58) - 0.1} \leq \lambda \leq \frac{1}{(1/5.71) - 0.1} \quad (\approx 95\%)$$

or

$$12.5 \leq \lambda \leq 13.3 \quad (\approx 95\%)$$

□

***Example 8.42 (Delayed Renewal Process).*** There is one variation of the regular renewal process that is of particular interest, where the distribution of the initial interarrival time $T_1$ may be different from the subsequent. Such a situation may emerge if we start the process in between two adjacent renewals rather than at a specific renewal time, hence the term delayed renewal process. Since only the first interarrival time differs, all previous limiting results can be shown to hold also for the delayed renewal process. In fact, it has some interesting theoretical properties that actually can be used to prove some of these results for the regular renewal process.

Let $F(t)$ denote the cdf of $T_2, T_3, \ldots$ as before and denote the cdf of $T_1$ by $\widetilde{F}(t)$. Furthermore, let $\widetilde{S}_n$ denote the renewals, $\widetilde{N}(t)$ the delayed renewal process, and $\widetilde{m}(t)$ the renewal function. Now, we can use the same method as in the proof of Proposition 8.16 to obtain

$$\begin{aligned}
\widetilde{m}(t) &= \int_0^t E[\widetilde{N}(t)|T_1 = u]\widetilde{f}(u)\,du = \int_0^t (1 + m(t-u))\widetilde{f}(u)\,du \\
&= \widetilde{F}(t) + \int_0^t m(t-u)\widetilde{f}(u)\,du
\end{aligned} \tag{8.14}$$

Note that we get the mean $m(t-u)$ in the integral since we get an ordinary renewal process after conditioning on $\{T_1 = u\}$. If we apply Proposition 8.16 to $m(t-u)$, we can write (8.14) as

$$\begin{aligned}
&\widetilde{F}(t) + \int_0^t \left( F(t-u) + \int_0^{t-u} m(t-u-v) f(v)\,dv \right) \widetilde{f}(u)\,du \\
&= \widetilde{F}(t) + \int_0^t F(t-u)\widetilde{f}(u)\,du + \int_0^t \left( \int_0^{t-v} m(t-v-u)\widetilde{f}(u)\,du \right) f(v)\,dv
\end{aligned}$$

The first integral can be transformed, using integration by parts, into

$$\int_0^t F(t-u)\widetilde{f}(u)\,du = \left[F(t-u)\widetilde{F}(u)\right]_0^t + \int_0^t f(t-u)\widetilde{F}(u)\,du$$
$$= \int_0^t \widetilde{F}(t-u)f(u)\,du$$

and we can use (8.14) again to rewrite the inner integral as

$$\int_0^{t-v} m(t-v-u)\widetilde{f}(u)\,du = \widetilde{m}(t-v) - \widetilde{F}(t-v)$$

Together, these results imply that

$$\widetilde{m}(t) = \widetilde{F}(t) + \int_0^t \widetilde{F}(t-u)f(u)\,du + \int_0^t (\widetilde{m}(t-v) - \widetilde{F}(t-v))f(v)\,dv$$
$$= \widetilde{F}(t) + \int_0^t \widetilde{m}(t-v)f(v)\,dv \tag{8.15}$$

which is the renewal equation for the delayed renewal process.

As for the ordinary renewal process, this equation uniquely determines the renewal function. If we look at the asymptotic properties as $t \to \infty$, we realize that they depend largely on $F(t)$ and that the initial interarrival time eventually becomes irrelevant. This means that we may choose whatever $\widetilde{F}(t)$ we like without affecting any limiting results. Proposition 8.18 says that the scaled renewal function $m(t)/t$ for a regular renewal process converges toward $1/\mu$ and, as mentioned above, the same can be shown to hold for $\widetilde{m}(t)/t$. Let us see if we can choose $\widetilde{F}(t)$ so that

$$\widetilde{m}(t) = \frac{t}{\mu}$$

for *all* $t \geq 0$. In that case, (8.15) implies that

$$\widetilde{F}(t) = \widetilde{m}(t) - \int_0^t \widetilde{m}(t-v)f(v)\,dv = \frac{t}{\mu} - \int_0^t \frac{t-v}{\mu} f(v)\,dv$$
$$= \frac{t}{\mu} - \left[\frac{t-v}{\mu}F(v)\right]_0^t - \int_0^t \frac{1}{\mu}F(v)\,dv = \frac{1}{\mu}\int_0^t (1-F(v))\,dv$$

as long as $\mu < \infty$. □

## 8.7 BROWNIAN MOTION

The simple random walk presented in Section 8.3.1 describes a discrete process that at unit time points jumps one unit up or one unit down with fixed probabilities $p$ and $1-p$, respectively, independent of all previous jumps. Let us see if we can construct a continuous version of this that makes infinitesimally small jumps infinitesimally often. To achieve this, we can consider random walks with jump sizes $\Delta x$ and time intervals of length $\Delta t$ and then gradually decrease $\Delta x$ and $\Delta t$ toward zero. In the following, we will consider only the symmetric random walk where $p = 1 - p = \frac{1}{2}$.

Let $X_1, X_2, \ldots$ be independent random variables such that

$$X_k = \begin{cases} +1 & \text{with probability } 1/2. \\ \\ -1 & \text{with probability } 1/2 \end{cases} \tag{8.16}$$

Then we can define the process

$$S_n(t) = \Delta x\, X_1 + \Delta x\, X_2 + \cdots + \Delta x\, X_{[t/\Delta t]} = \Delta x \sum_{i=1}^{n} X_i \tag{8.17}$$

where $n = [t/\Delta t]$ denotes the largest integer less than or equal to the real number $t/\Delta t$.

Since $E[X_k] = 0$ and $\text{Var}[X_k] = 1$, we can calculate the mean and variance of (8.17) as

$$E[S_n(t)] = \Delta x \sum_{i=1}^{n} E[X_k] = 0$$

$$\text{Var}[S_n(t)] = (\Delta x)^2 \sum_{i=1}^{n} \text{Var}[X_k] = (\Delta x)^2 n = (\Delta x)^2 \left[\frac{t}{\Delta t}\right] \tag{8.18}$$

If we let $\Delta t \to 0$, then $n \to \infty$ and we can use the central limit theorem to conclude that the sum in (8.17), properly standardized, converges to the standard normal distribution.

To obtain a similar result for $S_n(t)$, we also have to let $\Delta x \to 0$ in some orderly fashion. It is clear from (8.18) that $\Delta x$ would have to decrease at a slower rate than $\Delta t$, of the order $\sqrt{\Delta t}$, if we are to obtain a positive and finite variance in the limit. For simplicity, let $\Delta x = \sigma\sqrt{\Delta t}$ to obtain the variance limit $\text{Var}[S_n(t)] \to \sigma^2 t$ and let $\Delta t \to 0$. Then, the central limit theorem implies that

$$\frac{S_n(t) - E[S_n(t)]}{\sqrt{\text{Var}[S_n(t)]}} \xrightarrow{d} N(0, 1) \tag{8.19}$$

as $n \to \infty$.

Since both $E[S_n(t)]$ and $\text{Var}[S_n(t)]$ converge to finite and, for the variance, positive limits, we can reformulate (8.19) as

$$S_n(t) \xrightarrow{d} B(t) \sim N(0, \sigma^2 t) \tag{8.20}$$

Another important property of the limiting process $B(t)$ is that it has independent increments, that is, the changes $B(t_3) - B(t_2)$ and $B(t_2) - B(t_1)$ are independent random variables for $t_1 < t_2 < t_3$, which should be intuitively clear from the random

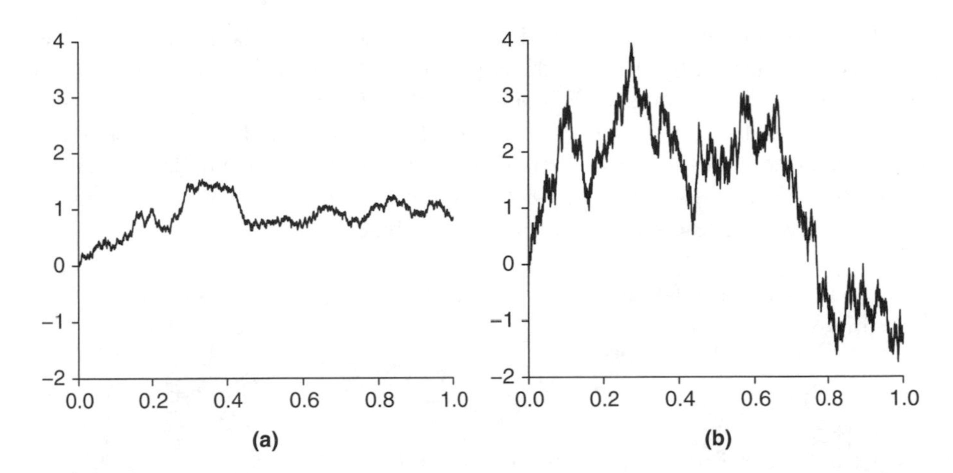


**FIGURE 8.3** Two realizations of Brownian motions in the unit interval for (a) $\sigma = 1$ and (b) $\sigma = 4$.

walk definition above. In fact, these are the two fundamental defining properties of the process.

> ***Definition 8.15.*** The *Brownian Motion*[12] $B(t)$ is a real valued stochastic process in real time $t \geq 0$ that satisfies
>
> (i) $B(0) = 0$
>
> (ii) If $t_1 < t_2 < \ldots < t_n$, then $B(t_1), B(t_2) - B(t_1), \ldots, B(t_n) - B(t_{n-1})$ are independent
>
> (iii) $B(t+s) - B(s) \sim N(0, \sigma^2 t)$ for $t, s \geq 0$

Note that part (iii) of Definition 8.15 is slightly more general than (8.20) since it says that all increments are also normally distributed. It does behave very erratically, which is illustrated in Figure 8.3 where two simulated Brownian motions are shown for $\sigma = 1$ and $\sigma = 4$.

This is one of the most studied stochastic processes in the mathematical literature partly because it has a lot of fascinating theoretical properties and because it has been found to be very useful in quite different areas like physics (quantum mechanics), electronics (filtering theory), and economics (option pricing). We will not have time to explore the Brownian motion in detail here except for some of the most fundamental and useful properties.

[12]Named after the Scottish botanist Robert Brown (1773–1858) who studied pollen grains submerged in liquid. To his amazement he observed that particles ejected from the pollen grains moved in a very erratic random fashion. He was the first to observe this phenomenon but he did not manage to explain it. Later, the American mathematician Norbert Wiener (1894–1964) developed the mathematical theory behind it and therefore the process $B(t)$ is sometimes also called the *Wiener process*

One interesting feature of $B(t)$ is that it is scale invariant in both time and space. It is quite easy to see, by checking the conditions in Definition 8.15, that if $B(t)$ is a Brownian motion with variance $\sigma^2$, then $B(t)/\sigma$ is a Brownian motion with variance 1. This is called the *standard Brownian motion* and since any Brownian motion can be standardized in this way, it is common to assume that $\sigma^2 = 1$. Similarly, it can be shown that if $B(t)$ is a standard Brownian motion, then $B(\sigma^2 t)$ is a Brownian motion with variance $\sigma^2$. These scalings can also be combined.

**Proposition 8.21 (Self-Similarity).** Let $B(t)$ be a standard Brownian motion. Then the process

$$\widetilde{B}(t) = \frac{B(a^2 t)}{a}$$

is also a standard Brownian motion for any $a \neq 0$.

Essentially, what this says is that if we would "zoom in" on the Brownian motion we would see a process that would look quite similar to what we started with, that is, the Brownian motion can be characterized as a *random fractal*. However, in order for this to work, we see that the rescaling of the horizontal time axis has to be the square of the rescaling of the vertical space axis. Also note that the rescaling constants does not have to be positive. A negative $a$ means that we reverse the vertical axis and, by symmetry of the normal distribution, we still get a Brownian motion.

### 8.7.1 Hitting Times

One quantity of interest is the time until the Brownian motion attains some predetermined level. As a practical example, let us say that we buy a share of stock whose value can be described by a Brownian motion and decide to sell this after its value has increased by a certain amount. Then, we would like to know how long it takes until our investment pays off. We start with the following definition.

***Definition 8.16.*** The time until the Brownian motion $B(t)$ hits $a$ is called the *hitting time* and is defined

$$T_a = \min\{t : B(t) = a\}$$

Clearly, $T_a$ is a nonnegative continuous random variable whose distribution is given below.

**Proposition 8.22.** The hitting time $T_a$ of a standard Brownian motion has pdf

$$f_{T_a}(t) = \frac{|a|}{\sqrt{2\pi t^3}} e^{-a^2/2t} \quad t > 0$$

for any $a \neq 0$.

*Proof.* Let us first assume that $a > 0$ and consider the event that the Brownian motion exceeds $a$ at time $t$. The law of total probability gives us

$$P(B(t) \geq a) = P(B(t) \geq a | T_a > t) P(T_a > t) + P(B(t) \geq a | T_a \leq t) P(T_a \leq t) \quad (8.21)$$

The first conditional probability above is clearly 0 since the event $\{T_a > t\}$ means that the Brownian motion has not hit $a$ at time $t$ and since $B(t)$ is continuous, it cannot be above $a$. If we turn to the second conditional probability, the condition says that we have hit $a$ before $t$. Then we can split $B(t)$ into two parts

$$B(t) = B(T_a) + (B(t) - B(T_a)) = a + (B(t) - a)$$

Now, the increment $B(t) - a$ is normally distributed with mean 0, which means that

$$P(B(t) \geq a | T_a > t) = P(B(t) - a \geq 0) = \frac{1}{2}$$

by symmetry. This means that (8.21) can be written as

$$P(T_a \leq t) = 2P(B(t) \geq a) = 2\left(1 - \Phi\left(\frac{a-0}{\sqrt{t}}\right)\right)$$

For $a < 0$, we consider the event $\{B(t) \leq a\}$ and, again by symmetry, we get that

$$P(T_a \leq t) = 2P(B(t) \leq a) = 2\Phi\left(\frac{a-0}{\sqrt{t}}\right) = 2\left(1 - \Phi\left(-\frac{a}{\sqrt{t}}\right)\right)$$

Hence, the cdf can be written

$$F_{T_a}(t) = 2\left(1 - \Phi\left(\frac{|a|}{\sqrt{t}}\right)\right) \quad (8.22)$$

and taking the derivative with respect to $t$ yields the pdf

$$f_{T_a}(t) = -2\phi\left(\frac{|a|}{\sqrt{t}}\right) \times \left(-\frac{|a|}{2t^{3/2}}\right) = \frac{|a|}{\sqrt{2\pi t^3}} e^{-a^2/2t}$$

■

The properties of this distribution are not that easy to investigate. Even if $T_a$ is finite with probability 1, implying that we will hit $a$ eventually, it turns out that the mean is infinite, in an analogy with Corollary 8.7, suggesting that it might take awhile. The median, however, turns out to be just $1.67a^2$, so for small $a$ there is a fair chance that we will reach $a$ fairly soon.

***Example 8.43.*** Since the Brownian motion can be characterized as the scaled limit of a symmetric random walk, we can use its properties to say something, at least approximately, about random walks. Therefore, let

$$S_n = \sum_{k=1}^{n} X_k$$

where $X_1, X_2, \ldots$ are defined in (8.16), be a symmetric random walk. As an example, let us consider the probability that the random walk will reach $S_n = 100$ within 1000 steps.

If we let $\Delta t = 0.001$ and $\Delta x = \sqrt{\Delta t} = \sqrt{0.001}$, by proper scaling, we see that the event that $S_n$ reaches 100 within 1000 steps is equivalent to $S_n(t)$ defined by (8.17) reaching $100\Delta x = \sqrt{10}$ in the interval [0, 1]. Since $\Delta t$ is comparatively small, we can use (8.20) to conclude that $S_n(t) \overset{d}{\approx} B(t)$, which means that we can express the event approximately as $\{T_{\sqrt{10}} \leq 1\}$ and (8.22) yields

$$P(\max_{0 \leq n \leq 1000} S_n \geq 100) \approx P(T_{\sqrt{10}} \leq 1) = 2(1 - \Phi(3.16)) = 0.0016$$

This is quite unlikely, so let us consider 10,000 steps instead. We could do this by another rescaling where $\Delta t = 10^{-4}$, but it suffices to use the scaling we have and instead look at the time $t = n\Delta t = 10$. Then we get that

$$P(\max_{0 \leq n \leq 10{,}000} S_n \geq 100) \approx P(T_{\sqrt{10}} \leq 10) = 2(1 - \Phi(1)) = 0.32$$

□

Another related quantity of interest is the maximum of a Brownian motion in a fixed interval. Therefore, we define the random variable

$$M_t = \max_{0 \leq s \leq t} B(s)$$

for the standard Brownian motion $B(t)$. This variable has the following distribution.

**Proposition 8.23.** The maximum $M_t$ of a standard Brownian motion in the interval $[0, t]$ has pdf

$$f_{M_t}(x) = \frac{2}{\sqrt{2\pi t}} e^{-x^2/2t} \quad x > 0$$

*Proof.* The cdf of $M_t$ can be written

$$F_{M_t}(x) = P(M_t \leq x) = 1 - P(M_t > x) = 1 - P(T_x < t)$$

since we have to hit the level $x$ in order to exceed it before the time point $t$. Now, Equation (8.22) yields

$$F_{M_t}(x) = 1 - F_{T_x}(t) = 1 - 2\left(1 - \Phi\left(\frac{|x|}{\sqrt{t}}\right)\right) = 2\Phi\left(\frac{x}{\sqrt{t}}\right) + 1$$

since $x > 0$, and taking the derivative with respect to $x$ gives us the pdf

$$f_{M_t}(x) = 2\phi\left(\frac{x}{\sqrt{t}}\right) \times \frac{1}{\sqrt{t}} = \frac{2}{\sqrt{2\pi t}} e^{-x^2/2t}$$

■

Since $t$ is fixed and $x$ is variable, this distribution is easier to understand. Essentially, it consists of the positive part of a normal distribution with mean 0 and variance $t$, which among other things means that the mean and variance can be calculated analytically (see Problem 66).

### 8.7.2 Variations of the Brownian Motion

The standard Brownian motion $B(t)$ is quite interesting in itself, but it is also used as a component in other, more realistic processes applied in various fields. In this section, we are going to look at some of the most common variations of $B(t)$.

***Example 8.44 (Brownian Motion with Drift).*** Sometimes, it is not enough to look at a Brownian motion with zero mean but rather a process that has a tendency to drift in a particular direction. We say that a Brownian motion with drift parameter $\mu$ and variance $\sigma^2$ is defined as

$$X(t) = \mu t + \sigma B(t)$$

For a fixed $t$, we get that $X(t)$ is normally distributed with mean $\mu t$ and variance $\sigma^2 t$. This means that the mean is increasing (or decreasing for $\mu < 0$) toward infinity, but if the variance also increases toward infinity, can we really be sure that the process will have a tendency to drift upward? This is indeed the case, which can be seen by considering the limit of

$$P(X(t) > a) = 1 - \Phi\left(\frac{a - \mu t}{\sigma\sqrt{t}}\right) = 1 - \Phi\left(\frac{a}{\sigma}\frac{1}{\sqrt{t}} - \frac{\mu}{\sigma}\sqrt{t}\right)$$

as $t \to \infty$. If $\mu > 0$, we see that the argument of $\Phi$ above tends to $-\infty$, which means that $P(X(t) > a) \to 1$ for any $a$ showing that $X(t)$ will eventually exceed any finite level. □

***Example 8.45 (Geometric Brownian Motion).*** Let $X(t)$ be Brownian motion with drift as defined in Example 8.44. Geometric Brownian motion is then defined as

$$Y(t) = e^{X(t)} = e^{\mu t + \sigma B(t)}$$

Since ordinary Brownian motion can be characterized as a sum of independent normal increments, this gives us a process that can handle products of independent increments. Furthermore, since $X(t)$ is normally distributed for any fixed $t$, we get that $Y(t)$ follows a lognormal distribution (see Section 2.8.1). This means that we can get the mean and variance as

$$E[Y(t)] = e^{\mu t + \sigma^2 t/2} = e^{(\mu + \sigma^2/2)t} \tag{8.23}$$
$$\mathrm{Var}[Y(t) = e^{2\mu t + \sigma^2 t}\left(e^{\sigma^2 t} - 1\right)$$

It is interesting to note that we can have an increasing mean even if $\mu$ is negative as long as $\mu > -\sigma^2/2$. It also holds that we can have an increasing variance if $\mu > -\sigma^2$. This yields an interesting interval $-\sigma^2 < \mu < -\sigma^2/2$ where we get a process whose mean decreases exponentially to zero but whose variance increases exponentially to infinity. Basically, it means that the drift downward is not strong enough to prevent brief excursions upward and the exponential functions amplifies the effect rather strongly.

One of the most common applications of the geometric Brownian motion is to use it as a model for stock prices and even whole stock exchange indices. Since price fluctuations are relative to the stock, we get a natural situation where increments are multiplicative rather than additive. In this context, $\mu$ is usually called the *expected logarithmic return* and $\sigma$ the *volatility* of the stock. The *risk* is usually interpreted as the probability that the stock will decrease in value over a fixed time period, that is,

$$P(Y(t) < Y(0) = 1) = P(X(t) < 0) = \Phi\left(\frac{0 - \mu t}{\sigma\sqrt{t}}\right) = \Phi\left(-\frac{\mu}{\sigma}\sqrt{t}\right)$$

This shows that in order to minimize the risk, we should choose a stock with high expected return but low volatility. This is unfortunately not always possible in reality, stocks with high expected return often suffer from high volatility and vice versa. On the other hand, a high volatility also means a higher expected value, as shown in (8.23), so it might still be worth to accept the higher risk. The geometric Brownian motion has also proved to be a valuable tool in the field of *option pricing*, but we will not go further into that area here. □

***Example 8.46 (Brownian Bridge).*** In some situations, we may be interested in the properties of a Brownian motion that returns to its starting point. Without loss of generality, we will consider a standard Brownian motion $B(t)$ that returns to 0 at time $t = 1$. Brownian bridges with arbitrary variance $\sigma^2$ and arbitrary time points $t$ can be obtained by appropriate time and space scaling. One way to do that is to condition on the event $\{B(1) = 0\}$, but an equivalent and more convenient construction is

$$B^\circ(t) = B(t) - tB(1)$$

The Brownian bridge $B^\circ(t)$ is a linear combination of two normally distributed quantities and, hence, also normally distributed. The mean is clearly 0 and to calculate the variance, we first note that $B(t)$ and $B(1) - B(t)$ are independent and normally distributed increments, which yields

$$\begin{aligned}\mathrm{Var}[B^{\circ}(t)] &= \mathrm{Var}[B(t) - tB(1)] = \mathrm{Var}[(1-t)B(t) - t(B(1) - B(t))]\\ &= (1-t)^2\mathrm{Var}[B(t)] + t^2\mathrm{Var}[B(1) - B(t)]\\ &= (1-t)^2 t + t^2(1-t) = t(1-t)\end{aligned}$$

One of the most important applications of the Brownian bridge is as a large sample approximation of the empirical distribution function

$$\widehat{F}_n(x) = \frac{1}{n}\sum_{k=1}^{n} I_{\{X_k \le x\}}$$

introduced in Section 6.9.1. We note that, for a fixed $x$, the total number $Y$ of events $\{X_k \le x\}$ that occur is binomially distributed with parameters $n$ and $p = F(x)$. Then, the central limit theorem implies that

$$\frac{Y - np}{\sqrt{np(1-p)}} \stackrel{d}{\to} N(0, 1)$$

as $n \to \infty$ and, since $\widehat{F}_n(x) = Y/n$, we get

$$\sqrt{n}(\widehat{F}_n(x) - F(x)) \stackrel{d}{\to} N(0, F(x)(1 - F(x))) \stackrel{d}{=} B^{\circ}(F(x))$$

Now, this result holds pointwise for every fixed $x$, but can actually be generalized to hold uniformly over the whole interval and is then usually expressed as

$$\sqrt{n}(\widehat{F}(\cdot) - F(\cdot)) \stackrel{d}{\to} B^{\circ}(F(\cdot)) \tag{8.24}$$

as $n \to \infty$.[13] To show this requires more elaborate mathematical tools than we have time to go into here. Suffice it to say that it can be used to derive distributions of more complex quantities like, for instance, the test statistic $D_n = \max_x |\widehat{F}_n(x) - F_0(x)|$ for the one-sample Kolmogorov–Smirnov test introduced in Section 6.9.1. Then (8.24) implies that

$$\sqrt{n}D_n \stackrel{d}{\to} \max_x |B^{\circ}(F_0(x))|$$

which in turn can be used to derive (6.7). □

## PROBLEMS

### Section 8.2. Discrete-Time Markov Chains

1. The weather at a coastal resort is classified each day simply as "sunny" or "rainy." A sunny day is followed by another sunny day with probability 0.9,

[13] This result is usually called *Donsker's Theorem.*

and a rainy day is followed by another rainy day with probability 0.3. **(a)** Describe this as a Markov chain. **(b)** If Friday is sunny, what is the probability that Sunday is also sunny? **(c)** If Friday is sunny, what is the probability that both Saturday and Sunday are sunny?

2. At another resort, it is known that the probability that any two consecutive days are both sunny is 0.7 and that the other three combinations are equally likely. Find the transition probabilities.
3. A machine produces electronic components that may come out defective and the process is such that defective components tend to come in clusters. A defective component is followed by another defective component with probability 0.3, whereas a nondefective component is followed by a defective component with probability 0.01. Describe this as a Markov chain and find the long-term proportion of defective components.
4. An insurance company classifies its auto insurance policyholders in the categories "high," "intermediate," or "low" risk. In any given year, a policyholder has no accidents with probability 0.6, one accident with probability 0.2, two accidents with probability 0.1, and more than two accidents with probability 0.1. If you have no accidents, you are moved down one risk category; if you have one, you stay where you are; if you have two accidents, you move up one category; and if you have more than two, you always move to high risk. **(a)** Describe the sequence of moves between categories of a policyholder as a Markov chain. **(b)** If you start as a low-risk customer, how many years can you expect to stay there? **(c)** How many years pass on average between two consecutive visits in the high-risk category?
5. Consider the ON/OFF system from Example 8.4. Let $X_n$ be the state after $n$ steps, and define $Y_n = (X_n, X_{n+1})$. Show that $\{Y_n\}$ is a Markov chain on the state space $\{0, 1\} \times \{0, 1\}$, find its transition matrix and stationary distribution.
6. Suppose that state $i$ is transient and that $i \to j$. Can $j$ be recurrent?
7. Consider the state space $S = \{1, 2, \ldots, n\}$. Describe a Markov chain on $S$ that has only one recurrent state.
8. Consider an irreducible Markov chain on a finite state space, such that the transition matrix is symmetric ($p_{ij} = p_{ji}$ for all $i, j \in S$). Find the stationary distribution.
9. Markov chains are named for Russian mathematician A. A. Markov, who in the early twentieth century examined the sequence of vowels and consonants in the 1833 poem *Eugene Onegin* by Alexander Pushkin. He empirically verified the Markov property and found that a vowel was followed by a consonant 87% of the time and a consonant was followed by a vowel 66% of the time. **(a)** Give the transition graph and the transition matrix. **(b)** If the first letter is a vowel, what is the probability that the third is also a vowel? **(c)** What are the proportions of vowels and consonants in the text?
10. A text is such that a vowel is followed by a consonant 80% of the time and a consonant is followed by a vowel 50% of the time. In the following cases, how

should you guess in order to maximize your probability to guess correctly: **(a)** a letter is chosen at random, **(b)** a letter is chosen at random and the next letter in the text is recorded, **(c)** five letters are chosen at random with replacement, and **(d)** a sequence of five consecutive letters is chosen at random?

11. Consider a text composed of consonants, vowels, blank spaces, and punctuation marks. When a letter is followed by another letter, which happens 80% of the time, the probabilities are as in the previous problem. If a letter is not followed by a letter, it is followed by a blank space 90% of the time. A punctuation mark is always followed by a blank space, and a blank space is equally likely to be followed by a vowel or a consonant. **(a)** State the transition matrix and find the stationary distribution. **(b)** If a symbol is chosen at random and turns out to be a punctuation mark, what is the expected number of blank spaces before the next punctuation mark? **(c)** If this is a literary text in English, what in the model do you find unrealistic?

12. Customers arrive at an ATM where there is room for three customers to wait in line. Customers arrive alone with probability $\frac{2}{3}$ and in pairs with probability $\frac{1}{3}$ (but only one can be served at a time). If both cannot join, they both leave. Call a completed service or an arrival an "event" and let the state be the number of customers in the system (serviced and waiting) immediately after an event. Suppose that an event is equally likely to be an arrival or a completed service. **(a)** State the transition graph and transition matrix and find the stationary distribution. **(b)** If a customer arrives, what is the probability that he finds the system empty? Full? **(c)** If the system is empty, the time until it is empty again is called a "busy period." During a busy period, what is the expected number of times that the system is full?

13. Show that a limit distribution is a stationary distribution. The case of finite $S$ is easier, so you may assume this.

14. Consider the Markov chain with the following transition graph:

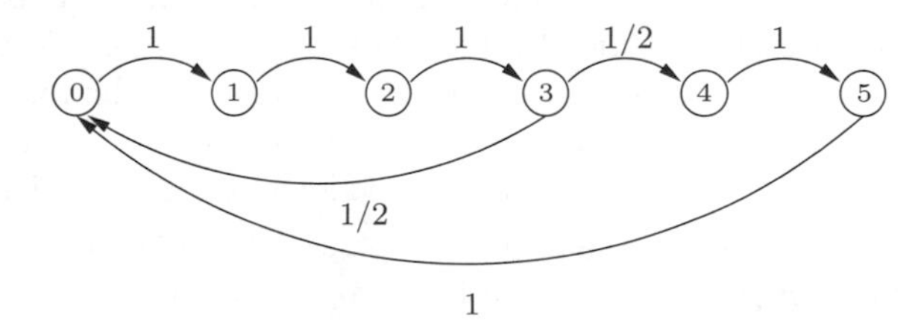


**(a)** What is the smallest number of steps (excluding 0) in which a state can reach itself? **(b)** What is the period of the chain? **(c)** Find the stationary distribution. Is it the limit distribution?

15. Consider the success run chain in Example 8.16. Suppose that the chain has been running for a while and is in state 10. **(a)** What is the expected number of steps until the chain is back at state 10? **(b)** What is the expected number of times the chain visits state 9 before it is back at 10?

16. Consider a version of the success run chain in Example 8.16 where we disregard sequences of consecutive tails, in the sense that, for example $T$, $TT$, $TTT$, and so on, all simply count as $T$. Describe this as a Markov chain and examine it in terms of irreducibility, recurrence, and periodicity. Find the stationary distribution and compare with Example 8.16. Is it the limit distribution?
17. *Reversibility.* Consider an ergodic Markov chain, observed at a late timepoint $n$. If we look at the chain *backward*, we have the backward transition probability $q_{ij} = P(X_{n-1} = j|X_n = i)$. **(a)** Express $q_{ij}$ in terms of the forward transition probabilities and the stationary distribution $\pi$. **(b)** If the forward and backward transition probabilities are equal, the chain is called *reversible*. Show that this occurs if and only if $\pi_i p_{ij} = \pi_j p_{ji}$ for all states $i, j$ (this identity is usually taken as the definition of reversibility). **(c)** Show that if a probability distribution $\boldsymbol{\pi}$ satisfies the equation $\pi_i p_{ij} = \pi_j p_{ji}$ for all $i, j$, then $\boldsymbol{\pi}$ is stationary.
18. The intuition behind reversibility is that if we are given a sequence of consecutive states under stationary conditions, there is no way to decide whether the states are given in forward or backward time. Consider the ON/OFF system in Example 8.4, use the definition in the previous problem to show that it is reversible, and explain intuitively.
19. For which values of $p$ is the following matrix the transition matrix of a reversible Markov chain? Explain intuitively.

$$P = \begin{pmatrix} 0 & p & 1-p \\ 1-p & 0 & p \\ p & 1-p & 0 \end{pmatrix}$$

20. *Ehrenfest model of diffusion.* Consider two containers containing a total of $N$ gas molecules, connected by a narrow aperture. Each time unit, one of the $N$ molecules is chosen at random to pass through the aperture from one container to the other. Let $X_n$ be the number of molecules in the first container. **(a)** Find the transition probabilities for the Markov chain $\{X_n\}$. **(b)** Argue intuitively why the chain is reversible and why the stationary distribution is a certain binomial distribution. Then use Problem 17 to show that it is indeed the stationary distribution. **(c)** Is the stationary distribution also the limit distribution?
21. Consider an irreducible and positive recurrent Markov chain with stationary distribution $\boldsymbol{\pi}$ and let $g : S \to R$ be a real-valued function on the state space. It can be shown that

$$\frac{1}{n}\sum_{k=1}^{n} g(X_k) \xrightarrow{P} \sum_{j \in S} g(j)\pi_j$$

for any initial distribution, where we recall convergence in probability from Section 4.2. This result is reminiscent of the law of large numbers, but the summands are not i.i.d. We have mentioned that the interpretation of the stationary distribution is the long-term proportion of time spent in each state. Show how a particular choice of the function $g$ above gives this interpretation (note that we do not assume aperiodicity).

### Section 8.3. Random Walks and Branching Processes

22. Consider the symmetric simple random walk at two timepoints: $n$ and $n+m$. Find $\rho(S_n, S_{n+m})$. What happens as $m \to \infty$ for fixed $n$ and as $n \to \infty$ for fixed $m$? Explain intuitively.
23. Consider the simple random walk with $p \neq \frac{1}{2}$. Use the law of large numbers to argue that $S_n$ goes to $-\infty$ if $p < \frac{1}{2}$ and to $\infty$ if $p > \frac{1}{2}$.
24. Consider a symmetric simple random walk with *reflecting barriers* 0 and $a$, in the sense that $p_{0,0} = p_{0,1} = \frac{1}{2}$ and $p_{a,a-1} = p_{a,a} = \frac{1}{2}$. **(a)** Describe this as a Markov chain and find its stationary distribution. Is it the limit distribution? **(b)** If the walk starts in 0, what is the expected number of steps until it is back? **(c)** Suppose instead that reflection is immediate, so that $p_{0,1} = 1$ and $p_{a,a-1} = 1$, everything else being the same. Describe the Markov chain, find its stationary distribution $\boldsymbol{\pi}$, and compare it with (a). Explain the difference. Is $\boldsymbol{\pi}$ the limit distribution?
25. Consider a variant of the simple random walk where the walk takes a step up with probability $p$, down with probability $q$, or stays where it is with probability $r$, where $p+q+r=1$. Let the walk start in 0, and let $\tau_1$ be the time of the first visit to 1. Find $P_0(\tau_1 < \infty)$ and $E_0[\tau_1]$.
26. Consider the simple random walk starting in 0 and let $\tau_r$ be the time of the first visit to state $r$, where $r \geq 1$. Find the expected value of $\tau_r$ if $p > \frac{1}{2}$.
27. Consider the simple random walk with $p \neq \frac{1}{2}$, starting in 0 and let

$$\tau_0 = \min\{n \geq 1 : S_n = 0\}$$

the time of the first return to 0. Use Corollary 8.5 to show that $P_0(\tau_0 < \infty) = 2\min(p, 1-p)$.

28. Consider the simple random walk with $p > \frac{1}{2}$ started in state 1. By Corollary 8.5 "reversed," the probability that the walk ever visits 0 is $(1-p)/p$. Now let the initial state $S_0$ be random, chosen according to a distribution on $\{0, 1, \ldots\}$ that has pgf $G$. **(a)** Show that the probability that 0 is ever visited (which could occur in step 0 if $S_0 = 0$) is $G((1-p)/p)$. **(b)** Now instead consider the probability that 0 is ever visited at step 1 or later. Show that this equals $G((1-p)/p) - 2p + 1$. **(c)** Let $p = \frac{2}{3}$ and $S_0 \sim \text{Poi}(1)$. Compute the probabilities in (a) and (b) and also compare with the corresponding probability if $S_0 \equiv 1$.
29. Consider a three-dimensional random walk $\mathbf{S}_n$ where in each step, one of the six neighbors along the axes is chosen with probability $\frac{1}{6}$ each. Let the walk start in the origin and show that

$$P(\mathbf{S}_{2n} = (0,0,0)) = \left(\frac{1}{6}\right)^{2n} \sum_{i+j+k=n} \frac{(2n)!}{(i!j!k!)^2}$$

and use Stirling's formula to conclude that the walk is transient.

30. Consider a branching process with mean number of offspring $\mu$, letting $Y_n$ be the total number of individuals up to and including the $n$th generation and letting $Y$ be the total number of individuals ever born. **(a)** For what values of $\mu$ is $Y$ finite? **(b)** Express $Y_n$ in terms of $Z_0, \ldots, Z_n$ and find $E[Y_n]$. What happens as $n \to \infty$? In particular, compare the two cases $\mu = 1$ and $\mu < 1$.
31. Consider a branching process where the offspring distribution is given by

$$P(X = k) = \frac{1}{3}\left(\frac{2}{3}\right)^k, \quad k = 0, 1, 2, \ldots$$

(geometric distribution including 0). Find **(a)** $E[Z_n]$, **(b)** $P(Z_2 = 0)$, and **(c)** the extinction probability $q$.
32. *Branching with immigration.* Consider a branching process where the offspring distribution is $\{p_0, p_1, \ldots\}$, with pgf $G(s)$ and mean $\mu$. Suppose that in each generation there is immigration into the population according to a sequence of i.i.d. random variables $Y_0, Y_1, \ldots$ with range $\{0, 1, 2, \ldots\}$ and the population is thus started by the first nonzero $Y_k$. Let the $Y_k$ have distribution $\{q_0, q_1, \ldots\}$, pgf $H(s)$, and mean $\nu$, and let $Z_n$ be the $n$th-generation size. **(a)** Show that $P(Z_1 = 0) = q_0 H(p_0)$. **(b)** Show that $Z_n$ has pgf given by $\prod_{j=0}^{n} H(G_{n-j}(s))$, where $G_{n-j}$ is the pgf of the $(n-j)$th generation in a branching process without immigration. Use this to find an expression for $E[Z_n]$. **(c)** Suppose that the $Y_k$ are Poisson with mean $\lambda$ and the offspring distribution is $p_0 = 1 - p$, $p_1 = p$. What is the distribution of $Z_n$?

## Section 8.4. Continuous-Time Markov Chains

33. Consider a continuous-time Markov chain where state $i$ is absorbing. How should the $i$th row of the generator $G$ be defined?
34. Consider the ON/OFF system in Example 8.21. State the backward and forward equations and solve the forward equations. (Why are these easier to solve than the backward equations?) What happens when $t \to \infty$?
35. Birds arrive at four bird feeders according to a Poisson process with rate one bird per minute. If all feeders are occupied, an arriving bird leaves, but otherwise it occupies a feeder and eats for a time that has an exponential distribution with mean 1 min. Consider this as a Markov chain where a "state" is the number of occupied feeders. The rate diagram is given below. **(a)** Explain the rates in the diagram. **(b)** Find the generator $G$. **(c)** If three feeders are occupied, what is the expected time until this changes? **(d)** If all feeders are occupied, what is the probability that a bird arrives before a feeder becomes free?

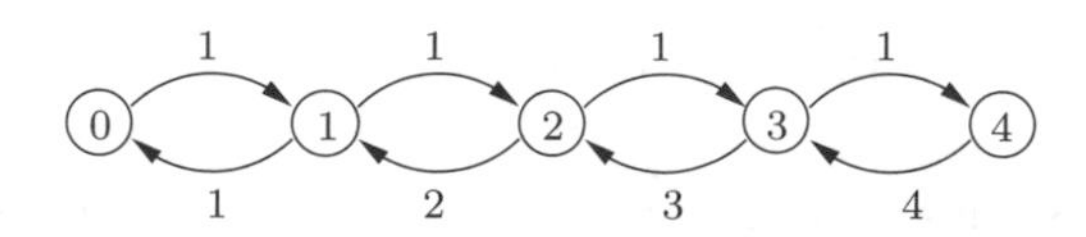

36. Consider a continuous-time Markov chain that allows jumps from states to themselves, after an exponentially distributed time. Although this sounds more general than our construction, it is not. Explain! *Hint:* Problem 147 in Chapter 3.

37. Consider a continuous-time Markov chain whose jump chain is the simple random walk with reflecting barriers 0 and $m$ from Problem 24(c). Suppose that the holding times in states 0 and $m$ are $\exp(a)$ and in all other states $\exp(b)$. **(a)** Describe this in a rate diagram and give the generator $G$. **(b)** Find the stationary distribution. For which values of $a$ and $b$ is it uniform? Compare with the stationary distribution for the jump chain.

38. An *invariant measure* for a discrete-time Markov chain is a nonnegative vector $\boldsymbol{\nu}$ such that $\boldsymbol{\nu} P = \boldsymbol{\nu}$. Thus, an invariant measure is more general than a stationary distribution since its entries need not sum to one. **(a)** Let $\boldsymbol{\nu}$ be invariant and suppose that $0 < \sum_{j \in S} \nu_j < \infty$. Show that a stationary distribution exists. **(b)** Let $\{X(t)\}$ be a continuous-time Markov chain with stationary distribution $\boldsymbol{\pi}$, and let its jump chain $\{X_n\}$ have invariant measure $\boldsymbol{\nu}$. Show that

$$\pi_k = \frac{c\nu_k}{\lambda(k)}, \quad k \in S$$

where $\lambda(k)$ is the holding-time parameter in state $k$ and $c$ is a constant (show that $\boldsymbol{\pi} G = 0$ and explain the role of the constant $c$). Note that this gives an alternative to solving $\boldsymbol{\pi} G = 0$ in order to find $\boldsymbol{\pi}$. **(c)** Now let the jump chain $\{X_n\}$ be the simple symmetric random walk. Show that $\boldsymbol{\nu}$ defined by $\nu_k = 1$ for all $k$ is invariant for $\{X_n\}$. **(d)** Let $\lambda(0) = 1$ and $\lambda(k) = k^2$ for $k \neq 0$, and show that $\{X(t)\}$ has a stationary distribution but that the jump chain $\{X_n\}$ does not. This shows that the jump chain is null recurrent but the continuous-time chain is positive recurrent.

39. Consider a continuous-time Markov chain $\{X(t)\}$ whose jump chain is the success run chain from Problem 16. Give a condition on the holding time parameters $\lambda(k), k = 0, 1, \ldots$ guaranteeing that $\{X(t)\}$ has a stationary distribution (remember the previous problem).

40. Consider a linear birth–death process where the individual birth rate is $\lambda = 1$, the individual death rate is $\mu = 3$, and there is constant immigration into the population according to a Poisson process with rate $\alpha$. **(a)** State the rate diagram and the generator. **(b)** Suppose that there are 10 individuals in the population. What is the probability that the population size increases to 11 before it decreases to 9? **(c)** Suppose that $\alpha = 1$ and that the population just became extinct. What is the expected time until it becomes extinct again?

41. In the previous problem suppose that an immigrating individual joins the population only if it is extinct, and otherwise leaves. Find the rate diagram, generator, and stationary distribution.

42. Consider state 0 in a birth–death process with stationary distribution $\boldsymbol{\pi}$. Under stationary conditions, we ought to have the *balance equation* $\pi_0 \lambda_0 = \pi_1 \mu_1$ ("rate in equals rate out"), which is also precisely the first equation of $\boldsymbol{\pi} G = 0$.

**(a)** Suggest how to formulate balance equations for any three states $k-1, k$, and $k+1$, and show that these equations are the same as $\pi G = 0$. **(b)** Describe how the equation $\pi G = 0$ has an interpretation as balance equations for any continuous-time Markov chain, not just birth–death processes.

43. Consider an $M/M/1/r$ queue in equilibrium where $r = 5$ and the service rate equals the arrival rate. **(a)** What is the proportion of lost customers? **(b)** How does this change if the service rate is doubled?
44. Consider a queueing system where there is one server and no room to wait in line (i.e., an $M/M/1/1$ queue). Further suppose that the arrival rate $\lambda$ and the service rate $\mu$ are equal. Under stationary conditions, find the proportion of customers who are lost **(a)** in this system, **(b)** if the service rate is doubled, **(c)** if one customer can wait in line, and **(d)** if a second server is added.
45. Consider the following queueing system. Customers arrive in pairs according to a Poisson process with rate $\lambda = 1$ customer pair/min. There is one server and room for two customers to wait in line. Service times are exponential with mean 30 s. If there is not room for both arriving customers, they both leave. **(a)** Describe the system in a rate diagram and find the stationary distribution. **(b)** Now suppose that pairs may split up. If there is not room for both, then with probability $\frac{1}{2}$ they both leave and with probability $\frac{1}{2}$ one stays and the other leaves. Do (a) again under these assumptions.
46. Customer groups arrive to a service station according to a Poisson process with rate $\lambda$ groups/min. With probability $p$, such a group consists of a single individual, and with probability $1-p$, it consists of a pair. There is a single server and room for two to wait in line. Service times are exponential with rate $\mu$. If a pair arrives and they cannot both join, they both leave. **(a)** Give the state space and describe the system in a rate diagram. **(b)** Suppose $\lambda = \mu$ and $p = \frac{1}{2}$. Find the stationary distribution $(\pi_0, \pi_1, \pi_2, \pi_3)$.
47. Phone calls arrive at a company according to two independent Poisson processes, one of female callers with rate 2 and one of male callers with rate 1 (calls/min). There is one server and room for one to wait in line. If the server is busy, a female caller stays to wait in line with probability 0.8; a male caller, with probability 0.5. Service times are i.i.d. exponential with mean length 2 min. Let the state be the number of customers in the system. **(a)** Describe the system in a rate diagram and find the stationary distribution. **(b)** What proportion of callers is lost?
48. Consider an $M/M/1$ queue with arrival rate $\lambda$ and service rate $\mu$ and where an arriving customer who finds $k$ individuals in the system joins with probability $1/(k+1)$. When does a stationary distribution exist and what is it?
49. *Reneging*. Consider an $M/M/1$ queue with arrival rate $\lambda$ and service rate $\mu$, where a customer who is waiting in line *reneges* and leaves the line after a time that is $\exp(\nu)$ (unless service has started), independent of the queue length. Describe this system in a rate diagram and state the generator.
50. Consider an $M/M/1$ queue in equilibrium and let $W$ be the waiting time of an arriving customer. Find the cdf of $W$. What type of distribution is this?

*Hint:* First find $P(W = 0)$ and then compute $P(W > x)$ by conditioning on the number of customers in the system.

51. Consider an $M/M/1$ queue in equilibrium and let $T$ be the total time an arriving customer spends in the system. Find the distribution of $T$ (condition on the number of customers in the system at arrival).

## Section 8.5. Martingales

52. Let $X_1, X_2, \ldots$ be positive i.i.d. random variables with mean $\mu$ and let $P_n = X_1 \times \ldots \times X_n$ for $n = 1, 2, \ldots$. Find a function $g(x)$ so that $Y_n = g(P_n)$ becomes a martingale with respect to $X_1, X_2, \ldots$.
53. Let $X_1 = 1$ and $X_{n+1}$ be uniformly distributed on the interval $[0, X_n]$ for $n = 1, 2, \ldots$. Find a function $g(x)$ so that $Y_n = g(X_n)$ becomes a martingale with respect to $X_1, X_2, \ldots$.
54. Consider the branching process in Example 8.35. Show that $Y_n = q^{Z_n}$ for $n = 1, 2, \ldots$ is a martingale with respect to $Z_1, Z_2, \ldots$.
55. *Polya's urn model.* Let us assume that we have an urn containing one white ball and one black ball initially. A ball is selected completely at random from the urn and put back together with a ball of the same color. Repeat this operation infinitely and let $Y_n$ denote the ratio of white balls in the urn after $n$ steps. Show that $Y_1, Y_2, \ldots$ is a martingale with respect to itself and use Proposition 8.14 to conclude that $Y_n \to Y$ with probability 1. What distribution does $Y$ have?
56. Consider a simple random walk with $p > \frac{1}{2}$. Show that $Y_n = S_n - (p - q)n$ is a martingale with respect to $S_1, S_2, \ldots$ and use Proposition 8.15 to prove Corollary 8.7.

## Section 8.6. Renewal Processes

57. Consider a renewal process $N(t)$ where $T_1, T_2, \ldots$ are i.i.d. and uniformly distributed on $[0, 1]$. Derive the renewal function $m(t)$ for $t \leq 1$.
58. Consider a lamp where the lightbulbs are replaced either when they burn out or when they have burned for 336 h. The lifelengths of lightbulbs are assumed to be i.i.d. and exponentially distributed with mean 300 h. **(a)** How often are the lightbulbs replaced in the long run? **(b)** What is the probability that a supply of 50 lightbulbs will last for a year (8760 h)?
59. *M/G/1/1.*[14] Assume that customers arrive to a service station with one server according to a Poisson process with rate $\lambda$. When an arriving customer finds the service station empty, he enters and starts being served, whereas if the server is busy, he leaves and never returns. **(a)** If we denote the mean service time $\mu_S$, at what rate do customers enter the service station? **(b)** What proportion of customers is actually served by the service station?

[14]The $G$ stands for *General*, denoting that we allow any service time distribution.

60. Let us assume that $\lambda = 5$ customers per minute in the previous problem and that the mean and variance of the service times are $\mu_S = 0.25$ and $\sigma_S^2 = 0.1$, respectively. What is the probability that at least 120 customers will be served during 1 h?
61. Find the distribution $\widetilde{F}(t)$ of the initial interarrival time in a delayed renewal process if the subsequent interarrival times are **(a)** exponentially distributed (delayed Poisson process), **(b)** uniformly distributed on $[0, 1]$, and **(c)** Gamma distributed with $\alpha = 2$

## Section 8.7. Brownian Motion

62. Calculate **(a)** $\text{Cov}[B(t), B(s)]$ and **(b)** $\text{Cov}[B^\circ(t), B^\circ(s)]$.
63. Derive the distribution of $B(t) + B(s)$ for $t \neq s$.
64. Calculate $E[B(t)B(s)]$.
65. Show that $B^*(t) = tB(1/t)$ is a standard Brownian motion.
66. Calculate the mean and variance of the maximum $M_t$ of a standard Brownian motion in the interval $[0, t]$.
67. Derive the conditional distribution of $B(t)$ for $t_1 < t < t_2$ conditioned on the event $\{B(t_1) = x_1, B(t_2) = x_2\}$
68. Consider a Brownian motion with drift parameter $\mu$ and variance $\sigma^2$. Derive the distribution of $B(t)$ conditioned on the event $\{B(s) = c\}$ for **(a)** $t > s$ and **(b)** $t < s$.
69. The discounted value of a share of stock can be described as a geometric Brownian motion with drift parameter $\mu = 0$ and variance $\sigma^2 = 0.2$. The time unit is 1 year. Let us assume that we decide to sell the share when it has increased by 20%. What is the probability that we sell the share within 6 months?
70. *Two-dimensional Brownian motion.* Let $B_x(t)$ and $B_y(t)$ be two independent standard Brownian motion describing the horizontal and vertical location of a particle moving in two dimensions. Derive the pdf of the distance $R(t) = \sqrt{B_x(t)^2 + B_y(t)^2}$ from the origin after $t$ time units.

# APPENDIX A

# TABLES

**TABLE A.1 Values of the cdf $\Phi(x)$ of the Standard Normal Distribution [e.g., $\Phi(1.41) = 0.921$]**

| $x$ | 0 | 1 | 2 | 3 | 4 | 5 | 6 | 7 | 8 | 9 |
|---|---|---|---|---|---|---|---|---|---|---|
| 0.0 | 0.500 | 0.504 | 0.508 | 0.512 | 0.516 | 0.520 | 0.524 | 0.528 | 0.532 | 0.536 |
| 0.1 | 0.540 | 0.544 | 0.548 | 0.552 | 0.556 | 0.560 | 0.564 | 0.568 | 0.571 | 0.575 |
| 0.2 | 0.579 | 0.583 | 0.587 | 0.591 | 0.595 | 0.599 | 0.603 | 0.606 | 0.610 | 0.614 |
| 0.3 | 0.618 | 0.622 | 0.626 | 0.629 | 0.633 | 0.637 | 0.641 | 0.644 | 0.648 | 0.652 |
| 0.4 | 0.655 | 0.659 | 0.663 | 0.666 | 0.670 | 0.674 | 0.677 | 0.681 | 0.684 | 0.688 |
| 0.5 | 0.692 | 0.695 | 0.698 | 0.702 | 0.705 | 0.709 | 0.712 | 0.716 | 0.719 | 0.722 |
| 0.6 | 0.726 | 0.729 | 0.732 | 0.736 | 0.739 | 0.742 | 0.745 | 0.749 | 0.752 | 0.755 |
| 0.7 | 0.758 | 0.761 | 0.764 | 0.767 | 0.770 | 0.773 | 0.776 | 0.779 | 0.782 | 0.785 |
| 0.8 | 0.788 | 0.791 | 0.794 | 0.797 | 0.800 | 0.802 | 0.805 | 0.808 | 0.811 | 0.813 |
| 0.9 | 0.816 | 0.819 | 0.821 | 0.824 | 0.826 | 0.829 | 0.832 | 0.834 | 0.836 | 0.839 |
| 1.0 | 0.841 | 0.844 | 0.846 | 0.848 | 0.851 | 0.853 | 0.855 | 0.858 | 0.860 | 0.862 |
| 1.1 | 0.864 | 0.867 | 0.869 | 0.871 | 0.873 | 0.875 | 0.877 | 0.879 | 0.881 | 0.883 |
| 1.2 | 0.885 | 0.887 | 0.889 | 0.891 | 0.892 | 0.894 | 0.896 | 0.898 | 0.900 | 0.902 |
| 1.3 | 0.903 | 0.905 | 0.907 | 0.908 | 0.910 | 0.912 | 0.913 | 0.915 | 0.916 | 0.918 |
| 1.4 | 0.919 | 0.921 | 0.922 | 0.924 | 0.925 | 0.926 | 0.928 | 0.929 | 0.931 | 0.932 |
| 1.5 | 0.933 | 0.934 | 0.936 | 0.937 | 0.938 | 0.939 | 0.941 | 0.942 | 0.943 | 0.944 |
| 1.6 | 0.945 | 0.946 | 0.947 | 0.948 | 0.950 | 0.951 | 0.952 | 0.952 | 0.9545 | 0.954 |

*(Continued)*

*Probability, Statistics, and Stochastic Processes*, Second Edition. Peter Olofsson and Mikael Andersson.

**TABLE A.1 (*Continued*)**

| $x$ | 0 | 1 | 2 | 3 | 4 | 5 | 6 | 7 | 8 | 9 |
|---|---|---|---|---|---|---|---|---|---|---|
| 1.7 | 0.955 | 0.956 | 0.957 | 0.958 | 0.959 | 0.960 | 0.961 | 0.962 | 0.962 | 0.963 |
| 1.8 | 0.964 | 0.965 | 0.966 | 0.966 | 0.967 | 0.968 | 0.969 | 0.969 | 0.970 | 0.971 |
| 1.9 | 0.971 | 0.972 | 0.973 | 0.973 | 0.974 | 0.974 | 0.975 | 0.976 | 0.976 | 0.977 |
| 2.0 | 0.977 | 0.978 | 0.978 | 0.979 | 0.979 | 0.980 | 0.980 | 0.981 | 0.981 | 0.982 |
| 2.1 | 0.982 | 0.983 | 0.983 | 0.983 | 0.984 | 0.984 | 0.985 | 0.985 | 0.985 | 0.986 |
| 2.2 | 0.986 | 0.986 | 0.987 | 0.987 | 0.988 | 0.988 | 0.988 | 0.988 | 0.989 | 0.989 |
| 2.3 | 0.989 | 0.990 | 0.990 | 0.990 | 0.990 | 0.991 | 0.991 | 0.991 | 0.991 | 0.992 |
| 2.4 | 0.992 | 0.992 | 0.992 | 0.992 | 0.993 | 0.993 | 0.993 | 0.993 | 0.993 | 0.994 |
| 2.5 | 0.994 | 0.994 | 0.994 | 0.994 | 0.995 | 0.995 | 0.995 | 0.995 | 0.995 | 0.995 |
| 2.6 | 0.995 | 0.996 | 0.996 | 0.996 | 0.996 | 0.996 | 0.996 | 0.996 | 0.996 | 0.996 |
| 2.7 | 0.996 | 0.997 | 0.997 | 0.997 | 0.997 | 0.997 | 0.997 | 0.997 | 0.997 | 0.997 |
| 2.8 | 0.997 | 0.998 | 0.998 | 0.998 | 0.998 | 0.998 | 0.998 | 0.998 | 0.998 | 0.998 |
| 2.9 | 0.998 | 0.998 | 0.998 | 0.998 | 0.998 | 0.998 | 0.998 | 0.998 | 0.999 | 0.999 |

**TABLE A.2 Values of $\Phi(x)$ Commonly used in Confidence Intervals and Tests, and the Corresponding $x$ Values**

| $\Phi(x)$ | 0.90 | 0.95 | 0.975 | 0.99 | 0.995 |
|---|---|---|---|---|---|
| $x$ | 1.28 | 1.64 | 1.96 | 2.33 | 2.58 |

**TABLE A.3 Percentiles of the $t$ Distribution with DF Degrees of Freedom [e.g., $F_{t_7}(1.89) = 0.95$]**

| DF | 0.95 | 0.975 | 0.99 | 0.995 | DF | 0.95 | 0.975 | 0.99 | 0.995 |
|---|---|---|---|---|---|---|---|---|---|
| 1 | 6.31 | 12.71 | 31.82 | 63.66 | 16 | 1.75 | 2.12 | 2.58 | 2.92 |
| 2 | 2.92 | 4.30 | 6.96 | 9.92 | 17 | 1.74 | 2.11 | 2.58 | 2.90 |
| 3 | 2.35 | 3.18 | 4,54 | 5.84 | 18 | 1.73 | 2.10 | 2.55 | 2.88 |
| 4 | 2.13 | 2.78 | 3.74 | 4.60 | 19 | 1.73 | 2.09 | 2.54 | 2.86 |
| 5 | 2.02 | 2.57 | 3.36 | 4.03 | 20 | 1.72 | 2.09 | 2.53 | 2.85 |
| 6 | 1.94 | 2.45 | 3.14 | 3.71 | 21 | 1.72 | 2.08 | 2.52 | 2.83 |
| 7 | 1.89 | 2.36 | 3.00 | 3.50 | 22 | 1.72 | 2.07 | 2.51 | 2.82 |
| 8 | 1.86 | 2.31 | 2.90 | 3.36 | 23 | 1.71 | 2.07 | 2.50 | 2.81 |
| 9 | 1.83 | 2.26 | 2.82 | 3.25 | 24 | 1.71 | 2.06 | 2.49 | 2.80 |
| 10 | 1.81 | 2.23 | 2.76 | 3.17 | 25 | 1.71 | 2.06 | 2.49 | 2.79 |
| 11 | 1.80 | 2.20 | 2.72 | 3.11 | 26 | 1.71 | 2.06 | 2.48 | 2.78 |
| 12 | 1.78 | 2.18 | 2.68 | 3.05 | 27 | 1.70 | 2.05 | 2.47 | 2.77 |
| 13 | 1.77 | 2.16 | 2.65 | 3.01 | 28 | 1.70 | 2.05 | 2.47 | 2.76 |
| 14 | 1.76 | 2.14 | 2.62 | 2.98 | 29 | 1.70 | 2.05 | 2.46 | 2.76 |
| 15 | 1.75 | 2.13 | 2.60 | 2.95 | 30 | 1.70 | 2.04 | 2.46 | 2.75 |

**TABLE A.4 Percentiles of the Chi-square Distribution with DF Degrees of Freedom [e.g., $F_{\chi^2_{20}}(10.85) = 0.05$]**

| DF | 0.025 | 0.05 | 0.95 | 0.975 | DF | 0.025 | 0.05 | 0.95 | 0.975 |
|---|---|---|---|---|---|---|---|---|---|
| 1 | 0.001 | 0.004 | 3.84 | 5.02 | 16 | 6.91 | 7.96 | 26.30 | 28.84 |
| 2 | 0.05 | 0.10 | 5.99 | 7.38 | 17 | 7.56 | 8.67 | 27.59 | 30.19 |
| 3 | 0.22 | 0.35 | 7.82 | 9.34 | 18 | 8.23 | 9.39 | 28.87 | 31.53 |
| 4 | 0.48 | 0.71 | 9.49 | 11.14 | 19 | 8.91 | 10.12 | 30.14 | 32.85 |
| 5 | 0.83 | 1.14 | 11.07 | 12.83 | 20 | 9.59 | 10.85 | 31.41 | 34.17 |
| 6 | 1.24 | 1.64 | 12.59 | 14.45 | 21 | 10.28 | 11.60 | 32.67 | 35.48 |
| 7 | 1.69 | 2.17 | 14.07 | 16.01 | 22 | 10.98 | 12.34 | 33.92 | 36.78 |
| 8 | 2.18 | 2.73 | 15.51 | 17.54 | 23 | 11.69 | 13.09 | 35.17 | 38.08 |
| 9 | 2.70 | 3.32 | 19.92 | 19.02 | 24 | 12.40 | 13.85 | 36.42 | 39.36 |
| 10 | 3.25 | 3.94 | 18.31 | 20.48 | 25 | 13.12 | 14.61 | 37.65 | 40.65 |
| 11 | 3.82 | 4.58 | 19.68 | 21.92 | 26 | 13.84 | 15.38 | 38.88 | 41.92 |
| 12 | 4.40 | 5.23 | 21.03 | 23.34 | 27 | 14.57 | 16.15 | 40.11 | 43.19 |
| 13 | 5.01 | 5.89 | 22.36 | 27.74 | 28 | 15.31 | 16.93 | 41.34 | 44.46 |
| 14 | 5.63 | 6.57 | 23.68 | 26.12 | 29 | 16.05 | 17.71 | 42.56 | 45.72 |
| 15 | 6.26 | 7.26 | 25.00 | 27.49 | 30 | 16.79 | 18.49 | 43.77 | 46.98 |

**TABLE A.5 Percentiles of the *F* Distribution with *r* and *s* Degrees of Freedom [e.g., $F_{F_{8,20}}(2.45) = 0.95$]**

| | | | | 2.5% percentile | | | | | |
|---|---|---|---|---|---|---|---|---|---|
| *s* | *r* = 2 | 3 | 4 | 5 | 6 | 7 | 8 | 9 | 10 |
| 2 | 0.026 | 0.062 | 0.094 | 0.119 | 0.138 | 0.153 | 0.165 | 0.175 | 0.183 |
| 3 | 0.026 | 0.065 | 0.100 | 0.129 | 0.152 | 0.170 | 0.185 | 0.197 | 0.207 |
| 4 | 0.025 | 0.066 | 0.104 | 0.135 | 0.161 | 0.181 | 0.198 | 0.212 | 0.224 |
| 5 | 0.025 | 0.067 | 0.107 | 0.140 | 0.167 | 0.189 | 0.208 | 0.223 | 0.236 |
| 6 | 0.025 | 0.068 | 0.109 | 0.143 | 0.172 | 0.195 | 0.215 | 0.231 | 0.246 |
| 7 | 0.025 | 0.068 | 0.110 | 0.146 | 0.176 | 0.200 | 0.221 | 0.238 | 0.253 |
| 8 | 0.025 | 0.069 | 0.111 | 0.148 | 0.179 | 0.204 | 0.226 | 0.244 | 0.259 |
| 9 | 0.025 | 0.069 | 0.112 | 0.150 | 0.181 | 0.207 | 0.230 | 0.248 | 0.265 |
| 10 | 0.025 | 0.069 | 0.113 | 0.151 | 0.183 | 0.210 | 0.233 | 0.252 | 0.269 |
| 12 | 0.025 | 0.070 | 0.114 | 0.153 | 0.186 | 0.214 | 0.238 | 0.259 | 0.276 |
| 15 | 0.025 | 0.070 | 0.116 | 0.156 | 0.190 | 0.219 | 0.244 | 0.265 | 0.284 |
| 16 | 0.025 | 0.070 | 0.116 | 0.156 | 0.191 | 0.220 | 0.245 | 0.267 | 0.286 |
| 18 | 0.025 | 0.070 | 0.116 | 0.157 | 0.192 | 0.222 | 0.248 | 0.270 | 0.290 |
| 20 | 0.025 | 0.071 | 0.117 | 0.158 | 0.193 | 0.224 | 0.250 | 0.273 | 0.293 |
| 21 | 0.025 | 0.071 | 0.117 | 0.158 | 0.194 | 0.225 | 0.251 | 0.274 | 0.294 |
| 24 | 0.025 | 0.071 | 0.117 | 0.159 | 0.195 | 0.227 | 0.253 | 0.277 | 0.297 |
| 25 | 0.025 | 0.071 | 0.118 | 0.160 | 0.196 | 0.227 | 0.254 | 0.278 | 0.298 |
| 27 | 0.025 | 0.071 | 0.118 | 0.160 | 0.197 | 0.228 | 0.255 | 0.279 | 0.300 |
| 28 | 0.025 | 0.071 | 0.118 | 0.160 | 0.197 | 0.228 | 0.256 | 0.280 | 0.301 |
| 30 | 0.025 | 0.071 | 0.118 | 0.161 | 0.197 | 0.229 | 0.257 | 0.281 | 0.302 |

*(Continued)*

**TABLE A.5** (*Continued*)

| 95% percentile | | | | | | | | | |
|---|---|---|---|---|---|---|---|---|---|
| $s$ | $r = 2$ | 3 | 4 | 5 | 6 | 7 | 8 | 9 | 10 |
| 2 | 19.00 | 19.16 | 19.25 | 19.30 | 19.33 | 19.35 | 19.37 | 19.38 | 19.40 |
| 3 | 9.55 | 9.28 | 9.12 | 9.01 | 8.94 | 8.89 | 8.85 | 8.81 | 8.79 |
| 4 | 6.94 | 6.59 | 6.39 | 6.26 | 6.16 | 6.09 | 6.04 | 6.00 | 5.96 |
| 5 | 5.79 | 5.41 | 5.19 | 5.05 | 4.95 | 4.88 | 4.82 | 4.77 | 4.74 |
| 6 | 5.14 | 4.76 | 4.53 | 4.39 | 4.28 | 4.21 | 4.15 | 4.10 | 4.06 |
| 7 | 4.74 | 4.35 | 4.12 | 3.97 | 3.87 | 3.79 | 3.73 | 3.68 | 3.64 |
| 8 | 4.46 | 4.07 | 3.84 | 3.69 | 3.58 | 3.50 | 3.44 | 3.39 | 3.35 |
| 9 | 4.26 | 3.86 | 3.63 | 3.48 | 3.37 | 3.29 | 3.23 | 3.18 | 3.14 |
| 10 | 4.10 | 3.71 | 3.48 | 3.33 | 3.22 | 3.14 | 3.07 | 3.02 | 2.98 |
| 12 | 3.89 | 3.49 | 3.26 | 3.11 | 3.00 | 2.91 | 2.85 | 2.80 | 2.75 |
| 15 | 3.68 | 3.29 | 3.06 | 2.90 | 2.79 | 2.71 | 2.64 | 2.59 | 2.54 |
| 16 | 3.63 | 3.24 | 3.01 | 2.85 | 2.74 | 2.66 | 2.59 | 2.54 | 2.49 |
| 18 | 3.55 | 3.16 | 2.93 | 2.77 | 2.66 | 2.58 | 2.51 | 2.46 | 2.41 |
| 20 | 3.49 | 3.10 | 2.87 | 2.71 | 2.60 | 2.51 | 2.45 | 2.39 | 2.35 |
| 21 | 3.47 | 3.07 | 2.84 | 2.68 | 2.57 | 2.49 | 2.42 | 2.37 | 2.32 |
| 24 | 3.40 | 3.01 | 2.78 | 2.62 | 2.51 | 2.42 | 2.36 | 2.30 | 2.25 |
| 25 | 3.39 | 2.99 | 2.76 | 2.60 | 2.49 | 2.40 | 2.34 | 2.28 | 2.24 |
| 27 | 3.35 | 2.96 | 2.73 | 2.57 | 2.46 | 2.37 | 2.31 | 2.25 | 2.20 |
| 28 | 3.34 | 2.95 | 2.71 | 2.56 | 2.45 | 2.36 | 2.29 | 2.24 | 2.19 |
| 30 | 3.32 | 2.92 | 2.69 | 2.53 | 2.42 | 2.33 | 2.27 | 2.21 | 2.16 |

| 97.5% percentile | | | | | | | | | |
|---|---|---|---|---|---|---|---|---|---|
| $s$ | $r = 2$ | 3 | 4 | 5 | 6 | 7 | 8 | 9 | 10 |
| 2 | 39.00 | 39.17 | 39.25 | 39.30 | 39.33 | 39.36 | 39.37 | 39.39 | 39.40 |
| 3 | 16.04 | 15.44 | 15.10 | 14.88 | 14.73 | 14.62 | 14.54 | 14.47 | 14.42 |
| 4 | 10.65 | 9.98 | 9.60 | 9.36 | 9.20 | 9.07 | 8.98 | 8.90 | 8.84 |
| 5 | 8.43 | 7.76 | 7.39 | 7.15 | 6.98 | 6.85 | 6.76 | 6.68 | 6.62 |
| 6 | 7.26 | 6.60 | 6.23 | 5.99 | 5.82 | 5.70 | 5.60 | 5.52 | 5.46 |
| 7 | 6.54 | 5.89 | 5.52 | 5.29 | 5.12 | 4.99 | 4.90 | 4.82 | 4.76 |
| 8 | 6.06 | 5.42 | 5.05 | 4.82 | 4.65 | 4.53 | 4.43 | 4.36 | 4.30 |
| 9 | 5.71 | 5.08 | 4.72 | 4.48 | 4.32 | 4.20 | 4.10 | 4.03 | 3.96 |
| 10 | 5.46 | 4.83 | 4.47 | 4.24 | 4.07 | 3.95 | 3.85 | 3.78 | 3.72 |
| 12 | 5.10 | 4.47 | 4.12 | 3.89 | 3.73 | 3.61 | 3.51 | 3.44 | 3.37 |
| 15 | 4.77 | 4.15 | 3.80 | 3.58 | 3.41 | 3.29 | 3.20 | 3.12 | 3.06 |
| 16 | 4.69 | 4.08 | 3.73 | 3.50 | 3.34 | 3.22 | 3.12 | 3.05 | 2.99 |
| 18 | 4.56 | 3.95 | 3.61 | 3.38 | 3.22 | 3.10 | 3.01 | 2.93 | 2.87 |
| 20 | 4.46 | 3.86 | 3.51 | 3.29 | 3.13 | 3.01 | 2.91 | 2.84 | 2.77 |
| 21 | 4.42 | 3.82 | 3.48 | 3.25 | 3.09 | 2.97 | 2.87 | 2.80 | 2.73 |
| 24 | 4.32 | 3.72 | 3.38 | 3.15 | 2.99 | 2.87 | 2.78 | 2.70 | 2.64 |
| 25 | 4.29 | 3.69 | 3.35 | 3.13 | 2.97 | 2.85 | 2.75 | 2.68 | 2.61 |
| 27 | 4.24 | 3.65 | 3.31 | 3.08 | 2.92 | 2.80 | 2.71 | 2.63 | 2.57 |
| 28 | 4.22 | 3.63 | 3.29 | 3.06 | 2.90 | 2.78 | 2.69 | 2.61 | 2.55 |
| 30 | 4.18 | 3.59 | 3.25 | 3.03 | 2.87 | 2.75 | 2.65 | 2.57 | 2.51 |

**TABLE A.6 95% Percentiles of the Studentized Range Distribution with $r$ and $s$ Degrees of Freedom [e.g., $F_{R_{8,20}}(4.77) = 0.95$]**

| $s$ | $r = 3$ | 4 | 5 | 6 | 7 | 8 | 9 | 10 |
|---|---|---|---|---|---|---|---|---|
| 2 | 8.33 | 9.80 | 10.88 | 11.73 | 12.43 | 13.03 | 13.54 | 13.99 |
| 3 | 5.91 | 6.82 | 7.50 | 8.04 | 8.48 | 8.85 | 9.18 | 9.46 |
| 4 | 5.04 | 5.76 | 6.29 | 6.71 | 7.05 | 7.35 | 7.60 | 7.83 |
| 5 | 4.60 | 5.22 | 5.67 | 6.03 | 6.33 | 6.58 | 6.80 | 6.99 |
| 6 | 4.34 | 4.90 | 5.30 | 5.63 | 5.90 | 6.12 | 6.32 | 6.49 |
| 7 | 4.16 | 4.68 | 5.06 | 5.36 | 5.61 | 5.82 | 6.00 | 6.16 |
| 8 | 4.04 | 4.53 | 4.89 | 5.17 | 5.40 | 5.60 | 5.77 | 5.92 |
| 9 | 3.95 | 4.41 | 4.76 | 5.02 | 5.24 | 5.43 | 5.59 | 5.74 |
| 10 | 3.88 | 4.33 | 4.65 | 4.91 | 5.12 | 5.30 | 5.46 | 5.60 |
| 12 | 3.77 | 4.20 | 4.51 | 4.75 | 4.95 | 5.12 | 5.27 | 5.39 |
| 15 | 3.67 | 4.08 | 4.37 | 4.59 | 4.78 | 4.94 | 5.08 | 5.20 |
| 16 | 3.65 | 4.05 | 4.33 | 4.56 | 4.74 | 4.90 | 5.03 | 5.15 |
| 18 | 3.61 | 4.00 | 4.28 | 4.49 | 4.67 | 4.82 | 4.96 | 5.07 |
| 20 | 3.58 | 3.96 | 4.23 | 4.45 | 4.62 | 4.77 | 4.90 | 5.01 |
| 21 | 3.56 | 3.94 | 4.21 | 4.42 | 4.60 | 4.74 | 4.87 | 4.98 |
| 24 | 3.53 | 3.90 | 4.17 | 4.37 | 4.54 | 4.68 | 4.81 | 4.92 |
| 25 | 3.52 | 3.89 | 4.15 | 4.36 | 4.53 | 4.67 | 4.79 | 4.90 |
| 27 | 3.51 | 3.87 | 4.13 | 4.33 | 4.50 | 4.64 | 4.76 | 4.86 |
| 28 | 3.50 | 3.86 | 4.12 | 4.32 | 4.49 | 4.62 | 4.74 | 4.85 |
| 30 | 3.49 | 3.85 | 4.10 | 4.30 | 4.46 | 4.60 | 4.72 | 4.82 |

**TABLE A.7 Critical Values $c$ for the Wilcoxon Signed Rank Test, Where $n$ is the Sample Size and $C = n(n+1) - c$ [e.g., if $n = 20$, then $P(W \leq 61) = P(W \geq 149) \approx 0.05$]**

| $n$ | 0.025 | 0.05 | $n(n+1)/2$ | $n$ | 0.025 | 0.05 | $n(n+1)/2$ |
|---|---|---|---|---|---|---|---|
| 5 | 0 | 1 | 15 | 18 | 41 | 48 | 171 |
| 6 | 1 | 3 | 21 | 19 | 47 | 54 | 190 |
| 7 | 3 | 4 | 28 | 20 | 53 | 61 | 210 |
| 8 | 4 | 6 | 36 | 21 | 59 | 68 | 231 |
| 9 | 6 | 9 | 45 | 22 | 67 | 76 | 253 |
| 10 | 9 | 11 | 55 | 23 | 74 | 84 | 276 |
| 11 | 11 | 14 | 66 | 24 | 82 | 92 | 300 |
| 12 | 14 | 18 | 78 | 25 | 90 | 101 | 325 |
| 13 | 18 | 22 | 91 | 26 | 99 | 111 | 351 |
| 14 | 22 | 26 | 105 | 27 | 108 | 120 | 378 |
| 15 | 26 | 31 | 120 | 28 | 117 | 131 | 406 |
| 16 | 30 | 36 | 136 | 29 | 127 | 141 | 435 |
| 17 | 35 | 42 | 153 | 30 | 138 | 152 | 465 |

**TABLE A.8 Critical Values $c$ for the Wilcoxon Rank Sum Test, Where $m$ is the Size of the Smaller Sample and $C = m(m + n + 1) - c$ [e.g., if $m = 4$ and $n = 8$, then $P(W \leq 16) = P(W \geq 36) \approx 0.05$]**

| $n$ | $P(W \leq c)$ | $m = 2$ | 3 | 4 | 5 | 6 | 7 | 8 | 9 | 10 | 11 |
|---|---|---|---|---|---|---|---|---|---|---|---|
| 2 | 0.025 | 3 | | | | | | | | | |
| | 0.05 | 3 | | | | | | | | | |
| 3 | 0.025 | 3 | 3 | | | | | | | | |
| | 0.05 | 6 | 7 | | | | | | | | |
| 4 | 0.025 | 3 | 6 | 11 | | | | | | | |
| | 0.05 | 3 | 7 | 12 | | | | | | | |
| 5 | 0.025 | 3 | 7 | 12 | 18 | | | | | | |
| | 0.05 | 4 | 8 | 13 | 20 | | | | | | |
| 6 | 0.025 | 3 | 8 | 13 | 19 | 27 | | | | | |
| | 0.05 | 4 | 9 | 14 | 21 | 29 | | | | | |
| 7 | 0.025 | 3 | 8 | 14 | 21 | 28 | 37 | | | | |
| | 0.05 | 4 | 9 | 15 | 22 | 30 | 40 | | | | |
| 8 | 0.025 | 4 | 9 | 15 | 22 | 30 | 39 | 50 | | | |
| | 0.05 | 5 | 10 | 16 | 24 | 32 | 42 | 52 | | | |
| 9 | 0.025 | 4 | 9 | 15 | 23 | 32 | 41 | 52 | 63 | | |
| | 0.05 | 5 | 11 | 17 | 25 | 34 | 44 | 55 | 67 | | |
| 10 | 0.025 | 4 | 10 | 16 | 24 | 33 | 43 | 54 | 66 | 79 | |
| | 0.05 | 5 | 11 | 18 | 27 | 36 | 46 | 57 | 70 | 83 | |
| 11 | 0.025 | 5 | 10 | 17 | 25 | 35 | 45 | 56 | 69 | 82 | 97 |
| | 0.05 | 5 | 12 | 19 | 28 | 38 | 48 | 60 | 73 | 87 | 101 |

**TABLE A.9 Critical Values $c$ for the One-Sample Kolmogorov–Smirnov Test Where $n$ is the Sample Size [e.g. $F_{D_{10}}(0.409) = 0.95$]**

| $n$ | 0.01 | 0.05 | $n$ | 0.01 | 0.05 |
|---|---|---|---|---|---|
| 1 | 0.995 | 0.975 | 16 | 0.392 | 0.327 |
| 2 | 0.929 | 0.842 | 17 | 0.381 | 0.318 |
| 3 | 0.829 | 0.708 | 18 | 0.371 | 0.309 |
| 4 | 0.734 | 0.624 | 19 | 0.361 | 0.301 |
| 5 | 0.669 | 0.563 | 20 | 0.352 | 0.294 |
| 6 | 0.617 | 0.519 | 21 | 0.344 | 0.287 |
| 7 | 0.576 | 0.483 | 22 | 0.337 | 0.281 |
| 8 | 0.542 | 0.454 | 23 | 0.330 | 0.275 |
| 9 | 0.513 | 0.430 | 24 | 0.323 | 0.269 |
| 10 | 0.489 | 0.409 | 25 | 0.317 | 0.264 |
| 11 | 0.468 | 0.391 | 26 | 0.311 | 0.259 |
| 12 | 0.449 | 0.375 | 27 | 0.305 | 0.254 |
| 13 | 0.432 | 0.361 | 28 | 0.300 | 0.250 |
| 14 | 0.418 | 0.349 | 29 | 0.295 | 0.246 |
| 15 | 0.404 | 0.338 | 30 | 0.290 | 0.242 |

**TABLE A.10 Critical Values $c$ for the Two-Sample Kolmogorov–Smirnov Test on the 5% Level Where $m$ is the Size of the Smaller Sample [e.g. $P(D_{7,10} \geq 0.614) \approx 0.05$]**

| $n$ | $m = 2$ | 3 | 4 | 5 | 6 | 7 | 8 | 9 | 10 |
|---|---|---|---|---|---|---|---|---|---|
| 4 | – | – | 1 | | | | | | |
| 5 | – | 1 | 1 | 1 | | | | | |
| 6 | – | 1 | 0.833 | 0.8 | 0.833 | | | | |
| 7 | – | 1 | 0.857 | 0.8 | 0.714 | 0.857 | | | |
| 8 | 1 | 0.875 | 0.875 | 0.75 | 0.708 | 0.714 | 0.75 | | |
| 9 | 1 | 0.889 | 0.778 | 0.778 | 0.722 | 0.667 | 0.639 | 0.667 | |
| 10 | 1 | 0.9 | 0.75 | 0.8 | 0.667 | 0.657 | 0.6 | 0.589 | 0.7 |
| 11 | 1 | 0.909 | 0.75 | 0.709 | 0.652 | 0.623 | 0.602 | 0.596 | 0.545 |
| 12 | 1 | 0.833 | 0.75 | 0.717 | 0.667 | 0.631 | 0.625 | 0.583 | 0.55 |
| 13 | 1 | 0.846 | 0.75 | 0.692 | 0.667 | 0.615 | 0.596 | 0.556 | 0.538 |
| 14 | 1 | 0.857 | 0.75 | 0.657 | 0.643 | 0.643 | 0.571 | 0.556 | 0.529 |
| 15 | 0.933 | 0.8 | 0.733 | 0.733 | 0.633 | 0.590 | 0.558 | 0.556 | 0.567 |
| 16 | 0.938 | 0.812 | 0.75 | 0.675 | 0.625 | 0.571 | 0.625 | 0.542 | 0.525 |
| 17 | 0.941 | 0.824 | 0.706 | 0.647 | 0.608 | 0.571 | 0.566 | 0.536 | 0.524 |
| 18 | 0.944 | 0.833 | 0.694 | 0.667 | 0.667 | 0.571 | 0.556 | 0.556 | 0.511 |
| 19 | 0.947 | 0.790 | 0.697 | 0.642 | 0.614 | 0.571 | 0.540 | 0.520 | 0.495 |
| 20 | 0.95 | 0.8 | 0.75 | 0.65 | 0.6 | 0.564 | 0.55 | 0.517 | 0.55 |

| $n$ | $m = 11$ | 12 | 13 | 14 | 15 | 16 | 17 | 18 | 19 | 20 |
|---|---|---|---|---|---|---|---|---|---|---|
| 11 | 0.636 | | | | | | | | | |
| 12 | 0.546 | 0.583 | | | | | | | | |
| 13 | 0.524 | 0.519 | 0.538 | | | | | | | |
| 14 | 0.532 | 0.512 | 0.489 | 0.571 | | | | | | |
| 15 | 0.509 | 0.517 | 0.492 | 0.467 | 0.533 | | | | | |
| 16 | 0.506 | 0.5 | 0.486 | 0.473 | 0.475 | 0.5 | | | | |
| 17 | 0.497 | 0.490 | 0.475 | 0.466 | 0.455 | 0.456 | 0.471 | | | |
| 18 | 0.490 | 0.5 | 0.470 | 0.460 | 0.456 | 0.444 | 0.435 | 0.5 | | |
| 19 | 0.488 | 0.474 | 0.462 | 0.455 | 0.446 | 0.438 | 0.436 | 0.415 | 0.474 | |
| 20 | 0.486 | 0.483 | 0.462 | 0.45 | 0.45 | 0.438 | 0.429 | 0.422 | 0.421 | 0.45 |

**TABLE A.11 Critical Values $c$ for the Kruskal–Wallis Test on the 5% Level, Where $k$ is the Number of Groups and $n$ is the Sample Size [e.g., if $k = 4$ and $n = 5$, then $P(K \geq 7.37) \approx 0.05$]**

| $n$ | $k = 3$ | 4 | 5 | 6 | 7 | 8 | 9 | 10 |
|---|---|---|---|---|---|---|---|---|
| 2 | – | 6.00 | 7.31 | 8.85 | 10.17 | 11.47 | 12.74 | 14.03 |
| 3 | 5.42 | 6.90 | 8.30 | 9.75 | 11.13 | 12.49 | 13.83 | 15.18 |
| 4 | 5.65 | 7.21 | 8.73 | 10.16 | 11.59 | 12.98 | 14.33 | 15.68 |
| 5 | 5.66 | 7.37 | 8.91 | 10.38 | 11.84 | 13.21 | 14.62 | 15.95 |
| 6 | 5.72 | 7.45 | 9.00 | 10.49 | 11.96 | 13.38 | 14.77 | 16.10 |
| 7 | 5.77 | 7.48 | 9.09 | 10.60 | 12.08 | 13.47 | 14.87 | 16.27 |
| 8 | 5.80 | 7.53 | 9.12 | 10.68 | 12.10 | 13.56 | 14.96 | 16.34 |
| 9 | 5.85 | 7.58 | 9.18 | 10.71 | 12.18 | 13.60 | 15.02 | 16.40 |
| 10 | 5.86 | 7.61 | 9.22 | 10.76 | 12.22 | 13.66 | 15.08 | 16.44 |
| 11 | 5.84 | 7.59 | 9.25 | 10.79 | 12.26 | 13.71 | 15.12 | 16.51 |
| 12 | 5.88 | 7.65 | 9.26 | 10.80 | 12.31 | 13.73 | 15.12 | 16.54 |
| 13 | 5.86 | 7.65 | 9.27 | 10.82 | 12.30 | 13.77 | 15.17 | 16.57 |
| 14 | 5.86 | 7.63 | 9.30 | 10.84 | 12.34 | 13.77 | 15.21 | 16.58 |
| 15 | 5.91 | 7.67 | 9.33 | 10.86 | 12.38 | 13.81 | 15.22 | 16.62 |
| 16 | 5.91 | 7.71 | 9.29 | 10.86 | 12.38 | 13.82 | 15.26 | 16.66 |
| 17 | 5.90 | 7.70 | 9.31 | 10.88 | 12.40 | 13.85 | 15.22 | 16.65 |
| 18 | 5.91 | 7.69 | 9.34 | 10.92 | 12.43 | 13.85 | 15.25 | 16.66 |
| 19 | 5.93 | 7.72 | 9.35 | 10.91 | 12.37 | 13.89 | 15.28 | 16.69 |
| 20 | 5.93 | 7.72 | 9.35 | 10.94 | 12.40 | 13.89 | 15.31 | 16.69 |

# APPENDIX B

# ANSWERS TO SELECTED PROBLEMS

## CHAPTER 1

1 **(a)** $\{3, 4, \ldots, 18\}$ **(b)** $[0, 1] \times [0, 1]$ **(c)** $\{M, F\} \times \{0, 1, 2, \ldots\}$ **(d)** $\{(i, j) : 1 \leq i < j \leq 10\}$ **(e)** $[0, 1]$.

2 $S = \{rr, rs, sr, ss\}$

3 **(c)** and **(e)**

4 **(a)** $(A \cap B^c \cap C^c) \cup (A^c \cap B \cap C^c) \cup (A^c \cap B^c \cap C)$ **(b)** $A^c \cap B^c \cap C^c$ **(c)** $A \cup B \cup C$

5 **(a)** $B_1$ **(b)** $B_1^c \cap B_2 \cap B_3$ **(c)** $B_1 \cap B_2 \cap B_3 \cap B_4$ **(d)** $(B_1^c \cap B_2 \cap B_3 \cap B_4 \cap B_5) \cup (B_1 \cap B_2^c \cap B_3 \cap B_4 \cap B_5) \cup (B_1 \cap B_2 \cap B_3^c \cap B_4 \cap B_5)(B_1 \cap B_2 \cap B_3 \cap B_4^c \cap B_5)$ **(e)** $B_1 \cap B_2 \cap B_3 \cap B_4^c \cap B_5^c \cap B_6^c \cap B_7^c$

6 0.6

7 **(a)** $P(B) = 0.4$ **(b, c)** $P(A \cap B^c) = 0.1$ **(d)** $P(A^c) = 0.7$ **(e)** $P(B^c) = 0.6$ **(f)** $P(A^c \cap B^c) = 0.5$ (which also equals $P((A \cup B)^c)$).

8 **(a)** $0.5 - p$ **(b)** $1 - 2p$ **(c)** $p$

10 (b)

12 **(a)** 3/8 **(b)** 1/8 **(c)** 7/8 **(d)** 1/8 **(e)** 3/8

*Probability, Statistics, and Stochastic Processes*, Second Edition. Peter Olofsson and Mikael Andersson.

16 **(a)** $0.5 + 0.33 + 0.2 - (0.16 + 0.2 + 0.08) + 0.08 = 0.67$
**(b)** $P(A_i) + P(A_j) + P(A_k) - (P(A_{\text{lcm}(i,j)}) + P(A_{\text{lcm}(i,k)}) + P(A_{\text{lcm}(j,k)})) + P(A_{\text{lcm}(i,j,k)})$ (see Example 1.10)

17 **(a)** $(7.1 + 15 - 0.75)/143 = 0.149$
**(b)** $(7.1 + 15 + 10 - (0.75 + 0.5 + 0) + 0)/143 = 0.216$
**(c)** $(71 + 15 + 71 + 1 - (7.5 + 35.5 + 0 + 7.5 + 0 + 0) + (3.75 + 0 + 0 + 0) - 0)/143 = 0.778$

19 **(a)** $\binom{7}{2} \times 26^5 \times 10^2 = 2.5 \times 10^{10}$ **(b)** $\binom{7}{2} \times (26)_5(10)_2 = 1.5 \times 10^{10}$

20 **(a)** $23 \times 22 \times 21/23^3 = 0.87$ **(b)** $10 \times 9 \times 8/10^3 = 0.72$ **(c)** $23/23^3 = 0.002$
**(d)** $5^3/10^3 = 0.125$ **(e)** $23 \times 22 \times 21 \times 10/(23^3 10^3) = 0.009$

21 **(a)** $5 \times 26^4/26^{10} = 1.6 \times 10^{-8}$ **(b)** $5 \times (20)_4/(26)_{10} = 3.0 \times 10^{-8}$

22 **(a)** 1/4 **(b)** 11/24 **(c)** 10/24

20 **(a)** 6, 12

25 **(a)** $\binom{8}{3}/\binom{64}{3} = 1/744$ **(b)** $\binom{32}{3}/\binom{64}{3} = 5/42$ **(c)** $8 \times \binom{8}{3}/\binom{64}{3} = 1/93$
**(d)** $8 \times 2 \times \binom{4}{3}/\binom{64}{3} = 1/651$

26 **(a)** $1/\binom{n}{j}$ **(b)** $2/\binom{n}{j}$ **(c)** $(n - j + 1)/\binom{n}{j}$

27 $3\left(\binom{2n}{k} - \binom{n}{k}\right)/\binom{3n}{k}$

28 $\left(4 \times \binom{39}{13} - 6 \times \binom{26}{13} + 4\binom{13}{13}\right)/\binom{52}{13} = 0.05$

29 **(a)** $\binom{5}{4}\binom{39}{1}\binom{43}{1}/(\binom{44}{5}\binom{44}{1}) = 0.00018$ **(b)** $\binom{5}{3}\binom{39}{2}\binom{1}{1}/(\binom{44}{5}\binom{44}{1}) = 0.00016$

30 **(a)** $4/\binom{52}{5} = 1.5 \times 10^{-6}$ **(b)** $9 \times 4/\binom{52}{5} = 0.000014$
**(c)** $13 \times 48/\binom{52}{5} = 0.00024$ **(d)** $13 \times 12 \times \binom{4}{3}\binom{4}{2}/\binom{52}{5} = 0.0014$
**(e)** $4(\times\binom{13}{5} - 10)/\binom{52}{5} = 0.002$ **(f)** $(10 \times 4^5 - 40)/\binom{52}{5} = 0.0039$
**(g)** $13 \times \binom{12}{2} \times 4^3/\binom{52}{5} = 0.021$ **(h)** $\binom{13}{2}\binom{4}{2}\binom{4}{2} \times 44/\binom{52}{5} = 0.0475$
**(i)** $13 \times \binom{4}{2}\binom{12}{3} \times 4^3/\binom{52}{5} = 0.4226$

31 **(a)** $\binom{6}{2}/\binom{10}{3}$ **(b)** $\binom{3}{1}/\binom{10}{3}$ **(c)** $(k - j - 1)/\binom{n}{3}$

32 $m(n)_{k-1}/(n + m)_k$

34 $\binom{k-1}{n-1}$

33 false, $0 < P(A) < 1$, $B = A$

35 0.63

36 **(a)** $n_0 = n! \sum_{k=0}^{n} \frac{(-1)^k}{k!}$ **(b)** $P(j \text{ matches}) = \frac{n_j}{n!} = \frac{1}{j!} \sum_{k=0}^{n-j} \frac{(-1)^k}{k!}$, $j = 0, \ldots, n$, which is $\approx \frac{e^{-1}}{j!}$ for large $n$

40 **(a)** $P(A \cup B) = \frac{2}{3}$, $P(A \cap B) = \frac{1}{6}$ **(b)** $P(A \cup B) = \frac{5}{6}$, $P(A \cap B) = 0$
**(c)** $P(A \cup B) = \frac{2}{3}$, $P(A \cap B) = \frac{1}{6}$ **(d)** $P(A \cup B) = \frac{1}{2}$, $P(A \cap B) = \frac{1}{3}$

41 **(a)** $P(A \cap B) = P(A)P(B|A) = 0.60 \cdot 0.75 = 0.45$
**(b)** $P(A \cap B^c) = P(A)P(B^c|A) = 0.60 \cdot 0.25 = 0.15$
**(c)** $P(A|B) = P(A \cap B)/P(B) = 0.45/0.50 = 0.90$
**(d)** $P(B|A^c) = \frac{P(A^c \cap B)}{P(A^c)} = \frac{P(A^c|B)P(B)}{P(A^c)}$

42 **(a)** true **(b)** false **(c)** true **(d)** true (if $P(A) < 1$)

45 $p = 0, 1/2, 1$

48 No, for any events $A$ and $B$, $P(A \cap B) \leq P(A)$ and $P(A \cap B) \leq P(B)$

49 $P(B|A) = 1$, $P(A|B) = 3/5$

50 **(a)** $\frac{1}{6}$ **(b)** $\frac{1}{2}$

51 $n = 7$

52 **(a)** $(0.4 \times 0.16)^2 = 0.0041$ **(b)** $2 \times 0.378 \times (1 - 0.378) = 0.47$
**(c)** $1 - (1 - 0.45 \times 0.84)^2 = 0.61$
**(d)** $0.96 \times 0.84 + 0.96 \times 0.84 \times 0.16 = 0.94$
**(e)** $0.45^2 + 0.40^2 + 0.11^2 + 0.04^2 = 0.38$ **(f)** $0.38 \times 2 \times 0.84 \times 0.16 = 0.10$

54 No, the probability that you win is $\approx 0.45$ (compare the birthday problem).

55 $p^2 + (1-p)^2$, $p = 1/2$

56 **(a)** $\sum_{k=1}^{365} p_k^2$

57 **(a)** $(1/2)^5$ **(b)** $(5/9)^5$ **(c)** $(2/3)^5$

59 **(a)** $P(A \cap B \cap C) = 1/36$, $P(A) = 1/2$, $P(B) = 2/3$, $P(C) = 1/12$ **(b)** No

61 **(a)** $(1 - 1/143)^3 = 0.98$ **(b)** 0.02 **(c)** $(71/143)^3 = 0.1224$
**(d)** $3(10 \times 118^2 + 10 \times 15 \times 118)/143^3 = 0.161$

62 $(4/5)^3 = 0.512$

63 **(a)** $8/15$ **(b)** $((2n-2)(2n-4)\cdots 2)/((2n-1)(2n-3)\cdots 1)$

64 $n/2^{n-1}$

65 $3((2/3)^5 - (1/3)^5) = 0.38$

67 $(1 - (1-p)^2)^2$, $1 - (1 - p^2)^2$

68 $\sum_{j=k}^{n} \binom{n}{j} p^j (1-p)^{n-j}$

69 4 rounds, $P(\text{win}) = 1 - (5/6)^4 \approx 0.52$

70 **(a)** $1 - (1 - 1/1,000,000)^{2,000,000} = 0.86$ **(b)** $n \geq 693147$

71 $P(A_1) = 6(1/6)^n$, $P(A_2) = \binom{6}{2}((2/6)^n - 2(1/6)^n)$

72 99.4%

73 $\sum_{k=1}^{6} (1/2)^k (1/6) = 0.164$

74 **(a)** $2 \times 0.40 \times 0.11 = 0.09$ **(b)** $0.45(1 - 0.45) + (0.40 + 0.11)(0.45 + 0.04)$
$+0.04(1 - 0.04) = 0.54$ **(c)** $1 - 0.09 = 0.91$ **(d)** 0.38 [same as 52(e)]

76 $(1 \times (1/56) + (1/3)(15/56 + 6/56) + (1/6)(34/56)) = 0.24$

77 $(1 \times (1/8) + 3 \times (1/2) + 6 \times (1/4))/10 = 5/16$

78 $(10 \times 1 + 3 \times \binom{10}{2} \times 3 + 855 \times 6)/1000^2 = 0.0055$

79 $\left(\binom{4}{k}\binom{48}{4-k}\binom{4}{n}\binom{48}{k-n}\right) / \left(\binom{52}{4}\binom{52}{k}\right)$, $n \leq k = 0, 1, \ldots, 4$

80 **(a)** $1/2$ **(b)** $1/(1 + 2p)$

83 $A$

84 $0.8 \times 0.05/0.135 = 0.296$

87 **(a)** $1/3$ **(b)** $1/2$

88 $376/459 \approx 0.82$

89 **(a)** $0.90^2 = 0.81$ **(b)** $(9/11)(18/19) = 0.775$

90 **(a)** $10/11$ **(b)** $1/2$

92 $0.99 \times 0.001/(0.99 \times 0.001 + 0.01 \times 0.999) = 0.09$

93 $\frac{1 \cdot 0.002}{1 \cdot 0.002 + 0.05 \cdot 0.998} \approx 0.04$

94 $\frac{1/3 \cdot 1/3}{1/3 \cdot 1/3 + 2/3 \cdot 2/3} = \frac{1}{5}$

97 $1/(1+p)$

98 **(a)** $p^2 + r(1-p)$ **(b)** $1-(1-r)(1-p)$ **(c)** $(1-p)p/((1-p)(p+1-r))$

99 $2p/(1+p)$

100 **(a)** $1/2$ **(b)** $2/3$ **(c)** $2^j/(2^j + 2^{n-j})$

101 **(a)** $2401/2500 = 0.96$ **(b)** $1/2499 = 0.04$

102 **(a)** 0 and 1 **(b)** 0 and 1/2

103 **(a)** $1/250,000$ **(b)** $1/30$ **(c)** $N = 4n^2$

104 **(a)** $(1-p)/(2-p)$ **(b)** $p_A/(1-p_B+p_Ap_B)$

105 $p_A(1-p_B)/(p_A+p_B-2p_Ap_B)$

106 0.47, 0.45, 0.37, and 0.26, respectively.

107 $(1-p)/p$

108 **(a)** $P(TH \text{ before } HH) = 3/4$ **(b)** $P(THHH \text{ before } HHHH) = 15/16$

109 $244/495 \approx 0.49$

## CHAPTER 2

1 **(a)** $1/3$ **(b)** $2/3$ **(c)** 0 **(d)** $2/3$

2 **(a)** $P(X=1) = 1/4$ **(b)** $P(X=2) = 1/2$ **(c)** $P(X=2.5) = 0$
**(d)** $P(X \leq 2.5) = 3/4$

3 **(a)** $p(k) = (13-2k)/36, k = 1, \ldots, 6$
**(b)** $p(0) = 6/36, p(k) = (12-2k)/36, k = 1, \ldots, 5$

4 **(a)** $c = 1/6$ **(c)** $1/2$ **(d)** $5/6$

5 **(a)** $c = 1/2$ **(b)** $1/2$ **(c)** $2/3$

6 **(a)** $p(k) = \binom{5}{k}(1/13)^k(12/13)^{5-k}, k = 0, \ldots, 5$ **(b)** $\binom{4}{k}\binom{48}{5-k}/\binom{52}{5}, k = 0, \ldots, 4$

7 $p(-2) = p(2) = 1/4, p(0) = 1/2$

8 **(a)** $p(k) = (51/52)^{k-1}(1/52), k = 1, 2, \ldots$ **(b)** $p(k) = 1/52, k = 1, 2, \ldots, 52$

9 If $x = 1/n$ for some integer $n$, then $F_Y(x) = (1/2)^{n-1}$, for other $x \in (0, 1]$, $F_Y(x) = (1/2)^{[1/x]}$ ($[\cdot]$ denotes integer part)

12 No and no.

13 **(a)** $c = 3$ **(b)** $F(x) = x^3, P(X > 0.5) = 1 - 0.5^3 = 0.875$
**(c)** $F_Y(x) = 6x^5, 0 \leq x \leq 1$

14 **(a)** $F(x) = (1-x^2)/2, -1 \leq x \leq 0, F(x) = (1+x^2)/2, 0 \leq x \leq 1$
**(b)** not possible
**(c)** $F(x) = x+1, -1 \leq x \leq 0$ **(d)** $F(x) = 1 - 1/x, x \geq 1$

15 $b = 1 - a/2, -2 \leq a \leq 2$

16 **(a)** $a = 1/\sqrt{2}$ **(b)** $1/18$ **(c)** $x = \sqrt{2}$

17 **(a)** $a = 1/100$ **(b)** $f_W(x) = (43 - x)/400, 13 \leq x \leq 33$

18 **(c)** Same pdf as $X$

19 unif (0, 1)

20 **(a)** $f_Y(x) = 1/x^2, x > 1$ **(b)** $f_Y(x) = 2/x^3, x > 1$ **(c)** $f_Y(x) = \exp(x), x < 0$

21 unif [0, 1]

22 $f_Y(x) = 2x \exp(-x^2), x \geq 0$

23 No.

24 **(a)** $91/36 = 2.53$ **(b)** $70/36 = 1.94$

25 **(a)** 1.5 **(b)** 1

26 3/4

27 $E[X] = 101 - 100(1 - p)^{100} \leq 100$ if $p \leq 0.045$

28 $E[X] = -\frac{2}{38}$, $\text{Var}[X] = \frac{398}{38} - (-\frac{2}{38})^2 = 10.5$

29 **(a)** \$6.20 **(b)** $-\frac{3}{38}$ cents

30 7.9 cents

34 \$85, \$3.75

35 3/4

36 **(a)** $-0.027, 0.9993$ **(b)** $-0.027, 34.08$

37 $E[A] = 55/6$, $\text{Var}[A] = 275/36$, $E[W] = 64/3$, $\text{Var}[W] = 275/9$

38 **(a)** $E[X] = 3/4$, $\text{Var}[X] = 3/80$ **(b)** $E[Y] = 6/7$, $\text{Var}[Y] = 3/196$

40 $E[Y] = \infty$, $E[Z] = 2$, $\text{Var}[Z] = \infty$

41 $E[V] = \pi/3$, $\text{Var}[V] = \pi^2/7$

42 $E[V] = 2$, $\text{Var}[V] = \frac{36}{7}$

43 **(a)** $E[X] = \frac{2}{3}$ $\text{Var}[X] = \frac{1}{18}$ **(b)** $\frac{8\pi}{15}$

44 **(a)** $c = \frac{1}{2}$ **(b)** $F(x) = \frac{1-\cos x}{2}, \quad 0 \leq x \leq \pi$ **(c)** $\frac{\pi}{2}$

45 **(a)** $E[-X] = -\mu$, $\text{Var}[-X] = \sigma^2$ **(b)** $a = \frac{1}{\sigma}, b = -\frac{\mu}{\sigma}$

46 **(a)** $c = (b - a)/(\sqrt{3}(b + a))$ **(b)** $c \to 0$

50 **(a)** Binomial with $n = 10$ and $p = 0.8$. **(b)** Not binomial as trials are not independent (one rainy day makes another more likely). **(c)** Not binomial as $p$ changes between months. **(d)** Binomial with $n = 10$ and $p = 0.2$.

51 $P(X > 0) = 1 - 0.9^{10} \approx 0.65$

52 **(a)** 0.1615 **(b)** 0.5155 **(c)** 0.9303

53 **(a)** $E[X] = 1$, $E[Y] = 2$ **(b)** $P(X > E[X]) = 0.6651$, $P(Y > E[Y]) = 0.6187$

54 **(a)** $\binom{n-1}{k-1}(1/2)^{n-1}$ **(b)** $\binom{n-1}{k}(1/2)^{n-1}$ **(c)** $\binom{n}{k}(1/2)^n/(1 - (1/2)^n)$

55 **(a)** 5 – 5 **(b)** somebody wins 6 – 4
**(c)** 0.7734 ($P(X \geq 3)$ where $X \sim \text{bin}(7, 1/2)$)
**(d)** $1 - (1/2)^5 = 0.9688$ **(e)** $(1/2)^8 = 0.004$

56 **(a)** 77/2048 **(b)** 37/64 **(c)** 0.596 [distribution is bin(6, 5/8)] **(d)** $p \approx 0.8$ [distribution is bin(6, $(1 + 3p)/4$)] **(e)** (a): 1.5, 1.125 (b): 3.75, 0.5625 (c): 3.75, 1.4062

57 17

58 **(b)** $P(Y \leq n) = P(X \geq 1)$

59 $P(X = 0) = (15/16)^n$, $E[X] = n/16$

60 $P(D > S) = 1/2$, $P(D = S) = 1/4$, $P(D < S) = 1/4$

62 **(a)** $(4/5)^4(1/5) = 0.08$ **(b)** $\binom{9}{4}(1/5)^5(4/5)^5 = 0.01$

63 **(a)** $1 - \exp(-2) = 0.86$ **(b)** $p_Y(k) = \exp(-2)2^k/(k!(1 - \exp(-2)))$, $E[Y] = 2/(1 - \exp(-2)) = 2.3$ **(c)** $\exp(2) - 1 = 6.4$

64 **(a)** $0.9 + 0.1\exp(-1) = 0.94$ **(b)** $0.1\exp(-1)/(0.9 + 0.1\exp(-1)) = 0.04$ **(c)** $1 - (0.9/(0.9 + 0.1\exp(-1)))^{10} = 0.33$

66 **(a)** 0.0107 (exact) 0.0404 (Poisson) **(b)** 0.8741 (exact) 0.8698 (Poisson)

70 **(a)** 100 **(b)** 0.3935 **(c)** 0.3935 **(d)** $\exp(-1) = 0.37$

71 No, memoryless property does not seem to hold.

73 **(a)** 0.70 **(b)** 0.30 **(c)** $E[X] = 58/\log 2 = 83.7$, $\mathrm{Var}[X] = 7002$

76 1/4

77 **(a)** $P(X \leq 220) = \Phi((220 - 200)/10) = \Phi(2) = 0.98$
**(b)** $P(X \leq 190) = \Phi((190 - 200)/10) = \Phi(-1) = 1 - \Phi(1) = 0.16$
**(c)** $P(X > 185) = \Phi(1.5) = 0.93$
**(d)** $P(X > 205) = 1 - \Phi(0.5) = 0.31$
**(e)** $P(190 \leq X \leq 210) = \Phi(1) - \Phi(-1) = 0.68$
**(f)** $P(180 \leq X \leq 210) = \Phi(1) - \Phi(-2) = 0.81$

78 $2\Phi(c) - 1$

79 66 and 74

80 $Z = c$

81 A 24-pound A-fish (the $Z$ scores are 2 and 1, respectively).

82 180 points

83 $2\Phi(1.16) - 1 = 0.754$

84 **(a)** $T \sim N(t + 0.1, 0.01)$ **(b)** $\Phi(-1) = 0.16$ **(c)** $\Phi(-0.5) - \Phi(-1.5) = 0.24$

85 **(a)** $\varphi(x), x \in R$ **(b)** $2\varphi(x), x \geq 0$ **(c)** $\varphi(\sqrt{x})/\sqrt{x}, x \geq 0$ **(d)** $\varphi(\log x)/x, x > 0$

87 **(a)** $g(X) = X$ if $X < 100$, $g(X) = X - 100$ if $X \geq 100$.
**(b)** $E[g(X)] = \mu - 100(1 - \Phi((100 - \mu)/\sqrt{2}))$ has minimum for $\mu = 103.7$

88 $\sqrt{\pi}/2$

90 $9.6 \times 10^{-18}$

92 $X^k$ is lognormal with parameters $(k\mu, k^2\sigma^2)$

94 **(a)** 0.76 **(b)** 0.64 **(c)** 0.31 **(d)** 1

96 **(a)** $1/2 + \arctan(x)/\pi, x \in R$

97 $F(x) = 0.2 + 0.8x/30, 0 \leq x \leq 30$

98 **(a)** $F(x) = 1 - (\exp(-x) + \exp(-2x))/2$
**(b)** $P(\text{type} I | X \geq t) = 1/(1 + \exp(-t))$

99 **(a)** $\infty$, 2, 1 **(b)** 1.5, any number in $[1, 2]$, any number in $[0, 1] \cup [2, 3]$ **(c)** $1/3$, $1 - 1/\sqrt{2}$, 0 **(d)** 0, 0, and $-1$ or 1 **(e)** does not exist, 0, 0

101 **(a)** (i), (iii), (v)

103 **(a)** 2 **(b)** $1/\sqrt{\lambda}$

105 **(a)** 3 **(b)** 9 **(c)** 1.8 **(d)** $3 + 1/\lambda$

107 **(b)** $r(k) = 1/(7-k)$, $k = 1, \ldots, 6$

109 **(a)** $2t/(1-t^2)$ **(b)** $1/t$ **(c)** $2t$

110 **(a)** $F(t) = 1 - 1/(1+t)$ **(b)** $F(t) = 1 - \exp(-t^2)$ **(c)** $F(t) = 1 - \exp(t^4/4)$ **(d)** $1 - \exp(\exp(-t) - 1)$

111 $c = \log 0.4/\log 0.2 = 0.57$

112 **(a)** $a = 0.26$ **(b)** 0.096

113 **(a)** 0.15 **(b)** $f(t) = (1-t)\exp(-t + t^2/2)$, $0 \leq t \leq 1$, $f(t) = (t-1)\exp(t - t^2/2 - 1)$, $t \geq 1$, $m = 1.6$

114 **(a)** 0.22 **(b)** 0.54 **(c)** 0.53 **(d)** $2\log 2 = 1.39$ **(e)** 1.1 years

115 **(a)** $r(t) = c$, $0 \leq t \leq 100$, $r(t) = c2^k$, $100 + 2k \leq t < 100 + 2(k+1)$, $k = 0, 1, 2, \ldots$ **(b)** $c = \log 2/162 = 0.043$ **(c)** 0.35

## CHAPTER 3

3 **(a)** 1 **(b)** 0 **(c)** 0

4 **(a)** true **(b)** false

6 **(a)** $F_X(x) = 1 - \exp(-x)$, $F_Y(y) = 1 - \exp(-y)$ **(b)** $F_X(x) = x^2$, $F_Y(y) = \sqrt{y}$ **(c)** $F_X(x) = 2x$, $F_Y(y) = y$ **(d)** $F_X(x) = (x^2 + 2x)/3$, $F_Y(y) = (y + 2y^2)/3$

7 **(a)** $p_X(0) = 1/4$, $p_X(1) = 1/2$, $p_X(2) = 1/4$, $p_Y(0) = p_Y(1) = 1/2$ **(b)** $p(0,0) = p(2,1) = 1/3$, $p(1,0) = p(1,1) = 1/6$

9 $p(0,3) = p(3,0) = 1/8$, $p(1,2) = p(2,1) = 3/8$

10 **(a)** $p(j,k) = 1/36$, $j = 1, \ldots, 6$, $k = j+1, \ldots, j+6$ **(b)** $p(j, 2j) = 1/36$, $j = 1, \ldots, 6$, $p(j,k) = 1/18$, $j = 1, \ldots, 6$, $k = 2j+1, \ldots, j+6$

11 **(a)** $p(j,k) = \binom{13}{j}\binom{13}{k}\binom{26}{3-j-k}/\binom{52}{3}$, $0 \leq j+k \leq 3$ **(b)** $P(H = S) = p(0,0) + p(1,1) = 0.32$

12 **(a)** $p(0,0) = 36/52$, $p(0,1) = 3/52$, $p(1,0) = 12/52$, $p(1,1) = 1/52$ **(b)** $40/52$

13 **(a)** $1/4$ **(b)** $3/32$ **(c)** $25/64$

15 (c), (d)

16 **(a)** $c = 6$ **(b)** $3/4$ **(c)** $1/4$ **(d)** $f_X(x) = 3x^2$, $0 \leq x \leq 1$, $f_Y(y) = 3y^2 - 6y + 3$, $0 \leq y \leq 1$

17 $f(x,y) = x + y$, $f_X(x) = x + 1/2$, $f_Y(y) = y + 1/2$

19 unif [0, 1]

20 **(a)** $c = 2/\pi$ **(b)** $d^2(2 - d^2)$

22 0.68

23 $\exp(-2) = 0.14$

28 $f(x) = 1/2, 0 \leq x \leq 2$

29 $8x/3, 0.5 \leq x \leq 1$

31 **(a)** $f_Y(y|x) = \exp(x - y), y \geq x$ **(b)** $f_Y(y|x) = x\exp(-xy), y \geq 0$
**(c)** $f_Y(y|x) = 1/y^2, y \geq 1$ **(d)** $f_Y(y|x) = 1/(2\sqrt{y}), 0 \leq y \leq 1$

32 **(a)** $f(x, y) = x, 0 \leq y \leq 1/x, 0 \leq x \leq 1$ **(b)** $f_Y(y) = 1/2, 0 \leq y \leq 1$,
$f_Y(y) = 1/(2y^2), y \geq 1$ **(c)** 1/3

33 $p_Y(k) = 1/(n + 1), k = 0, 1, \ldots, n$

34 **(a)** $450/1350 = 1/3$ **(b)** $800/1350 = 0.59$

35 **(b)** $\lambda_1/(\lambda_1 + \lambda_2)$

36 **(a)** yes **(b)** no **(c)** yes **(d)** yes **(e)** yes

38 $(\pi/6)^{1/3}/2 = 0.40$

39 $1/2 + n/24$ for $n \leq 4$ and $1 - 2/(3\sqrt{n})$ for $n \geq 4$

40 **(a)** 1/4 **(b)** 0.53 **(c)** 1/4

43 **(a)** $F(x) = 1 - (1 - x)^2, 0 \leq x \leq 1$ **(b)** $F(x) = x/(2(1 - x)), 0 \leq x \leq 1/2$,
$F(x) = 1 - (1 - x)/(2x), 1/2 \leq x \leq 1$

44 $f_W(x) = -\log x, 0 < x \leq 1$

45 **(a)** 10/3 min **(b)** 85/6 min

46 7/12 Ml ($E[\max(X - Y, 0)]$)

47 4/9

48 **(a)** 1/4 **(b)** $\infty$ **(c)** $-2$ **(d)** 1/3

49 **(a)** $E[A] = a$, $\text{Var}[A] = \sigma^2/2 + \tau^2/4$

50 $E\bar{X}] = \mu$, $\text{Var}[X] = \sigma^2/n$

51 **(a)** $n(\mu^2 + \sigma^2)$ **(b)** $n(n\mu^2 + \sigma^2)$

53 **(c)** $(1 - 2p)/\sqrt{np(1 - p)}$ and $3 + (1 - 6p + 6p^2)/(np(1 - p))$

55 **(a)** $n(1 - (1 - 1/n)^k)$

58 $(2N - n)(2N - n - 1)/(4N - 2)$ (let $I_j$ be the indicator for the event that the $j$th married couple remains and find $E[I_j]$)

61 **(a)** $f_{(X,Z)}(x, z) = xf(x, xz)$
**(b)** $f_{(X,Z)}(x, z) = x\exp(-x(1 + z))$, $f_Z(z) = 1/(1 + z)^2$

62 $f_{(U,V)}(u, v) = u, 0 \leq uv \leq 1, 0 \leq u(1 - v) \leq 1$

63 $f_{(U,V)}(u, v) = \exp(-(u^2 + v^2)/4)/(4\pi)$ (note that $U = X - Y$ and $V = X + Y$ are independent; this only holds for the normal distribution)

64 $f_{(X,Y)}(x, y) = 1/(2\pi\sqrt{x^2 + y^2})$, $x^2 + y^2 \leq 1$

65 $f_{(W,I)}(w, i) = 2/i$, $0 \leq i \leq 1, i^2 \leq w \leq i$ (note the range), $f_I(i) = 2(1 - i), 0 \leq i \leq 1$, $f_W(w) = -\log w, 0 \leq w \leq 1$

66 $f_{(U,V)}(u, v) = v/2,\ 0 \le u \le 2, u/2 \le v \le 1 + u/2$ (note the range)

67 $f_{(U,V)}(u, v) = \exp(-(u^2 + 3v^2)/6)/(2\pi\sqrt{3}), u, v \in R$

69 **(a)** \$360 ($E[g(X)]$ where $g(X) = X - 0.8$ if $X \ge 0.8$ and $g(X) = 0$ otherwise)
**(b)** \$600 ($E[X|X > 0.8] - 0.8$)

70 **(a)** $E[Y|X = k] = k + 3.5$
**(b)** $E[Y|X = k] = 52/11, 56/9, 54/7, 46/5, 32/3, 12$
for $k = 1, 2, \dots, 6$

71 **(a)** $E[Y|X = x] = x + 1$ **(b)** $E[Y|X = x] = 1/x$ **(c)** $E[Y|X = x] = \infty$
**(d)** $E[Y|X = x] = 1/3$

72 **(a)** $1/6$ **(b)** $1/\pi$ **(c)** $\infty$ **(d)** $1/2$

73 **(a)** $1/2$ **(b)** 2 **(c)** $\infty$

74 **(a)** $\infty$ **(b)** $n/2$

78 $2/((1 - p)^2 + p^2)$

79 $(2 - p)/(1 - p(1 - p))$

81 10 and 20

83 $2^{n+1} - 2$

84 **(a)** 36 **(b)** 258

85 $1/36$

86 1 [$\mathrm{Var}[I_j] = (n - 1)/n^2, \mathrm{Cov}[I_j, I_k] = 1/(n^2(n - 1))$]

87 $-1/3$

95 None are independent **(a)** correlated **(b)** uncorrelated **(c)** uncorrelated **(d)** uncorrelated

96 $\sqrt{6/7} = 0.93$

97 $E[X] + E[Y] = 0$

98 $E[Y|X] = X^2/2, l(X) = 2X/3 - 1/5$

102 a,d,e,g

103 **(a)** $E[C] = 30, E[C|X = x] = 5.2x + 4$
**(b)** $E[A] = 50.016, E[A|X = x] = 1.6x^2 + 2x$
**(c)** 0.08 **(d)** 0.02 **(e)** $c \le 0.46$

104 0.37

105 **(a)** 0.08 **(b)** 0.09 **(c)** $n \ge 3$

107 $U$ and $V$ are independent with $U \sim N(0, 2(1 - \rho)), V \sim N(0, 2(1 + \rho))$.

108 **(a)** $f_Z(x) = \varphi((x - 900)/\sqrt{181/4})/\sqrt{181/4}$
**(b)** $f_W(x) = (\varphi((x - 1000)/10)/20 + \varphi((x - 800)/9)/18$
**(c)** $E[Z] = 900, \mathrm{Var}[Z] = 181/4\ E[W] = 900, \mathrm{Var}[W] = 10090.5$

109 $E[N] = e$

110 $E[N] = 1/(1 - F(c))$

111 32.5 min

114 **(a)** $f_M(x) = \Phi(x - 5)\varphi(x - 7) + \Phi(x - 7)\varphi(x - 5)$ **(b)** $\Phi(1)\Phi(-1) = 0.13$

117 **(a)** 1/2 **(b)** 1/24 **(c)** 1/8 **(d)** 1/8
118 $100 + 100 \times 0.9^{1/4} = 197.40$
120 *Hint:* the lifetime is $\max(T_1, T_2)$ where $T_1, T_2$ are i.i.d. exp(2)
121 $(\exp(t) + 2)/(\exp(t) + 1)$
122 $f(x, y) + f(y, x)$
123 0.0001
125 0.38
126 $f_T(x) = 1 - \exp(-x), 0 \leq x \leq 1$ and $f_T(x) = \exp(-x)(e - 1), x \geq 1$
128 $f_{X+Y}(x) = (x - 2)\exp(2 - x), x \geq 2$
131 $f_{X+Y}(x) = \exp(-x)x^2/2, x \geq 0$
134 **(a)** $G_X(s) = (s + s^2 + \cdots s^6)/6$ **(b)** $G_S(s) = G_X(s)^5$
135 **(a)** $ps/(1 - s + ps)$ **(b)** $(ps/(1 - s + ps))^r$
137 $P(X_k = 0) = 1 - \mu/\lambda$, $P(X_k = 1) = \mu/\lambda$
138 \$8.75
139 **(a)** 0.83 **(b)** 9 ms
140 **(a)** $G_Y(s) = \exp(s + s^2 + s^3 + s^4 - 4)$
**(b)** $\exp(-4) = 0.02$ **(c)** $E[Y] = 10$, $\text{Var}[Y] = 30$
142 **(b)** $\approx$ \$95
145 0 and 3
148 Not normal
149 **(a)** $G_X(s) = G(s, 1)$ **(b)** $\text{Cov}[X, Y] = G_{st}(1, 1) - G_s(1, 1)G_t(1, 1)$
150 **(a)** $\exp(-10/13) = 0.46$ **(b)** $\exp(-5/13) = 0.68$ **(c)** $\exp(-1) = 0.37$
**(d)** $((10/13)\exp(-10/13))^3 = 0.045$
151 **(a)** $\exp(-2/7) = 0.75$ **(b)** $1 - \exp(-8/10) = 0.55$ **(c)** bin(52, exp(−2))
155 **(a)** $\exp(-5) \times 5^2/2 = 0.08$ **(b)** $2\exp(-2) \times 3\exp(-3) = 0.04$ **(c)** 9/25
**(d)** $\binom{4}{2}(1/2)^4 = 0.375$
156 **(a)** no **(b)** $3\exp(-2) = 0.41$ **(c)** $P$(even number of accidents) = 0.51

## CHAPTER 4

3 $\int_0^1 g(x)dx$
4 2/3
9 $\Phi(2) = 0.977$
10 $1 - \Phi(10/\sqrt{200}) = 0.24$
11 78
12 Exact: 0.67, 1, 1, approximately: 0.50, 0.87, 0.99
13 $P(X \geq 80) = 1 - P(X \leq 79) \approx 0.72$ $(n = 100, p = 13/16)$
14 0.44

15 254

16 **(a)** $1/2$ and $1/2$ (exact: $1/2$)
**(b)** $1/4$ and $1/3$ (exact: $1/3$) **(c)** $1/16$ and $3/16$ (exact: $1/5$)
**(d)** 1.65 and 1.72 (exact: $\exp(1) - 1 \approx 1.72$)
**(e)** 0.61 and 0.63 (exact: $1 - \exp(-1) \approx 0.63$)
**(f)** 1 and 0.59 (exact: $2/\pi \approx 0.64$) **(g)** 0 and 0 (exact: 0)

17 $\approx 216,000$ times

18 $N(2/3, 4/27n)$

24 $\lambda = 1, \alpha = 2$

25 For example $a_n = n^{1/3}$, Weibull with $\lambda = 1, \alpha = 3$

## CHAPTER 5

3 Let $S_n = \sum_1^n X_k, n = 1, 2, \ldots$ and let $Y = \min\{n : S_n > \lambda\} - 1$.

6 $F^{-1}(u) = (b - a)u + a$

7 $X = U^{1/3}$

8 **(a)** $X = \tan(\pi(U - 1/2))$

9 $X = (-\log U/\lambda)^{1/\alpha}$

10 $X = 1 + [\log U/\log p]$ ($[\cdot]$ denotes integer part)

17 $X_1 = \mu_1 + \sigma_1 X, Y_1 = \mu_2 + \rho\sigma_2(X_1 - \mu_1)/\sigma_1 + \sigma_2\sqrt{1 - \rho^2}Y$

18 $X = \sqrt{V}\sin(2\pi U), Y = \sqrt{V}\cos(2\pi U)$

## CHAPTER 6

2 $\bar{Y} - \bar{X}, \sqrt{\sigma_1^2/n + \sigma_2^2/m}$

5 **(a)** $\widehat{N} = kn/X$, not unbiased **(b)** $\widehat{N} = kX$, unbiased

7 $\widehat{\lambda} = X/t, \sqrt{\lambda/t}$

8 $I(\lambda)^{-1} = \lambda/t$

12 $P(X \leq 5) \leq 0.59$ and 0.06, respectively

13 $\widehat{c} = 0.11$, no

14 **(a)** $\widehat{p} = 0.0113$ **(b)** $\widehat{x} = 17.8$

15 $\widehat{\mu} = 98.3, \widehat{\sigma} = 24.7$

17 $A - B = 4.7 \pm 17.8$

18 $\mu = \bar{X} \pm 1.96/\sqrt{5} = 1000.3 \pm 0.88$

19 $p = 0.54 \pm 0.03$

20 $\approx 0.25$

21 **(a)** $n \geq 246$ **(b)** $n \geq 385$

22 $p = 0.53 \pm 0.06$

24 $p_B - p_G = 0.03 \pm 0.01$

25 $\mu = \bar{X} \pm z\frac{\sigma}{\sqrt{n}}$ $(\approx q)$

26 $\sigma = \frac{s}{1 \pm z/\sqrt{2n}}$ $(\approx q)$

28 $p_d - p_p \geq -0.03$

30 **(c)** $1086 \pm 215$

32 $\widehat{\lambda} = n/T$

33 $\widehat{p} = X/n$

34 $\widehat{b} = X_{(n)}$ (MLE) and $\widehat{b} = 2\bar{X} - a$

35 $\widehat{a} = X_{(1)}, \widehat{b} = X_{(n)}$ (MLE's) and
$\widehat{a} = \bar{X} - \sqrt{3(\widehat{\mu}_2 - \bar{X}^2)}, \widehat{b} = \bar{X} + \sqrt{3(\widehat{\mu}_2 - \bar{X}^2)}$

36 $\widehat{\theta} = \max(-X_{(1)}, X_{(n)})$

37 $\widehat{\theta} = X_{(1)}$ (MLE) and $\widehat{\theta} = \bar{X} - 1$

38 **(a)** $\widehat{\theta} = \bar{x}/(1 - \bar{X})$ **(b)** $\widehat{\theta} = -n/\sum_k \log X_k$ **(c)** $\theta = \widehat{\theta}(1 \pm z/\sqrt{n})$ $(\approx q)$

39 $\widehat{\theta} = \sum_k X_k^2/n$ (both)

40 **(a)** $\widehat{a} = \pi/(2\bar{X}^2)$ **(b)** $\widehat{a} = 2n/\sum_k X_k^2$ **(c)** $a = \widehat{a}(1 \pm z/\sqrt{n})$ $(\approx q)$

43 $H_0 : p = \frac{1}{2}, H_A : p \neq \frac{1}{2}, \alpha = 0.0078, C = \{0, 1, 7, 8\}$

44 $H_0 : \lambda = 5, H_A : \lambda > 5, C = \{\sum_{k=1}^{4} X_k \geq 26\}$, We can reject $H_0$.

45 **(a)** Base the test on $T = (\bar{X} - \mu_0)/(\sigma/\sqrt{n}) \sim N(0, 1)$ **(b)** accept $H_0$

47 **(a)** $T = \sqrt{n}(1 - \lambda_0\bar{X})$ **(b)** $T = \sqrt{n}(1/(\lambda_0\bar{X}) - 1)$ The second test is better.

48 $H_0 : p = 0.21$ vs $H_A : p > 0.21$, accept $H_0$

49 $H_0 : p_W = 2p_T$ vs $H_A : p_W > 2p_T$, accept $H_0$

50 Assume independent Poisson variables. Test statistic
$T = (X_2 - X_1)/\sqrt{X_1 + X_2} \overset{d}{\approx} N(0, 1)$.

51 $$\frac{n\widehat{p} + z^2/2}{n - z^2} \pm \sqrt{\frac{nz^2\widehat{p}(1 - \widehat{p}) + z^4/4}{(n - z^2)^2}}$$

56 $1 - 0.95^k$

60 **(a)** $n \geq 210$ **(b)** $\Phi((p - 0.5)\sqrt{n} - 1.64/2)/\sqrt{p(1 - p)} = q$

63 Reject $H_0$, beware of data-snooping!

64 Accept $H_0$, $\chi^2 = 4.06$, lump together into five classes

65 Geometric: $\chi^2 = 1.31$, four classes. Poisson: $\chi^2 = 10.48$, four classes

68 Accept $H_0$, $\chi^2 = 0.26$, DF=1

70 Reject $H_0$, $\chi^2 = 15.2$, five classes

71 Reject $H_0$, $\chi^2 = 7.1$

72 Reject $H_0$, $\chi^2 = 76.6$, DF $= 1$

74 Reject $H_0$, $\chi^2 = 207$, DF $= 3$

75 $p$-value $= 0.14$. Do not reject $H_0$.

76 $P(p = 1/3|D) = q/(4 - 3q)$, $E[p|D] = (8 - 7q)/(12 - 9q)$

78 $P(p|D) = 30p^4(1-p)$, mean $= 5/7$, mode $= 4/5$, MLE $= 3/4$

80 **(a)** $f(\theta|D) = 32\theta(\theta-1)/(99 \cdot 2^\theta) \quad \theta = 2, \ldots, 6$ **(b)** 3.94 and 1.69 **(c)** $3 \le \theta \le 5$

83 **(a)** $1/5$ and $1/75$ **(b)** $2/15$ and $2/225$

87 **(a)** Beta$(\alpha, \beta)$ **(b)** 0.2 and 0.072 **(c)** 0.175 and 0.070

88 Parametric: $101.2 \le \mu \le 116.4$, nonparametric (sign test): $96 \le m \le 116$, nonparametric (Wilcoxon test) $101 \le m \le 116$

89 Parametric: $22.3 \le \mu \le 283.6$, nonparametric: $27.3 \le \mu \le 263.9$

90 0.06 and 0.03, respectively, if $N_+$ is either 0 or 5

91 Accept $H_0 : m = 100$

92 Reject $H_0 : m = 1$ on the 5% level

95 $1/16$

96 Reject $H0 : \mu = 100$ in favor of $H_A : \mu > 100$ ($W = 59$, $C = 66 - 14 = 52$)

97 Pairwise differences. Reject $H_0 : \mu = 0$ in favor of $H_A : \mu \neq 0$ ($W = 54$, $c = 9$, $C = 46$)

98 $W = 3, 4, \ldots, 13$, $P(W = 7) = 1/5$

99 Accept $H_0 : \mu_A = \mu_B$ ($W = 46.5$, $c = 22$, $C = 48$)

100 Accept $H_0 : D_{5,8} = 0.475$ ($c = 0.8$)

101 Accept $H_0$ ($W = 43$, $c = 33$, $C = 69$)

102 Accept $H_0$ ($W = 106$, $W \stackrel{d}{\approx} N(90, \sqrt{345})$)

## CHAPTER 7

8 $f(v) = \sqrt{2/\pi}\sigma^3 v^2 e^{-v^2/(2\sigma^2)} \quad v \ge 0$

9 $\mu = 0.11 \pm 0.05$

11 $0.034 \le \sigma \le 0.104$

12 $15.2 \le \sigma \le 23.6$ (0.90), $14.7 \le \sigma \le 24.8$ (0.95)

13 $F_{\chi^2_{n-1}}(n\epsilon^2/(4t^2))$

14 $2\sigma^4/(n-1)$

15 $H_0 : \mu = 12$ vs $H_A : \mu \neq 12$, $|T| = (12.1 - 12)/(\sqrt{0.96}/10) = 1.02 < 1.96$, accept $H_0$

16 $H_0 : \mu = 100$ vs $H_A : \mu > 100$, $T = (\bar{X} - 100)/(s/\sqrt{7}) = 3.1 > 1.94$, reject $H_0$

17 $H_0 : \mu = 0$ vs $H_A : \mu \neq 0$, $|T| = \bar{X}/(s/\sqrt{5}) = 1.0 < 2.78$, accept $H_0$

18 $H_0 : \mu = 0.30$ vs $H_A : \mu > 0.30$, reject $H_0$

19 $H_0 : \mu = 0$ vs $H_A : \mu > 0$, accept $H_0$

20 $n \ge 62$

22 $H_0 : \sigma^2 = 1$ vs $H_A : \sigma^2 > 1$, reject $H_0$

23 $H_0 : \sigma = 1$ vs $H_A : \sigma > 1$, accept $H_0$

24 $H_0 : \sigma = 15$ vs $H_A : \sigma \neq 15$, accept $H_0$

25 Reject $H_0$ ($T = 1.875$, $t_10 = 1.812$)

26 $d = -2.2 \pm 16.3$

27 $73 \pm 22$ (95%)

28 Assume normally distributed samples. Test $\sigma_1 = \sigma_2$. Reject $H_0$ on 5% level. ($F = 7.23$, $x_1 = 0.14$, $x_2 = 7.15$)

29 $2\mu_1 + 2\mu_2 = 489 \pm 27$

32 $H_0 : \mu = 0$ vs $H_A : \mu > 0$, accept $H_0$

33 Each $t$ test will be conditioned on the outcome of the first test, which means that the test statistic no longer is $t$ distributed.

34 **(a)** Reject $H_0$. ($F = 20.85$, $F_{3,16} = 3.24$) **(b)** $\bar{X}i_1\cdot - \bar{X}_{i_2\cdot} \pm 17.5$. **(c)** Reject $H_0$. ($K = 15.16 > 7.37$)

41 $a = 36 \pm 12$, $b = 9.7 \pm 3.4$

42 Accept $H_0 : a = 0$ ($T = -0.48$)

44 **(a)** $\widehat{b} = (S_{xY} - aS_x)/S_{xx}$ **(b)** $\widehat{a} = \bar{Y} - b\bar{x}$ **(c)** $\widehat{a} = 24.6$, $\widehat{b} = 11.3$

45 $\widehat{a} = 0.85$, $0 \leq a \leq 25.5$, $\widehat{b} = 0.15$, $b = 0.15 \pm 1.12$ (consider $\log Y$)

46 **(a)** $y = -0.039x + 105$, 57 s

47 $Y = 133 \pm 27$

50 $180 \leq \sigma \leq 330$

51 $H_0 : \rho = 0$ vs $H_A : \rho \neq 0$, $R = 0.25$, accept $H_0$

52 $R_S = 0.418$, accept $H_0$ ($T = 1.30$).

53 $-0.45 \leq \rho \leq 0.76$ ($\approx 0.95$)

54 $H_0 : \rho = 0.9$ vs $H_A : \rho > 0.9$, $L = 4.66$, reject $H_0$

57 $\beta_0 = 2.71 \pm 0.70$, $\beta_1 = 10.20 \pm 0.78$, $\beta_2 = 2.07 \pm 0.16$

60 $\text{SSA} = \sum_{i,j,k}(\bar{X}_{i\cdot\cdot} - \bar{X})^2$, $\text{SSB} = \sum_{i,j,k}(\bar{X}_{\cdot j\cdot} - \bar{X})^2$, $\text{SSE} = \sum_{i,j,k}(X_{ijk} - \bar{X}_{ij\cdot})^2$, $\text{SST} = \sum_{i,j,k}(X_{ijk} - \bar{X})^2$, and $\text{SSAB} = \sum_{i,j,k}(\bar{X}_{ij\cdot} - \bar{X}_{i\cdot\cdot} - \bar{X}_{\cdot j\cdot} + \bar{X})^2$. (Derive them in this order.)

## CHAPTER 8

1 **(b)** 0.88 **(c)** 0.81

2 $p_{ss} = 7/8$, $p_{rr} = 1/2$

3 1/71

4 **(b)** 4 (geom(0.2) including 0) **(c)** $1/\pi_{\text{high}} = 5.8$

9 **(b)** 0.59 **(c)** 43% vowels

10 **(a)** c(onsonant) **(b)** c **(c)** ccccc **(d)** cvcvc

11 **(b)** 10

12 **(a)** $\boldsymbol{\pi} = (1/10, 2/10, 4/15, 5/18, 7/45)$ (Note: $p_{32} = 3/5$, $p_{34} = 2/5$ and $p_{43} = 1$)
**(b)** $1/10, 7/45$ **(c)** $14/9$

14 **(a)** 4 **(b)** 2 **(c)** $\pi_0 = \pi_1 = \pi_2 = \pi_3 = 1/5, \pi_4 = \pi_5 = 1/10$, no

15 **(a)** 2048 **(b)** 2

16 $\pi_0 = \pi_1 = 1/3, \pi_k = (1/3)(1/2)^{k-1}, k \geq 2$

17 **(a)** $q_{ij} = p_{ji}\pi_j/\pi_i$

19 $p = 1/2$

20 **(a)** $p_{0,1} = p_{N,N-1} = 1$, $p_{k,k-1} = k/N$, $p_{k,k+1} = 1 - k/N, k = 1, 2, \ldots, N-1$
**(b)** bin($N$, 1/2) **(c)** no

22 $1/\sqrt{1 + m/n}$

24 **(a)** $\pi_k = 1/(a+1), k = 0, 1, \ldots a$ **(b)** $a + 1$ **(c)** $\pi_0 = \pi_a = 1/(2a)$,
$\pi_k = 1/a, k = 1, 2, \ldots, a-1$

25 $P_0(\tau_1 < \infty) = p/q$ if $p < q$ and 1 otherwise. $E_0[\tau_1] = 1/(p-q)$ if $p > q$ and $\infty$ otherwise

26 $r/(p-q)$

28 **(c)** $\exp(-1/2) = 0.61$ in (a), $2\exp(-1)(1 - 2/3) + \exp(-1/2) - \exp(-1) = 0.48$ in (b), and $1/2$ if $S_0 \equiv 1$

30 **(a)** $\mu < 1$ **(b)** $Y_n = 1 + Z_1 + \cdots Z_n$, $E[Y_n] = n + 1$
if $\mu = 1$ and $(1 - \mu^{n+1})/(1 - \mu)$ if $\mu \neq 1$

31 **(a)** $2^n$ **(b)** 3/7 **(c)** 1/2

32 **(b)** Note that $Z_n = Z_n^{(0)} + \cdots Z_n^{(n)}$ where $Z_n^{(j)}$ is the number of individuals in the $n$th generation stemming from the $Y_j$ immigrants in generation $j$, $j = 0, 1, \ldots, n$. Differentiation of the pgf gives $E[Z_n] = \nu \sum_{k=0}^{n} \mu^k$ **(c)** Poi($\lambda(1 - p^{n+1})/(1-p)$)

34 For example, $p_{00}(t) = \lambda/(\lambda + \mu)e^{-(\lambda+\mu)t} + \mu/(\lambda + \mu)$

35 **(c)** 1/4 minute **(d)** 1/5

37 **(b)** uniform if $b = 2a$

40 **(b)** $(10 + \alpha)/(40 + \alpha)$ **(c)** 1.5

41 $\pi_0 = 1/(1 - \log(2/3)), \pi_n = \pi_0/(n3^n)$

43 **(a)** 17% **(b)** 1.6%

44 **(a)** 50% **(b)** 33% **(c)** 33% **(d)** 20%

45 **(a)** $\pi_0 = 4/10, \pi_1 = 2/10, \pi_2 = 3/10, \pi_3 = 1/10$
**(b)** $\pi_0 = 16/43, \pi_1 = 8/43, \pi_2 = 12/43, \pi_3 = 7/43$

46 **(b)** $\pi_0 = \pi_1 = 4/19, \pi_2 = 6/19, \pi_3 = 5/19$

47 **(a)** $\pi_0 = 0.03, \pi_1 = 0.19, \pi_2 = 0.78$
**(b)** $\pi_2 + \pi_1((2/3) \times 0.2 + (1/3) \times 0.5) = 0.84$

48 Poi($\lambda/\mu$)

49 $\lambda_n \equiv \lambda, \mu_n = \mu + (n-1)\nu, n \geq 1$

50 $P(W \leq x) = 1 - \rho \exp(-x(\mu - \lambda)), x \geq 0$ $(P(W = 0) = 1 - \rho)$ and conditioned on $k \geq 1$ customers in the system, $W \sim \Gamma(k, \mu)$)

51 $T \sim \exp(\mu(1 - \rho))$

52 $Y_n = P_n/\mu^n$

53 $Y_n = 2^n X_n$

55 $Y \sim \text{unif}[0, 1]$

57 $m(t) = e^t - 1, \quad t \leq 1$

58 **(a)** Once in 190 h **(b)** 0.79

59 **(a)** $\lambda/(1 + \lambda\mu_S)$ **(b)** $1/(1 + \lambda\mu_S)$

60 0.918

61 **(a)** $\widetilde{F}(t) = 1 - e^{-\lambda t}$
**(b)** $\widetilde{F}(t) = t(2 - t) \quad 0 \leq t \leq 1$ **(c)** $\widetilde{F}(t) = 1 - e^{-\lambda t}(1 + \lambda t/2)$

62 **(a)** $\min(s, t)$ **(b)** $t(1 - s)$ if $t \leq s$ and $s(1 - t)$ if $t \geq s$

63 $N(0, t + s + 2\min(t, s))$

64 $\min(t, s)$

66 $E[M_t] = \sqrt{2t/\pi}$, $\text{Var}[M_t] = t(t - 2/\pi)$

67 $N\left(\frac{t_2 - t}{t_2 - t_1}x_1 + \frac{t - t_1}{t_2 - t_1}x_2, \frac{(t_2 - t)(t - t_1)}{t_2 - t_1}\right)$

68 **(a)** $N(c + \mu(t - s), \sigma^2(t - s))$ **(b)** $N(ct/s, \sigma^2 t(s - t)/s)$

69 0.362

70 $f(r) = \frac{r}{t}e^{-r^2/(2t)} \quad r \geq 0$

# FURTHER READING

1. F. Mosteller, *Fifty Challenging Problems in Probability with Solutions*, Addison-Wesley, Reading, MA London, 1965 [the classic, and charmingly dated (mentions logarithm tables!), collection of probability problems].
2. G. Blom, L. Holst, and D. Sandell, *Problems and Snapshots from the World of Probability*, Springer-Verlag, New York, 1994 (the modern, comprehensive, and mathematically stringent successor of Mosteller. A wealth of interesting problems).
3. J. Haigh, *Taking Chances: Winning with Probability*, Oxford University Press, 2003 (an entertaining nontechnical book on the probability considerations of everything from roulette to "Who wants to be a millionaire?").
4. J. A. Paulos, *Innumeracy: Mathematical Illiteracy and Its Consequences*, Hill & Wang, 2001 (the bestseller that debunks myths and discusses misunderstandings of mathematics, probability, and statistics in everyday life).
5. D. Kahneman, P. Slovic, and A. Tversky, *Judgment under Uncertainty: Heuristics and Biases*, Cambridge University Press, 1982 (extensive treatment of the psychology of probability considerations in decision making).
6. R. Meester, *A Natural Introduction to Probability Theory*, Birkhäuser Basel, 2003 (a clear and focused introduction to probability, with an aim to prepare the reader for more advanced theory).
7. G. Grimmett and D. Stirzaker, *Probability and Random Processes*, 3rd edition, Oxford University Press, New York, 2001 (a comprehensive introduction to intermediate probability theory, Markov chains, and other stochastic processes).

*Probability, Statistics, and Stochastic Processes*, Second Edition. Peter Olofsson and Mikael Andersson.

8. O. Häggström, *Finite Markov Chains and Algorithmic Applications*, Cambridge University Press, 2002 (an excellent introduction to the theory of Markov chains with finite state space, and their use in computer simulation).
9. J. Stoyanov, *Counterexamples in Probability*, 2nd edition, Wiley, Chichester, UK, 1997 (the book that somebody just had to write, with all those results that ought to be true but are not; fast-paced on a mostly advanced level).
10. D. Freedman, R. Pisani, and R. Purves, *Statistics*, 3rd edition, Norton, New York, 1998 (it is hard to believe that a book with this title can be highly entertaining, but it is; the text is clear and exact without being technical, and contains plenty of interesting real-world examples).

# INDEX



*Probability, Statistics, and Stochastic Processes*, Second Edition. Peter Olofsson and Mikael Andersson.